Electronic Principles

OTHER BOOKS BY ALBERT P. MALVINO

Transistor Circuit Approximations

Electronic Instrumentation Fundamentals

Digital Principles and Applications (with D. Leach)

Electronic Principles

Albert Paul Malvino, Ph.D.

McGraw-Hill Book Company

NEW YORK	KUALA LUMPUR	PANAMA
ST. LOUIS	LONDON	RIO DE JANEIRO
SAN FRANCISCO	MEXICO	SINGAPORE
DÜSSELDORF	MONTREAL	SYDNEY
JOHANNESBURG	NEW DELHI	TORONTO

Library of Congress Cataloging in Publication Data

Malvino, Albert Paul.
  Electronic principles.

  1. Electronics.   1. Title.
TK7816.M25   621.381   72-10292
ISBN 0-07-039841-0

ELECTRONIC PRINCIPLES

1 2 3 4 5 6 7 8 9 MURM 7 9 8 7 6 5 4 3

*The editors for this book were Alan W. Lowe and Cynthia
Newby, the designer was Marsha Cohen, and its production was supervised by Jim Lee. It was set in Palatino
by Progressive Typographers.*
*It was printed by The Murray Printing Company and
bound by Rand McNally & Company.*

TO JOANNA,

My brilliant and beautiful wife
without whom I would be nothing.
She always comforts and consoles,
never complains or interferes,
asks nothing and endures all,
and writes my dedications.

The entire truth about anything is almost impossible to tell within practical limits. No desert is absolutely dry, no liquid absolutely wet. No plain completely flat; no ball-bearing truly round.

Shidle

Contents

Preface

You can find two ways to explain any idea: a simple way and a complex way. This book is the *simple way*. Do not misunderstand me. Nothing is quite so complicated as simplicity; to attain it, you have to force all kinds of distractions and irrelevancies into perspective. I am sure teachers and students will welcome this "simple" book in a world cluttered with the other kind.

This book is for a student taking a basic course in electronics, but it will help many advanced readers. The prerequisites are an ac-dc course, algebra, and some trigonometry. The Contents and illustrations will give you a concrete idea of the topics and level. In addition, here are some general ideas.

In-depth coverage of fewer topics is better than shallow treatment of many. Bombard a student with too many devices and circuits, and he becomes confused and discouraged; even worse, he will get wrong ideas. A good example is negative feedback. A shallow coverage deals only with the best-known type of negative feedback; invariably, the student generalizes this one type to all types. It may take him years to realize *there are four kinds of negative feedback*. This book is not for shallow courses; it is for courses with "bite," enough in-depth discussion to learn key principles in the right way.

Where is the student going after he finishes his schooling? Most likely to industry. To succeed, he needs the right blend of theory and practice; industry wants a graduate who has *learned his theory in a practical way*. Twenty years as a technician and engineer helped me capture what I think is the optimum blend of theory and practice. Throughout this book, topics, problems, and approximations are related to widespread industrial practice.

Readers despise books that go through mountains of mathematics with no practical conclusions. This book uses rather than abuses mathematics. The mathematics is not intended as a substitute for thinking; it is a *compressed way of thinking.* And the final formulas are compact summaries of that thinking. As an aid, the easy-to-remember important formulas have a triple asterisk *** attached to them.

You will find many new ideas in this book not discussed elsewhere. Occasionally, you may wonder if a circuit works as described. Let me assure you I have *built and tested all circuits* to prove the theories behind them. In fact, *Experiments for Electronic Principles* (a correlated laboratory manual) contains 47 experiments that demonstrate the principles of this book in the laboratory.

At the end of each chapter are many good problems, some purely theoretical, others the bread-and-butter kind you find in industry. The student should do *all* these problems; they are essential for learning the electronic principles in this book.

Most schools now give more than 90 percent coverage to solid-state devices and less than 10 percent to vacuum tubes. The reason is clear enough: the tube has disappeared from new designs, except for a few isolated areas. To keep in step with modern schools, this book devotes more than 90 percent of the discussion to solid-state devices and less than 10 percent to tubes.

Many thanks to my good friend Gregory Johnson, Chairman of Electrical Technology, Ulster Community College, Stone Ridge, New York. He reviewed the final manuscript and gave me more good ideas than I will admit in print.

Albert Paul Malvino

Electronic Principles

1. Introduction

This chapter is about idealization and approximation, ways to simplify analysis. We also review Thevenin's theorem.

1-1.   IDEALIZATION

Most electronic devices are complicated. To get to the main ideas, we often idealize devices; this means we strip away all unnecessary detail. The device that remains is then ideal or perfect.

Need for idealization

You have used idealization many times already. A good example is a piece of wire. In most cases, you treat it as a perfect conductor; but this is far from the truth. Take a piece of copper wire 1 ft long as shown in Fig. 1-1a. If it is AWG 22, this piece of wire has a resistance of 0.016 Ω and an inductance of 0.24 μH. If the wire is 1 in above a metallic plane (see Fig. 1-1b), it has a capacitance of 3.3 pF. Therefore, the piece of wire acts like the circuit of Fig. 1-1c.

Even this is not exact. The truth is you cannot draw a circuit at all because the R, L, and C are *distributed* over the length of the wire rather than *lumped* between nodes A, B, C, and D. For exact analysis you need to use an approach based on advanced formulas called Maxwell's equations.

Fortunately, you don't need Maxwell's equations below 300 MHz and can use lumped-constant circuits like Fig. 1-1c. (Lumped constant means the resistance is between one pair of nodes, the inductance between another pair, and the capacitance

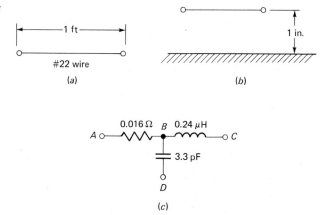

Figure 1-1. Idealizing a piece of wire.

between a third pair.) Better yet, for frequencies below 1 MHz the inductive and capacitive reactances are usually negligible, and the wire's resistance is so small compared with other circuit resistance that it too is negligible. In other words, most of the time we *idealize* a piece of wire by neglecting its R, L, and C.

Approximations

Make no mistake about it. At high enough frequencies you do need an equivalent circuit like Fig. 1-1c for a piece of wire. This should give you a clue as to how complicated the exact circuits are for some electronic devices. For most everyday needs, we can approximate; otherwise, we'd be bogged down in unnecessary detail. Throughout this book, we use approximations as much as possible.

The *ideal,* or first approximation, strips away everything but the key ideas behind a device. In this way, we get to the bones of device operation. With ideal devices, circuit analysis is easier. When necessary, we can improve the analysis by using a second approximation, sometimes a third approximation, and once in a while, an exact equivalent circuit. In the chapters to come, look for these levels of approximation:

Ideal or first approximation: the simplest equivalent circuit. Retains only one or two ideas of how the device works.

Second approximation: includes extra features to improve analysis. Usually, this is as far as many engineers and technicians go in daily work.

Third approximation: includes other effects of lesser importance. In some circuits, we will need this approximation.

Exact circuit: this will be as complete as possible using lumped constants. We almost never use this circuit.

1-2.   COMPONENTS

Most of the time, we idealize a resistor by taking only its resistance into account. But a resistor also has capacitance and inductance. Likewise, pure capacitors and pure inductors exist only as idealizations. Because R's, L's, and C's are so much used in electronic circuits, you should know their approximations.

Resistor approximations

Figure 1-2a shows the ideal and second approximations for a resistor. At low frequencies, only the R matters. But at higher frequencies, the capacitance from one end of the resistor to the other becomes important because it provides a path for ac current.

For lower-frequency work, you can forget about end-to-end capacitance. But above 100 kHz, you sometimes have to include its effect in your analysis. As a guide, the end-to-end capacitance of typical resistors (¼ to 2 W) is in the vicinity of 1 pF, with the exact value determined by the length of the leads, the size of the resistor body, and other factors. When you need the exact capacitance value, you can measure it on an RLC bridge.

The end-to-end capacitance is negligible when the current through it is much smaller than the ac current through the resistance. Many people use this guide: neglect end-to-end capacitance when

$$\frac{X_C}{R} > 10 \tag{1-1}$$

As an example, if a 10-kΩ resistor has 1 pF of capacitance, the X_C at 1 MHz equals

$$X_C = \frac{1}{2\pi fC} = \frac{1}{2\pi \,(10^6)\,10^{-12}}$$
$$= 159 \text{ k}\Omega$$

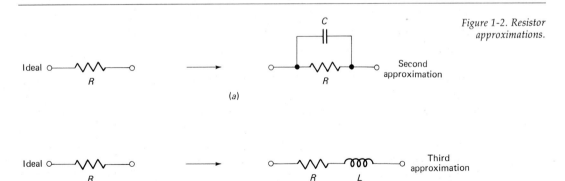

Figure 1-2. Resistor approximations.

The ratio of reactance to resistance is

$$\frac{X_C}{R} = \frac{159 \text{ k}\Omega}{10 \text{ k}\Omega} = 15.9$$

This is greater than 10; therefore, we can neglect the capacitance of a 10-kΩ resistor operating at 1 MHz.

Occasionally, we need the third approximation of a resistor (see Fig. 1-2*b*). This approximation includes the *lead inductance,* approximately 0.02 μH/in for wire sizes used in low-power electronic circuits.[1] When you need the exact value of lead inductance, you can measure it on a high-frequency *RLC* bridge. The typical rule for lead inductance is this: neglect lead inductance when

$$\frac{R}{X_L} > 10 \tag{1-2}$$

Lead inductance is less important than end-to-end capacitance in most cases. For example, suppose the leads of a 1-kΩ resistor are cut to ½ in on each end. Then, the total lead length is 1 in. Estimating the inductance by 0.02 μH/in, we have a lead inductance of 0.02 μH. At 300 MHz, the reactance is

$$X_L = 2\pi fL = 2\pi (300) 10^6 (0.02) 10^{-6}$$
$$= 37.7 \ \Omega$$

and the ratio of resistance to reactance is

$$\frac{R}{X_L} = \frac{1000}{37.7} = 26.5$$

Therefore, even at 300 MHz we can neglect the lead inductance of a 1-kΩ resistor.

By setting $X_C/R = 10$ and $R/X_L = 10$, we can plot frequency versus resistance as shown in Fig. 1-3. This graph gives us the dividing lines between the ideal, second, and third approximations, assuming an end-to-end capacitance of 1 pF and a lead inductance of 0.02 μH.

Here is how we use the graph. For any point below the two lines, we can idealize the resistor, that is, neglect its capacitance and inductance. On the other hand, if a point falls above either line, we may want to use the second or third approximation. For instance, a 10-kΩ resistor operating at 1 MHz can be idealized because it falls in the ideal region of Fig. 1-3. But if this 10-kΩ resistor operates at 5 MHz, we may include end-to-end capacitance in precise calculations. Similarly, a 20-Ω resistor acts ideally up to 16 MHz, but beyond this frequency, the lead inductance becomes important.

Don't lean too heavily on Fig. 1-3; it is only a guide to help you estimate when to include end-to-end capacitance or lead inductance in your calculations. If working at high frequencies where precise analysis is required, you may have to measure the end-to-end capacitance and the lead inductance on a high-frequency *RLC* bridge.

---

[1] These are wire sizes from around AWG 18 to 26. AWG 18 has 0.018 μH/in; AWG 26 has 0.024 μH/in.

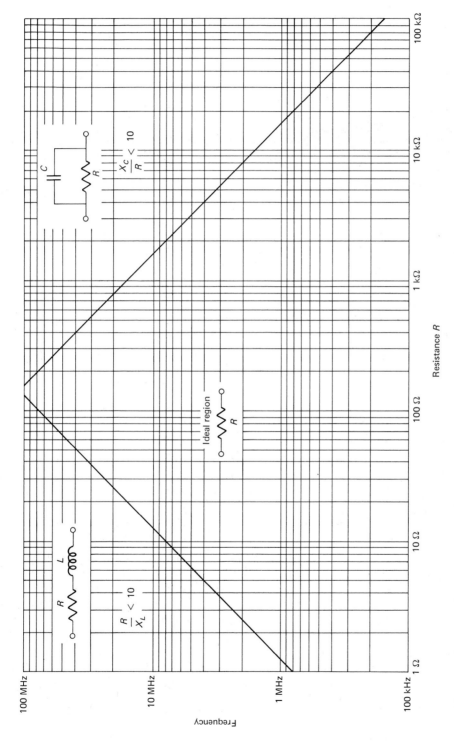

Figure 1-3. Approximation locator.

Inductor and capacitor approximations

Figure 1-4a shows the ideal, second, and third approximations of an inductor. For frequencies where only the X_L matters, we use the ideal approximation. But at lower frequencies where X_L is not large compared with R, we must use the second approximation. At higher frequencies, the distributed capacitance between the windings of the inductor becomes important; in this case, we need the third approximation shown in Fig. 1-4a. We will use these inductor approximations later.

Figure 1-4b shows the approximations for a capacitor. Most of the time, we can idealize a capacitor. For large capacitors, especially the electrolytic type, we may include the leakage resistance (the second approximation). And at the highest frequencies, the lead inductance becomes important; therefore, occasionally, we need the third approximation shown in Fig. 1-4b.

EXAMPLE 1-1.

How much inductance does a 5-in piece of wire have? What does its X_L equal at 10 MHz?

SOLUTION.

The inductance is approximately 0.02 μH/in; therefore, 5 in of wire has an inductance of approximately 0.1 μH.

At 10 MHz, the inductive reactance equals

$$X_L = 2\pi fL = 2\pi (10) 10^6 (0.1) 10^{-6}$$
$$= 6.28 \ \Omega$$

This inductive reactance may be important; it depends on the circuit. All we can say is this: for the 6.28 Ω to be negligible, the circuit impedance in series with it must be much larger.

Figure 1-4. (a) *Inductor approximations.* (b) *Capacitor approximations.*

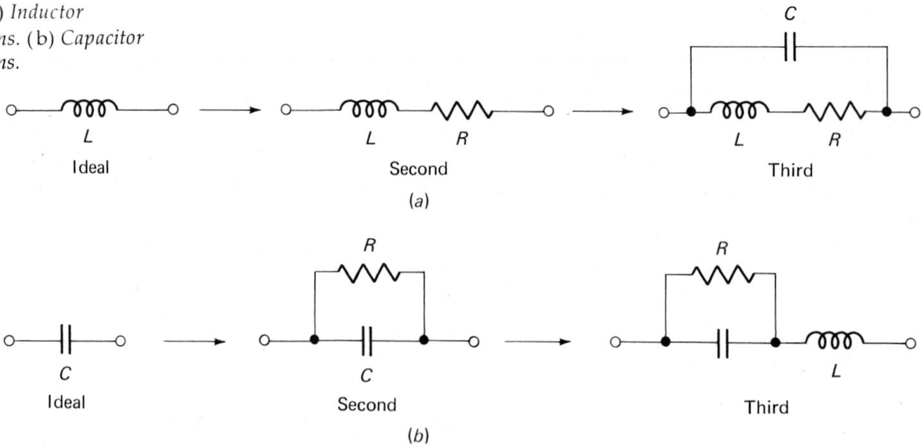

Besides having inductance, a piece of wire forms a capacitance with any nearby metal. With increasing frequency, this capacitance becomes important because it provides a path for current.

The inductance and capacitance of connecting wires may disturb circuit operation. It depends on the resistances and frequencies involved. As a general rule, keep all connecting wires as short as possible for circuits operating above 100 kHz.

EXAMPLE 1-2.
Capacitors have lead inductance. For this reason, they become self-resonant at higher frequencies. Calculate the resonant frequency for a 1000-pF capacitor with ½-in leads on each end.

SOLUTION.
Assume a lead inductance of 0.02 μH/in. Then,

$$f = \frac{1}{2\pi \sqrt{LC}} = \frac{1}{2\pi \sqrt{0.02\,(10^{-6})\,1000\,(10^{-12})}}$$
$$= 35.6 \text{ MHz}$$

This is the resonant frequency of the capacitor. Beyond this frequency, the inductive reactance becomes greater than the capacitive reactance. In effect, the capacitor no longer acts like a capacitor above 35.6 MHz; it acts like a small inductor.

1-3. SOURCES

Besides *passive* components like resistors, capacitors, and inductors, we have *active* devices like voltage and current sources. They supply the energy needed to operate the circuit.

Voltage sources

Dc voltage sources deliver a constant voltage when the load resistance is fixed. Figure 1-5 shows the ideal and second approximation of a dc voltage source. In the ideal approximation, the voltage stays fixed no matter how much current flows. The second approximation is more realistic; it includes an internal resistance R. Only when very large currents flow do we worry about the voltage drop across R.

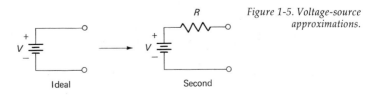

Figure 1-5. Voltage-source approximations.

Ideal   Second

Figure 1-6. Loaded source.

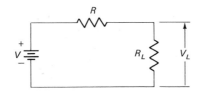

If R_L represents the load resistance (see Fig. 1-6), the load voltage equals

$$V_L = \frac{R_L}{R + R_L} V$$

Our rule for using the ideal approximation is this: neglect internal drop when

$$\frac{R_L}{R} > 100 \qquad (1\text{-}3)$$

When load resistance is at least 100 times greater than internal resistance, we get less than 1 percent error by neglecting R.

Figure 1-7 shows the three approximations for an ac voltage source. When internal impedance is negligible, we will show only the ideal symbol for the source. When internal resistance is important, we will use the second approximation. Every so often, the internal impedance of an ac source is not constant at different frequencies; in a case like this, we represent the internal impedance by a complex number Z. As a guide, we neglect internal impedance when the load resistance is at least 100 times greater than the magnitude of Z.

Current sources

Figure 1-8a shows the approximations for a dc current source. The ideal dc current source delivers a constant load current regardless of the value of load resistance. For instance, if $I = 5$ A, then 5 A flows whether you have a 1-Ω load or a 50-MΩ load.

The second approximation includes an internal resistance R. Because of this, some current flows through R; this reduces the load current. In Fig. 1-8b, the larger R is compared with R_L, the smaller its share of total current. When R is at least 100 times greater

Figure 1-7. Ac-voltage-source approximations.

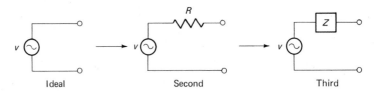

Ideal   Second   Third

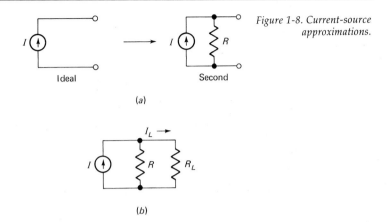

Figure 1-8. Current-source approximations.

than R_L, we will neglect R. In symbols, we disregard R when

$$\frac{R}{R_L} > 100 \qquad (1\text{-}4)$$

This leads to less than 1 percent error.

In many analyses, we don't need this kind of accuracy. Often, 10 percent error or less is adequate and we can use the less severe condition: $R/R_L > 10$. For instance, if R_L equals 10 kΩ, we can neglect internal resistance when R is greater than 100 kΩ; the resulting error is less than 10 percent.

Ac current sources can have reactance as well as resistance. For this reason, Fig. 1-9 shows three approximations for the ac current source. Often, we can use the ideal source. Sometimes, we need the internal resistance R. And occasionally, the reactance of the source must be taken into account. That is, in the third approximation, Z is a complex number.

The big use we make of current sources is with transistors. If we tried to analyze the transistor with exact methods, we would be swamped in detail. When we idealize a transistor, however, it is simple; it becomes a current source. More about this later.

Incidentally, capital letters V and I are normally used for dc voltage and dc current, shown in Figs. 1-5 and 1-8. Lowercase letters v and i are for ac voltage and ac current (Figs. 1-7 and 1-9).

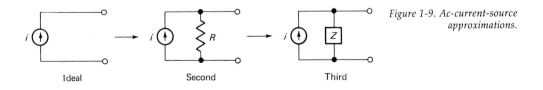

Figure 1-9. Ac-current-source approximations.

EXAMPLE 1-3.
A PP9 dry-cell battery has a voltage of 9 V and an internal resistance of 4 Ω. For what value of load resistance may we neglect internal resistance?

SOLUTION.
Use the 100-to-1 rule: R_L must be at least 100 times greater than R. With an R of 4 Ω, the smallest R_L is 400 Ω. If R_L is less than 400 Ω, we may want to calculate the internal voltage drop to improve our answers.

EXAMPLE 1-4.
Calculate V_{out} in Fig. 1-10. The current source points in the direction of *conventional* current.

Figure 1-10. Example 1-4.

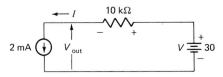

SOLUTION.
The ideal current source forces 2 mA to flow, regardless of what is connected across it. Therefore, 2 mA flows through the 10-kΩ resistor. The IR drop across this resistor equals

$$IR = 2\,(10^{-3})\,10\,(10^{3}) = 20 \text{ V}$$

with the plus-minus polarity shown in Fig. 1-10. And V_{out} equals

$$V_{out} = V - IR$$
$$= 30 - 20 = 10 \text{ V}$$

1-4.   THEVENIN'S THEOREM

You have already studied Thevenin's theorem elsewhere. It is a *must* in circuit analysis. What follows is a fast recap of this great theorem.

Briefly, here is the Thevenin theorem for dc circuits. Suppose you have a linear circuit where only dc currents and voltages exist. (Linear means the resistance values do not change with increasing voltage.) Pick any pair of terminals A and B. The Thevenin theorem says these terminals act as if a single battery and resistor is behind them, as shown in Fig. 1-11a. The original circuit may be complicated with many sources and loops. But the Thevenin equivalent has only one source and one loop when you connect R_L (see Fig. 1-11b). This one-loop simplicity is the power of the Thevenin circuit; anyone can calculate currents and voltages in a circuit like this.

To find V_{TH}, you calculate or measure the *open-circuit* voltage in Fig. 1-12a. To get R_{TH}, you reduce all sources to zero, and calculate or measure the resistance from A to B.

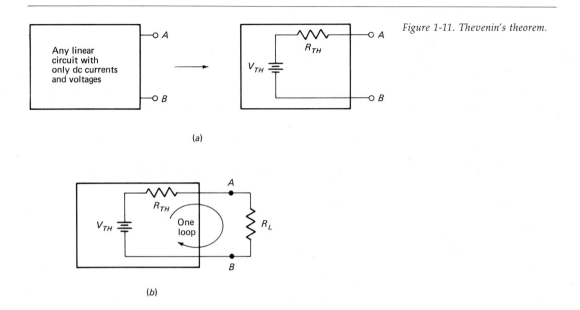

Figure 1-11. Thevenin's theorem.

When finding R_{TH} by measurement, it may be inconvenient to reduce all sources to zero. In this case, you can connect a variable load resistance as shown in Fig. 1-12b; when V_L equals half of V_{TH}, R_L equals R_{TH}; you then can measure R_L with an ohmmeter. For instance, suppose we measure an open-circuit voltage of 12 V (see Fig. 1-12c). Then if a 10-kΩ resistor drops the load voltage to 6 V, we will know the Thevenin resistance is 10 kΩ (see Fig. 1-12d).

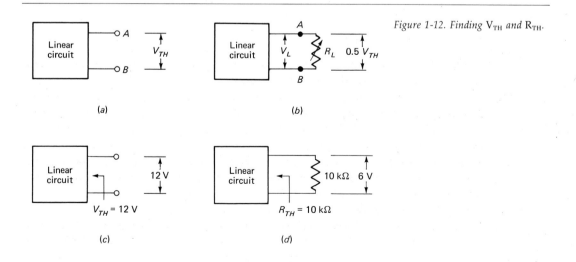

Figure 1-12. Finding V_{TH} and R_{TH}.

Thevenin's theorem also applies to ac circuits. In fact, combined with the superposition theorem, Thevenin's theorem is the key to understanding many different electronic circuits. We discuss this combined use of Thevenin's theorem and the superposition theorem in Chap. 9.

EXAMPLE 1-5.
Figure 1-13a shows a circuit with a dc voltage source and a dc current source. *Thevenize* the circuit at the *AB* terminals. (From now on *Thevenize* means "find the Thevenin equivalent of.")

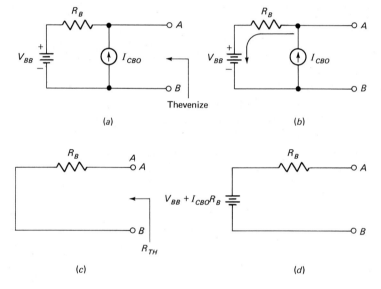

Figure 1-13. Example 1-5.

SOLUTION.
The I_{CBO} current source forces a conventional current of I_{CBO} to flow back through the resistor and down through the voltage source as shown in Fig. 1-13b. The voltage across the resistor equals $I_{CBO}R_B$. Therefore,

$$V_{TH} = V_{BB} + I_{CBO}R_B$$

To get R_{TH}, you reduce all sources to zero. Reducing a *voltage source* to zero is identical to replacing it with a *short circuit* as shown in Fig. 1-13c. Reducing a *current source* to zero is identical to replacing it with an *open* circuit because zero current flows through an open circuit. When a current source is open, we can remove it altogether from the circuit as shown in Fig. 1-13c. Looking into the *AB* terminals, we see a resistance of R_B.

Figure 1-13d shows the Thevenin equivalent circuit. (You will see this circuit again in Chap. 7.)

Problems

1-1. Channel 4 on a TV receiver operates at a center frequency of 69 MHz. How much inductive reactance does a 3-in piece of wire have at this frequency? (Use 0.02 μH/in.)

1-2. A 1-MΩ resistor has an end-to-end capacitance of 1 pF. At what frequency does the capacitive reactance equal 1 MΩ? At what frequency will X_C equal 10 MΩ?

1-3. A 1-Ω resistor has a lead inductance of 0.02 μH. What does its X_L equal at 10 MHz? At what frequency does X_L equal 0.1 Ω?

1-4. Using Fig. 1-3 as a guide, estimate the maximum frequency for which the ideal approximation is valid for each of these: 5-Ω resistor, 100-Ω resistor, 1-kΩ resistor, and 56-kΩ resistor.

1-5. You are going to build a circuit that will operate up to 10 MHz. Using Fig. 1-3 as a guide, you can idealize all resistors if they are within what range?

1-6. A 0.01-μF capacitor has a lead inductance of 0.04 μH. What is its resonant frequency?

1-7. A battery has an internal resistance of 2 Ω. To have better than 1 percent accuracy in calculations, how large should the load resistance be?

1-8. What is the Thevenin circuit for Fig. 1-14a?

1-9. Thevenize the circuit of Fig. 1-14b.

1-10. Thevenize Fig. 1-14c looking into the AB terminals.

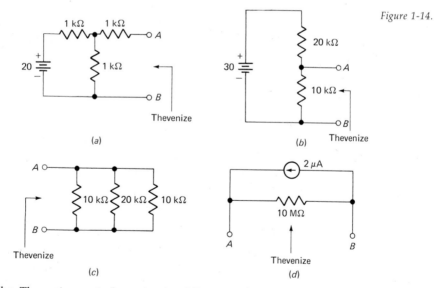

Figure 1-14.

1-11. Work out the Thevenin equivalent circuit of Fig. 1-14d at the AB terminals.

1-12. If the 10-MΩ resistor of Fig. 1-14d opens up, what will the R_{TH} of the circuit be? What is the R_{TH} of any ideal current source?

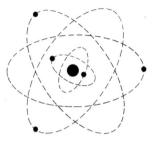

Figure 2-1. Bohr model.

2-2.   ORBITAL RADIUS

In the silicon atom (Fig. 2-2a), the electrons travel with the right velocity to balance the inward pull of the nucleus and the outward push of centrifugal force. When an electron is in a large orbit, the nuclear attraction is low; therefore, the velocity needed to balance is low. For smaller orbits, the nuclear attraction is greater, which means the electron must travel faster to keep the balance.

You may think an electron can travel in an orbit of any radius, provided its velocity has the right value. Modern physics says otherwise; the radii must have *discrete* values (discrete means separated by a gap). For instance, the smallest orbit in a hydrogen atom has a radius of

$$r_1 = 0.53\,(10^{-10})\ \text{m}$$

The next orbit has a radius of

$$r_2 = 2.12\,(10^{-10})\ \text{m}$$

All values between r_1 and r_2 are *forbidden*; an electron cannot find a stable orbit between r_1 and r_2.

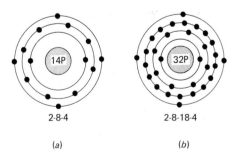

Figure 2-2. (a) Silicon atom. (b) Germanium atom.

2-8-4   2-8-18-4

(a)   (b)

2. Semiconductor Theory

You already know something about atoms, electrons, and protons. This chapter adds to your knowledge. Our treatment will be simple, sometimes idealized; otherwise, we would be caught up in highly advanced mathematics and physics.

Some topics in this chapter may seem far out and of little value in practical electronics. However, these topics are *musts* for understanding devices like diodes, transistors, and integrated circuits.

2-1. ATOMIC STRUCTURE

Bohr idealized the atom. He saw it as a nucleus surrounded by orbiting electrons (Fig. 2-1). The nucleus has a positive charge and attracts the electrons. The electrons would fall into the nucleus except for the centrifugal force of their motion. When an electron travels in a stable orbit, it has just the right velocity for centrifugal force to balance nuclear attraction. The nearer an electron is to the nucleus, the faster it must travel to offset nuclear attraction.

Three-dimensional drawings like Fig. 2-1 are difficult to show for complicated atoms. Because of this, we often symbolize the atom in two dimensions. An isolated silicon atom (Fig. 2-2a) has 14 protons in its nucleus. Two electrons travel in the first orbit, eight electrons in the second, and four in the outer or *valence* orbit. The 14 revolving electrons neutralize the charge of the nucleus so that from a distance the atom acts electrically neutral.

Germanium is another element used in electronics work. It resembles silicon in this respect: an isolated atom of germanium has a valence orbit with four electrons (see Fig. 2-2b). Germanium differs, however, because it has 32 protons in its nucleus. A neutral germanium atom has 32 electrons spread into orbits with 2, 8, 18, and 4 electrons.

Why? Strange as it seems, an electron appears to be two different things at the same time. In some experiments, it acts like a particle, something with mass lumped in one place. Yet in other experiments, it will act like a wave — a massless something with vibrations distributed in space. The orbit of an electron must satisfy not only equations for its particle nature, but also equations for its wave nature. We cannot go into this topic too far, but we can say this much: because it is a *wave,* an electron can *fit* only into an orbit whose circumference equals the electron *wavelength* or some multiple of it.

For a hydrogen atom, an electron can fit into orbits whose radii are given by

$$r_n = n^2 r_1 \qquad\qquad (2\text{-}1)$$

where $n = 1, 2, 3$, and so on, corresponding to the first, second, third orbit, etc. The quantity r_1 equals $0.53(10^{-10})$ m. Since atomic distances are so small, the *angstrom* unit is often used. It is defined as

$$\text{angstrom} = 10^{-10} \text{ m}$$

Therefore, the r_1 of hydrogen equals 0.53 angstrom (Å). And with Eq. (2-1), the first three permitted radii equal

$$r_1 = 0.53 \text{ Å}$$

$$r_2 = 4r_1 = 2.12 \text{ Å}$$

$$r_3 = 9r_1 = 4.77 \text{ Å}$$

Because a silicon atom has 14 protons, the r_1 of silicon is smaller than hydrogen's r_1; the larger positive charge of the silicon nucleus makes electrons travel in smaller orbits. Each electron in silicon feels the influence not only of the nuclear charge but of every other electron as well. For this reason, the radius formula of silicon is more complicated than Eq. (2-1). But as before, the electrons can travel only in orbits of discrete radii.

2-3. ENERGY LEVELS

Suppose we magnify the silicon atom of Fig. 2-2a until all we see is part of each orbit as shown in Fig. 2-3. For convenience in drawing, we can take this a step further and visualize the curved lines of Fig. 2-3 as horizontal lines like those of Fig. 2-4. We can also identify each orbit with an *energy level.* To move an electron from the first to the second orbit takes energy because the electron is pulled away from the nucleus. When an electron has been moved into a larger orbit, it acquires potential energy with respect to smaller orbits. If the electron is allowed to fall back to its original orbit, it will give up its potential energy.

An analogy will help. Think of the edge of the nucleus in Fig. 2-4 as the surface of the earth. It takes energy to lift a stone off the earth. The higher you lift it, the more energy you expend. Where does the energy go? The stone absorbs it in the form of potential energy with respect to the earth. If you release the stone, it will fall and give back its acquired energy when it strikes the earth.

Figure 2-3. Magnified orbits.

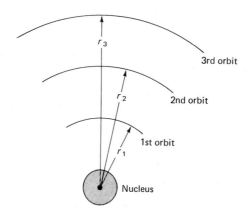

Figure 2-4. Energy levels.

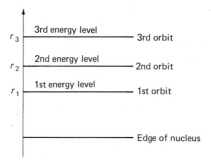

The total energy of an electron is the sum of its potential energy, its kinetic energy, and a few other forms of energy. To a good approximation (at least for our purposes), we can identify the total energy of an electron with the radius of its orbit. That is, we can think of each radius in Fig. 2-4 as equivalent to an energy level. Electrons in the smallest orbit are on the first energy level; electrons in the second orbit are on the second energy level, and so on. The larger the radius, the greater the energy of an electron. When two electrons differ slightly in energy, we will view this as a slight difference in radius. This viewpoint is an approximation, but it is adequate for our needs.[1]

The importance of energy levels is this. If external energy like heat, light, or other radiation bombards an atom, it can lift an electron to a higher energy level (larger orbit). The atom is then in an *excited* state. The excitement does not last long. Ordinarily, the electron remains in the larger orbit approximately 10^{-8} s before falling back to its smaller orbit. In falling from a higher to a lower energy level, the electron gives up energy; this energy leaves the atom as heat, light, or other radiation.[2]

[1] A more accurate viewpoint requires *quantum mechanics,* an abstract combination of advanced mathematics and physics.
[2] This is the principle behind neon signs, fluorescent lamps, ultraviolet lamps, etc.

2-4. CRYSTALS

Up to now, we have talked about the atomic structure and energy levels of an *isolated* atom. When you combine atoms, the picture changes. Take silicon. When silicon atoms combine to form a solid, they arrange themselves in an orderly pattern called a *crystal*. The forces holding the atoms together are the *covalent bonds*.

What is a covalent bond? An isolated atom has four electrons in its valence orbit. For reasons covered by advanced equations, silicon atoms combine so as to have eight electrons in the valence orbit. To manage this, each silicon atom positions itself between four other silicon atoms (see Fig. 2-5a). Each neighbor shares an electron with the central atom. In this way, the central atom has picked up four electrons, making a total of eight in its valence orbit. Actually, the electrons no longer belong to a single atom; they are shared by adjacent atoms. It is this sharing that sets up the covalent bond.

Here is how a covalent bond holds atoms together. In Fig. 2-5a, a core represents the nucleus plus the inner orbits. As a result, each core has a net positive charge equal to the difference between nuclear protons and inner-orbit electrons. In silicon, the core charge equals +4 (14 protons minus 10 inner-orbit electrons). Now, look at the central core and the one on the right. These two cores attract the pair of electrons between them with equal and opposite forces. This motionless pulling in opposite directions is what holds the atoms together. (Similar to tug-of-war teams pulling on a rope. As long as the teams pull with equal and opposite forces, they remain immobile and bonded together.)

Figure 2-5b symbolizes the mutual sharing and pulling on electrons. Each line represents a shared electron. Each shared electron establishes a bond between the central atom and a neighbor. For this reason, we call each line a covalent bond.

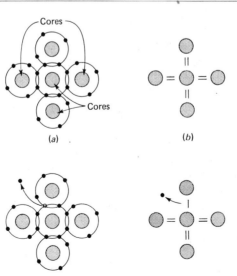

Figure 2-5. (a) *Covalent bonds.* (b) *Bonding diagram.* (c) *Hole.* (d) *Broken bond.*

(a) (b)

(c) (d)

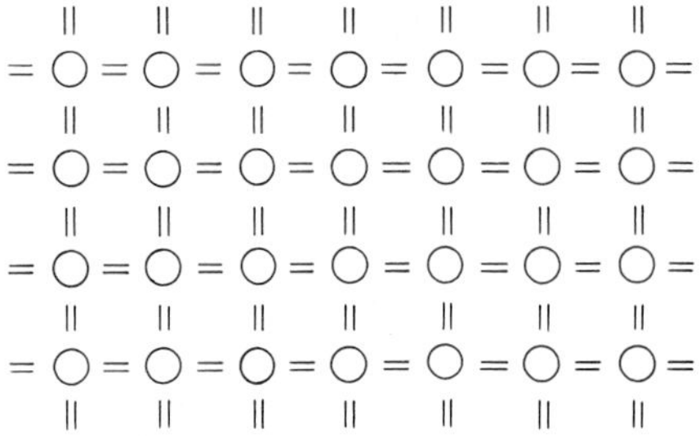

Figure 2-6. Bonding diagram.

$$
\begin{array}{c}
\|\quad\|\quad\|\quad\|\quad\|\quad\|\quad\| \\
= \bigcirc = \bigcirc = \bigcirc = \bigcirc = \bigcirc = \bigcirc = \bigcirc = \\
\|\quad\|\quad\|\quad\|\quad\|\quad\|\quad\| \\
= \bigcirc = \bigcirc = \bigcirc = \bigcirc = \bigcirc = \bigcirc = \bigcirc = \\
\|\quad\|\quad\|\quad\|\quad\|\quad\|\quad\| \\
= \bigcirc = \bigcirc = \bigcirc = \bigcirc = \bigcirc = \bigcirc = \bigcirc = \\
\|\quad\|\quad\|\quad\|\quad\|\quad\|\quad\| \\
= \bigcirc = \bigcirc = \bigcirc = \bigcirc = \bigcirc = \bigcirc = \bigcirc = \\
\|\quad\|\quad\|\quad\|\quad\|\quad\|\quad\|
\end{array}
$$

When outside energy lifts a valence-orbit electron to a higher energy level (larger radius), the departing electron leaves a vacancy in the outer orbit (see Fig. 2-5c). We call this vacancy a *hole*. The hole is equivalent to a broken covalent bond, symbolized by Fig. 2-5d. When a hole is present, the cores pull only on the remaining electron; as a result, the covalent force is cut in half.

Figure 2-6 shows a *bonding diagram* for many silicon atoms in a crystal. All covalent bonds are intact; each atom has eight electrons traveling in its valence orbit. For reasons explained only with advanced equations, this orbit is filled or saturated when it contains eight electrons. No more electrons can fit in the valence orbit; additional electrons have to go into higher energy levels, that is, larger orbits.

An isolated germanium atom has four outer-orbit electrons. For this reason, a germanium crystal also has a bonding diagram like Fig. 2-6.

2-5. ENERGY BANDS

When silicon atoms combine into a crystal, the orbit of an electron is influenced not only by charges in its own atom but by the nucleus and electrons of every other atom in the crystal. The nearest charges have the greatest influence, but even the faraway ones have some effect, however small. In other words, *the orbit of each electron is controlled to some extent by every charge in the crystal.*

Since each electron has a different position inside the crystal, no two electrons see exactly the same pattern of surrounding charges. Because of this, the energy of each electron is different. Figure 2-7 shows what happens to energy levels. All electrons traveling in first orbits have slightly different energy levels because no two see exactly the same charge environment. Since there are billions of first-orbit electrons, the slightly different energy levels form a cluster or band. Similarly, the billions of second-orbit electrons, all with slightly different energy levels, form the second energy band shown. And all third-orbit electrons form the third band.

Figure 2-7 shows the energy bands as dark regions. This will be our way of indicating filled or saturated bands; that is, every permitted orbit is occupied by an electron. If we leave part of a band unshaded, it means some orbits are empty; equivalently, some energy levels are vacant.

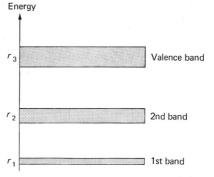

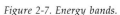

Figure 2-7. Energy bands.

Figure 2-7 shows the energy levels in a silicon crystal at absolute zero temperature (−273°C). The next section tells how this picture changes at higher temperatures.

2-6. CONDUCTION IN CRYSTALS

How well does a silicon crystal conduct? Figure 2-8*a* shows a bar of silicon with metal end surfaces. An external voltage sets up an electric field between the ends of the crystal.

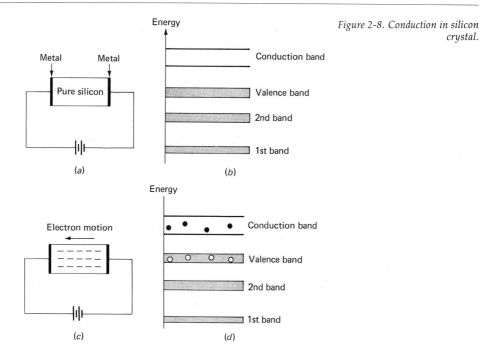

Figure 2-8. Conduction in silicon crystal.

Does current flow? It depends. On what? On whether there are any movable electrons inside the crystal.

Absolute zero

At absolute zero temperature, electrons cannot move through the crystal. All electrons are tightly held by the silicon atoms; inner-orbit electrons are buried deep within atoms; outer-orbit electrons are part of the covalent bonding and cannot break away without receiving outside energy. Therefore, at absolute zero temperature a silicon crystal acts like a perfect insulator.

In terms of energy, here is why silicon will not conduct at absolute zero temperature. Figure 2-8b shows the energy-band diagram. The first three bands are filled, and electrons cannot move easily in these bands. But beyond the valence band is a *conduction band.* This band represents the next larger group of radii that satisfy the particle-wave nature of an electron. At absolute zero temperature, the conduction band is empty; this means no electron has enough energy to travel in a conduction-band orbit. We get no current because no electrons are in the conduction band; furthermore, electrons in lower bands cannot move because there is no room to move.

Above absolute zero

Raise the temperature above absolute zero and things change. The incoming heat energy breaks some covalent bonds, that is, knocks valence electrons into the conduction band. In this way, we get a limited number of conduction-band electrons symbolized by the minus signs of Fig. 2-8c. Under the influence of the electric field, these electrons move to the left and set up a current.

Above absolute zero, we visualize the energy bands shown in Fig. 2-8d. Heat energy has lifted some electrons into the conduction band where they travel in orbits of larger radii than before. In these larger conduction-band orbits, the electrons are only loosely held by the atoms. Furthermore, so many vacant conduction-band orbits exist that an electron can easily move from one atom to the next with only a slight change in radius or energy.

In Fig. 2-8d, each time an electron is bumped up to the conduction band, a hole is created in the valence band. Therefore, the valence band is no longer saturated or filled: each hole represents an available orbit of rotation.

The higher the temperature, the greater the number of electrons kicked up to the conduction band, and the larger the current in Fig. 2-8c. At room temperature (around 25°C) the current is too small to be useful in most applications. At this temperature, a piece of silicon is neither a good insulator nor a good conductor. For this reason, it is called a *semiconductor.*

Silicon versus germanium

A germanium crystal also is a semiconductor at room temperature. But there is crucial difference between silicon and germanium. The energy gap between the valence

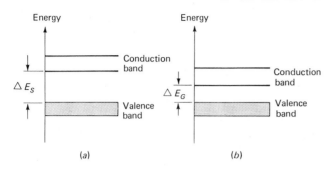

Figure 2-9. Silicon and germanium energy gaps.

band and the conduction band is greater for silicon than for germanium, as shown in Fig. 2-9*a* and *b*. In symbols,

$$\Delta E_S > \Delta E_G$$

where Δ stands for "the change in." This means it takes more heat energy to free valence electrons in silicon. Therefore, fewer conduction-band electrons exist in a silicon crystal than in a germanium crystal. Because of this, much less current flows in a pure silicon crystal than in a pure germanium crystal of the same size.

The fact that a pure silicon crystal has less current than a comparable germanium crystal turns out to be an overwhelming advantage for silicon. (The reason will be given when we discuss *diodes*.) Because of this advantage, silicon has become the main semiconductor material. Unless otherwise specified, all devices we discuss are made of silicon.

2-7.   HOLE CURRENT

Conduction-band electrons produce a current we are already familiar with. It is the same kind of current that flows in a copper wire. Conduction-band electrons are in large orbits and are only loosely held by atoms. With many unoccupied conduction-band orbits, an electron can move easily from one atom to the next.

Holes also can move and produce a current. This kind of current is new to us because it does not occur in copper wire. In other words, in a semiconductor there are two distinct kinds of current: conduction-band current and hole current.

How holes move

At the extreme right of Fig. 2-10*a* is a hole. This hole attracts the valence electron at *A*. With only a slight readjustment of energy, the valence electron at *A* can move into the hole. When this happens, the original hole vanishes and a new one appears at position *A*. The new hole at *A* can attract and capture the valence electron at *B*. When the valence electron moves from *B* to *A*, the hole moves from *A* to *B*. The motion of valence electrons can continue along the path shown by the arrows; the holes move in the opposite direc-

Figure 2-10. Three ways to visualize hole current.

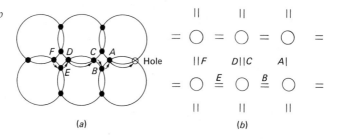

(a)　　　　　　　(b)

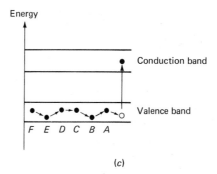

(c)

tion. What it boils down to is this: because holes are present in valence orbits, we have a second way for electrons to move through the crystal.

You can also visualize the hole current in terms of a bonding diagram (see Fig. 2-10b). Just to the right of A is a broken bond, or hole. When the valence electron moves into the hole, the bond at A is broken. Continuing in this way, the broken bond (the hole) can move from A to B to C and so on.

Here is what happens in terms of energy levels. To begin with, *thermal energy* (same as heat energy) bumps an electron from the valence band into the conduction band. This leaves a hole in the valence band as shown in Fig. 2-10c. With a slight readjustment of energy, the valence electron at A can move into the hole. When this happens, the original hole disappears and a new one appears at A. Next, the valence electron at B can move into the new hole with a slight change in energy. In this way, with minor changes in energy, valence electrons can move along the path shown by the arrows. This is equivalent to the hole moving through the valence band along the *ABCDEF* path.

In other words, a semiconductor gives electrons two ways to produce current because it has two energy bands in which the electrons can travel. Under the influence of an electric field, the conduction-band electrons move toward one end of the crystal; by counting the number that flow, we will have one component of the current. The valence electrons move toward the same end of the crystal; by counting the number that flow, we get the second component of current. The sum of the two components is the total current.

Electron-hole pairs

When we apply an external voltage across a crystal, it forces the electrons to move. In Fig. 2-11a, two kinds of movable electrons exist: conduction-band electrons and valence electrons. The motion of valence electrons to the right means holes are moving to the left.

Most of the time, we prefer to talk about holes rather than valence electrons. In a pure semiconductor, each conduction-band electron means a hole exists in the valence orbit of some atom. In other words, thermal energy produces *electron-hole pairs*. The holes act like positive charges and for this reason are shown as plus signs in Fig. 2-11b. As before, we visualize conduction-band electrons moving to the right. But now, we think of holes (positive charges) moving to the left.

Recombination

In Fig. 2-11b, you may wonder why the negative signs don't cancel the positive signs. To some extent, this does happen. Each minus sign is a conduction-band electron in a large orbit, and each plus sign is a hole in a smaller orbit. Occasionally, a conduction-band orbit of one atom may overlap the hole orbit of another. Because of this, a conduction-band electron can fall into a hole every so often. This merging of a conduction-band electron and a hole is called *recombination*. When recombination takes place, the hole does not move elsewhere; it disappears.

Recombination occurs continuously in a semiconductor. Because of this, every hole would eventually be filled except for one thing. Incoming heat energy continuously produces new electron-hole pairs. Therefore, at any instant in time, a number of electron-hole pairs exist. The average time between the creation and disappearance of an electron-hole pair is called the *lifetime*; it varies from a few nanoseconds to several microseconds, depending on how perfect the crystal structure is, and other factors.

2-8. DOPING

We will call a *pure* silicon crystal (every atom a silicon atom) an *intrinsic semiconductor*. The only current carriers in an intrinsic semiconductor are the electron-hole pairs. For most applications, not enough of these exist to produce a usable current.

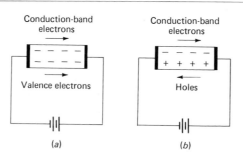

Figure 2-11. Two components of current.

Doping means adding impurity atoms (nonsilicon atoms) to a crystal to increase either the number of conduction-band electrons or the number of holes. When a crystal has been doped, it is called an *extrinsic semiconductor.*

n-type semiconductor

To get extra conduction-band electrons, we add *pentavalent* atoms; these have *five* electrons in the valence orbit. (Examples are arsenic, antimony, and phosphorus.) After adding pentavalent atoms to a pure silicon crystal, we still have mostly silicon atoms. But every now and then, we find a pentavalent atom between four neighbors as shown in Fig. 2-12a. The pentavalent atom originally had five electrons in its valence orbit. After forming covalent bonds with four neighbors, this central atom has an extra electron left over. Since the valence orbit can hold no more than eight electrons, the extra electron must travel in a conduction-band orbit.

For a crystal doped by a pentavalent impurity, here is what we find:

1. Many new conduction-band electrons produced by doping. Since each pentavalent atom contributes one conduction-band electron, we can control the number of conduction-band electrons by the amount of impurity added.

Figure 2-12. Pentavalent doping.

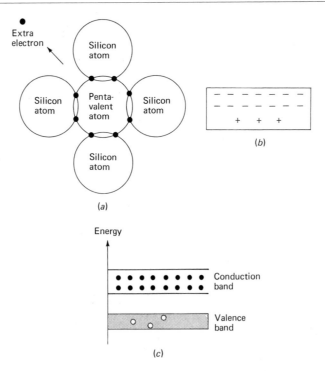

(a)

(b)

(c)

2. Thermal energy still generates a few electron-hole pairs. These are far fewer in number than the conduction-band electrons produced by doping.

Figure 2-12*b* shows a crystal that has been doped by a pentavalent impurity. We have a large number of conduction-band electrons produced mostly by doping. Only a few holes exist, created by thermal energy. For obvious reasons, we call the electrons the *majority carriers* and the holes the *minority carriers*. Doped silicon of this kind is known as *n*-type semiconductor, where *n* stands for negative.

Figure 2-12*c* shows the energy diagram of an *n*-type semiconductor. The valence band has a few holes; the conduction band has many electrons.

p-type semiconductor

How can we dope a crystal to get extra holes? By using a *trivalent* impurity (one with three electrons in the outer orbit). Examples are aluminum, boron, and gallium. After adding the impurity, we find each trivalent atom between four neighbors as shown in Fig. 2-13*a*. Since each trivalent atom brought only three valence-orbit elec-

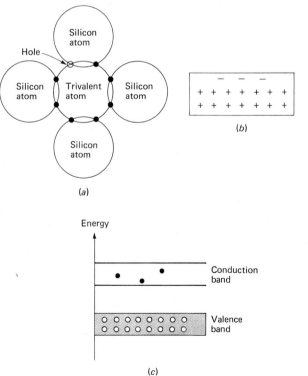

Figure 2-13. Trivalent doping.

trons with it, only seven electrons will travel in its valence orbit. In other words, one hole appears in each trivalent atom. By controlling the amount of impurity added, we can control the number of holes in the doped crystal.

A semiconductor doped by a trivalent impurity is known as a *p*-type semiconductor; the *p* stands for positive. As shown in Fig. 2-13*b*, the holes of a *p*-type semiconductor far outnumber the conduction-band electrons. For this reason, the holes are the majority carriers in a *p*-type semiconductor, while the conduction-band electrons are the minority carriers.

Figure 2-13*c* shows the energy diagram for a *p*-type semiconductor. The conduction band has a few electrons, produced by thermal energy. The valence band, however, has many holes produced by doping. These holes provide the room needed for the valence electrons to move easily.

A final point. A doped semiconductor still has resistance; we call this resistance the *bulk resistance*. The greater the doping, the lower the bulk resistance. On the other hand, a lightly doped semiconductor has a higher bulk resistance.

Problems

2-1. Calculate the radii of the fourth and fifth orbits in a hydrogen atom.

2-2. Figure 2-14*a* shows a hypothetical energy diagram. How much energy must an electron receive to go from the second to the third level? If an electron falls from the fourth to the first level, how much energy does it radiate?

Figure 2-14.

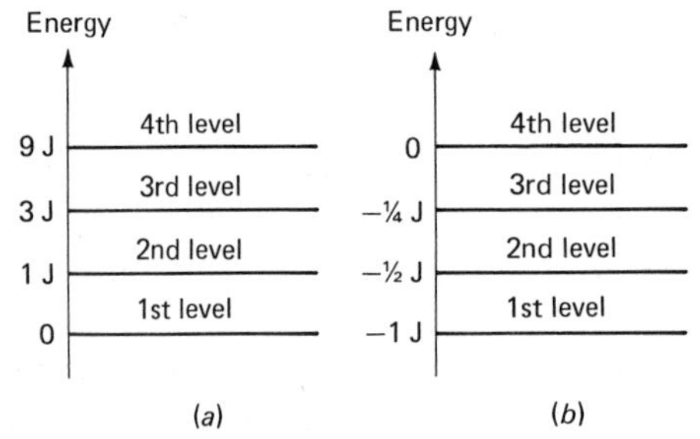

2-3. The fourth level of Fig. 2-14*b* has zero energy. The lower levels therefore have negative energy with respect to the fourth level. How much energy does an electron receive if it rises from the first to the third level? If an electron falls from the fourth to the third level, how much energy does it give up?

3. pn Junctions

By itself, an *n*-type semiconductor is about as useful as a carbon resistor; the same can be said for *p*-type material. But, dope a crystal so half is *n*-type and the other half *p*-type, and you have something. For one thing, a *pn* crystal conducts better in one direction than in the other. This one-way action has all kinds of applications.

The *junction* is where the *p*-type and *n*-type regions meet, and *junction diode* is another name for a *pn* crystal. (The word *diode* is a contraction of *two* electr*ode,* where *di* stands for two.) This chapter describes how junction diodes work and prepares us for the transistor, which combines two junction diodes.

3-1. THE UNBIASED DIODE

Figure 3-1*a* shows a junction diode. The *p* side has many holes and the *n* side many conduction-band electrons. To avoid confusion, no minority carriers are shown; bear in mind, though, a few conduction-band electrons are on the *p* side and a few holes on the *n* side.

The conduction-band electrons on the *n* side travel in large orbits. With minor changes in energy, a conduction-band electron can move from a large orbit around one atom to a large orbit around another. In other words, electrons on the *n* side can move easily through the *n* region.

The depletion layer

Because of their repulsion for each other, the electrons on the *n* side tend to *diffuse* (spread) in all directions. Some diffuse across the junction. When an electron enters the

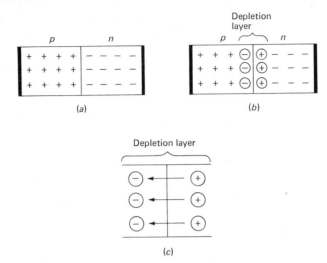

Figure 3-1. Charges in a diode. (a) Before diffusion. (b) After diffusion. (c) Depletion-layer field.

p region, it becomes a minority carrier. With so many holes around it, this minority carrier has a short lifetime; soon after entering the p region, the electron will fall into a hole. When this happens, the hole disappears and the conduction-band electron becomes a valence electron.

An atom with fewer electrons than protons has a positive charge; such an atom is called a *positive ion*. If an atom has more electrons than protons, it is *negative ion*. Each time an electron diffuses across the junction, it creates a pair of ions. When an electron leaves the n side, it leaves behind a pentavalent atom that is short one negative charge; this pentavalent atom becomes a positive ion. After the migrating electron falls into a hole on the p side, it makes a negative ion out of the trivalent atom that captures it.

Figure 3-1b shows these ions on each side of the junction. The circled plus signs are the positive ions, and the circled minus signs are the negative ions. The ions are fixed in the crystal structure because of covalent bonding and cannot move around like conduction-band electrons or holes.

Each pair of positive and negative ions in Fig. 3-1b is called a *dipole*. The creation of a dipole means one conduction-band electron and one hole have been taken out of circulation. As the number of dipoles builds up, the region near the junction is emptied of movable charges. We call this charge-empty region the *depletion layer* (see Fig. 3-1b).

Barrier potential

Each dipole has an electric field (Fig. 3-1c). The arrows show the direction of force on a positive charge. Therefore, when an electron enters the depletion layer, the field tries to push the electron back into the n region. The strength of the field increases with each crossing electron until equilibrium is reached. To a first approximation, this means the field eventually stops diffusion of electrons across the junction.

To a second approximation, we need to include minority carriers. Remember the *p* side has a few thermally produced conduction-band electrons. Those inside the depletion layer are pushed by the field into the *n* region. This slightly reduces the depletion layer, decreases the field strength, and lets a few majority carriers diffuse from right to left to restore the field to its original strength.

Here is our final picture of equilibrium at the junction:

1. A few minority carriers drift from left to right across the junction. They would reduce the field except that
2. A few majority carriers diffuse from right to left across the junction and restore the field to its original value.

In Fig. 3-1*b*, having a field between ions is equivalent to a difference of potential called the *barrier potential*. At 25°C, the barrier potential approximately equals 0.3 V for germanium diodes and 0.7 V for silicon diodes.

Temperature effects

The value of barrier potential depends on temperature. Higher temperature creates more electron-hole pairs. As a result, the drift of minority carriers across the junction increases. This forces equilibrium to occur at a slightly lower barrier potential.

As a guide, we will use this approximation for changes in barrier potential: for either germanium or silicon diodes, the barrier potential decreases 2 mV for each Celsius degree rise.[1] In symbols, the change in barrier potential equals

$$\Delta V = -0.002 \, \Delta T \qquad \text{(3-1)} ***$$

where Δ stands for "the change in." Also, the equation is marked with a triple asterisk ***; this means the formula is easy to remember but very important.

As an example, at 25°C the barrier potential of a silicon diode equals 0.7 V. Raise the junction temperature to 100°C and you get a change of

$$\Delta V = -0.002 \, (100 - 25)$$
$$= -0.150 \text{ V}$$

Therefore, the barrier potential at 100°C equals approximately 0.7 − 0.15, or 0.55 V.

Equation (3-1) is a useful guide; it helps us estimate the effects of temperature change on diode and transistor circuits.

EXAMPLE 3-1.
Calculate the barrier potential at 0°C for a silicon diode.

[1] This rule is derived in Millman, J., and H. Taub, *Pulse, Digital, and Switching Waveforms*, McGraw-Hill Book Company, New York, 1965, p. 183.

SOLUTION.

At 25°C, the barrier potential of a silicon diode equals 0.7 V, approximately. When the temperature drops to 0°C, the barrier potential increases because fewer minority carriers drift across the junction. With Eq. (3-1),

$$\Delta V = -0.002\,(0 - 25)$$
$$= 0.05\text{ V}$$

Therefore, at 0°C the barrier potential equals approximately 0.75 V.

3-2.   THE ENERGY HILL

Let's take a close look at the unbiased diode from the energy viewpoint. This will give us more insight into the action at a junction.

Before diffusion

Assuming an abrupt junction (one that suddenly changes from p to n material), what does the energy diagram look like? Figure 3-2a shows the energy bands before electrons have diffused across the junction. The p side has many holes in the valence band, and the n side many electrons in the conduction band. But why are the p bands slightly higher than the n bands?

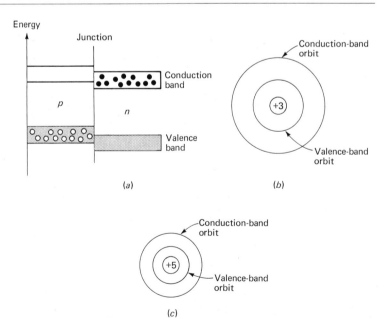

Figure 3-2. (a) *Energy bands before diffusion.* (b) p *orbits.* (c) n *orbits.*

The p side has trivalent atoms with a core charge of +3, shown in Fig. 3-2b. On the other hand, the n side has pentavalent atoms with a core charge of +5 (Fig. 3-2c). A +3 core attracts an electron less than a +5 core. Therefore, the orbits of a trivalent atom (p side) are slightly larger than those of a pentavalent atom (n side). This is why the p bands of Fig. 3-2a are slightly higher (more energy and larger radius) than the n bands.

A truly abrupt junction is an idealization; the p side does not suddenly end where the n side begins. A better picture allows for a gradual change from one material to the other. To a second approximation, Fig. 3-3a shows the energy diagram for a junction *before* diffusion has occurred.

At equilibrium

What happens when conduction-band electrons diffuse across the junction? The electrons near the top of the n conduction band move across the junction as previously described. Not only does this create the depletion layer, it also changes the energy levels in the junction area.

Figure 3-3b shows the energy diagram after equilibrium is reached. The p bands have moved up with respect to the n bands. In fact, the bottom of each p band is even with the top of the corresponding n band. This means electrons on the n side no longer have enough energy to get across the junction. What follows is a simplified explanation of the reasons why.

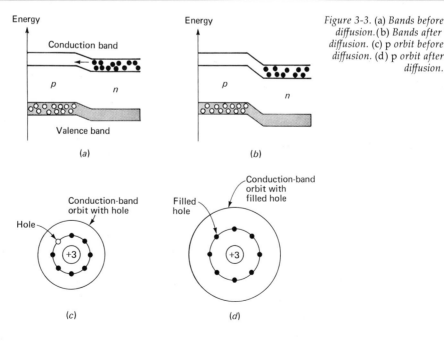

Figure 3-3. (a) *Bands before diffusion.* (b) *Bands after diffusion.* (c) p *orbit before diffusion.* (d) p *orbit after diffusion.*

Figure 3-3c shows a conduction-band orbit around one of the trivalent atoms before diffusion has occurred. When an electron diffuses across the junction, it fills the hole of a trivalent atom (see Fig. 3-3d). The extra electron will push the conduction-band orbit farther away from the trivalent atom as shown in Fig. 3-3d. Therefore, any other electrons coming into the area will need more energy than before to travel in a conduction-band orbit. This is equivalent to saying the p bands move up with respect to the n bands after the depletion layer has built up.

At equilibrium, conduction-band electrons on the n side travel in orbits not quite large enough to match the p-side orbits. In other words, electrons on the n side do not have enough energy to get across the junction. To an electron trying to diffuse across the junction, the path it must travel looks like a hill, an *energy hill* (see Fig. 3-3b). The electron cannot climb this hill unless it receives energy from an outside source.

3-3. FORWARD BIAS

Figure 3-4a shows a dc source across a diode. The negative source terminal connects to the n-type material, and the positive terminal to the p-type material. We call this connection *forward bias*. How does it affect the diode?

Figure 3-4. Forward bias. (a)
Charges. (b) *Fields.* (c) *Bands.*

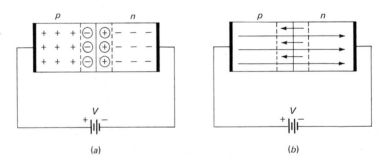

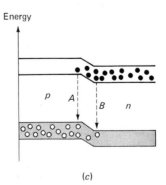

(c)

Depletion layer narrows

The dc source sets up the electric field shown in Fig. 3-4*b*. This field opposes the field of the depletion layer. Therefore, it pushes electrons and holes toward the junction. As the holes and electrons move toward the junction, they *deionize* the edges of the depletion layer. This narrows the depletion layer. The greater the external voltage, the narrower the depletion layer becomes.

Large forward current

Current flows easily in a circuit like Fig. 3-4*a*. Why? To begin with, when those conduction-band electrons move toward the junction, the right end of the crystal becomes slightly positive. This happens because electrons at the right end of the crystal move toward the junction and leave positively charged atoms behind. The positively charged atoms then pull electrons into the crystal from the negative source terminal.

When electrons on the *n* side get near the junction, they recombine with holes. These recombinations occur at varying distances from the junction, depending on how long a conduction-band electron can avoid falling into a hole. The odds are high that recombinations occur very close to the junction. To a first approximation, we can visualize all the conduction-band electrons recombining when they reach the junction.

When we look at Fig. 3-4*a*, here is what we see. Electrons are pouring into the right end of the crystal, while the bulk of electrons in the *n* region move toward the junction. The left edge of this moving group disappears as it hits the junction (the electrons fall into holes). In this way, there is a continuous stream of electrons from the negative source terminal toward the junction.

Those conduction-band electrons disappearing at the junction—what happens to them? They become valence electrons. As valence electrons, they can move through holes in the *p* region. We can visualize the valence electrons on the *p* side moving en masse toward the left end of the crystal; this is equivalent to holes moving to the right. When the valence electrons reach the left end of the crystal, they leave the crystal and flow into the positive terminal of the source.

Here are the highlights of what happens to an electron in Fig. 3-4*a*:

1. After leaving the negative source terminal, it enters the right end of the crystal.
2. It travels through the *n* region as a conduction-band electron.
3. Near the junction it recombines and becomes a valence electron.
4. It travels through the *p* region as a valence electron.
5. After leaving the left end of the crystal, it flows into the positive source terminal.

Energy bands

Forward bias lowers the energy hill (see Fig. 3-4*c*). Because of this, conduction-band electrons have enough energy to invade the *p* region. Soon after entering the *p* region, each electron falls into a hole (path *A*). As a valence electron, it continues its journey toward the left end of the crystal.

A conduction-band electron may fall into a hole even before it crosses the junction. In Fig. 3-4*c* a valence electron may cross the junction from right to left; this leaves a hole just to the right of the junction. This hole does not live long. A conduction-band electron soon falls into it (path *B*).

Regardless of where the recombination takes place, the result is the same. A steady stream of conduction-band electrons moves toward the junction and falls into holes near the junction.[2] The captured electrons (now valence electrons) move left in a steady stream through the holes in the *p* region. In this way, we get a continuous flow of electrons through the diode.

3-4. REVERSE BIAS

Turn the dc source around and you *reverse-bias* the diode as shown in Fig. 3-5*a*. Does the diode still conduct? What happens to the depletion layer? To the energy bands?

Depletion layer widens

The externally produced field is in the same direction as the depletion-layer field (Fig. 3-5*b*). Because of this, holes and electrons move toward the ends of the crystal (away from the junction). The fleeing electrons leave positive ions behind, and departing holes leave negative ions. Because of this, the depletion layer gets wider. The greater the reverse bias, the wider the depletion layer becomes.

How wide does the depletion layer get? In Fig. 3-5*a*, when the holes and electrons move away from the junction, the newly created ions increase the difference of potential across the depletion layer. The wider the depletion layer, the greater the difference of potential. The depletion layer stops growing when its difference of potential equals the applied reverse voltage. When this happens, electrons and holes stop moving away from the junction.

Transient current

How much current is there in a reverse-biased diode? While the depletion layer is adjusting to its new width, holes and electrons move away from the junction. This means electrons are flowing from the negative source terminal into the left end of the crystal; at the same time, electrons are leaving the right end of the crystal and flowing into the positive source terminal. Therefore, a current flows in the external circuit while the depletion layer is adjusting to its new width. This *transient* current drops to zero after the depletion layer stops growing.

---

[2] When the electron falls into a hole, it gives off energy in the form of heat, light, etc. This is the principle behind the light-emitting diode (LED).

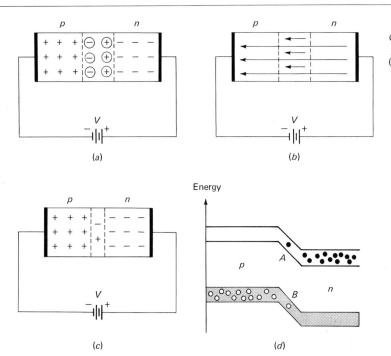

Figure 3-5. Reverse bias. (a) Charges. (b) Fields. (c) Thermal production of minority carriers. (d) Electron-hole pair in energy bands.

The amount of time this current flows depends on the RC time constant and other factors. All we need to know now is that the depletion layer adjusts its width very rapidly. To a first approximation, we can neglect the small transient current that flows.[3]

Minority-carrier current

Is there any current at all after the depletion layer settles down? Yes. A small current flows. Thermal energy creates electron-hole pairs. In other words, a few minority carriers exist on both sides of the junction. Most of these recombine with the majority carriers. But those inside the depletion layer may live long enough to get across the junction. When this happens, a small current flows in the external circuit.

Figure 3-5c illustrates the idea. When an electron-hole pair is created inside the depletion layer, the depletion-layer field pushes the electron to the right, forcing one electron to leave the right end of the crystal. The hole in the depletion layer is pushed to the left. This extra hole on the p side lets one electron enter the left end of the crystal and fall into a hole. Since thermal energy is continuously producing electron-hole pairs near the junction, we get a small continuous current in the external circuit.

[3] Above 10 MHz this transient current prevents normal diode action. Often a special diode, the hot-carrier (Schottky-barrier) diode, is used at very high frequencies.

The reverse current caused by the minority carriers is called the *saturation current,* designated I_S. The name *saturation* reminds us we cannot get more minority-carrier current than is produced by thermal energy. In other words, increasing the reverse voltage will not increase the number of thermally created minority carriers.

Thermal energy produces saturation current; the higher the temperature, the greater the saturation current. A good practical rule to remember is this: I_S approximately doubles for each 10°C rise; this is equivalent to a 7 percent increase per degree rise.[4] For example, if I_S equals 5 nA (nanoamperes) at 25°C, it will approximately equal 10 nA at 35°C, 20 nA at 45°C, 40 nA at 55°C, and so on.

As mentioned earlier, the energy gap between the valence band and conduction band is greater in silicon than in germanium. Because of this, thermal energy produces fewer minority carriers in silicon diodes than in germanium diodes. In other words, a silicon diode has a much smaller I_S than a germanium diode with the same junction area. This is an immense advantage for silicon and is one of the reasons silicon dominates the semiconductor field.

Figure 3-5d illustrates saturation current in terms of energy bands. Assume an electron-hole pair appears near the junction at A and B. The electron at A falls down the energy hill, pushing an electron out the right end of the n conduction band. Similarly, a valence electron falls down the hill into the hole at B. The falling valence electron leaves a hole behind. This extra hole on the p side lets an electron enter the left end of the crystal.

The greater the reverse voltage, the steeper the energy hill. A conduction-band electron falling down this hill can gain a high velocity. The importance of this is brought out when we discuss *breakdown voltage.*

Surface-leakage current

Besides transient current and minority-carrier current, does any other current flow in a reverse-biased diode? Yes. A small current flows on the *surface* of the crystal. We call this component the *surface-leakage current.* The exact mechanism for this current is not completely understood, although most people believe it is caused by surface impurities and imperfections.

Here is one way imperfections may produce a surface current. Suppose the atoms at the top of Fig. 3-6a are atoms on the surface of the crystal. With no neighbors on top of them, these atoms have broken covalent bonds, that is, holes. Visualize these holes along the surface of the crystal as shown in Fig. 3-6b. In effect, the skin of a crystal is like a p-type semiconductor. Because of this, electrons can enter the left end of the crystal, travel through the surface holes, and leave the right end of the crystal. In this way, we get a small reverse current along the surface; this current increases when you increase the reverse voltage.

[4] This rule is derived in Millman, J., and H. Taub, *Pulse, Digital, and Switching Waveforms,* McGraw-Hill Book Company, New York, 1965, pp. 182–183.

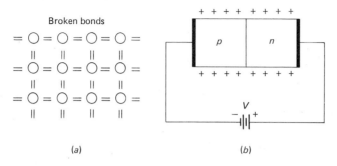

Broken bonds

(a)   *(b)*

Figure 3-6. Surface leakage current.

Breakdown voltage

Keep increasing the reverse voltage and you eventually reach the *breakdown voltage*. For rectifier diodes (those manufactured to conduct better one way than the other), the breakdown voltage is usually greater than 50 V. Once the breakdown voltage is reached, a large number of minority carriers appear in the depletion layer and the diode conducts heavily.

Where do the carriers suddenly come from? Figure 3-7a shows a thermally produced electron-hole pair inside the depletion layer. With reverse bias, the electron is

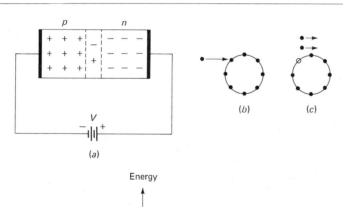

Figure 3-7. Breakdown effect. (a) Diode charges. (b) Collision. (c) Two moving charges. (d) Avalanche.

(b)   *(c)*

(a)

Energy

(d)

pushed right and the hole left. As it moves, the electron gains speed. The stronger the depletion-layer field, the faster the electron moves. For large reverse voltages, the electron reaches high velocities. This high-speed electron may collide with a valence electron (see Fig. 3-7b). If the high-speed electron has enough energy, it can bump the valence electron into a conduction-band orbit. This results in two conduction-band electrons as shown in Fig. 3-7c. Both of these will now accelerate and may go on to dislodge two more electrons. In this way, the number of minority carriers may become quite large and the diode can conduct heavily.

In terms of the energy hill, the larger the reverse bias, the steeper the hill becomes (Fig. 3-7d). A few thermally produced conduction-band electrons gain very high speeds as they fall down the energy hill. When they collide with valence electrons in the depletion layer, they bump the valence electrons loose. Because the action reminds us of an avalanche, we call it the *avalanche effect.*

A final point. With a rectifier diode, you should not exceed the breakdown voltage. The manufacturer specifies the largest reverse voltages his diodes can handle; we discuss these maximum ratings in the next chapter.

3-5. BIPOLAR AND UNIPOLAR DEVICES

The junction diode is a *bipolar device.* (Bipolar is a contraction of *two polar*ities with bi used for two). A bipolar device needs holes *and* conduction-band electrons to work properly.

Some devices are *unipolar;* they need holes *or* electrons to operate normally. A carbon resistor, for instance, is a unipolar device; it uses conduction-band electrons only. If we wanted, we could make a resistor out of *p*-type semiconductor; it would be a unipolar device.

In later chapters, we discuss transistors. Bipolar transistors are older and more widely used; they are also less expensive. Unipolar transistors (called field-effect transistors, or FETs) are important in special applications.

Figure 3-8 shows a chip, a small piece of semiconductor material. The dimensions are representative; chips are often smaller than this, occasionally larger. By advanced photographic techniques, a manufacturer can produce circuits on the surface of this chip, circuits containing many diodes, resistors, transistors, etc. The finished network is so small you need a microscope to see the connections. We call a circuit like this an *integrated circuit* (IC).

Figure 3-8. Semiconductor chip.

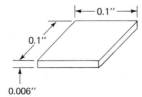

0.1″

0.1″

0.006″

We will discuss ICs in Chap. 17. For now, we should know they can be bipolar or unipolar. Bipolar ICs are older and more widely used; the unipolar ICs are becoming important because of their low power needs and other advantages.

A final point. A *discrete circuit* is the kind of circuit you build when you connect separate resistors, capacitors, transistors, etc. Each component you add to the circuit is discrete, that is, distinct or separate from the others. This differs from an integrated circuit where components are atomically part of the semiconductor chip.

Problems

3-1. Calculate the approximate barrier potential of a silicon diode at 60°C, 90°C, and 150°C.

3-2. What is the approximate barrier potential for a germanium diode at −20°C?

3-3. A silicon diode has a saturation current of 2 nA at 25°C. What is the value of I_S at 75°C? At 125°C?

3-4. At 25°C, a silicon diode has a reverse current of 25 nA. The surface leakage component equals 20 nA. If the surface current still equals 20 nA at 75°C, what does the total reverse current equal at 75°C?

4. Diode Equivalent Circuits

This chapter is about approximations for a rectifier diode. With these approximations, you can analyze diode circuits quickly and easily. These approximations are also a big help in transistor-circuit analysis.

4-1. THE RECTIFIER DIODE

In Fig. 4-1a, the ac source pushes electrons up the resistor during the positive half cycle of input voltage, and down the resistor during the negative half cycle. The up and down currents are equal.

Put a diode in the circuit and things change. In Fig. 4-1b, the positive half cycle of input voltage will forward-bias the diode; therefore, the diode conducts during the positive half cycle. But on the negative half cycle, the diode is reverse-biased and only a small reverse current can flow. The arrows symbolize the large upward flow of electrons and the small downward flow. The diode has rectified the ac current, that is, changed it from an alternating current to a unidirectional current.

When a manufacturer optimizes a diode for use as a rectifier, the diode is called a rectifier diode. This is the oldest and most widely used type. In the next few chapters, we concentrate on the rectifier diode.

Figure 4-1c shows the schematic symbol of a rectifier diode. The p side is called the *anode,* and the n side the *cathode.* The diode symbol looks like an arrow that points from the p side to the n side. Because of this, it is a reminder that *conventional* current flows easily from the p side to the n side.

A forward-biased diode conducts well, and a reverse-biased one poorly. Therefore, when analyzing diode circuits, one of the things to decide is whether the diode is

Figure 4-1. (a) Equal currents. (b) Rectified current. (c) Schematic symbol.

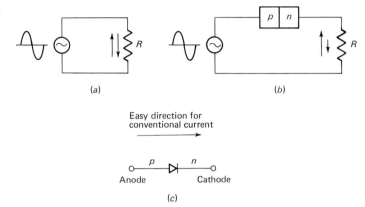

(a)　　　　　　　(b)

Easy direction for
conventional current

Anode　　　　　　Cathode

(c)

forward- or reverse-biased. This is not always easy to do. But here is something that helps. Ask yourself this question: Is the external circuit trying to push conventional current in the direction of the diode arrow or in the opposite direction? If conventional current is in the same direction as the diode arrow, the diode is forward-biased. On the other hand, if conventional current tries to flow opposite the arrowhead, the diode is reverse-biased.

EXAMPLE 4-1.
In each diode circuit of Fig. 4-2, determine whether the diodes are forward- or reverse-biased.

Figure 4-2. Example 4-1.

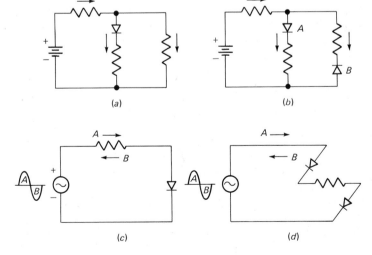

(a)　　　　　　　(b)

(c)　　　　　　　(d)

SOLUTION.

Fig. 4-2a. Conventional current flows out of the positive source terminal and down through each vertical branch as shown. Since conventional current flows in the direction of the diode arrow, the diode is forward-biased.

Fig. 4-2b. As before, conventional current flows down through each vertical branch. Diode *A* is forward-biased, but diode *B* is reverse-biased. Diode *A* conducts well, diode *B* poorly.

Fig. 4-2c. In this problem, time is important. During the positive half cycle, the source voltage is plus-minus as shown. Therefore, during time interval *A*, conventional current flows in the same direction as the diode arrow and the diode is forward-biased.

During the negative half cycle (time interval *B*), the polarity of the source voltage is opposite that shown. As a result, conventional current tries to flow up through the diode. Since this is opposite the diode arrow, the diode is reverse-biased.

Fig. 4-2d. During the positive half cycle, conventional current flows in the same direction as the diode arrows; therefore, both diodes are forward-biased. On the other hand, the negative half cycle of input voltage tries to push conventional current opposite the diode arrows. So, both diodes are reverse-biased during the negative half cycle.

4-2. THE FORWARD DIODE CURVE

We have the main idea behind a rectifier diode; it conducts better one way than the other. This is a start. We can now sharpen our understanding by graphing diode current versus diode voltage.

Figure 4-3a shows a circuit you can set up in the laboratory. Since the dc source pushes conventional current in same direction as the diode arrow, the diode is forward-biased. The greater the applied voltage, the larger the diode current. By varying the applied voltage, you can measure the diode current (use a series ammeter) and the diode voltage (a voltmeter in parallel with the diode). By plotting the corresponding currents and voltages, you get a graph of diode current versus diode voltage.

Knee voltage

Figure 4-3b shows how the graph looks for a forward-biased silicon diode. It is customary to plot voltage along the horizontal axis because voltage is the independent variable. Each value of diode voltage produces a particular current; the current is a dependent variable and is plotted along the vertical axis.

What does the graph tell us? To begin with, the diode doesn't conduct well until we overcome the barrier potential. This is why the current is small for the first few-tenths volt. As we approach 0.7 V, conduction-band electrons and holes start crossing

Figure 4-3. (a) *Diode circuit.* (b)
Forward curve. (c) *Bulk
resistance.* (d) *Forward currents.*

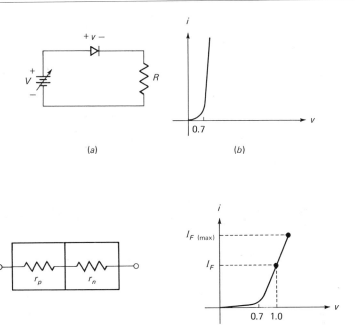

(a)   (b)

(c)   (d)

the junction in larger numbers. This is the reason the current starts increasing rapidly. Above 0.7 V, each 0.1-V increase produces a sharp increase in current.

The voltage where the current starts to increase rapidly is called the *knee voltage* of the diode. For a silicon diode, the knee voltage equals the barrier potential, approximately 0.7 V. A germanium diode, on the other hand, has a knee voltage of about 0.3 V.

Bulk resistance

Above the knee voltage, the diode current increases rapidly; small increases in diode voltage cause large increases in diode current. The reason is this: after overcoming the barrier potential, all that impedes current is the resistance of the p and n regions, symbolized by the r_p and r_n in Fig. 4-3c. Since any conductor has some resistance, so too does the p region have a certain amount of resistance. The n region also has some resistance. The sum of these resistances is called the *bulk resistance* of the diode. In symbols,

$$r_B = r_p + r_n$$

The value of bulk resistance r_B depends on the doping and size of the p and n regions; typically, r_B is from 1 to 25 Ω.

How to calculate bulk resistance

It helps to have the approximate value of r_B when analyzing a diode circuit. Here is a way to estimate r_B. A manufacturer's data sheet often gives you the amount of forward current I_F at 1 V (see Fig. 4-3d). For a silicon diode, the first 0.7 V is dropped across the depletion layer; the final 0.3 V is dropped across the r_B of the diode. Therefore,

$$I_F \cong \frac{0.3}{r_B}$$

or
$$r_B \cong \frac{0.3}{I_F} \qquad \text{for silicon diodes} \qquad (4\text{-}1)\,^{***}$$

where I_F is the dc forward current at 1 V. (The symbol $\cong$ means "approximately equal to.")

For germanium diodes, 0.3 V appears across the depletion layer and 0.7 V across the bulk resistance. Therefore, the bulk resistance of a germanium diode equals $0.7/I_F$, where I_F is the forward current at 1 V.

As an example, a 1N456 is a silicon diode with an I_F of 40 mA at 1 V. With Eq. (4-1),

$$r_B \cong \frac{0.3}{40\,(10^{-3})} = 7.5 \ \Omega$$

In a circuit using a 1N456, the first 0.7 V appears across the depletion layer. Any additional diode voltage is dropped across the 7.5 Ω of bulk resistance.

Maximum dc forward current

If the current in a diode is too large, excessive heat will destroy the diode. Even approaching the burnout value without reaching it can shorten diode life and degrade other properties. For this reason, a manufacturer's data sheet specifies the maximum current a diode can safely handle without shortening its life or degrading its characteristics.

The maximum dc forward current is one of the maximum ratings usually given on a data sheet. In Fig. 4-3d, this current is designated $I_{F(\max)}$. It represents the largest dc forward current the diode can safely handle. For instance, the 1N456 has an $I_{F(\max)}$ rating of 135 mA. This means it can safely handle a continuous forward current of 135 mA.

Current-limiting resistor

Which brings us to why a resistor is almost always used in series with a diode. In Fig. 4-3a, R is called a *current-limiting resistor*. The larger we make R, the smaller the diode current. The exact size of R will depend on what we are trying to do with the circuit. We will give examples later. For now it is enough to know the diode depends on R to keep the current less than the maximum rated current.

Maximum power dissipation

Closely related to the maximum dc forward current is the maximum power dissipation. Like a resistor, a diode has a power rating. This rating tells you how much power the diode can safely dissipate without shortening its life or degrading its properties.

When the diode current is a steady forward current, the product of diode voltage and diode current equals the power dissipated by the diode.

In many diode circuits, the current is not steady. Rather, it fluctuates (like the rectifier circuit of Fig. 4-1b). Because of this, you have to use the rms value of current to calculate the power dissipated by the diode. Under no condition should the maximum power rating of the diode be exceeded. We give examples in Chap. 5.

EXAMPLE 4-2.
Estimate the bulk resistance of each of these silicon diodes:

1. A 1N662 with an I_F of 10 mA at 1 V.
2. A 1N3070 with an I_F of 100 mA at 1 V.

SOLUTION.

1. The 1N662 has a bulk resistance of approximately

$$r_B = \frac{0.3}{0.010} = 30 \ \Omega$$

2. A 1N3070 has an approximate bulk resistance of

$$r_B = \frac{0.3}{0.100} = 3 \ \Omega$$

4-3. THE DIODE CURVE

When you reverse-bias a diode (Fig. 4-4a), you get only a small current. By measuring diode current and voltage, you can plot the reverse curve; it will look something like Fig. 4-4b. There are no surprises here; diode current is very small for all reverse voltages less than the breakdown voltage BV; at breakdown the current increases rapidly for small increases in voltage. As already mentioned, with a rectifier diode you always try to keep the applied voltage less than BV.

By using positive values for forward current and voltage, and negative values for reverse current and voltage, we can plot the forward and reverse curves in a single graph (see Fig. 4-5). This graph summarizes the action of a rectifier diode; it tells us how much diode current flows for each value of diode voltage. Conversely, if we know how much current is flowing, the graph tells us what the diode voltage is.

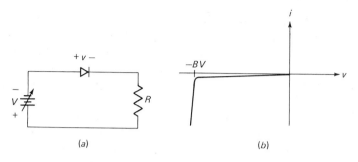

Figure 4-4. (a) Reverse bias. (b) Reverse curve.

Another point. In getting the graph, we used a dc source and measured dc current and voltage. For this reason, the graph is a plot of dc quantities, where dc means steady or continuous. Figure 4-5 does *not* include the small transient current produced when the depletion layer changes its width. (The depletion layer changes its width to match each value of applied voltage.) The adjustment in depletion-layer width is so rapid that by the time we measure current and voltage, the transient is gone. For this reason, Fig. 4-5 is sometimes called a *static curve,* where static means "at rest" or "standing still."

Even though Fig. 4-5 is a static curve, you can use it when the applied voltage is an ac signal, provided the frequency is not too high. If the frequency is too high, the depletion layer cannot adjust fast enough; in this case, the transient current becomes noticeable. At a later time, we will examine this high-frequency effect.

For now, here is what counts. At normal frequencies (those for which the rectifier diode is intended), Fig. 4-5 applies. This curve tells us the value of diode current for a particular diode voltage. Even though we measure the current and voltage in the circuits of Figs. 4-3a and 4-4a, the curve of Fig. 4-5 can be used in other circuits as well. In other words, no matter what circuit we use a diode in, a given diode voltage produces exactly the same current indicated by Fig. 4-5. This is why we can use Fig. 4-5 in all rectifier circuits we encounter.

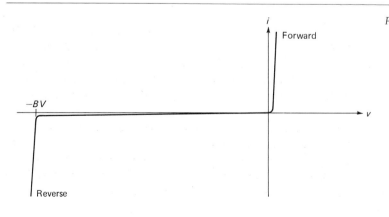

Figure 4-5. Complete diode curve.

4-4. THE IDEAL DIODE

Chapter 1 talked about the difficulty of getting exact answers. When even a piece of wire may become a back-breaking obstacle, we abandon all hope for exact analysis and settle for the degree of precision that suits the problem. In other words, to get anywhere in circuit analysis, we often must accept approximate answers.

The *ideal-diode* approximation strips away all but the bones of diode operation. What does a rectifier diode do? It conducts well in the forward direction and poorly in the reverse direction. Boil this down to its essence, and this is what you get: ideally, a rectifier diode acts like a perfect conductor (zero voltage) when forward-biased and like a perfect insulator (zero current) when reverse-biased as shown in Fig. 4-6a.

In circuit terms, an ideal diode acts like an *automatic switch*. When conventional current tries to flow in the direction of the diode arrow, the switch is closed (see Fig. 4-6b). If conventional current tries flowing the other way, the switch is open (Fig. 4-6c). This is rock bottom; we cannot simplify beyond this point without losing the main idea of a diode.

Extreme as the ideal-diode approximation seems at first, it gives good answers for most diode circuits. There will be times when the approximation breaks down; for this reason, we need a second and third approximation. But for all preliminary analysis of rectifier-diode circuits, the ideal diode is an excellent approximation.

EXAMPLE 4-3.
Use the ideal-diode approximation to find the output waveform in Fig. 4-7a.

Figure 4-6. (a) *Ideal-diode curve.* (b) *Closed-switch analogy.* (c) *Open-switch analogy.*

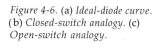

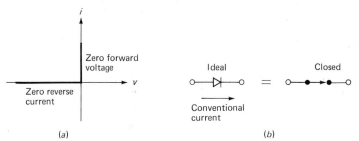

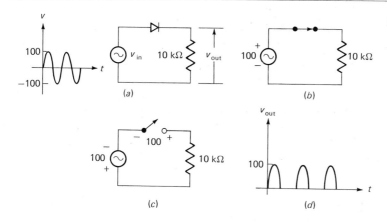

Figure 4-7. Examples 4-3 and 4-4.

SOLUTION.

The input waveform is a sine wave with a positive peak of 100 V and a negative peak of −100 V. During the positive half cycle, the diode is forward-biased because conventional current tries to flow in the direction of the diode arrow. Figure 4-7b shows the circuit right at the positive peak of the input voltage.

During the negative half cycle, the source tries pushing conventional current against the diode arrow. For this reason, the diode looks like an open switch during the negative half cycle. Figure 4-7c shows the circuit at the negative peak of input voltage.

We can summarize the circuit action like this: each positive half cycle of input voltage appears across the output; each negative half cycle is blocked from the output. Figure 4-7d shows the output waveform.

EXAMPLE 4-4.

Calculate the peak current through the diode in Fig. 4-7a. Also, what is the maximum voltage across the diode?

SOLUTION.

The largest current we can get through the diode occurs when the input voltage reaches its positive peak. Figure 4-7b shows the circuit for this peak-current condition. The diode current equals the current through the 10-kΩ resistor:

$$i_{peak} = \frac{100}{10,000} = 10 \text{ mA}$$

During forward bias, the ideal-diode voltage drop equals zero. But during reverse bias, the diode voltage gets quite large. How large? Look at Fig. 4-7c. Kirchhoff's voltage law implies that 100 V appears across the open switch (the reverse-biased diode). Since this occurs at the negative peak, it is the maximum voltage that appears across the diode.

4-5. THE SECOND APPROXIMATION

We need about 0.7 V before a silicon diode really conducts well. When we have a large input voltage, this 0.7 V is too small to matter. But when the input voltage is not large, we may want to take the knee voltage into account.

Figure 4-8*a* shows the graph for the *second approximation*. The graph says no current flows until 0.7 V appears across the diode. At this point the diode turns on. No matter what forward current flows, we allow only 0.7 V drop across a silicon diode. (Use 0.3 V for germanium diodes.)

Figure 4-8*b* shows the equivalent circuit for the second approximation. We think of the diode as a switch in series with a 0.7-V battery. If the external circuit can force conventional current in the direction of the diode arrow, the switch is closed and the diode voltage equals 0.7 V; otherwise, the switch is open.

EXAMPLE 4-5.

Use the second approximation in Fig. 4-9*a* to find the output waveform. Also, calculate the peak forward current and maximum reverse voltage across the diode.

SOLUTION.

During the positive half cycle of input voltage, the first 0.7 V is wasted in overcoming the barrier potential; thereafter, the diode can conduct.

How do we find the current? Figure 4-9*b* shows the circuit when the applied voltage equals +10 V. By Kirchhoff's voltage law, the voltage across the resistor equals 9.3 V. Therefore, the current through the resistor equals

$$i_{peak} = \frac{9.3}{10,000} = 0.93 \text{ mA}$$

Because the diode drops 0.7 V when conducting, the peak output voltage equals 9.3 V rather than the full 10 V. Figure 4-9*c* shows the output waveform.

What about the maximum reverse voltage? When the diode is reverse-biased, it is open. No current flows through the resistor; therefore, to satisfy Kirchhoff's voltage law, all of the applied voltage must appear across the diode. At the peak, this voltage equals −10 V.

Figure 4-8. (a)
Second-approximation curve. (b)
Switch-battery analogy.

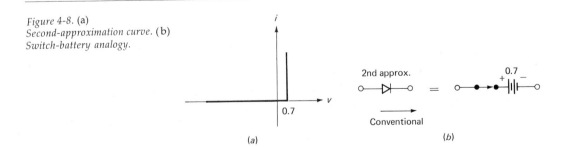

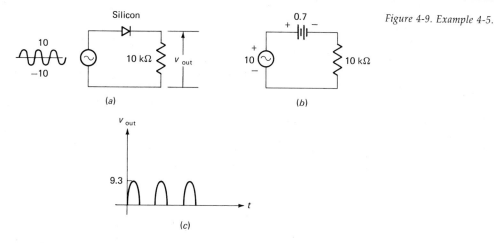

Figure 4-9. Example 4-5.

4-6 THE THIRD APPROXIMATION

In the third approximation of a diode, we include bulk resistance r_B. Figure 4-10a shows the effects of r_B. After the silicon diode turns on, the current produces a voltage across r_B. The greater the current, the larger the voltage.

The equivalent circuit for the third approximation is a switch in series with a 0.7-V battery and a resistance of r_B (see Fig. 4-10b). After the external circuit has overcome the barrier potential, it forces conventional current in the direction of the diode arrow. Therefore, the total voltage across the silicon diode equals

$$V_F = 0.7 + I_F r_B \qquad (4\text{-}2)^{***}$$

(For a germanium diode, use 0.3 V instead of 0.7 V.)

As an example, if a silicon diode has an r_B of 20 Ω, the forward voltage with 1 mA of current equals

$$V_F = 0.7 + 0.001\,(20)$$
$$= 0.72 \text{ V}$$

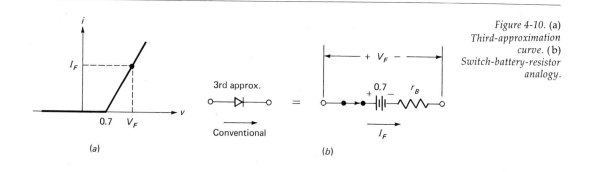

Figure 4-10. (a) Third-approximation curve. (b) Switch-battery-resistor analogy.

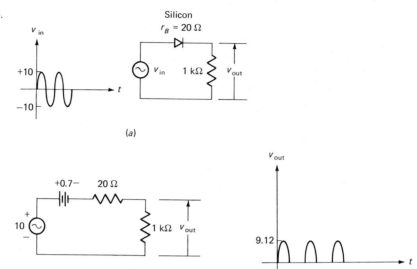

Figure 4-11. Example 4-6.

For the same diode, 10 mA of forward current produces a voltage of

$$V_F = 0.7 + 0.01\,(20)$$
$$= 0.9 \text{ V}$$

EXAMPLE 4-6.
Use the third approximation with an r_B of 20 Ω to calculate the peak current in Fig. 4-11a. Also, show the output waveform.

SOLUTION.
Figure 4-11b shows the circuit at the instant the input voltage reaches its positive peak of 10 V. Therefore, the peak current equals

$$i_{\text{peak}} = \frac{10 - 0.7}{1020}$$
$$= 9.12 \text{ mA}$$

When this flows through the 1-kΩ resistor, it produces an output voltage of

$$v_{\text{out(peak)}} = 9.12\,(10^{-3})\,10^3$$
$$= 9.12 \text{ V}$$

Figure 4-11c shows the output waveform. (If we had used an ideal diode, the output waveform would be a half-wave-rectified signal with a peak of 10 V.)

4-7. SELECTING AN APPROXIMATION

Which approximation should you use? If you are troubleshooting or making a preliminary analysis, large errors are often acceptable. No need to waste time if all you want is
a basic idea of how a circuit works. On the other hand, there may be times when you
want at least slide-rule accuracy, errors no greater than 1 or 2 percent.

In a one-diode circuit, you can visualize the diode between a pair of AB terminals
as shown in Fig. 4-12a. You can Thevenize the external circuit driving the diode to get
a one-loop circuit like Fig. 4-12b. With the values of V and R, you can now decide
which approximation you need as follows.

With our best approximation (the third), a forward-biased diode appears as shown
in Fig. 4-12c. In this circuit, the forward current equals

$$I_F = \frac{V - 0.7}{R + r_B} \tag{4-3}$$

This equation pinpoints the effects of the 0.7 V and r_B as far as the value of I_F is concerned. If V is more than 7 V, we get less than 10 percent error by neglecting the knee
voltage. Similarly, if R is at least 10 times greater than r_B, we get less than 10 percent
error by neglecting r_B.

When you need high accuracy like 1 percent or so, V must be at least 70 V before
you can neglect the 0.7 V, and R must be at least $100r_B$ before r_B is negligible.

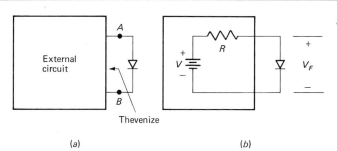

(a)   (b)

*Figure 4-12. Effect of knee voltage
and bulk resistance.*

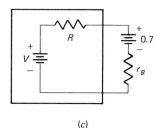

(c)

Figure 4-13. Example 4-7.

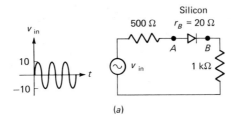

(a)

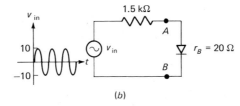

(b)

EXAMPLE 4-7.

In Fig. 4-13a, we want less than 10 percent error in our analysis. Can we neglect the knee voltage? The bulk resistance?

SOLUTION.

We first work out the Thevenin equivalent circuit driving the diode; this is shown in Fig. 4-13b. The input voltage has a peak of 10 V; therefore, we can neglect the 0.7 V. We also can neglect r_B because it is less than one-tenth of 1500 Ω.

For 1 percent work, you must take knee voltage and bulk resistance into account. That is, to find peak current, you would need Eq. (4-3).

4-8. REVERSE RESISTANCE

When reverse-biased, a diode has a small reverse current. One way to estimate the importance of this current is with the *reverse resistance* of a diode.

Reverse resistance defined

This is the definition for reverse resistance:

$$R_R = \frac{V_R}{I_R} \qquad (4\text{-}4)^{***}$$

where I_R is the reverse current at a reverse voltage V_R. As an example, at 25°C the 1N456 has an I_R of 25 nA for a V_R of 25 V.[1] Therefore, the reverse resistance equals

[1] A good reference for diode and transistor data is *The Semiconductor Data Book*, Motorola Semiconductor Products, Inc., 1970.

$$R_R = \frac{25}{25(10^{-9})} = 1000 \text{ M}\Omega$$

At 150°C, the data sheet of the 1N456 gives an I_R of 5 μA for a V_R of 25 V. So, the reverse resistance at 150°C is

$$R_R = \frac{25}{5(10^{-6})} = 5 \text{ M}\Omega$$

Neglecting reverse resistance

How high does R_R need to be before we can neglect it? Figure 4-14a shows a Thevenin circuit driving the diode. Since it is reverse-biased, the diode looks like a resistance R_R as shown in Fig. 4-14b. With the voltage-divider formula, we get a diode voltage of

$$V_R = \frac{R_R}{R + R_R} V \qquad (4\text{-}5)$$

When R_R is 100 times greater than R, you get less than 1 percent error by neglecting R_R (treating it as infinite). If R_R is at least 10 times greater than R, the error is less than 10 percent when you neglect R_R.

EXAMPLE 4-8.
Sketch the output waveform for the circuit of Fig. 4-15a, taking reverse resistance into account. Neglect knee voltage and bulk resistance.

SOLUTION.
The reverse resistance is given as 4 MΩ, and the Thevenin resistance facing the diode is 1 MΩ. Figure 4-15b shows the equivalent circuit at the negative peak of the input voltage. The voltage divider delivers 6 V to the output. Therefore, the output waveform is the poorly rectified signal of Fig. 4-15c. To improve the circuit, either you must reduce the 1-MΩ resistor, or you must use a diode with a higher reverse resistance.

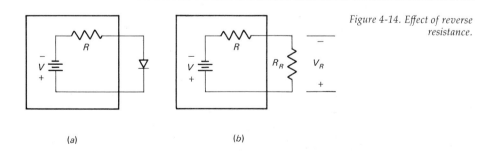

Figure 4-14. Effect of reverse resistance.

(a)   (b)

Figure 4-15. Example 4-8.

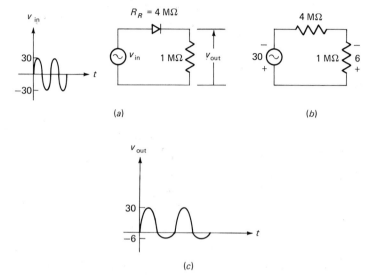

(a)

(b)

(c)

4-9. DIODE CAPACITANCE

Like any component with leads, a diode has end-to-end capacitance that may affect high-frequency operation; this external capacitance is usually less than 1 pF. More important than this external capacitance, however, is the internal capacitance built into the junction of a diode. We call this internal capacitance the *transition capacitance,* designated C_T. The word "transition" refers to the transition from *p*-type to *n*-type material. Transition capacitance is also known as depletion-layer capacitance, barrier capacitance, and junction capacitance.

What is transition capacitance? Figure 4-16*a* shows a reverse-biased diode. As previously discussed, the depletion layer widens until its difference of potential equals the applied reverse voltage. The greater the reverse voltage, the wider the depletion layer. Because the depletion layer has almost no charge carriers, it acts like an insulator or dielectric. The doped *p* and *n* regions, on the other hand, act like fairly good conductors. With a little imagination, we can visualize the *p* and *n* regions separated by the depletion layer as a parallel-plate capacitor. This parallel-plate capacitance is the same as the transition capacitance.

When you increase reverse voltage, you make the depletion layer wider. It's as though you have moved the parallel plates farther apart. In effect, the transition capacitance decreases when the reverse voltage increases. (When the diode is optimized for its variable-capacitance properties, it's called a *varactor*. Varactors are often used to tune resonant circuits.)

Each cycle of input voltage alternately forward- and reverse-biases the diode. Diode capacitance has little effect on forward current, but it will affect reverse current

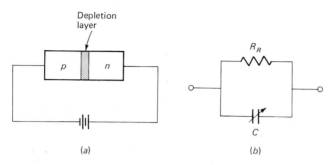

Depletion layer

p n

(a)

R_R

C

(b)

Figure 4-16. (a) Transition capacitance. (b) Equivalent circuit.

at higher frequencies. Figure 4-16b shows how diode capacitance appears during reverse bias; C includes transition capacitance and end-to-end capacitance. With increasing frequency, X_C gets smaller. Eventually, the shunting effect of X_C prevents normal rectification.

For preliminary analysis, we will adopt this rule: neglect diode capacitance when the reactance is at least ten times the Thevenin resistance driving the diode. In symbols, we neglect diode capacitance when

$$X_C \geq 10R \tag{4-6}$$

where R is the Thevenin resistance facing the diode (see Fig. 4-14a).

In applying Eq. (4-6), we will use the diode capacitance at 0 V because this is usually specified on a data sheet. For instance, the data sheet of a 1N456 gives diode capacitance as 10 pF at 0 V. A 1N3062 is better; it has a capacitance of only 1 pF at 0 V.

A final point. If you don't have a data sheet for a diode, you can measure its capacitance on an *RLC* bridge or any instrument that measures capacitance. The only precaution is to make sure the test voltage across the diode is not large enough to produce forward current. The peak test voltage across the diode should be less than 0.7 V for silicon diodes and less than 0.3 V for germanium diodes.

EXAMPLE 4-9.

Estimate the frequency beyond which diode capacitance becomes important in Fig. 4-17.

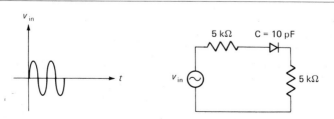

Figure 4-17. Example 4-9.

v_{in}

t

5 kΩ   C = 10 pF

V_{in} 5 kΩ

SOLUTION.

The Thevenin resistance facing the diode is 10 kΩ. With Eq. (4-6), the critical condition is

$$X_C = 10R$$

or

$$\frac{1}{2\pi fC} = 10R$$

$$\frac{1}{2\pi f(10^{-11})} = 10^5$$

Solving for f gives

$$f = 159 \text{ kHz}$$

At this frequency, the X_C of the diode just equals 100 kΩ. Beyond this frequency, X_C gets smaller and the rectification deteriorates.

If you wanted to rectify at much higher frequencies, you would have to select a diode with less capacitance. It also would help to reduce the resistance in series with the diode.[2]

Problems

4-1.   In Fig. 4-18*a*, are the diodes forward- or reverse-biased?

4-2.   Is the diode of Fig. 4-18*b* forward- or reverse-biased?

4-3.   At the peak of the positive half cycle in Fig. 4-18*c*, which diode is forward-biased? Which one is reverse-biased at the negative input peak?

4-4.   Figure 4-18*d* shows a circuit often used in a VOM to measure ac voltages. The ammeter has current through it when diode *B* is forward-biased. During which half cycle of input voltage does this happen? During the negative half cycle of input voltage, is diode *A* forward- or reverse-biased?

4-5.   The circuit of Fig. 4-18*e* has a sine wave induced across the secondary winding. When the upper end of the secondary is positive with respect to the lower end, which diode is forward-biased and which reverse-biased?

4-6.   In Fig. 4-18*f*, the diode is forward-biased at either the positive or negative input peak. Which of these is it?

4-7.   The resistance of the *p* region of a diode equals 5 Ω and the resistance of the *n* region is 3 Ω. What does the bulk resistance of the diode equal?

4-8.   A data sheet for a silicon diode says the forward current equals 50 mA at 1 V. Estimate the bulk resistance of the diode.

4-9.   A germanium diode has a forward current of 30 mA at 1 V. What is the approximate bulk resistance of this diode?

[2] For very high-frequency rectification, a *hot-carrier* (Schottky-barrier) diode is often used.

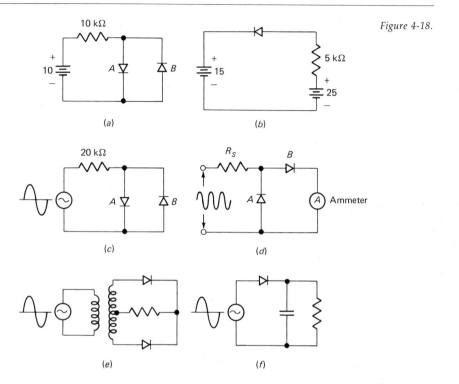

Figure 4-18.

4-10.   Figure 4-19*a* shows the forward curve of a silicon diode. Estimate the bulk resistance.

4-11.   Figure 4-19*b* gives two values of diode current for corresponding diode voltages. An alternative way to calculate bulk resistance is

$$r_B = \frac{\Delta v}{\Delta i}$$

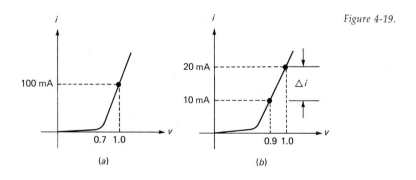

Figure 4-19.

where the symbol Δ means "the change in." As shown, Δi is the difference in the two values of current. The quantity Δv is the difference in the two values of corresponding voltages. Calculate r_B for Fig. 4-19b.

4-12. The diodes of Fig. 4-18a are ideal. How much current is there in each diode?

4-13. Treat the diode of Fig. 4-18b as ideal. How much current is there in the 5-kΩ resistor?

4-14. The sine wave of Fig. 4-18c has a positive peak voltage of 40 V and a negative peak of −40 V. What does the current in diode A equal at the positive peak? At the negative peak? (Use ideal diodes.)

4-15. Assuming an ideal diode in Fig. 4-20a, what is the peak forward current in the circuit? Sketch the output-voltage waveform.

4-16. Use the ideal-diode approximation and Thevenin's theorem to find the peak forward current through the diode of Fig. 4-20b.

4-17. If the diode in Fig. 4-20c is ideal, how much current is there in the 10-kΩ resistor?

4-18. Use Thevenin's theorem and the ideal-diode approximation to find the diode current in Fig. 4-20d.

4-19. In Fig. 4-20b, sketch the voltage waveform across the 20-kΩ resistor.

4-20. When the diode of Fig. 4-20a is reverse-biased, what is the maximum voltage across it?

Figure 4-20.

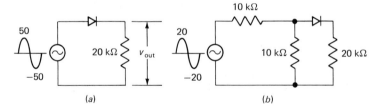

(a)   (b)

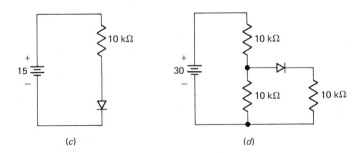

(c)   (d)

4-21. What is the maximum reverse voltage across the diode of Fig. 4-20b?

4-22. Use the second approximation in Fig. 4-21a to calculate the peak forward current. Sketch the output waveform.

4-23. Calculate the peak current in the 12.5-kΩ resistor using the second approximation for the diode (Fig. 4-21b).

4-24. How much current is there in Fig. 4-21c? (Use the second approximation.) And in Fig. 4-21d?

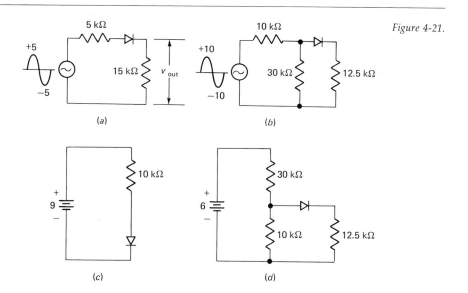

Figure 4-21.

4-25. Use the third approximation in Fig. 4-22a to calculate the peak forward current through the diode. What is the peak voltage across the 300-Ω resistor?

4-26. The 1N3062 of Fig. 4-22b has these specifications: $I_{F(max)} = 115$ mA, and $I_F = 20$ mA at 1 V. Use the third approximation of a silicon diode to calculate the current through the diode.

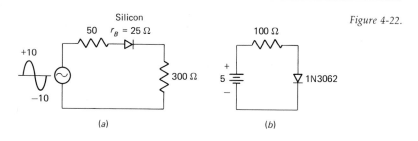

Figure 4-22.

4-27. A 1N5429 has a reverse current of 50 nA for a reverse voltage of 125 V (at a temperature of 25°C). Calculate the reverse resistance. At 150°C, $I_R = 50$ μA for a $V_R = 125$ V. What is the reverse resistance at this elevated temperature?

4-28. The 1N3595 is a low-leakage silicon diode. At 25°C it has an I_R of 1 nA for a V_R of 125 V. What does its reverse resistance equal? Even at 125°C, $I_R = 500$ nA for $V_R = 125$ V. Calculate the reverse resistance at 125°C.

5. Rectifier Circuits

This chapter covers diode circuits. In all of them, the diode acts like a rectifier because it changes alternating current to unidirectional current.

5-1. THE HALF-WAVE RECTIFIER

A diode conducts well only when forward-biased. When the input voltage is alternating, the diode is alternately forward- and reverse-biased. Because of this, the diode can conduct only during part of the input cycle. When current flows in the output resistor for half the input cycle, we call the circuit a *half-wave rectifier*.

The chassis

Figure 5-1 shows the schematic symbol for the chassis, a metal frame on which some components are mounted. The chassis is a good conductor and may be used as part of the circuit. We will refer to the chassis as *ground* because it is normally at the

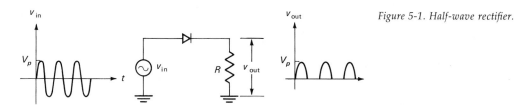

Figure 5-1. Half-wave rectifier.

same potential as the earth (the third prong of a three-wire plug usually connects the chassis to earth).

In Fig. 5-1, each positive half cycle of source voltage forward-biases the diode. During this half cycle, conventional current flows through the diode, through the resistor, and back to the source through the chassis. For an ideal diode, the output waveform is the half sine wave shown.

Figure 5-2. (a) Chassis may be live or neutral. (b) One live chassis, one neutral.

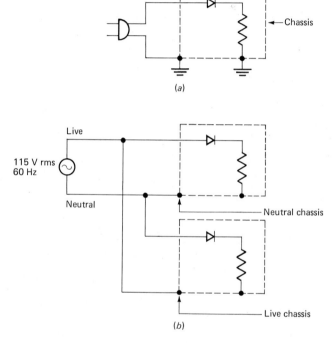

The isolation transformer

If either side of the power line connects to the chassis, the danger of electrical shock is high. Figure 5-2*a* shows the upper wire of a two-wire plug connected to a diode. The return path for current is through the chassis to the lower wire of the power plug. When you plug this equipment into a power socket, you will connect the chassis either to the *neutral* side of the power line or to the *live* side. If connected to the live side, the chassis will be 115 V rms with respect to the earth. In this case, you can get a shock in any of several ways.

Figure 5-2*b* shows the worst situation of all. Here you have two chassis, each with a two-wire plug. Since the plugs can connect in either of two ways, it is possible to connect the upper chassis to the neutral side of the power line, and the lower chassis to

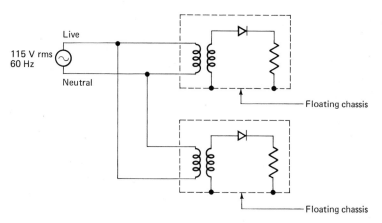

Figure 5-3. Isolation transformer.

Floating chassis

Floating chassis

the live side. Because of this, 115 V rms is between the two chassis. Anyone leaning against or otherwise touching both chassis will be directly across the power line.

To avoid chassis shocks, an *isolation transformer* is normally used between the power line and each piece of electronics equipment powered by the line. Figure 5-3 illustrates why this eliminates the voltage between chassis. No matter how you connect the power plugs, no voltage can exist between the two chassis because the power line connects only to the primary windings and not to the chassis.

An isolation transformer can step the voltage up or down, as needed. Figure 5-4 shows a 5:1 step-down transformer between the line and the half-wave rectifier. Peak line voltage is approximately 160 V. For this reason, the voltage across the secondary winding is a sine wave with a peak of about 32 V. Assuming an ideal diode, the output voltage is a half-wave signal with a peak of 32 V as shown.

The output resistor of Fig. 5-4 is called the *load resistor* and is designated R_L; the voltage across this resistor is the load voltage.

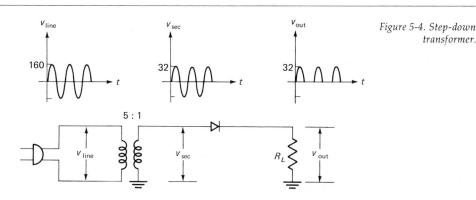

Figure 5-4. Step-down transformer.

5-2. THE FULL-WAVE RECTIFIER

In a *full-wave rectifier,* the load current flows during both half cycles, but it does not alternate. Rather, load current is unidirectional.

The dot convention

What do the dots on each winding of Fig. 5-5a mean? The voltage on one end of the secondary is positive with respect to the other end. But which end is which? To keep things straight, we use the *dot convention,* which says

1. When a dotted end of a winding goes positive, all dotted ends go positive.
2. When a dotted end goes negative, all dotted ends go negative.

An alternative way of stating the dot convention is all dotted ends have in-phase voltages.

For instance, during the *AB* half cycle of input voltage (Fig. 5-5a), the upper end of the primary goes positive; at this time, the upper end of the secondary goes positive.

The center-tap rectifier

The circuit of Fig. 5-5a has a center-tapped secondary winding. How does it work? During the *AB* half cycle of input voltage, the voltage is plus-minus across each half of

Figure 5-5. (a) Meaning of dot convention. (b) Positive half cycle. (c) Negative half cycle.

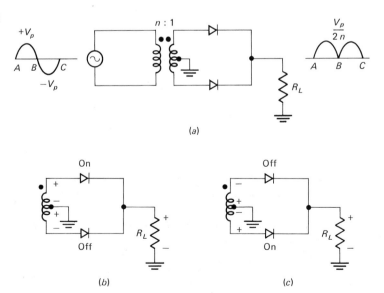

the secondary (Fig. 5-5b). Therefore, during the *AB* half cycle the upper diode is forward-biased and the lower diode reverse-biased. Note the load voltage is positive with respect to ground.

On the negative half cycle of input voltage, the polarities reverse as shown in Fig. 5-5c. Now the upper diode is off and the lower one is on. Load current is in the *same* direction as before; therefore, the upper end of the load resistor is positive with respect to ground.

Figure 5-5a shows the load-voltage waveform, assuming ideal diodes and negligible voltage drop in the transformer windings. The turns ratio of the transformer is *n*; therefore, each half of the secondary has a sine-wave voltage with a peak of $V_p/2n$. This is why the load voltage is a full-wave signal with a peak value of $V_p/2n$.

Usually, a center-tap rectifier has a small unbalance; the peak output voltage of two successive cycles is slightly different. The reason is the diodes may be slightly different, or the center tap may not be at the exact center of the secondary winding.

EXAMPLE 5-1.
Calculate the peak load current in Fig. 5-6a.

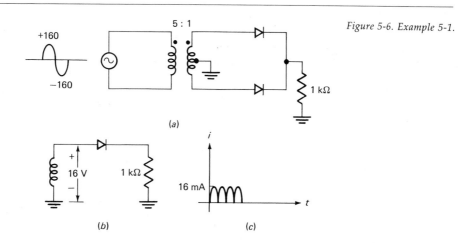

Figure 5-6. Example 5-1.

SOLUTION.
The total voltage across the secondary winding is one-fifth the input voltage; therefore, the secondary voltage is a sine wave with a peak of 32 V. Since the secondary is center-tapped, each half has a sine wave with a peak of 16 V.

Figure 5-6b shows the equivalent circuit for the *on* diode at the peak. With an ideal diode, the peak current equals

$$i_{peak} = \frac{16}{1000} = 16 \text{ mA}$$

Figure 5-6c is the load-current waveform, a full-wave signal with a peak value of 16 mA.

5-3.   THE BRIDGE RECTIFIER

Figure 5-7a shows a *bridge rectifier,* another way to full-wave-rectify. During the positive half cycle of input voltage, a plus-minus voltage is induced across the secondary winding as shown in Fig. 5-7b. Conventional current tries to flow down through the diodes. Because of this, diodes D_1 and D_2 are forward-biased, diodes D_3 and D_4 reverse-biased. In Fig. 5-7b, diodes D_3 and D_4 are shaded or dark; this will be our way of indicating a device is *off* or nonconducting.

During the negative half cycle of input voltage, the polarity reverses as shown in Fig. 5-7c. Conventional current tries flowing up through the diodes. Therefore, diodes D_1 and D_2 are off; D_3 and D_4 are on. Especially important, the load current is in the *same* direction in Fig. 5-7b and c. For this reason, the load voltage is unidirectional and looks like the full-wave signal of Fig. 5-7a. Since the transformer has a turns ratio of n, the load voltage has a peak value of V_p/n, assuming ideal diodes and an ideal transformer.

A full-wave signal has double the frequency of the input signal. Here is the reason. In Fig. 5-7a, the input signal must go through the positive and negative half cycles to complete one cycle. The output waveform is different; to complete one output cycle, the full-wave signal only has to trace out one of the cycles shown. (We are assuming a balanced rectifier with identical diodes.) For this reason, the period of the output signal is half the period of the input signal. This means the output frequency is twice the input frequency. As an example, if the input frequency is 60 Hz, the output frequency equals 120 Hz.

Figure 5-7. Bridge rectifier. (a) Circuit. (b) Positive half cycle. (c) *Negative half cycle.* (d) *Electrically equivalent circuit.*

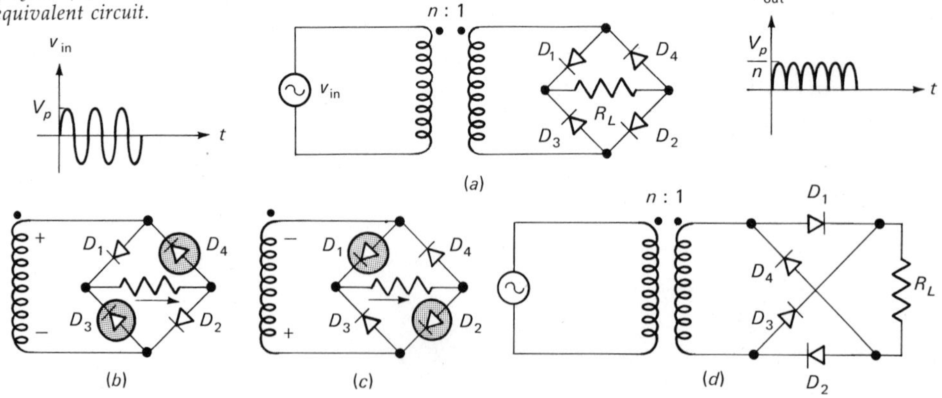

Figure 5-7*d* shows an alternative way to draw a bridge rectifier. By checking the current paths during positive and negative half cycles, you will see the equivalence of Fig. 5-7*d* to 5-7*a*.

5-4. POWER AND CURRENT RATINGS

Chapter 4 discussed $I_{F(max)}$, the maximum dc forward current the diode can safely conduct with a steady forward voltage. This rating does not apply to a rectifier circuit because the current and voltage are not constant. For this reason, we need another current rating called the *average rectified current* I_0.

Peak and average power dissipation

Whether a diode is used in a half-wave or full-wave rectifier, the current in each diode is half-wave like Fig. 5-8*a*. At the instant the current reaches the peak value, the power reaches its peak value. The diode has to dissipate the *peak power* only at the peak of each cycle. During the rest of the cycle, the instantaneous power is less than the peak power.

Because the diode has mass, its temperature does not fluctuate with the current through it (assuming an input frequency greater than 10 Hz, or thereabouts); the junction temperature will settle on an average value after a few cycles. Since it is the junction temperature that may damage a diode, we are interested in the *average power dissipation* rather than the peak power dissipation. The average power rating is sometimes given on a data sheet. For instance, a 1N3062 has a power rating of 250 mW. This means the 1N3062 can dissipate an average power of 250 mW.

Average rectified current

In Fig. 5-8*a*, the average current equals

$$I_0 = 0.318 I_{peak} \qquad \qquad (5\text{-}1)\,^{***}$$

That is, the average current equals 31.8 percent of the peak current.

You can prove Eq. (5-1) with calculus. Alternatively, you can prove this equation in the laboratory with a circuit like Fig. 5-8*b*. The rectified current through the ammeter

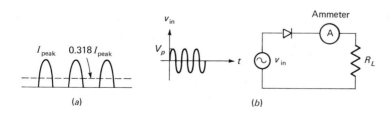

(a)   *(b)*

Figure 5-8. (a) *Diode current.* (b) *Measuring average current.*

will be a half-wave current like Fig. 5-8a. If the frequency is greater than 50 Hz or so (depends on ammeter construction), the ammeter cannot follow the rapid fluctuations because of its inertia. Instead, the ammeter reads the average value I_0 shown in Fig. 5-8a. In this way, you can experimentally prove Eq. (5-1).

The average rectified current through a diode is related to the average power dissipation of the diode. Current is more convenient to measure and work with than power. For this reason, many data sheets don't give the power rating; instead, they give you the maximum average rectified current the diode can safely handle. For instance, a 1N3062 has an $I_{0(max)}$ of 75 mA. As long as the average rectified current is less than 75 mA, the diode is dissipating less than its maximum power rating.

EXAMPLE 5-2.
The 1N3210 diodes of Fig. 5-9a have an $I_{0(max)}$ rating of 15 A. Work out the actual value of I_0 in the diodes and compare to $I_{0(max)}$.

SOLUTION.
With a turns ratio of 1:1, each half of the secondary has an 80-V-peak sine wave across it. Ideally, the peak load current equals

$$I_{peak} = \frac{80}{10} = 8 \text{ A}$$

The current through each diode is a half-wave current with an average value of

$$I_0 = 0.318 I_{peak}$$
$$= 0.318(8) = 2.54 \text{ A}$$

No problem here. The average rectified current is well within the manufacturer's maximum rating of 15 A.

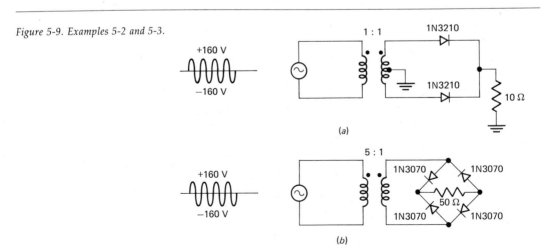

Figure 5-9. Examples 5-2 and 5-3.

EXAMPLE 5-3.

The 1N3070 has a maximum average rectified current rating of 100 mA. Check the bridge rectifier of Fig. 5-9b to see if the maximum current rating is exceeded.

SOLUTION.

The secondary voltage is a sine wave with a peak of 32 V. Ideally, the peak diode current equals

$$I_{\text{peak}} = \frac{32}{50} = 640 \text{ mA}$$

Therefore, the rectified current in each diode has an average value of

$$I_0 = 0.318(0.640) = 204 \text{ mA}$$

This is more than twice the maximum current rating for the 1N3070. In a case like this, you either have to increase the load resistance or select another diode type.

A final point. In a bridge rectifier, the two *on* diodes are in series. To a second approximation, we lose about 1.4 V across them (0.7 V for each). If the secondary voltage is low, you may want to take this 1.4 V into account.

5-5. PEAK INVERSE VOLTAGE

Besides checking the maximum current rating, we must check the *peak-inverse-voltage* (PIV) rating of the diode. The peak inverse voltage is the maximum voltage across a diode during the reverse-bias part of the cycle.

Figure 5-10a shows voltages in a half-wave rectifier at the instant the input voltage reaches its negative peak value. With an n-to-1 turns ratio, the induced secondary voltage equals V_p/n with the polarity shown. Since reverse current is negligibly small, the voltage across the load resistor equals zero; therefore, all the secondary voltage appears across the diode (Kirchhoff's voltage law must be satisfied). As shown in Fig. 5-10a, the PIV across the diode is V_p/n; this applies to any half-wave rectifier.

Figure 5-10b shows peak voltages for a center-tap rectifier. The load voltage has a peak value of $+V_p/2n$. As you can see, the lower half of the secondary applies a voltage of $-V_p/2n$ to the anode of the off diode. Therefore the voltage across this diode equals V_p/n. In other words, the PIV of a diode in a center-tap rectifier is V_p/n, the same as for a half-wave-rectifier diode.

Figure 5-10c shows peak voltages for a bridge rectifier. With the upper left diode *on* (approximately 0 V across it), the lower left diode is *off* and has V_p/n across it to satisfy Kirchhoff's voltage law. For a similar reason, the upper right diode has a maximum reverse voltage of V_p/n. Therefore, the diodes in a bridge rectifier must withstand a PIV of V_p/n, the same as the half-wave rectifier and the center-tap rectifier.

The PIV should not approach the breakdown voltage of the diode. The manufac-

Figure 5-10. Peak inverse voltage.
(a) Half-wave rectifier. (b)
Full-wave center-tap rectifier. (c)
Full-wave bridge rectifier.

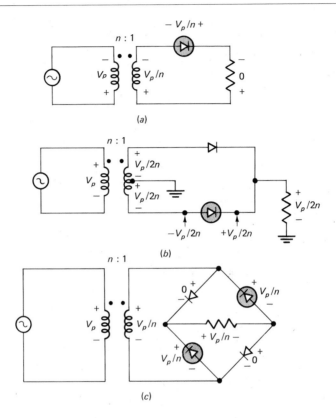

turer specifies maximum reverse voltage as the *working inverse voltage* WIV [also as $V_{RM(wkg)}$]. This rating tells us the maximum reverse voltage a diode can safely handle cycle after cycle. In no case should the PIV (peak inverse voltage across the diode) exceed the WIV (the manufacturer's maximum rating). In symbols: PIV < WIV.

The PIV across each diode in Fig. 5-10 equals V_p/n, whether the circuit is a half-wave, center-tap, or bridge rectifier. To stay within the manufacturer's specifications, we must satisfy this condition:

$$\frac{V_p}{n} < \text{WIV} \qquad\qquad (5\text{-}2)***$$

Which brings us to one advantage a bridge rectifier has over a center-tap rectifier. Figure 5-11 shows both rectifiers set up to deliver the same load voltage (a full-wave signal with a peak of 80 V). In the center-tap rectifier, the PIV equals 160 V. You can see this either by inspection of Fig. 5-11a or by using $n = 1$ to get

$$\text{PIV} = \frac{V_p}{n} = V_p = 160 \text{ V}$$

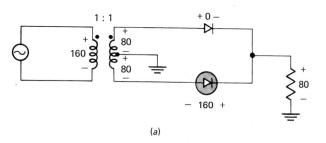

(a)

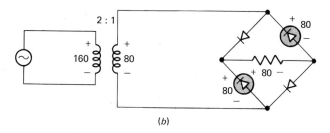

(b)

Figure 5-11. Bridge-rectifier diodes have lower PIV.

But with the bridge rectifier, the PIV across each diode is only 80 V, apparent in Fig. 5-11b, or by

$$PIV = \frac{V_p}{n} = \frac{V_p}{2} = 80 \text{ V}$$

This means the WIV rating of the bridge-rectifier diodes can be as low as 80 V, but the center-tap-rectifier diodes must have a WIV rating of at least 160 V.

Another advantage of the bridge rectifier is the lack of a center tap. It is difficult to locate the exact center of the secondary winding. For this reason, a center-tap rectifier will have a slight unbalance on alternate output cycles. In some applications, this small unbalance is unacceptable. With a bridge rectifier, you eliminate the need for a center tap. Then, the only unbalance is in diodes.

EXAMPLE 5-4.
The 1N3210 has WIV rating of 200 V. Make sure the PIV in Fig. 5-9a is less than this.

SOLUTION.
With a turns ratio of 1:1, we get

$$PIV = \frac{V_p}{n} = \frac{160}{1} = 160 \text{ V}$$

which is less than the WIV rating.

EXAMPLE 5-5.

A 1N3070 has a WIV rating of 175 V. Compare this to the PIV in Fig. 5-9*b*.

SOLUTION.

The PIV across the diodes is

$$\text{PIV} = \frac{V_p}{n} = \frac{160}{5} = 32 \text{ V}$$

which is much less than the WIV rating.

5-6. FILTERING TO GET A DC VOLTAGE

The average rectified current in a diode is 31.8 percent of the peak current. A similar idea applies to voltage.

Half-wave relation

Figure 5-12*a* shows a half-wave voltage. The average value of this voltage equals

$$V_{av} = 0.318 V_{peak} \qquad \text{(half wave)} \qquad (5\text{-}3)^{***}$$

You can prove this experimentally with a circuit like Fig. 5-12*b*. (You can also use calculus to prove it.) After rectification, you have a half-wave signal with a 100-V peak as shown. If the frequency is greater than 50 Hz, the voltmeter's inertia will prevent the needle from following the instantaneous voltage; in this case, the needle will settle on

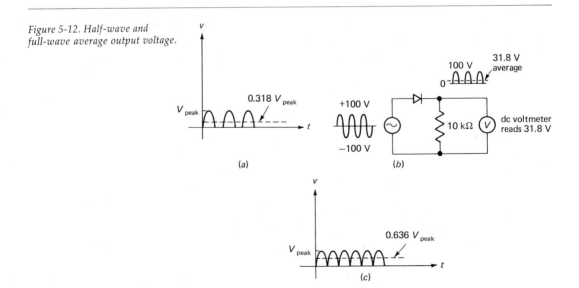

Figure 5-12. Half-wave and full-wave average output voltage.

an average or dc voltage; this average voltage equals 31.8 percent of the peak voltage (see Fig. 5-12b).

Full-wave relation

Since a full-wave signal has twice as many cycles as a half-wave signal, the full-wave signal has twice the average value, that is,

$$V_{av} = 0.636V_{peak} \qquad \text{(full wave)} \qquad (5\text{-}4)***$$

as shown in Fig. 5-12c. If a dc voltmeter were connected across the load resistor of a full-wave rectifier, it would read 63.6 percent of the peak voltage.

RC and LC filters

This brings us to the idea of *filtering* a rectified signal to get an average or dc voltage. In the *RC filter* of Fig. 5-13a, X_C at 60 Hz is much smaller than R. Because of this, the fluctuating part of the half-wave signal is dropped across the resistor. All that appears across the capacitor is the average value of the half-wave signal.

In a similar way, we can use an *LC filter* like the one in Fig. 5-13b. In this case, X_L is much greater than X_C at 60 Hz. Because of this, most of the fluctuating part of the half-wave signal is dropped across the inductor. All that appears across the capacitor is an average or dc voltage.

The analysis of filtering is difficult without *harmonics*. For this reason, we will postpone the details of how a filter works until after we have covered *harmonics* and *spectra*. Then, filter analysis will be simple.

All you need at this time is the general idea. Put a half-wave signal into a filter and out comes a dc voltage equal to 31.8 percent of the peak input voltage (see Fig. 5-13c). If you drive a filter with a full-wave signal, you get a dc voltage equal to 63.6 percent of the input peak voltage (Fig. 5-13d). In electronic equipment, dc voltages like these are needed to operate other electronic circuits.

EXAMPLE 5-6.
The 20-H inductor and the 500-μF capacitor of Fig. 5-13e form an *LC* filter similar to Fig. 5-13d Calculate the ideal dc voltage across the 10-kΩ load resistor.

SOLUTION.
Across each half of the secondary is a sine wave with a peak of 20 V. After rectification, a full-wave signal with approximately a 20-V peak appears at the input to the *LC* filter.

The average or dc voltage out of the filter equals

$$V_{av} = 0.636V_{peak} = 0.636(20)$$
$$= 12.7 \text{ V}$$

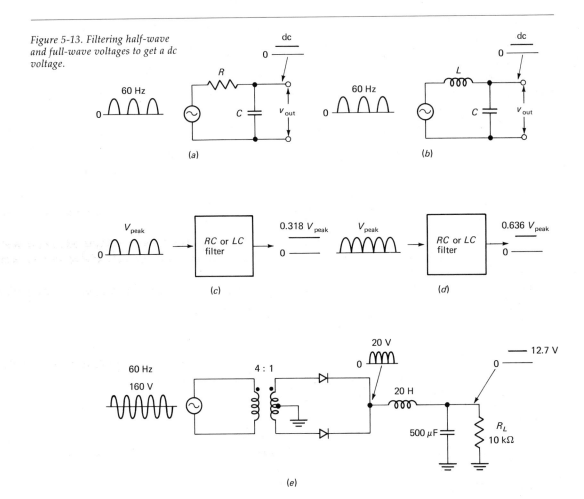

Figure 5-13. Filtering half-wave and full-wave voltages to get a dc voltage.

Ideally, this is a pure dc voltage, that is, it is a unidirectional voltage with a constant value of 12.7 V. Actually, it will have a small *ripple* or fluctuation because the filter does not completely remove the variations from the rectified signal.

5-7. THE CLIPPER

In radar, digital computers, and other electronic systems, we sometimes want to remove signal voltages above or below a specified voltage level. Diode *clippers* are one way to do this.

The positive clipper

Figure 5-14a shows a *positive clipper,* a circuit that removes positive parts of the signal. As shown, the output voltage has all positive half cycles clipped off. The circuit works as follows: During the positive half cycle of input voltage, the diode conducts heavily. To a first approximation, we visualize the diode as a closed switch (Fig. 5-14b). The voltage across a short must equal zero; therefore, the output voltage equals zero during each positive half cycle. (All the voltage is dropped across R.)

During the negative half cycle, the diode is reverse-biased and looks open (Fig. 5-14c). In effect, the circuit is a voltage divider with an output of

$$v_{\text{out}} = -\frac{R_L}{R + R_L} V_p$$

Normally, R_L is much greater than R, so that

$$v_{\text{out}} \cong -V_p \qquad \text{for } R_L \gg R$$

So, during each positive half cycle, the diode conducts heavily and most of the voltage is dropped across R; almost none of the voltage appears across R_L. During each

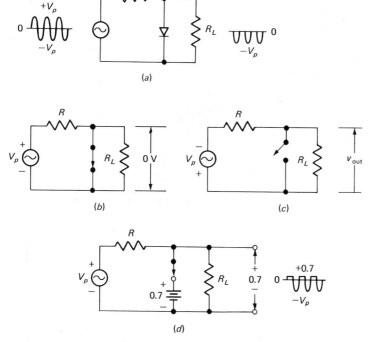

Figure 5-14. Positive clipper. (a) Circuit. (b) *At positive peak.* (c) *At negative peak.* (d) *Second approximation.*

negative half cycle, the diode is off. Because R_L is much greater than R, most of the negative half cycle appears across R_L.

Figure 5-14a shows the output waveform. All signal above the 0-V level has been clipped off. The positive clipper is also called a *positive limiter* because the output voltage is limited to a maximum of 0 V.

The clipping is not perfect. To a second approximation, a conducting silicon diode drops 0.7 V (see Fig. 5-14d). Because the first 0.7 V is used to overcome the barrier potential, the output signal is clipped near +0.7 V rather than 0 V. When top accuracy is required, you can include bulk resistance r_B, reverse resistance R_R, and diode reactance X_C in the analysis.

In Fig. 5-14a, what happens if we turn the diode around? Right, it conducts better on negative half cycles. Stated simply, if you reverse the polarity of the diode in Fig. 5-14a, you get a negative clipper that removes all signal below 0 V.

The biased clipper

In some applications, you may want clipping levels different from 0 V. With a *biased clipper* you can move the clipping level to a desired positive or negative level.

Figure 5-15a shows a biased clipper. For the diode to turn on, the input voltage must be greater than +V. When v_{in} is greater than +V, the circuit looks like Fig. 5-15b. Since the diode acts like a closed switch (ideally), the voltage across the output equals +V. This output voltage stays at +V as long as the input voltage exceeds +V.

When the input voltage is less than +V, the diode opens and the circuit reverts

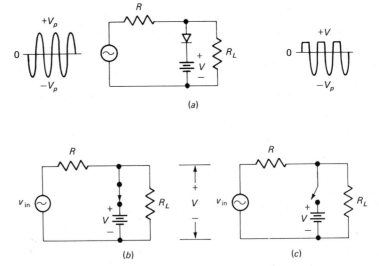

Figure 5-15. Biased positive clipper.

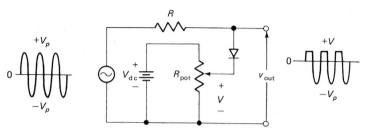

Figure 5-16. Adjustable clipping level.

to a voltage divider (Fig. 5-15c). As usual, R_L should be much greater than R; in this way, most of the input voltage appears across the output.

The output waveform of Fig. 5-15a summarizes the circuit action. The biased clipper removes all signal above the $+V$ level.

You can have an adjustable clipping level by using a circuit like Fig. 5-16. With the potentiometer, you can change the ideal clipping level from 0 to $+V_{dc}$. To operate properly, the resistance of the potentiometer should be negligibly small, that is,

$$R_{pot} \ll R$$

A combination clipper

You can combine biased positive and negative clippers as shown in Fig. 5-17a. Diode D_1 turns on when the input voltage is greater than $+V_1$. Therefore, the output voltage equals $+V_1$ when v_{in} is greater than $+V_1$ (see Fig. 5-17b).

On the other hand, when v_{in} is more negative than $-V_2$, diode D_2 turns on (Fig. 5-17c). With D_2 ideally shorted, the output voltage equals $-V_2$ as long as the input voltage is more negative than $-V_2$.

When v_{in} lies between $-V_2$ and $+V_1$, neither diode is on, and the circuit reverts to the voltage divider of Fig. 5-17d. With R_L much greater than R, most of the input voltage appears across the output.

When the input signal is large, that is, when V_p is much greater than the clipping levels, the output signal resembles a square wave like that of Fig. 5-17a.

Summary

For all clippers shown in Figs. 5-14 through 5-17, we get almost ideally clipped output signals when these conditions are satisfied: $V_p \gg 0.7$ V, $R_L \gg R$, $r_B \ll R$, $R_R \gg R$, $R_{pot} \ll R$, and $X_C \gg R$.

A final point. Clippers work just as well with nonsinusoidal waveforms. Figure 5-18a shows a negative clipper removing the negative half cycles of a *sawtooth* waveform. Figure 5-18b shows a biased positive clipper limiting a triangular wave at the $+V$ level.

Figure 5-17. Combining positive and negative clippers.

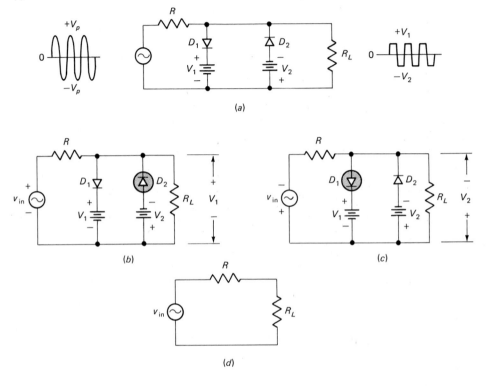

(a)

(b)

(c)

(d)

Figure 5-18. Clipping nonsinusoidal waveforms.

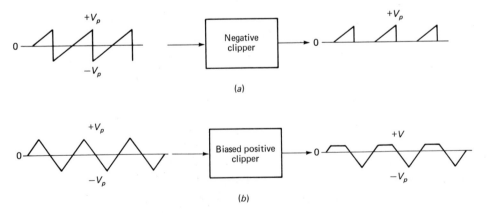

(a)

(b)

EXAMPLE 5-7.

The 1N3064 of Fig. 5-19a has these specifications:

$$I_F = 10 \text{ mA at } 1 \text{ V}$$

$$I_R = 0.1 \ \mu\text{A at } -50 \text{ V}$$

$$C = 2 \text{ pF at } 0 \text{ V}$$

1. What is the ideal output waveform with the wiper at mid-position?
2. Comment on the effects of R_L, R_R, X_C, r_B, and R_{pot}.

SOLUTION.

1. With the wiper at mid-position we get a clipping level of +5 V. Ideally, the output waveform looks like Fig. 5-19b.
2. The usual circuit conditions are that R_L, R_R, and X_C should be much greater than R, 10 kΩ in this particular circuit. R_L (1 MΩ) satisfies this condition.

From the given data we can calculate R_R as

$$R_R = \frac{V_R}{I_R} = \frac{50}{10^{-7}} = 500 \text{ M}\Omega$$

and X_C as

$$X_C = \frac{1}{2\pi f C} = \frac{1}{2\pi (10^3) \, 2 \, (10^{-12})} = 79.5 \text{ M}\Omega$$

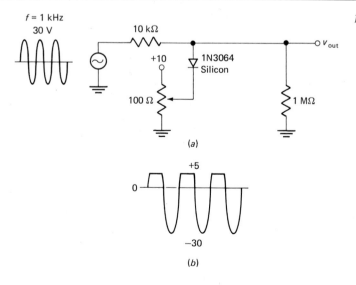

Figure 5-19. Example 5-6.

Both R_R and X_C are much larger than R and have negligible effect on circuit action.

The other circuit conditions to satisfy are r_B and R_{pot} much less than R (10 kΩ). From the given data, the bulk resistance equals

$$r_B = \frac{0.3}{10^{-2}} = 30 \ \Omega$$

The potentiometer has a total resistance of 100 Ω; what actually matters is the Thevenin resistance facing the diode. When the wiper is at mid-position (worst case), the diode sees a resistance of

$$R_{TH} = 50||50 = 25 \ \Omega$$

(The vertical lines $||$ mean "in parallel with.") Therefore, both r_B and R_{TH} are much smaller than R and have negligible effect on circuit action.

About the only significant nonideal effect is the knee voltage of the silicon diode. If we allow 0.7 V across the diode when conducting, we get a clipping level of approximately +5.7 V instead of the ideal 5-V level shown in Fig. 5-19b.

5-8.   THE PEAK RECTIFIER

The *peak rectifier* (also known as a peak detector) produces a dc voltage equal to the peak value of the input signal. For instance, put a sine wave with a peak of +50 V into a peak rectifier and out comes a dc voltage equal to +50 V.

How it works

Figure 5-20a shows a positive peak rectifier. During the first quarter cycle of input voltage, the diode is forward-biased. Ideally, it looks like a closed switch (see Fig. 5-20b). Since the diode connects the source directly across the capacitor, the capacitor charges to peak voltage V_p when the input voltage reaches its peak value.

Just past the positive peak, the diode stops conducting, that is, the switch opens as shown in Fig. 5-20c. Why? Because the capacitor has $+V_p$ across it. With the source voltage slightly less than $+V_p$, current tries flowing opposite the diode arrow. For this reason, the diode is reverse-biased.

With the diode off, the capacitor starts to discharge through load resistance R_L. But here is the key idea behind a peak rectifier: the RC discharge time constant is much greater than the period T of the input signal. Because of this, the capacitor will lose only a small amount of its charge. Near the next positive input peak, the diode turns on again, and recharges the capacitor.

Figure 5-20d shows the kind of output waveform we get from a peak rectifier. The maximum voltage equals $+V_p$. When the diode is off, the capacitor discharges into the load resistance. With a long RC time constant, the output voltage drops only slightly as shown. In other words, we see only the earliest part of an exponential discharge. Near

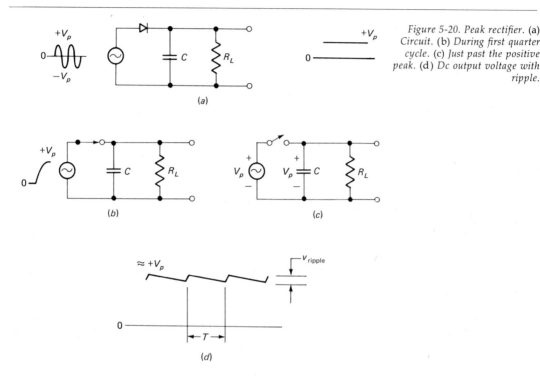

(a)

Figure 5-20. Peak rectifier. (a) Circuit. (b) During first quarter cycle. (c) Just past the positive peak. (d) Dc output voltage with ripple.

the next positive peak, the diode turns on briefly; this replaces the lost capacitor charge, and the output voltage rises to $+V_p$.

Ripple voltage

The signal of Fig. 5-20d is a unidirectional, almost constant voltage. The only deviation from a pure dc voltage is the small fluctuation caused by charging and discharging the capacitor. This fluctuation is called the *ripple*. We will designate the peak-to-peak ripple voltage as v_{rip}. The smaller v_{rip}, the better. In an ideal peak detector, v_{rip} equals zero.

Here are some useful guides for estimating v_{rip}:

1. v_{rip} is less than 1 percent of V_p when

$$R_LC \geqslant 100T \qquad (5\text{-}5)$$

where T is the period of the input signal.
2. v_{rip} is less than 10 percent of V_p when

$$R_LC \geqslant 10T \qquad (5\text{-}6)$$

These guides are based on the equation for exponential discharge of a capacitor and have been proved elsewhere.[1]

As an example of using these guides, suppose $C = 1\ \mu\text{F}$, $R_L = 100\ \text{k}\Omega$, and $f = 1$ kHz. The input signal has a period of

$$T = \frac{1}{f} = \frac{1}{1000} = 0.001 \text{ s}$$

The discharge time constant equals

$$R_L C = 10^5 (10^{-6}) = 0.1 \text{ s}$$

$R_L C$ is 100 times greater than T; therefore, v_{rip} is less than 1 percent of V_p. For a V_p of 30 V, v_{rip} is less than 0.3 V.

If the frequency is less than 1 kHz, the period T is greater than 0.001 s. This gives the capacitor more time to discharge and produces a larger ripple. In other words, *the ripple increases when the frequency decreases.*

Other nonideal effects

What other deviations are there from ideal peak-detector action? As usual, a few-tenths volt is dropped across the diode; this subtracts from the output voltage. Also, the source may have internal resistance R_S as shown in Fig. 5-21a. This prevents the capacitor voltage from reaching $+V_p$. As long as R_S is much smaller than R_L, the effect of source resistance is negligible.

When source resistance is too large to neglect, we account for it as follows: At the positive peak, the circuit looks like Fig. 5-21b. As far as the capacitor is concerned, it

Figure 5-21. Nonideal effects in peak rectifier.

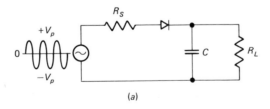

(a)

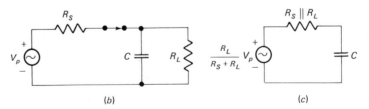

(b) (c)

[1] See Appendix.

sees a Thevenin circuit like Fig. 5-21c. The maximum voltage the capacitor can charge to is

$$V_{TH} = \frac{R_L}{R_S + R_L} V_p \tag{5-7}$$

If $R_S = 10$ kΩ, $R_L = 90$ kΩ, and $V_p = 50$ V,

$$V_{TH} = \frac{90}{10 + 90} 50 = 45 \text{ V}$$

With these values, the capacitor voltage can reach at most 45 V (less a small diode drop).

Inspection of Eq. (5-7) brings out these two rules:

1. V_{TH} is within 1 percent of V_p when

$$R_S < \frac{R_L}{100}$$

2. V_{TH} is within 10 percent of V_p when

$$R_S < \frac{R_L}{10}$$

Summary

The peak rectifier of Fig. 5-21a produces a dc voltage approximately equal to the peak voltage of the input signal. The circuit conditions to satisfy are these:

$$R_L C \gg T$$

$$R_S \ll R_L$$

$$V_p \gg 0.7 \text{ V}$$

Peak detectors are used in radio, television, radar, etc.

EXAMPLE 5-8.
At what frequency does the $R_L C$ time constant of Fig. 5-22a equal 100T?

SOLUTION.
The $R_L C$ time constant equals

$$R_L C = 50(10^3) 10^{-5} = 0.5 \text{ s}$$

Now we proceed as follows:

$$R_L C = 100T$$

$$0.5 = 100T$$

or

$$T = 5 \text{ ms}$$

This is equivalent to a frequency of

$$f = \frac{1}{T} = \frac{1}{5(10^{-3})} = 200 \text{ Hz}$$

This tells us the peak detector works down to 200 Hz, at which point v_{rip} equals 1 per-cent of V_p, in this case 0.3 V. If the input frequency is lower than 200 Hz, the ripple voltage is greater than 0.3 V.

EXAMPLE 5-9.
Figure 5-22b shows a full-wave peak rectifier. The capacitor charges twice per input cycle rather than once. Because of this, the ripple is half what it would be in a half-wave peak rectifier. Estimate the ripple voltage.

SOLUTION.
Each half of the secondary has a sine wave with a peak of 32 V. Therefore, the capacitor charges to around 32 V, less the diode drop and possibly some drop in the secondary winding.

Figure 5-22. Example 5-7.

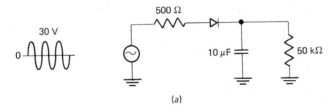

(a)

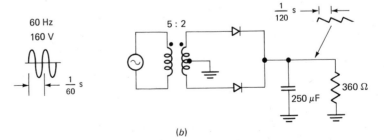

(b)

Since the capacitor charges twice per input cycle, the time between charging is 1/120 s as shown; this is the period of the output ripple. As a decimal fraction, this equals

$$\frac{1}{120} \text{ s} = 0.00833 \text{ s}$$

The $R_L C$ time constant equals

$$R_L C = 360\,(250)\,10^{-6} = 0.090 \text{ s}$$

This is slightly more than 10 times the period of the ripple; with Eq. (5-6), we conclude the ripple voltage is less than 10 percent of V_p, or 3 V peak to peak. Therefore, the output voltage is a unidirectional, almost constant voltage of about 30 V with a ripple of about 3 V.

5-9. SURGE CURRENT

Because the peak rectifier is so widely used, most data sheets include another diode rating called the *surge-current rating*. The initial charging of the capacitor in a peak rectifier causes a large transient current called the *surge current*.

What it is

The basic capacitor law says

$$C = \frac{Q}{V}$$

equivalent to

$$Q = CV$$

In Fig. 5-23a, when the capacitor charges to the peak voltage, the total number of coulombs equals

$$Q = CV_p$$

Where did these coulombs come from? From the source, of course. If R_S is zero, all of these coulombs flow during the first quarter cycle. Because of this, the charging current during this quarter cycle may be quite large. This sudden gush of current is known as the *surge current*.

In Fig. 5-23a, if R_S equals zero, the average surge current during the first quarter cycle equals

$$I_{\text{surge}} = \frac{Q}{t} = \frac{CV_p}{T/4}$$

$$= 4fCV_p \qquad \text{when } R_S \text{ is zero} \tag{5-8}$$

Figure 5-23. Surge current.

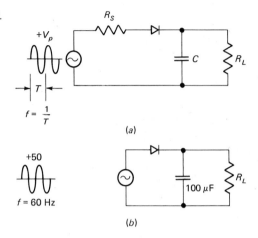

(a)

(b)

As an example suppose $f = 60$ Hz, $C = 100$ μF, and $V_p = 50$ volts as shown in Fig. 5-23b. Then, the surge current equals

$$I_{\text{surge}} = 4(60)\,10^{-4}(50)$$
$$= 1.2 \text{ A}$$

Real voltage sources have some internal resistance, however small. Because of this, the charging time constant $R_S C$ often prevents the capacitor from reaching the full charge during the first cycle; it may take several cycles for the capacitor voltage to reach V_p. As a result, the average surge current may not be as great as given by Eq. (5-8). But we can use Eq. (5-8) to estimate the worst-case surge current in a peak rectifier.

Surge-current rating

Data sheets normally include a surge-current rating designated $I_{F(\text{surge})}$ or something similar. Because surge current flows only briefly, the surge-current rating is greater than the average current rating $I_{0(\text{max})}$. For instance, a 1N456 has an $I_{0(\text{max})}$ of 90 mA and an $I_{F(\text{surge})}$ of 700 mA. A heavier-duty diode like a 1N3208 has an $I_{0(\text{max})}$ of 15 A and an $I_{F(\text{surge})}$ of 250 A.[2] When the surge current in a peak rectifier exceeds the surge-current rating, the diode may be destroyed or may have its rectifying properties degraded. To avoid this, you must select a diode with an adequate surge-current rating.

---

[2] Manufacturers test surge current under different conditions. For highly accurate analysis, you will have to know what these conditions are. If the data sheet does not include them, you may have to contact the manufacturer.

5-10. THE POWER SUPPLY

Transistors, ICs, and other devices need dc voltages to work properly. Sometimes, batteries supply these dc voltages (as in a transistor radio). Often, a power supply rectifies and filters ac line voltage to get the required dc voltages. Figure 5-13c and d shows filtering of half-wave and full-wave signals. But the common approach is to filter the output of a peak rectifier.

Figure 5-24a shows a full-wave peak rectifier followed by an RC filter. The diodes and C_1 are the peak rectifier; therefore, the voltage across C_1 is unidirectional and almost equal to the peak voltage across half the secondary winding. The size of the ripple depends on the discharge time constant $(R + R_L)C_1$. By deliberate design, we keep this time constant much greater than the period of the ripple. In this way, the ripple is not too large.

R and C_2 form an RC filter that reduces the ripple even further. (When necessary, you can use more than one RC filter.) Therefore, the final output voltage is almost ripple-free. On an oscilloscope, this output looks like a dc voltage (a horizontal line above the 0-V level).

Figure 5-24a has this disadvantage: resistor R drops some of the dc voltage because the dc current through R_L must also flow through R. In some applications, the drop across R may be too large. An LC filter gets around this problem.

The LC filter of Fig. 5-24b filters the signal from the peak rectifier. The dc current through R_L also flows through L, but the dc resistance of the inductor is usually small; therefore, the dc voltage across L is much smaller than across the R of Fig. 5-24a.

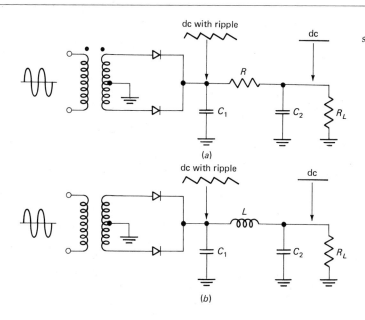

Figure 5-24. Typical power supply. (a) RC *filtering.* (b) LC *filtering.*

5-11. THE CLAMPER

All a clamper does is add a dc component to the signal. In Fig. 5-25a the input signal is a sine wave with a peak-to-peak value of 20 V. The clamper pushes the signal upward, so that the negative peaks fall on the 0-V level. As you can see, the shape of the original signal is preserved; all that happens is a vertical shift of the signal. We describe an output signal like that of Fig. 5-25a as *positively clamped.*

The key idea

Before getting to a particular clamper circuit, realize this: a clamper has to add a dc voltage to the incoming signal. For instance, in Fig. 5-25b the lower source represents the incoming signal. The clamper adds 10 V dc to the signal. Therefore, the sum of the incoming signal and the 10 V dc is the positively clamped sine wave shown in Fig. 5-25b.

Figure 5-25. Positive clamping.

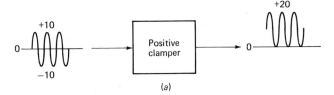

(a)

(b)

(c)

Given an incoming signal like the one in Fig. 5-25c, the clamper will add a dc voltage of V_p. As a result, the output signal is positively clamped as shown.

The positive clamper

Figure 5-26a shows a positive-clamper circuit. Ideally, here is how it works. On the first *negative* half cycle of input voltage, the diode turns on as shown in Fig. 5-26b. At the negative peak, the capacitor must charge to V_p with the polarity shown (how else can Kirchhoff's voltage law be satisfied?).

Slightly beyond the negative peak, the diode shuts off as shown in Fig. 5-26c. The R_LC time constant is deliberately made much greater than the period T of the incoming signal. For this reason, the capacitor remains almost fully charged during the off time of the diode. To a first approximation, the capacitor acts like the battery of Fig. 5-25c. This is why the output voltage in Fig. 5-26a is a positively clamped signal.

Figure 5-26d shows the circuit as it is usually drawn. Since the diode drops a few-tenths volt when conducting, the capacitor voltage does not quite reach V_p. For this

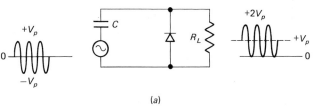

Figure 5-26. (a) *Positive-clamper circuit.* (b) *At the negative peak.* (c) *Diode off.* (d) *Effect of diode knee voltage.*

(a)

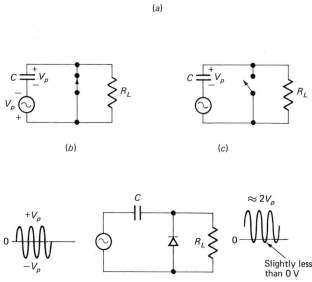

(b)   (c)

(d)

reason, the clamping is not perfect, and the output generally dips slightly below the 0-V level as shown.

What happens if we turn the diode in Fig. 5-26d around? The polarity of capacitor voltage reverses, and the circuit becomes a *negative clamper*. Both positive and negative clampers are widely used. Television receivers, for instance, use a clamper to add a dc voltage to the video signal. In television work, the clamper is usually called a *dc restorer*.

The biased clamper

Occasionally, you may want an adjustable clamping level. You can have it with a circuit like Fig. 5-27. The wiper picks off $+V$ and applies this to the anode of the diode. At the negative peak of the input signal, the capacitor charges to a voltage of V_p $+V$. Therefore, the output signal is positively clamped above the $+V$ level as shown.

Nonideal effects

The charging and discharging of the capacitor is similar to the action in a peak rectifier. For this reason, the deviations from ideal clamping action are similar. To get almost ideal clamping action, you need to make

$$R_L C \gg T$$
$$R_S \ll R_L$$
$$V_p \gg 0.7 \text{ V}$$

which are the same conditions as for a peak rectifier.

5-12. THE PEAK-TO-PEAK DETECTOR

If you cascade a positive clamper and a peak detector, you get a peak-to-peak detector (see Fig. 5-28a). The input sine wave is positively clamped; therefore, the input to the peak detector has a peak value of $2V_p$. This is why the output of the peak detector is a dc voltage equal to $2V_p$.

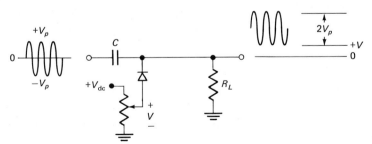

Figure 5-27. Adjustable clamping level.

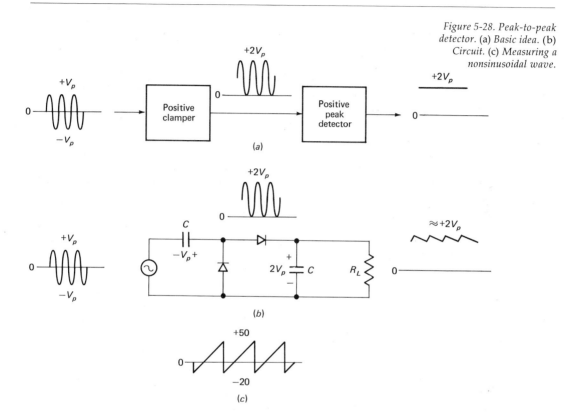

Figure 5-28. Peak-to-peak detector. (a) Basic idea. (b) Circuit. (c) Measuring a nonsinusoidal wave.

Figure 5-28b shows a practical peak-to-peak detector. As usual, the discharge time constant $R_L C$ must be much greater than the period of the incoming signal. By satisfying this condition, you get good clamping action and good peak detection. The output ripple will therefore be small.

Where are peak-to-peak detectors used? Sometimes, the output of a peak-to-peak detector is applied to a dc voltmeter. In this way, the combination acts like a peak-to-peak ac voltmeter. For instance, if the sawtooth wave of Fig. 5-28c is measured with such a voltmeter, the reading would be 70 V pp. Another use for the peak-to-peak detector is in power supplies to get twice as much dc output voltage as with an ordinary peak rectifier. When used in this way, the peak-to-peak detector is called a *voltage doubler*.

5-13. THE DC RETURN

One of the most baffling things that may happen in the laboratory is this: you connect a signal source to a circuit; for some reason, the circuit will not work; yet nothing is

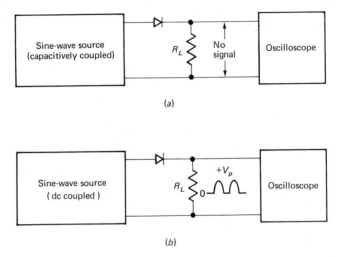

Figure 5-29. The dc-return problem.

(a)

(b)

defective in the circuit or in the signal source. As a concrete example, Fig. 5-29a shows a sine-wave source driving a half-wave rectifier. When you look at the output with an oscilloscope, you see no signal at all; the rectifier refuses to work. To add to the confusion, you may try another sine-wave source and find a normal half-wave signal across the load (Fig. 5-29b).

The phenomenon just described is a classic in electronics; it occurs again and again in practice. It may happen with diode circuits, transistor circuits, integrated circuits, etc. Unless you understand why one kind of source works and another doesn't, you will be confused and possibly discouraged every time the problem arises.

Types of coupling

The signal source of Fig. 5-30a is capacitively coupled; this means it has a capacitor in the signal path. Many commercial signal generators use a capacitor to *dc-isolate* the source from the load, that is, to prevent dc current between the source and load. The idea of a capacitively coupled source is to let only the ac signal pass from source to load.

The dc-coupled source of Fig. 5-30b is different. It has no capacitor; therefore, it provides a path for both ac and dc currents. When you connect this kind of source to a load, it is possible for the load to force a dc current through the source. As long as this dc current is not too large, no damage occurs to the source. Many commercial signal generators are dc-coupled like this.

Sometimes, a signal source is transformer-coupled like Fig. 5-30c. The advantage is that it passes the ac signal from source to load, and at the same time provides a dc path through the secondary winding.

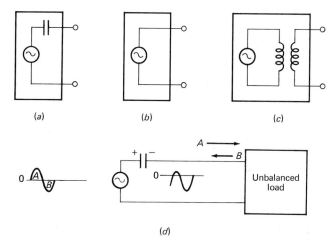

Figure 5-30. (a) *Capacitively coupled source.* (b) *Direct-coupled source.* (c) *Transformer-coupled source.* (d) *Unbalanced load causes unequal currents.*

All circuits discussed earlier in this chapter work fine with dc-coupled and transformer-coupled sources. It is only with the *capacitively coupled sources* that trouble may arise.

Unbalanced diode circuits

A capacitively coupled source will cause unwanted clamping action when the load is *unbalanced*. Figure 5-30d shows what we mean by an unbalanced load; it is any load that causes *unequal* currents during alternate half cycles. Because of this, the amount of charge deposited on the capacitor plates is different on one half cycle from the next. When current A is greater than current B, the capacitor voltage increases during each cycle. As we saw with clamper circuits, a charged capacitor like this will push the ac signal up or down depending on the polarity of the capacitor voltage. In other words, whether we like it or not, *an unbalanced load will cause clamping of the signal out of a capacitively coupled source.*

Now we know why a half-wave rectifier won't work if connected to a capacitively coupled source. In Fig. 5-31a, the capacitor charges to V_p during the first few cycles. Because of this, the signal coming from the source is negatively clamped and the diode cannot turn on after the first few cycles. This is why we see no signal on the oscilloscope.

Among the diode circuits discussed earlier, the following are unbalanced loads: half-wave rectifier, clipper, peak detector, clamper, and peak-to-peak detector. The last two are supposed to clamp the signal; therefore, they work fine with a capacitively coupled source. But the half-wave rectifier, the clipper, and the peak detector of Fig. 5-31b, c, and d will not work with a capacitively coupled source because of unwanted clamping action.

Figure 5-31. Capacitively coupled source produces unwanted clamping in some circuits.

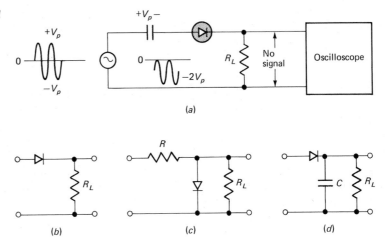

(a)

(b)   (c)   (d)

Dc-return resistor

Is there a remedy for unwanted clamping action? Yes. You can add a *dc-return resistor* across the input to the unbalanced circuit (see Fig. 5-32a). This resistor R_D allows the capacitor to discharge during the *off* time of the diode. In other words, any charge deposited on the capacitor plates is removed during the alternate half cycle.

Figure 5-32. Adding a dc return to eliminate unwanted clamping.

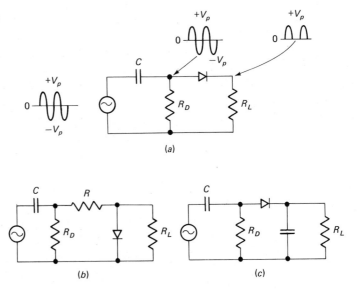

(a)

(b)   (c)

The size of R_D is not critical. The main idea in preventing clamping action is to keep the discharging resistance R_D less than or equal to the charging resistance in series with the diode. In Fig. 5-32a, this means

$$R_D \leq R_L$$

When this condition is satisfied, the capacitor voltage cannot build up significantly and only a small amount of clamping takes place. When possible, make R_D less than one-tenth of R_L.

The dc return for a peak detector (Fig. 5-32c) also requires keeping R_D less than R_L and, if possible, making R_D less than one-tenth R_L.

The clipper of Fig. 5-32b is slightly different. When the diode is conducting, the charging resistance in series with the diode equals R instead of R_L. Therefore, in Fig. 5-32b, the rule is

$$R_D \leq R$$

and when possible, R_D should be less than one-tenth of R.

Balanced diode circuits

Some diode circuits are *balanced loads.* Examples are the center-tap rectifier, the bridge rectifier, the full-wave peak rectifier, etc. These circuits work fine with a capacitively coupled source. In other words, no dc return is needed because the equal and opposite half-cycle currents produce an average capacitor voltage of zero. For instance, Fig. 5-33 shows a capacitively coupled source driving a bridge rectifier. Assuming identical diodes, currents A and B are equal and opposite. Because of this, the capacitor voltage does not build up and the bridge rectifier works normally.

In summary, whether you like it or not, you may get clamping action with a capacitively coupled source. This unwanted clamping may occur in diode circuits, transistor circuits, integrated circuits, etc. In general, whenever a capacitor drives a device that conducts for only *part of the ac cycle,* you may get unwanted clamping action. If this happens, you usually can eliminate the clamping by adding a dc return.

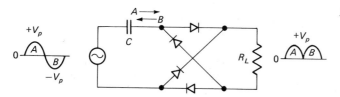

Figure 5-33. Full-wave bridge is a balanced load.

Problems

5-1. Line voltage typically is 115 V rms but may be as low as 105 V rms or as high as 125 V rms. Calculate the peak value for each of these extremes.

5-2. The transformer of Fig. 5-34a has a turns ratio of 4:1. Assuming an ideal diode, what is the peak output voltage? With the second approximation of a silicon diode, what is the peak output voltage?

5-3. If the transformer of Fig. 5-34a has a 2:1 turns ratio, what is the peak load current? (Use ideal diode.)

5-4. What is the peak output voltage for the center-tap rectifier of Fig. 5-34b? The peak load current?

5-5. In Fig. 5-34b, we want a peak load voltage of 40 V. What turns ratio do we need?

5-6. During the AB half cycle of primary voltage, which diodes in Fig. 5-34c conduct?

5-7. The turns ratio in Fig. 5-34c is 5:1. What value does the peak load current have?

Figure 5-34.

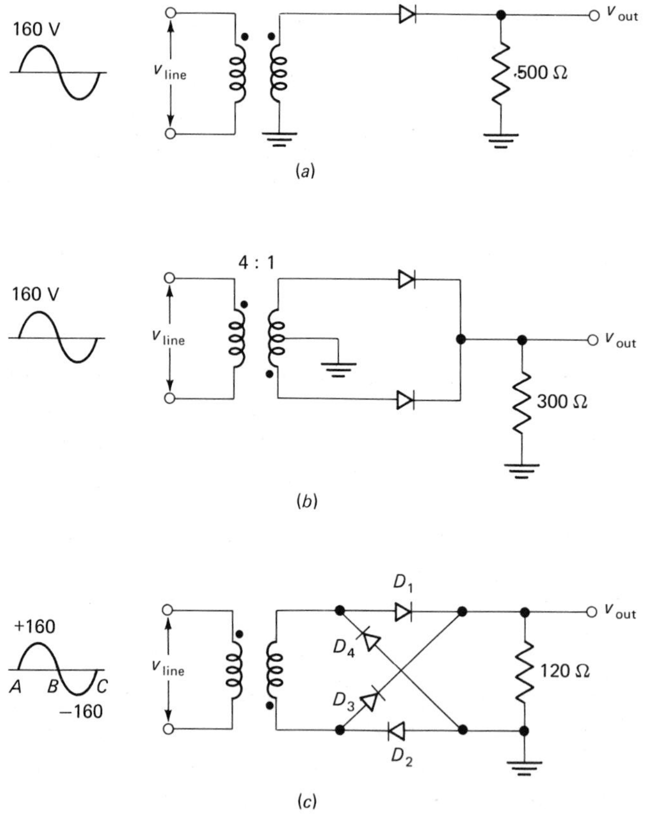

(a)

(b)

(c)

5-8. Using the second approximation of a silicon diode, what is the peak load voltage in Fig. 5-34c if the turns ratio equals 4:1?

5-9. The 1N3070 is a silicon diode with an I_F of 100 mA at 1 V. If the turns ratio in Fig. 5-34a equals 5:1, what is the peak load current with the third approximation of the 1N3070?

5-10. The bridge rectifier of Fig. 5-34c uses FD300s. These silicon diodes have an I_F of 200 mA at 1 V. Using the third approximation, calculate the peak load current for a turns ratio of 3:1.

5-11. Calculate the average value of rectified current through each diode in Fig. 5-34b. Does this equal the average load current, or is it half the average load current?

5-12. The diodes of Fig. 5-34c have an $I_{0(max)}$ rating of 150 mA. Calculate the average rectified current through each diode for a turns ratio of 2:1. Are the diodes safe, or do we need heavier-duty diodes?

5-13. For the half-wave rectifier of Fig. 5-34a, what is the PIV across the diode for each of these turns ratios:

    1. 1:1
    2. 2:1
    3. 3:1

5-14. What is the PIV across each diode in the center-tap rectifier of Fig. 5-34b?

5-15. The bridge rectifier of Fig. 5-34c uses FD600s; these have a WIV rating of 50 V. Which of the following turns ratios keep the PIV less than the WIV:

    1. 1:1
    2. 2:1
    3. 3:1

5-16. What turns ratio does the bridge rectifier of Fig. 5-34c need to produce the same peak load voltage as the center-tap rectifier of Fig. 5-34b? For this turns ratio, what is the PIV across the diodes in the bridge rectifier? Is this lower or higher than the PIV across the diodes in the center-tap rectifier?

5-17. A half-wave rectifier produces a peak load voltage of 40 V. What is the average or dc value of this load voltage? If a full-wave rectifier produces the same peak load voltage, what would the average load voltage equal?

5-18. If the full-wave output signal in Fig. 5-34b is filtered, how much dc voltage will there be at the output of the filter? (Forget about internal voltage drop in the filter.)

5-19. Neglecting all nonideal effects, what is the output signal in the clipper of Fig. 5-35a? If you allow 0.7 V across the diode when conducting, what happens to the output signal? If you include the drop across the 1-kΩ resistor when the diode is off, what happens to the output signal?

Figure 5-35.

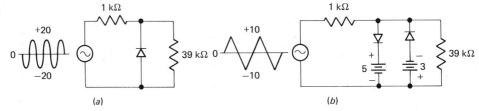

(a)   (b)

5-20. Neglect nonideal effects in the combination clipper of Fig. 5-35b. What does the output signal look like?

5-21. A sine wave with a positive peak of 100 V drives the peak rectifier of Fig. 5-36a. Approximately how much dc output voltage is there? The input frequency is 50 Hz. Is the output ripple less than 10 percent? Less than 1 percent?

5-22. In the peak rectifier of Fig. 5-36b, what is the maximum voltage the capacitor can charge to, assuming an ideal diode? If you allow 0.7-V drop across the diode, what is the maximum voltage the capacitor can charge to?

Figure 5-36.

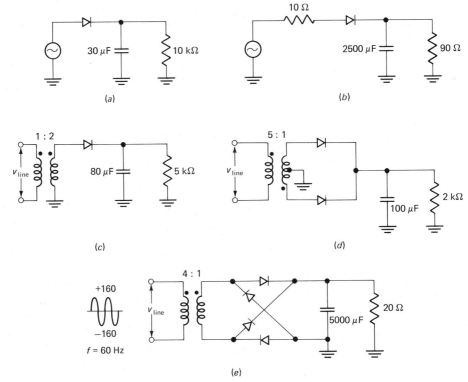

(a)   (b)

(c)   (d)

(e)

5-23.   Calculate the discharging time constant in Fig. 5-36b. Is the ripple less than 10 percent for an input frequency of 60 Hz?

5-24.   What is the approximate output voltage in Fig. 5-36c?

5-25.   Calculate the approximate value of dc output voltage in Fig. 5-36d. (Use a line peak voltage of 160 V.)

5-26.   Neglect all nonideal effects. What is the output dc voltage in Fig. 5-36e?

5-27.   In Fig. 5-36e, assume 1-V drop across each conducting diode. What is the maximum output voltage? If an additional 3 V is lost across the resistance of the secondary winding, what is the maximum output voltage?

5-28.   Use Eq. (5-8) to estimate the surge current in Fig. 5-36c. (Peak line voltage is 160 V.)

5-29.   The bridge peak rectifier of Fig. 5-36e uses 1N3208s. These diodes have a surge-current rating of 250 A. Calculate the surge current using Eq. (5-8). Can the 1N3208s withstand the surge current?

5-30.   What is the output voltage in Fig. 5-37a? At what input frequency does the $R_L C$ time constant equal $10T$, where T is the period of the input sine wave?

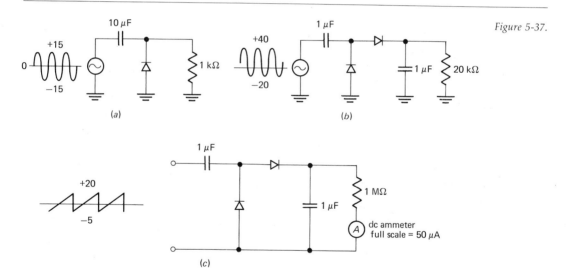

Figure 5-37.

(a)   (b)

(c)

5-31.   How much dc voltage is there across the 20-kΩ resistor of Fig. 5-37b?

5-32.   Figure 5-37c shows a peak-to-peak reading voltmeter. The dc ammeter has a full-scale current of 50 μA. Approximately how much voltage is there across the second capacitor? How much dc current is there through the ammeter? In a peak-to-peak reading voltmeter like this, we would mark the meter face to read peak-to-peak volts rather than current. What voltage should be marked at full scale? At half scale?

Figure 5-38.

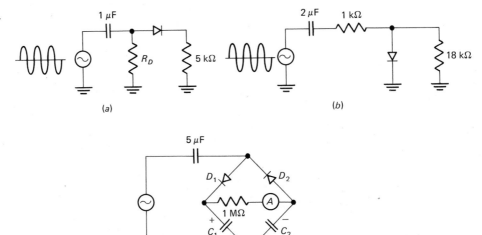

(a)

(b)

(c)

5-33. For the half-wave rectifier of Fig. 5-38a to work normally, what size should R_D be (use 10:1 rule)?

5-34. The clipper of Fig. 5-38b needs a dc return to work properly. Use the 10:1 rule to decide what size R_D should be.

5-35. We have not discussed the bridge circuit of Fig. 5-38c, but it is not difficult to understand. Diode D_1 and capacitor C_1 are a positive peak detector, while D_2 and C_2 are a negative peak detector. All we want to know in this problem is this: do you think the circuit is a balanced or unbalanced load?

6. Bipolar Transistors

You can dope a semiconductor to get an *npn* crystal or a *pnp* crystal. A crystal like this is called a *junction transistor*. The *n* regions have mostly conduction-band electrons, and the *p* regions mostly holes. For this reason, the junction transistor is often called a *bipolar transistor*.

Shockley worked out the theory of the junction transistor in 1949, and the first one was produced in 1951. The transistor's impact on electronics has been enormous. Besides starting the multibillion-dollar semiconductor industry, the transistor has led to all kinds of related inventions like integrated circuits, optoelectronic devices, and others.

The transistor far outperforms the vacuum tube in most applications. It allows us to do things that were either difficult or impossible with vacuum tubes. This is especially noticeable in the computer industry. To a first approximation, the transistor did not revise the computer industry; it created it.

6-1. THE THREE DOPED REGIONS

Figure 6-1*a* shows an *npn* crystal. The *emitter* is heavily doped; its job is to emit or inject electrons into the *base*. The base is lightly doped and very thin; it passes most of the emitter-injected electrons on to the *collector*. The doping of the collector is between the heavy doping of the emitter and the light doping of the base. The collector is so named because it collects or gathers electrons from the base. The collector is the largest of the three regions; it must dissipate more heat than the emitter or base.

The transistor of Fig. 6-1*a* has two junctions, a junction between the emitter and the base, and another between the base and the collector. Because of this, the transistor

Figure 6-1. The three transistor regions. (a) npn. (b) pnp.

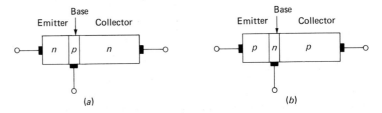

is like two diodes. We call the diode on the left the *emitter-base diode* or simply the *emitter diode.* The diode on the right is the *collector-base diode* or the *collector diode.*

Figure 6-1*b* shows the other possibility: a *pnp* transistor. The *pnp* transistor is the *complement* of the *npn* transistor; this means opposite currents and voltages are involved in the action of a *pnp* transistor. To avoid confusion, we will concentrate on the *npn* transistor during our early discussions. At the right time, we will extend the discussion to *pnp* transistors.

6-2.   THE UNBIASED TRANSISTOR

Figure 6-2*a* shows the majority carriers before any have moved across the junctions. The *like* charges repel each other; this causes diffusion as described in Sec. 3-1. In other words, some majority carriers move across the junctions; this results in two depletion layers (Fig. 6-2*b*). For each of these depletion layers, the barrier potential approximately equals 0.7 V at 25°C for a silicon transistor (0.3 V for a germanium transistor). As with diodes, we concentrate on silicon transistors because of their greater importance. In all

Figure 6-2. Charge distribution. (a) *Before diffusion.* (b) *After diffusion.* (c) *Depletion layers.* (d) *Energy bands.*

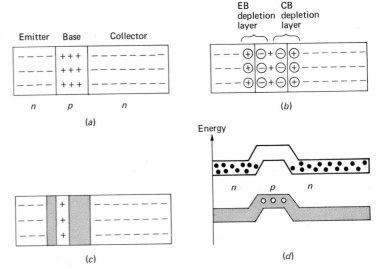

our discussions and figures, the transistors are silicon unless otherwise indicated.

For junction temperatures other than 25°C, we still use Eq. (3-1) to estimate the change in barrier potential. As you may recall, the barrier potential decreases approximately 2 mV for each degree rise in junction temperature.

Because the three regions have different doping levels the depletion layers do not have the same width. The more heavily doped a region is, the greater the concentration of ions near the junction. This means the depletion layer penetrates only slightly into the emitter region (it's heavily doped) but deeply into the base, which is lightly doped. The other depletion layer also extends well into the base, and penetrates the collector region to a lesser amount. Figure 6-2c shows what it boils down to. From now on, we will shade depletion layers to indicate they have no majority carriers. Notice the collector-base depletion layer is wider than the emitter-base depletion layer.

To complete the picture of the unbiased transistor, Fig. 6-2d shows the energy diagram. Because we have two depletion layers, we have two energy hills. Especially important, the conduction-band electrons in the emitter do not have enough energy to enter the base region. Or in terms of radius, these electrons are traveling in conduction-band orbits smaller than the smallest permitted conduction-band orbit in the base. Unless we forward-bias the emitter diode and lower the hill, the emitter electrons cannot enter the base.

6-3. FF AND RR BIAS

There are several ways to bias a transistor. This section discusses two methods we need for later discussions.

Forward-forward bias

Figure 6-3a illustrates *forward-forward* (FF) bias, so named because the emitter diode and the collector diode are forward-biased. The circuits driving the emitter and

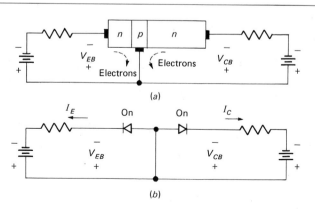

Figure 6-3. FF bias. (a) *Actual circuit.* (b) *Equivalent circuit.*

collector diodes may be lumped as shown or may represent Thevenin equivalent circuits. Either way, carriers cross the junctions and flow down through the base into the external base lead as shown.

Figure 6-3b shows the equivalent circuit for FF bias. Emitter-base voltage V_{EB} forward-biases the emitter diode and produces a conventional current I_E. Likewise, collector-base voltage V_{CB} forward-biases the collector diode, causing conventional current I_C to flow.

Reverse-reverse bias

Another possibility is to *reverse-reverse* (RR) bias the transistor as shown in Fig. 6-4a. Now both diodes are reversed-biased. For this condition, only small currents flow consisting of thermally produced saturation current and surface-leakage current. The thermally produced component is temperature-dependent, and approximately doubles for every 10° rise (discussed earlier in Sec. 3-4). The surface-leakage component, on the other hand, increases with voltage. To a first approximation, these reverse currents are negligible.

Figure 6-4b is the equivalent circuit for RR bias; both diodes are open unless V_{EB} or V_{CB} exceed the breakdown voltages.

6-4.   FORWARD-REVERSE BIAS

Forward-bias the emitter diode, reverse-bias the collector diode, and the unexpected happens. Figure 6-5a shows *forward-reverse* (FR) bias. We expect a large emitter current because the emitter diode is forward-biased. But we do not expect a large collector current because the collector diode is reversed-biased. Nevertheless, this is exactly what we get; this is precisely why the transistor is the great invention it is.

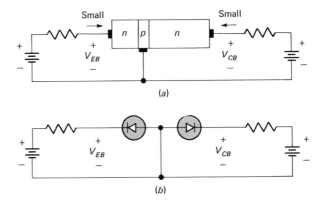

Figure 6-4. RR bias. (a) *Actual circuit.* (b) *Equivalent circuit.*

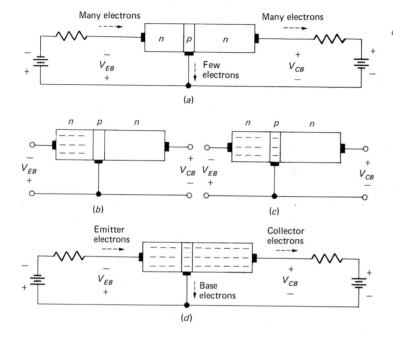

Figure 6-5. FR bias. (a) Actual circuit. (b) Electrons in emitter. (c) Electrons enter base. (d) Electrons diffuse into collector.

Preliminary explanation

Before getting into fine points, here is a brief explanation of why we get that large collector current in Fig. 6-5a. At the instant the forward bias is applied to the emitter diode, electrons in the emitter have not yet entered the base region (see Fig. 6-5b). If V_{EB} is greater than the barrier potential, many emitter electrons enter the base region as shown in Fig. 6-5c. These electrons in the base can flow in either of two directions: down the thin base into the external base lead, or across the collector junction into the collector region.

Which way will they go? For the electrons to flow down through the base region, they must first fall into holes, that is, recombine with base holes. Then, as valence electrons they can flow down through adjacent base holes and into the external base lead. This downward component of base current is called *recombination current*. It is small because the base is lightly doped, that is, has only a few holes.

A second crucial idea in transistor action is that the base is very thin. In Fig. 6-5c, the base is teeming with injected conduction-band electrons. Being *like* charges, they repel each other, causing diffusion into the collector depletion layer. Once inside this layer, they are pushed by the depletion-layer field into the collector region (see Fig. 6-5d). These collector electrons can then flow into the external collector lead as shown.

Our picture of FR bias is this. In Fig. 6-5d we visualize a steady stream of electrons leaving the negative source terminal and entering the emitter region. Simultaneously, emitter electrons enter the base region. The thin and lightly doped base gives almost all of these electrons enough lifetime to diffuse into the collector depletion layer. The depletion-layer field then pushes a steady stream of electrons into the collector region. Simultaneously, electrons leave the collector, enter the external collector lead, and flow into the positive terminal of the collector voltage source. In most transistors, more than 95 percent of the emitter-injected electrons flow to the collector; less than 5 percent fall into base holes and flow out the external base lead.

The energy viewpoint

An energy diagram is the next step to a deeper understanding of the transistor. Forward-biasing the emitter diode lowers its energy hill (see Fig. 6-6). Therefore, conduction-band electrons in the emitter now have enough energy to move into the base conduction band. In other words, the orbits of some emitter electrons are now large enough to match some of the available base orbits. Because of this, emitter electrons can diffuse from the emitter conduction band to the base conduction band.

On entering the base conduction band, the electrons become minority carriers because they are inside a p region. In almost any transistor, more than 95 percent of these minority carriers have a long enough lifetime to diffuse into the collector depletion layer and fall down the collector energy hill. As they fall, they give up energy, mostly in the form of heat. The collector must be able to dissipate this heat, and for this reason, it is usually the largest of the three doped regions. Less than 5 percent of the emitter-injected electrons fall along the recombination path shown in Fig. 6-6; those that become valence electrons flow through base holes into the external base lead.

There is a key point to make right now: the emitter diode is master and the collector diode is slave. Increase the forward bias on the emitter diode and you inject more electrons into the base. Almost all of these diffuse into the collector depletion layer. In Fig. 6-6, it doesn't matter much if the collector hill is steep or gentle. Any slope at all in the collector hill implies that any electrons entering the collector depletion layer will be swept into the collector region. We can summarize this by

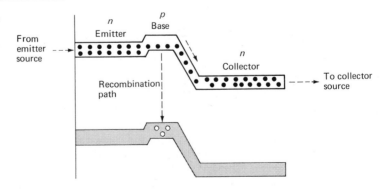

Figure 6-6. Transistor energy bands.

1. The forward bias on the emitter diode controls the number of electrons injected into the base. The larger V_{EB}, the larger the number of injected electrons.
2. The reverse bias on the collector diode has little influence on the number of electrons that enter the collector. Increasing V_{CB} steepens the collector hill but cannot significantly change the number of electrons arriving at the collector depletion layer.

Dc alpha

Saying that more than 95 percent of injected electrons reach the collector is the same as saying collector current almost equals emitter current. The *dc alpha* of a transistor indicates how close in value the two currents are; it is defined as

$$\alpha_{\text{dc}} = \frac{I_C}{I_E} \qquad\qquad (6\text{-}1) ***$$

For instance, if we measure an I_C of 4.9 mA and an I_E of 5 mA,

$$\alpha_{\text{dc}} = \frac{4.9}{5} = 0.98$$

The thinner and more lightly doped the base is, the higher the α_{dc}. Ideally, if all injected electrons went on to the collector, α_{dc} would equal unity. Many transistors have α_{dc} greater than 0.99, and almost all have α_{dc} greater than 0.95. Because of this, we can approximate α_{dc} as 1 in most preliminary analysis.

Base-spreading resistance

With two depletion layers penetrating the base, the base holes are confined to a thin channel of p-type semiconductor as shown in Fig. 6-7. Increasing the reverse bias on the collector diode (equivalent to increasing V_{CB}) widens the collector depletion layer; this reduces the width of the channel containing base holes. In other words, there will be fewer base holes available for recombination current. The resistance of the p channel in the base is called the *base-spreading resistance r_b'.*

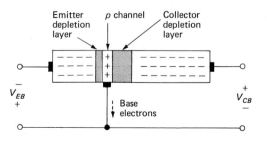

Figure 6-7. Base-spreading resistance r_b'.

Recombination current in the base must flow down through r_b'. When it does, it produces a voltage. We discuss the importance of this voltage later. For now, realize r_b' exists and depends on the width of the p channel in Fig. 6-7 as well as on the doping of the base. In rare cases, r_b' may be as high as 1000 Ω. Typically, it is in the range of 50 to 150 Ω. In most preliminary analysis, we neglect r_b'.

Breakdown voltages

Since the two halves of the transistor are diodes, too much reverse voltage on either diode can cause breakdown. With FR bias, we worry only about the collector diode. When V_{CB} is too large, the collector diode breaks down because of avalanche (previously described) or because of *reach-through effect* (also known as punch-through).

Reach-through means the collector depletion layer becomes so wide it reaches the emitter depletion layer. When this happens, emitter electrons are injected directly into the collector depletion layer. With even the slightest overlap of depletion layers, the collector current can become large enough to destroy the transistor.

Figure 6-8a gives us more insight into reach-through effect; reach-through has not yet occurred because the two depletion layers do not yet overlap. With an increase in

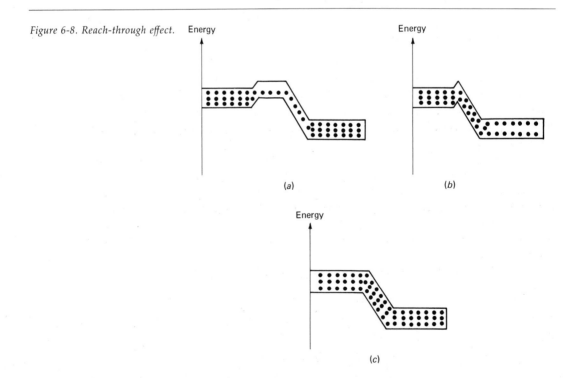

Figure 6-8. Reach-through effect.

collector voltage, the collector depletion layer widens. Figure 6-8b shows the condition when the two depletion layers overlap. Now, more electrons can flow from emitter to collector. Increasing the collector voltage further may wipe out the emitter hill altogether as shown in Fig. 6-8c. In this case, the enormous current would destroy the transistor.

Reach-through destroys a transistor only if the transistor cannot dissipate the heat produced by the reach-through current. In other words, it is possible to have a slight reach-through condition (something like Fig. 6-8b) where the overlap does not produce excessive currents. The transistor does not work too well under this condition, however, because the emitter hill has lost some of its control over collector current.

Avalanche and reach-through are undesirable in the ordinary transistor. We can avoid both by keeping the collector voltage less than the breakdown voltage specified on a manufacturer's data sheet; we discuss breakdown voltage in more detail later.

6-5. THE CE CONNECTION

In Fig. 6-9a, the emitter and the two voltage sources connect to the common point shown. Because of this, we call the circuit a *common-emitter* (CE) connection. The circuit first discussed (Fig. 6-5a) is a common-base (CB) connection because the base and two voltage sources are connected to a common point.

Voltage drop across emitter diode

In the CE circuit of Fig. 6-9a, V_{BB} forward-biases the emitter diode through R_B. Source V_{BB} and resistance R_B may represent a single voltage source and resistor as shown, or they may represent the Thevenin equivalent of a more complicated circuit; it makes no difference as far as the forward bias is concerned. All that matters is the value of V_{BE} appearing across the emitter diode. When this voltage is around 0.7 V, we get emitter current.

The use of 0.7 V for V_{BE} is the second approximation discussed earlier. To a third approximation, we get an additional voltage drop across the bulk resistance of the emitter diode. This bulk resistance is partly in the emitter region and partly in the base region. Because the emitter is heavily doped, its bulk resistance is negligible compared with the bulk resistance of the thin and lightly doped base. So, any additional voltage is mostly across the bulk resistance of the base, which is the same as the r_b' discussed earlier.

For our third approximation of the emitter diode, we will use

$$V_{BE} \cong 0.7 + I_B r_b'$$

To give you some idea of how $I_B r_b'$ adds to the total drop, suppose $I_B = 1$ mA and $r_b' = 100\ \Omega$, then

$$V_{BE} \cong 0.7 + (0.001)100 = 0.8 \text{ V}$$

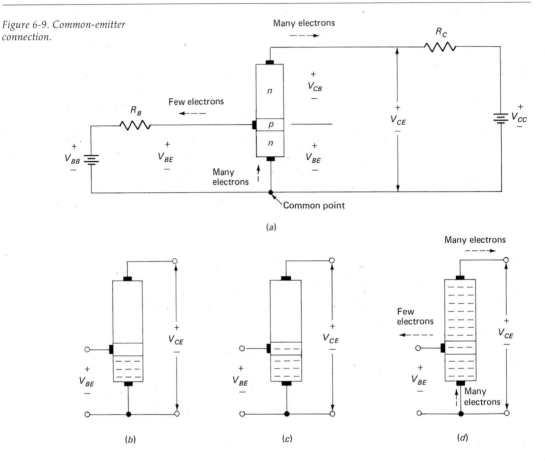

Figure 6-9. Common-emitter connection.

Many electrons

(a)

(b)   (c)   (d)

If the base current were as high as 3 mA, the total V_{BE} drop would be around 1 V. A base current of 3 mA is rather high and would be encountered only in power transistors (to be discussed in Chap. 10). The point is we won't worry about the additional $I_B r_b'$ drop unless we need highly accurate answers.

Reverse bias of collector diode

In the collector circuit of Fig. 6-9a, V_{CC} and R_C may be exactly as shown or may represent a Thevenin equivalent circuit. Again, it is immaterial what is connected to the collector as long as V_{CE} produces reverse bias of the collector diode. The actual voltage across the collector diode is V_{CB}; this voltage reverse-biases the collector diode. With Kirchhoff's voltage law, we can relate the three transistor voltages:

$$V_{CE} = V_{CB} + V_{BE}$$

or by rearranging,

$$V_{CB} = V_{CE} - V_{BE} \qquad (6\text{-}2)$$

To a second approximation, this equals

$$V_{CB} \cong V_{CE} - 0.7 \qquad (6\text{-}2a)$$

Examine Fig. 6-9a. For the collector diode to be reverse-biased, V_{CB} must be positive. According to Eq. (6-2a), V_{CB} can be positive only if V_{CE} is greater than 0.7 V. Or, at very high base currents when the $I_B r_b'$ drop is important, V_{CE} should be greater than a volt or thereabouts. In other words, if V_{CE} is less than 0.7 to 1 V, depending on the exact value of $I_B r_b'$, the collector diode will not be reverse-biased. In our discussions, therefore, we must keep V_{CE} greater than approximately a volt to make sure the collector diode stays reverse-biased.

Electron flow

Just because you change from a CB connection (Fig. 6-5a) to a CE connection (Fig. 6-9a), you do not change the way a transistor operates. Majority carriers move exactly the same as before. That is, the emitter is teeming with conduction-band electrons (Fig. 6-9b). When V_{BE} is greater than 0.7 V or so, the emitter injects these electrons into the base (Fig. 6-9c). As before, the thin and lightly doped base gives almost all these electrons enough lifetime to diffuse into the collector depletion layer. With a reverse-biased collector diode, the depletion-layer field pushes the electrons into the collector region where they flow out to the external voltage source (Fig. 6-9d).

The energy-band explanation for electron flow is identical to that given earlier. Also, α_{dc} still is the ratio of I_C to I_E, and is greater than 0.95 for almost any transistor; base current is less than 5 percent of emitter current.

Dc beta

We have related the collector current to the emitter current by using α_{dc}. We can also relate collector current to base current by defining the *dc beta* of a transistor as

$$\beta_{dc} = \frac{I_C}{I_B} \qquad (6\text{-}3)^{***}$$

For instance, if we measure a collector current of 5 mA and a base current of 0.05 mA, the transistor has a β_{dc} of

$$\beta_{dc} = \frac{5\ \text{mA}}{0.05\ \text{mA}} = 100$$

In almost any transistor, less than 5 percent of the emitter-injected electrons recombine with base holes to produce I_B; therefore, β_{dc} is almost always greater than 20.

Usually, it is from 50 to 200. And some transistors have β_{dc} as high as 1000. In another system of analysis called h parameters (Sec. 15-7), β_{dc} is called the dc current gain and is designated by h_{FE} (note the FE subscripts are capitalized). In other words, $\beta_{dc} = h_{FE}$. This is important to remember because data sheets usually give the value of h_{FE}.

Relation between α_{dc} and β_{dc}

Is there a relation between α_{dc} and β_{dc}? Of course there is. Kirchhoff's current law tells us

$$I_E = I_C + I_B$$

Dividing by I_C gives

$$\frac{I_E}{I_C} = 1 + \frac{I_B}{I_C}$$

or

$$\frac{1}{\alpha_{dc}} = 1 + \frac{1}{\beta_{dc}}$$

With simple algebra, we can rearrange to get

$$\beta_{dc} = \frac{\alpha_{dc}}{1 - \alpha_{dc}} \tag{6-4}$$

or as a good approximation for most transistors,

$$\beta_{dc} \cong \frac{1}{1 - \alpha_{dc}} \tag{6-4a}$$

because $\alpha_{dc} \cong 1$.

As an example, if $\alpha_{dc} = 0.98$, the exact value of β_{dc} is

$$\beta_{dc} = \frac{0.98}{1 - 0.98} = 49$$

and the approximate value is

$$\beta_{dc} \cong \frac{1}{1 - 0.98} = 50$$

For exact answers, therefore, we use Eq. (6-4), but for preliminary analysis, Eq. (6-4a) is accurate enough.

Occasionally, we may need a formula for α_{dc} in terms of β_{dc}. With simple algebra we can rearrange Eq. (6-4) to get

$$\alpha_{dc} = \frac{\beta_{dc}}{\beta_{dc} + 1} \tag{6-4b}$$

For instance, for a β_{dc} of 100,

$$\alpha_{dc} = \frac{100}{100 + 1} = 0.99$$

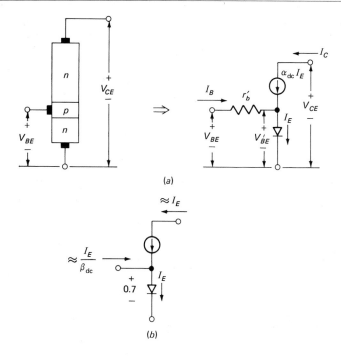

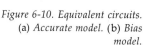

Figure 6-10. Equivalent circuits.
(a) Accurate model. (b) Bias
model.

(a)

(b)

Two equivalent circuits

To crystallize the main ideas of transistor action, we can use the equivalent circuit of Fig. 6-10a. The voltage V'_{BE} is the voltage across the emitter depletion layer. When this voltage is greater than 0.7 V or thereabouts, the emitter injects electrons into the base. As mentioned earlier, the emitter diode is master and the collector diode is slave; the current in the emitter diode controls the collector current. For this reason, the collector-current source forces a current of $\alpha_{dc}I_E$ to flow in the collector circuit. All of this assumes V_{CE} is greater than a volt or so; otherwise, the collector diode is not reverse-biased and the transistor cannot work normally.

The internal voltage V'_{BE} differs from the applied voltage V_{BE} by the drop across r'_b. That is,

$$V_{BE} = V'_{BE} + I_B r'_b$$

When the $I_B r'_b$ drop is small enough to neglect, $V_{BE} \cong V'_{BE}$. The equivalent circuit of Fig. 6-10a is the kind of circuit we would use for highly accurate analysis.

For most practical work, the equivalent circuit of Fig. 6-10b is accurate enough. First, we assume the voltage across the emitter diode is 0.7 V. Second, we disregard the $I_B r'_b$ voltage, equivalent to treating r'_b as negligibly small. Third, emitter current I_E sets up a slave current of approximately I_E in the collector circuit; again, this assumes the collector diode is reverse-biased, that is, a V_{CE} greater than a volt or so, but less than the breakdown value. Fourth, the base current approximately equals I_E/β_{dc}. This follows immediately from

$$\beta_{dc} = \frac{I_C}{I_B}$$

or

$$I_B = \frac{I_C}{\beta_{dc}} = \frac{\alpha_{dc}I_E}{\beta_{dc}} \cong \frac{I_E}{\beta_{dc}}$$

because α_{dc} is very close to unity.

We intend to use the equivalent circuit of Fig. 6-10b a great deal in Chap. 7. Remember a transistor doesn't care what kind of external circuitry you connect to it. As long as the emitter diode is forward-biased and the collector diode reverse-biased (less than breakdown), the transistor will operate as previously described. In other words, we can use the equivalent circuit of Fig. 6-10b for a CB circuit like Fig. 6-5a, or for a CE circuit like Fig. 6-9a, or for *any circuit* that FR-biases the transistor.

The transistor equivalent circuit of Fig. 6-10b is so important that we will give it a name. We call it the *bias model* of a transistor. Using the bias model, Chap. 7 will show you how to analyze the dc currents and voltages in practically any transistor circuit you will ever encounter.

EXAMPLE 6-1.

Figure 6-11a shows a transistor with a β_{dc} of 50 and an emitter current of 10 mA. The external circuitry producing this emitter current is not shown, but you may assume the

---

Figure 6-11. Example 6-1.

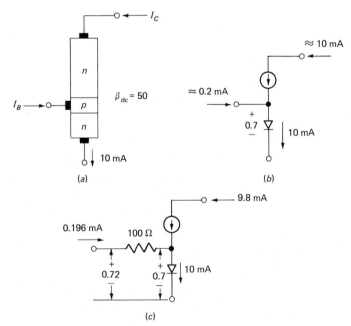

transistor is FR-biased. Show the *bias model* with as many currents and voltages as possible.

SOLUTION.
With an emitter current of 10 mA, the collector current is approximately 10 mA, and the base current equals

$$I_B \cong \frac{10 \text{ mA}}{50} = 0.2 \text{ mA}$$

The voltage across the emitter diode is approximately 0.7 V; V_{CE} is unknown except for being greater than approximately 1 V but less than the breakdown value.

Figure 6-11b shows the bias model with its currents.

EXAMPLE 6-2.
For the transistor of Fig. 6-11a, show the values in the accurate model of Fig. 6-10a. Use an $r_b' = 100 \ \Omega$, $V_{BE}' = 0.7$ V, and $\alpha_{dc} = 0.98$.

SOLUTION.
In this case, collector current equals

$$I_C = \alpha_{dc} I_E = 0.98 \,(10 \text{ mA}) = 9.8 \text{ mA}$$

and base current equals

$$I_B = \frac{I_C}{\beta_{dc}} = \frac{9.8 \text{ mA}}{50} = 0.196 \text{ mA}$$

When this base current flows through the base-spreading resistance, it produces a voltage drop of

$$I_B r_b' = 0.196\,(10^{-3})\,10^2 = 0.0196 \text{ V}$$

Therefore, the voltage across the base-emitter terminals is

$$V_{BE} = V_{BE}' + I_B r_b' = 0.7 + 0.0196$$
$$= 0.72 \text{ V}$$

Figure 6-11c summarizes these results. When we compare this model with its accurate values to the bias model of Fig. 6-11b, we see small differences in currents and voltages. These small differences are important only when making an exact analysis. For most practical analysis, the approximate values of the bias model are accurate enough.

6-6. TRANSISTOR CURVES

One way to see as many details as possible is with graphs that relate transistor currents and voltages.

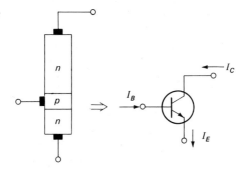

Figure 6-12. Transistor schematic symbol.

Schematic symbol

Figure 6-12 shows the schematic symbol for an *npn* transistor. The emitter has an arrowhead, but not the collector. Especially important, the arrowhead points in the direction of conventional emitter current. In other words, electrons flow into the emitter and out the base and collector. Conventional currents flow in opposite directions as shown in Fig. 6-12. Conventional emitter current flows out of the emitter; conventional base and collector current flow into the *npn* transistor.

Collector curves

You can get data for CE collector curves by setting up a circuit like Fig. 6-13*a* or by using a transistor curve tracer. Either way, the idea is to vary the V_{BB} and V_{CC} supplies to set up different transistor voltages and currents.

Figure 6-13. Getting collector curves.

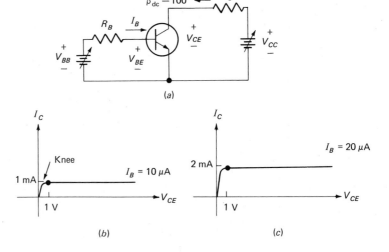

To keep things orderly, the usual procedure is to set a value of I_B and hold it fixed while you vary V_{CC}. By measuring I_C and V_{CE}, you can get data for graphing I_C versus V_{CE}. For instance, suppose we set up $I_B = 10$ μA in Fig. 6-13a. Next, we vary V_{CC} and measure the resulting I_C and V_{CE}. If we plot the data, we may get Fig. 6-13b. Notice we label this curve $I_B = 10$ μA because we got this curve by keeping I_B fixed during all measurements.

There is nothing mysterious about Fig. 6-13b. It echoes our explanation of transistor action. When V_{CE} is zero, the collector diode is not reverse-biased; therefore, the collector current is negligibly small. For V_{CE} between 0 and 1 V or thereabouts, the collector current rises sharply and then becomes almost constant. This is tied in with the idea of reverse-biasing the collector diode. It takes approximately 0.7 V or so to reverse-bias the collector diode; once you reach this level, the collector is gathering all electrons that reach its depletion layer.

Above the knee, the exact value of V_{CE} is not too important because making the collector hill steeper cannot appreciably increase the collector current. The small increase in collector current with increasing V_{CE} is caused by the collector depletion layer getting wider and capturing a few more base electrons before they fall into holes.

By repeating the measurements of I_C and V_{CE} for $I_B = 20$ μA, we can plot the graph of Fig. 6-13c. The curve is similar, except that above the knee the collector current approximately equals 2 mA. Again, an increase in V_{CE} produces a small increase in collector current because the widening depletion layer captures a few additional base electrons.

If several curves for different I_B are shown on the same graph, we get the collector curves of Fig. 6-14a. Since we used a transistor with a β_{dc} of approximately 100, the

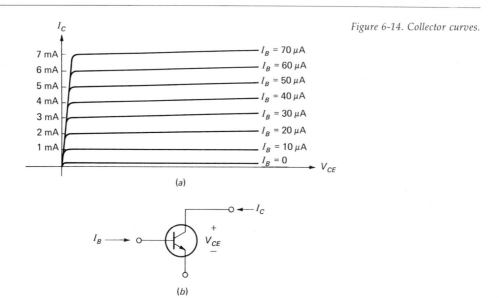

Figure 6-14. Collector curves.

collector current is approximately 100 times greater than the base current for any point above the knee of a given curve. Because collector current increases slightly with an increase in V_{CE}, β_{dc} increases slightly with an increase in V_{CE}.

Now here is something important. You must not think a set of collector curves like Fig. 6-14a applies only to the original circuit used to make the measurements. The relation between transistor currents and voltages is independent of the circuit connected to the transistor. Once you have a set of collector curves like Fig. 6-14a, you can use these collector curves no matter what you connect to the transistor.

In other words, refer to Fig. 6-13a. Now, mentally detach the transistor from this circuit and visualize it as shown in Fig. 6-14b. As long as this transistor is FR-biased, no matter what the circuitry connected to it, the collector curves of Fig. 6-14a tell us how the collector current, collector voltage, and base current are related. Remember this when we attach more complicated circuits to the base and collector. Remember this also when we draw the transistor sideways, upside down, or any way we please.

The base curve

In the measuring circuit of Fig. 6-13a, we can also get data for graphing I_B versus V_{BE}. Figure 6-15a shows the kind of graph we get. There is nothing surprising here. Since the base-emitter section of a transistor is a diode, we expect to see something resembling a diode curve. There is a fine point worth mentioning, however. As the collector depletion layer widens with increasing collector voltage, the base current decreases slightly because the collector depletion layer captures a few more base electrons. For this reason, if we plot another base curve for a different collector voltage, the new curve will look slightly different from the old.

Figure 6-15b shows what we mean. For the same value of V_{BE}, the curve with higher V_{CE} has a smaller base current because the collector depletion layer captures a few more base electrons before they fall into holes; as a result, the recombination base current I_B is reduced. The gap between the curves is small, and we take it into account when seeking highly accurate answers (Sec. 15-7 on h parameters). For all preliminary analysis, we disregard the effect a changing collector voltage has on base current.

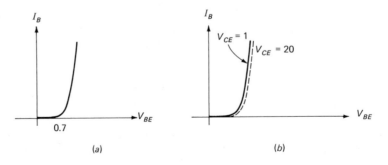

Figure 6-15. Base curves.

(a)

(b)

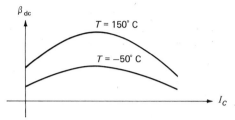

Figure 6-16. Variation in dc beta.

The current-gain curve

The β_{dc} of a transistor is an important quantity; we should be aware of how it changes with collector current and temperature. Figure 6-16 shows a typical variation in β_{dc}. At a fixed temperature, β_{dc} increases to a maximum when collector current increases. For further increases in I_C, the β_{dc} drops off. The variation in β_{dc} may be as much as 2:1 over the useful current range of the transistor; it depends on the type of transistor.

Increasing the temperature will increase β_{dc} at a given collector current. Over a large temperature range, the variations in β_{dc} may be greater than 3:1, depending on the transistor type. Chapter 7 will explain why such changes are undesirable and will show you how to eliminate their effects.

Cutoff current and breakdown voltage

In the collector curves of Fig. 6-14a, the lowest curve is for zero base current. The condition $I_B = 0$ is equivalent to having an *open base lead* (see Fig. 6-17a). The collector current with an open base lead is designated I_{CEO}, where subscripts CEO stand for collector to emitter with open base. I_{CEO} is caused partly by thermally produced carriers and partly by surface-leakage current.

Figure 6-17b shows the $I_B = 0$ curve. With a large enough collector voltage, we reach a breakdown voltage labeled BV_{CEO}, where the subscripts again stand for collector to emitter with open base. For normal transistor operation, we must keep V_{CE} less than BV_{CEO}. Most transistor data sheets list the value of BV_{CEO} among the maximum ratings. This breakdown voltage may be from less than 20 V to over 200 V, depending on the transistor type.

Collector saturation voltage

For normal transistor operation, the collector diode must be reverse-biased; this requires a V_{CE} greater than a volt or so, depending on how much collector current

Figure 6-17. Cutoff current and
breakdown voltage.

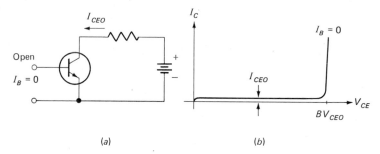

(a)   (b)

flows. As a guide to where the knee is, many data sheets list the $V_{CE(\text{sat})}$ of a transistor.

Figure 6-18 shows what we mean by $V_{CE(\text{sat})}$. It is the value of V_{CE} at some point below the knee with the exact position specified on the data sheet. For instance, a 2N4137 has a $V_{CE(\text{sat})}$ of 0.14 V for $I_C = 10$ mA and $I_B = 1$ mA. $V_{CE(\text{sat})}$ increases at higher collector currents. The same 2N4137 has a $V_{CE(\text{sat})}$ of 0.28 V when $I_C = 100$ mA and $I_B = 10$ mA. All we need to know for the next few chapters is that data sheets usually list $V_{CE(\text{sat})}$ values. Typically, $V_{CE(\text{sat})}$ is only a few-tenths volt, although at very large collector currents it may exceed a volt. Incidentally, the part of the curve below the knee of Fig. 6-18 is known as the *saturation region*.

Summary

Here are the main ideas of this section:

1. V_{CE} should be greater than $V_{CE(\text{sat})}$ but less than BV_{CEO}. This ensures a reverse-biased collector diode.
2. Increasing V_{CE} decreases I_B because the collector depletion layer penetrates deeper into the base. This is a small effect treated later using h parameters (Sec. 15-7).
3. β_{dc} varies with I_C and temperature. These variations are undesirable; Chap. 7 shows you how to get around them.

Figure 6-18. Saturation region.

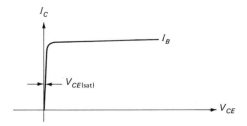

4. There is a small collector current even when $I_B = 0$. The importance of this current and other cutoff currents will be handled in Chap. 7.

All these points are important and will be discussed further at the right time. What we want to carry with us into Chap. 7 is the *bias model* of a transistor (Fig. 6-10*b*). Using the *bias model* of a transistor, Chap. 7 will show you how to calculate the dc currents and voltages in almost any transistor circuit you will encounter.

Problems

6-1. The barrier potential of the emitter depletion layer is approximately 0.7 V for a temperature of 25°C. Estimate the barrier potential at 0 and 50°C.

6-2. In Fig. 6-4*a*, a small reverse current flows in the collector circuit. The thermally produced component of this current equals 0.1 nA at 25°C. Estimate the values of this component at 75°C and at −15°C.

6-3. For the transistor circuit of Fig. 6-5*d*, suppose that only 2 percent of the electrons injected into the base recombine with base holes. If 10 million electrons enter the emitter in 1 μs, how many electrons come out the base lead in this period? How many come out the collector lead during this time?

6-4. When the electrons in Fig. 6-6 fall down the collector hill, they give up energy mostly in the form of heat. You can calculate the power by the product of VI, where V is voltage across the collector depletion layer and I is the current flowing through this depletion layer. If $V = 20$ V and $I = 10$ mA, how much power is being dissipated? If the voltage remains the same, but the current increases to 500 mA, how much power is there?

6-5. In a particular transistor, the collector current equals 5.6 mA and the emitter current equals 5.75 mA. What is the value of α_{dc}?

6-6. The base current in a transistor equals 0.01 mA and the emitter current equals 1 mA. What is the value of α_{dc}?

6-7. If $V_{CB} = 4$ V and $V_{BE} = 0.7$ V, what does V_{CE} equal?

6-8. In a transistor, we measure an I_C of 100 mA and an I_B of 0.5 mA. What does β_{dc} equal?

6-9. A transistor has a β_{dc} of 150. If the collector current equals 45 mA, what does the base current equal?

6-10. Suppose the β_{dc} of a particular transistor can vary from 20 to 100. Over what range does α_{dc} vary?

6-11. A transistor has an α_{dc} of 0.995. What does its β_{dc} equal? When the temperature is 100°C, the α_{dc} increases to 0.998. What value does β_{dc} have at this higher temperature?

6-12. In the transistor equivalent circuit of Fig. 6-19a, the α_{dc} equals 0.99. Calculate I_C, I_B, and V_{BE}.

6-13. In Fig. 6-19a, suppose we increase the emitter current from 20 mA to a very high value. If $\alpha_{dc} = 0.99$ and we measure a V_{BE} of 1 V, how much emitter current is there?

Figure 6-19.

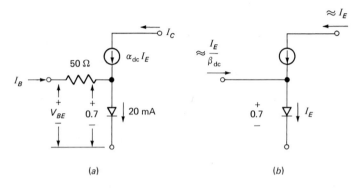

(a)   (b)

6-14. A 2N5067 is a power transistor (one that can dissipate more than approximately ½ W of power). Power transistors normally have small r_b'. The typical 2N5067 has an r_b' of 10 Ω. How much $I_B r_b'$ drop is there when $I_B = 1$ mA? When $I_B = 10$ mA? And when $I_B = 100$ mA?

6-15. A 2N3298 has a typical β_{dc} of 90. If the bias model of Fig. 6-19b represents the 2N3298, calculate the collector and base currents for an emitter current of 10 mA.

6-16. In the bias model of Fig. 6-19b, suppose the base current equals 1 mA and the emitter current equals 200 mA, what is the value of β_{dc}?

6-17. A transistor has a β_{dc} of 400. In the bias model of Fig. 6-19b, what does the base current equal when the emitter current equals 50 mA?

6-18. Figure 6-20a shows one of the collector curves for a particular transistor. Calculate the β_{dc} at point A and at point B. If the collector current changes at the same rate with increasing V_{CE}, what will β_{dc} equal at $V_{CE} = 15$ V?

6-19. Figure 6-20b shows a circuit to be analyzed in Chap. 7. Right now, you know enough to answer this question. If the transistor has the collector curve of Fig. 6-20a, what value does I_B have? Figure 9-20c shows the same circuit modified slightly. What does I_C equal?

6-20. A 2N5346 has the base curve shown in Fig. 6-21a. The voltage drop across r_b' causes V_{BE} to increase when the base current increases. Because of this, here is a way to estimate the value of r_b':

$$r_b' = \frac{\Delta V_{BE}}{\Delta I_B} \qquad \text{well above the knee}$$

Apply this formula to Fig. 6-21a to get an estimate for the r_b' of a 2N5346.

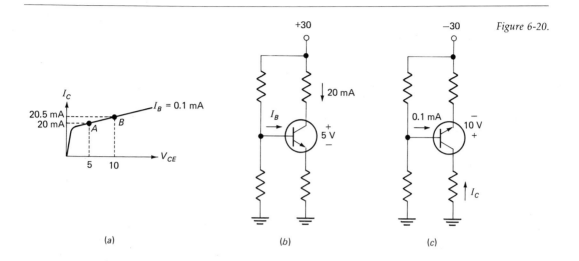

Figure 6-20.

(a)   (b)   (c)

6-21.   A 2N5346 has the β_{dc} variations shown in Fig. 6-21b. If the transistor is operating at collector currents much smaller than 1 A, what is the approximate value of β_{dc}? How much base current is there when $I_C = 1$ A? And when $I_C = 7$ A?

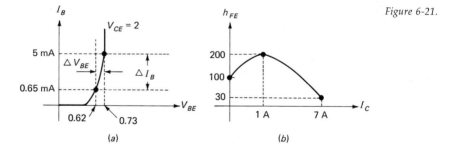

Figure 6-21.

(a)   (b)

6-22.   Figure 6-22a shows a transistor circuit with an open base lead. If we measure a V_{CE} of 9 V, what value does I_{CEO} have? If we change the collector resistor from 10 MΩ to 10 kΩ as shown in Fig. 6-22b, what is the new value of V_{CE}? (Assume I_{CEO} remains the same.)

6-23.   A transistor has the collector curves of Fig. 6-22c. If this transistor is used in the circuit of Fig. 6-22d, what will V_{CE} equal? What is the value of BV_{CEO}? Is the transistor of Fig. 6-22d in danger of breakdown?

Figure 6-22.

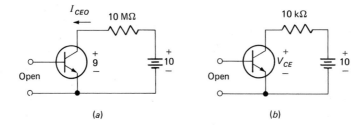

(a)   (b)

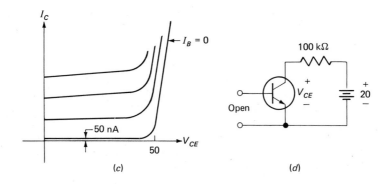

(c)   (d)

7. Transistor Biasing Circuits

Linear transistor circuits are intended to operate only with FR bias. *Digital* transistor circuits are different; they operate in any or all three modes: FF, RR, or FR. Digital circuits are sometimes called *switching* circuits because their biasing switches from FF to RR, or from FR to RR, or from one value of FR bias to another value of FR bias.

This book emphasizes linear transistor circuits. Circuits like these can do many useful things to an ac signal. But to work properly, linear circuits must stay FR-biased throughout the ac cycle. For this condition to be satisfied, we must first set up proper values of dc current and voltage in a transistor circuit.

7-1. BASE BIAS AND EMITTER BIAS

You will encounter many different biasing circuits. Fortunately, with Thevenin's theorem, you can reduce most practical biasing circuits to either of two basic forms: *base-biased* or *emitter-biased.*

Base-biased

Figure 7-1a shows a base-biased circuit. We call this base-biased because a voltage source V_{BB} drives the base through resistance R_B. Source V_{BB} and resistance R_B may be a single source and resistor as shown, or they may represent a Thevenin equivalent circuit. In Fig. 7-1a, base current I_B flows through R_B into the base of the transistor. Coming out the emitter is a conventional current I_E which flows through R_E. Likewise, the V_{CC} source and the R_C resistor in the collector circuit may be a single source and resistance, or they may represent a Thevenin equivalent circuit.

Figure 7-1. Bias prototypes. (a)
Base bias. (b) Emitter bias. (c)
Bias model.

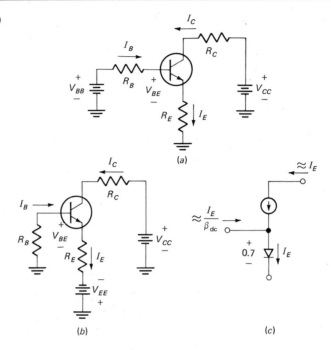

Emitter-biased

Emitter bias looks almost the same as base bias, except we move the dc source to the emitter circuit (see Fig. 7-1b). Now, the V_{EE} source and R_E resistor drive the emitter. Again, V_{EE} and R_E may be as shown, or may be a Thevenin equivalent circuit.

Here is our plan for most of this chapter. First, we will derive formulas for the base-biased circuit of Fig. 7-1a and the emitter-biased circuit of Fig. 7-1b. Second, we will practice reducing circuits to the base-biased form or the emitter-biased form and then apply the formulas. In this way, we will have a method for analyzing practical biasing circuits. Because the circuits of Fig. 7-1a and b are the foundations for this chapter, we will call them *prototypes*. (A prototype is a basic model or standard form.)

7-2. BASE-BIAS FORMULAS

Of the three transistor currents, *emitter current plays a central role*. If we have the value of I_E, we can get the value of I_C (approximately equal to I_E), or we can get I_B (divide by β_{dc}). Because of this, we will concentrate on emitter current.

Approximations

For preliminary analysis, accuracies within 5 percent are adequate. For this reason, we can save much time and work if we eliminate small effects *before* we start deriving formulas rather than after. For instance, since α_{dc} is almost always greater than 0.95, we get less than 5 percent error if we treat it as unity when calculating collector current from emitter current. Here is a list of the reasonable approximations we use freely in derivations (the symbol $\Rightarrow$ stands for "implies"):

$$\alpha_{dc} > 0.95 \Longrightarrow \alpha_{dc} \cong 1 \Longrightarrow I_C \cong I_E$$

$$I_C \cong I_E \Longrightarrow \frac{I_C}{I_B} \cong \frac{I_E}{I_B} \cong \beta_{dc}$$

$$\beta_{dc} > 20 \Longrightarrow \beta_{dc} \gg 1 \Longrightarrow \beta_{dc} + 1 \cong \beta_{dc}$$

$$\beta_{dc} \gg 1 \Longrightarrow I_C + I_B \cong I_C$$

$$V_{BE} \cong 0.7 \text{ V for silicon transistors}$$

$$V_{BE} \cong 0.3 \text{ V for germanium transistors}$$

When we use the bias model of a transistor (Fig. 7-1c), we automatically use most of the foregoing approximations.

The formula for dc emitter current

Figure 7-2a shows the base-biased prototype; Fig. 7-2b is an equivalent way to draw it. Let us work out the formula for emitter current.

To do this, we replace the transistor by its bias model as shown in Fig. 7-2c. Kirchhoff's voltage law is always true, no matter what kind of circuit you have. Therefore, all voltages in the base loop of Fig. 7-2c must add to zero. Starting with the V_{BB} source and going in a clockwise direction, we get

$$-V_{BB} + \frac{I_E}{\beta_{dc}} R_B + 0.7 + I_E R_E \cong 0$$

Solving for I_E gives

$$I_E \cong \frac{V_{BB} - 0.7}{R_E + R_B/\beta_{dc}} \qquad \text{(second)} \qquad\qquad (7\text{-}1)^{***}$$

This result is a formula worth memorizing; it applies to the base-biased prototype. Since many practical biasing circuits can be reduced to base-biased form, we will use Eq. (7-1) often.

If V_{BB} is more than 7 V, you can neglect the 0.7 V in Eq. (7-1) with less than 10 percent error. If V_{BB} is greater than 14 V, neglecting the 0.7 V gives less than 5 percent error. Depending on the application and desired accuracy, we often can use

$$I_E \cong \frac{V_{BB}}{R_E + R_B/\beta_{dc}} \qquad \text{(ideal)} \qquad\qquad (7\text{-}1a)$$

Equation (7-1a) is an ideal formula because it neglects the 0.7 V. Equation (7-1) is a second approximation because it allows 0.7 V across the emitter diode. We can get a

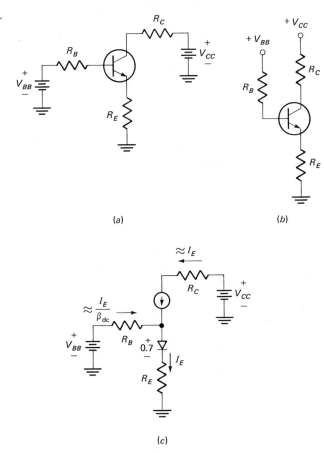

Figure 7-2. Deriving dc emitter current.

(a)

(b)

(c)

third approximation by using V_{BE} in the place of 0.7 V. That is,

$$I_E \cong \frac{V_{BB} - V_{BE}}{R_E + R_B/\beta_{dc}} \quad \text{(third)} \qquad (7\text{-}1b)$$

This is the formula to use when you want to include the additional drop across the r_b' of the base. As you recall from the preceding chapter, $V_{BE} = 0.7 + I_B r_b'$.

EXAMPLE 7-1.
Calculate the three transistor currents in Fig. 7-3a.

SOLUTION.
The circuit is in base-biased prototype form. With Eq. (7-1)

$$I_E \cong \frac{20 - 0.7}{10^4 + 10^6/100} \cong 1 \text{ mA}$$

Therefore,

$$I_C \cong I_E \cong 1 \text{ mA}$$

and
$$I_B \cong \frac{I_E}{\beta_{dc}} \cong \frac{1 \text{ mA}}{100} = 0.01 \text{ mA}$$

Figure 7-3b summarizes the three transistor currents.

EXAMPLE 7-2.
Voltages from any point to ground will be subscripted by the letter designating that point. For example, V_B is the base-to-ground voltage, V_E is the emitter-to-ground voltage, and V_C is the collector-to-ground voltage.

Calculate V_C, V_E, and V_{CE} in Fig. 7-3b.

SOLUTION.
The calculations are straightforward as follows:

$$V_E = I_E R_E \cong 10^{-3} 10^4 = 10 \text{ V}$$

$$V_C = V_{CC} - I_C R_C = 20 - 10^{-3}(5000) = 15 \text{ V}$$

and
$$V_{CE} = V_C - V_E = 15 - 10 = 5 \text{ V}$$

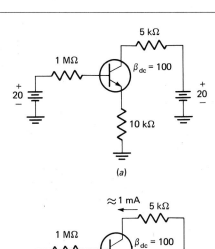

Figure 7-3. Examples 7-1 and 7-2.

Figure 7-4. Example 7-3.

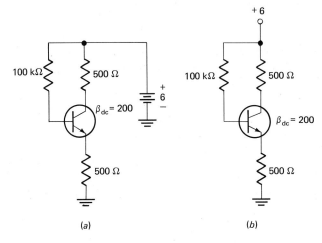

(a) (b)

EXAMPLE 7-3.
Calculate the three transistor currents in Fig. 7-4a.

SOLUTION.
Since V_{BB} and V_{CC} are positive supplies in the base-biased prototype, it is permissible to use only a single supply as shown in Fig. 7-4a. Figure 7-4b shows the same circuit drawn in a simpler way.

With prototype Eq. (7-1),

$$I_E \cong \frac{6 - 0.7}{500 + 10^5/200} = 5.3 \text{ mA}$$

From this, we get

$$I_C \cong I_E = 5.3 \text{ mA}$$

and

$$I_B \cong \frac{I_E}{\beta_{dc}} = \frac{5.3 \text{ mA}}{200} = 0.0265 \text{ mA}$$

7-3. SPECIAL CASE: $R_E = 0$

Figure 7-5a shows a special case of the base-biased prototype: $R_E = 0$. With prototype Eq. (7-1),

$$I_E \cong \frac{V_{BB} - 0.7}{R_B/\beta_{dc}} = \beta_{dc} \frac{V_{BB} - 0.7}{R_B}$$

Since I_E is directly proportional to β_{dc}, a 2:1 change in β_{dc} will cause a 2:1 change in I_E. This is always bad in linear transistor circuits (those that are supposed to stay in FR bias).

Here is the reason. The value of V_{CE} in Fig. 7-5a is

$$V_{CE} = V_{CC} - I_C R_C$$

When β_{dc} doubles with temperature or for other reasons, $I_C R_C$ doubles. This may force V_{CE} to drop under 1 V and bring the collector diode out of reverse bias. When this happens, normal transistor action is lost.

A circuit like Fig. 7-5a is the worst possible way to FR-bias a transistor when trying to set up a stable value of emitter current. Because of the large variations in β_{dc}, you almost never see the biasing circuit of Fig. 7-5a used in mass production of linear transistor circuits.

The circuit of Fig. 7-5a is sometimes called a *fixed-base-current* circuit because the base current remains almost constant when β_{dc} changes. Figure 7-5b shows the left end of R_B has a voltage V_{BB} with respect to ground; the right end of this resistor has a voltage of 0.7 V to ground. Therefore, the base current equals

$$I_B \cong \frac{V_{BB} - 0.7}{R_B} \cong \frac{V_{BB}}{R_B} \qquad \text{for large } V_{BB} \qquad (7-2)$$

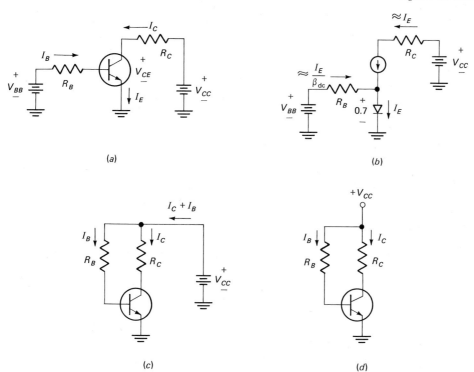

Figure 7-5. Fixed base current.

(a)

(b)

(c)

(d)

Since β_{dc} does not appear in this expression, I_B is independent of changes in β_{dc}. But,

$$I_E \cong \beta_{dc} I_B$$

which means I_E changes directly with β_{dc}.

We want to reverse the situation just described, at least in linear transistor circuits; we want circuits that set up a *fixed value of emitter current*. The next section shows you one classic circuit that does this.

Figure 7-5c shows how the circuit of Fig. 7-5a looks when a single supply is used. Note the source supplies a total current of $I_C + I_B$ to the transistor circuit. Figure 7-5d shows the simplest way to draw the circuit.

EXAMPLE 7-4.
Calculate all currents in Fig. 7-6a.

SOLUTION.
With prototype Eq. (7-1),

$$I_E \cong \frac{10 - 0.7}{0 + 10^6/100} \cong 1 \text{ mA}$$

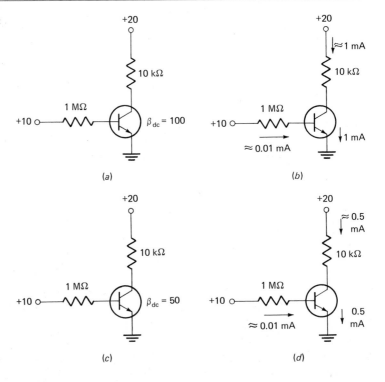

Figure 7-6. Examples 7-5 through 7-7.

With the emitter current, we can get the other two currents:

$$I_C \cong I_E = 1 \text{ mA}$$

and
$$I_B \cong \frac{I_E}{\beta_{dc}} = \frac{1 \text{ mA}}{100} = 0.01 \text{ mA}$$

Figure 7-6*b* shows these three currents.

EXAMPLE 7-5.
Same circuit as for the preceding example, but use a β_{dc} of 50 instead of 100 as shown in Fig. 7-6*c*.

SOLUTION.
With prototype Eq. (7-1),

$$I_E \cong \frac{10 - 0.7}{10^6/50} \cong 0.5 \text{ mA}$$

and
$$I_C \cong 0.5 \text{ mA}$$

$$I_B \cong \frac{0.5 \text{ mA}}{50} = 0.01 \text{ mA}$$

Figure 7-6*d* shows these transistor currents.

EXAMPLE 7-6.
Calculate the value of V_{CE} in Fig. 7-6*b* and *d*.

SOLUTION.
$V_E = 0$ because $R_E = 0$. Therefore,

$$V_{CE} = V_C = V_{CC} - I_C R_C$$
$$\cong 20 - 10^{-3}(10^4) = 10 \text{ V}$$

in Fig. 7-6*b*.
And for Fig. 7-6*d*,

$$V_{CE} \cong 20 - 0.5(10^{-3})(10^4) = 15 \text{ V}$$

You can see from these two calculations how a change in β_{dc} changes V_{CE}. You can easily verify that if $\beta_{dc} = 200$, $I_E \cong 2$ mA, and $V_{CE} \cong 0$, which means the collector diode is no longer reverse-biased. In a case like this, the transistor operates in the saturation region which is not suitable for linear transistor circuits.

Now is a good time to mention the following: The approximation $I_C \cong I_E$ is highly accurate when β_{dc} is greater than 20. When V_{CE} drops to less than a volt or so, the tran-

Figure 7-7. Example 7-7.

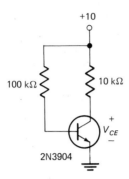

sistor operates in the saturation region. Near the knee of a collector curve, β_{dc} is still high; but as V_{CE} approaches zero, I_C approaches zero and so too does β_{dc}. Therefore, the approximation $I_C \cong I_E$ becomes invalid when V_{CE} equals zero.

This gives us a way to check whether a transistor has lost its FR bias or not. In solving problems, we start by assuming the transistor is FR-biased. If V_{CE} turns out to be *negative*, this contradictory answer tells us the transistor has lost its FR bias because V_{CE} can drop no lower than approximately zero. The next example illustrates this point.

EXAMPLE 7-7.
The 2N3904 of Fig. 7-7 has a β_{dc} of 100 when FR-biased. If you use a voltmeter to measure V_{CE}, approximately what value do you read?

SOLUTION.
We assume FR bias to start with. Then using prototype Eq. (7-1a),

$$I_E \cong \frac{10}{10^5/100} = 10 \text{ mA} \qquad \text{(assuming FR bias)}$$

and $$I_C \cong I_E = 10 \text{ mA} \qquad \text{(assuming FR bias)}$$
and $$V_{CE} = V_{CC} - I_C R_C = 10 - 10^{-2}(10^4) = -90 \text{ V}$$

Impossible! V_{CE} cannot be negative. Once V_{CE} drops to approximately zero, the relation $I_C \cong I_E$ becomes invalid.

Whenever a negative V_{CE} turns up in a calculation, we will know the *npn* transistor has lost its FR bias and is operating in the saturation region. We discuss problems like this further in Chap. 10. In this example, V_{CE} is approximately zero. This is what a voltmeter would read in the circuit of Fig. 7-7.

7-4. VOLTAGE-DIVIDER BIAS

The most important of all base-biased circuits is *voltage-divider bias*. Figure 7-8a shows this special case. R_1 and R_2 form a voltage divider in the base circuit. By Thevenizing the base circuit, we can reduce the actual circuit to the base-biased prototype.

The first step in applying Thevenin's theorem is to break the base lead as shown in Fig. 7-8*b*. The open-circuit voltage therefore equals

$$V_{TH} = \frac{R_2}{R_1 + R_2} V_{CC} \qquad (7\text{-}3)\,{}^{***}$$

The second step in Thevenizing is to reduce V_{CC} to zero as shown in Fig. 7-8*c*. Looking into the voltage divider, we see a Thevenin resistance of

$$R_{TH} = R_1 || R_2 \qquad (7\text{-}4)$$

where the vertical lines mean "in parallel with." We use this convenient notation in all our formulas. Given the value of R_1 and R_2, we calculate $R_1 || R_2$ with either

$$R_1 || R_2 = \frac{R_1 R_2}{R_1 + R_2} \qquad (7\text{-}4a)$$

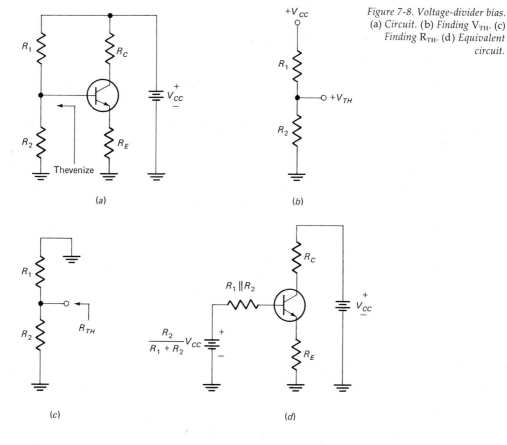

Figure 7-8. Voltage-divider bias. (a) Circuit. (b) Finding V_{TH}. (c) Finding R_{TH}. (d) Equivalent circuit.

or
$$R_1 \| R_2 = \frac{1}{1/R_1 + 1/R_2} \qquad (7\text{-}4b)$$

Now, we can draw the circuit in its equivalent form (Fig. 7-8d). This is the base-biased prototype where $V_{BB} = V_{TH}$ and $R_B = R_{TH}$. Therefore, prototype Eq. (7-1) can be used to calculate the emitter current.

EXAMPLE 7-8.
Calculate the emitter current in Fig. 7-9a.

SOLUTION.
Thevenize the base circuit by visualizing steps 1 and 2 of Thevenin's theorem as shown in Fig. 7-9b and c. From these we get $V_{TH} = 8$ V and $R_{TH} = 5$ kΩ.
   With prototype Eq. (7-1),

$$I_E \cong \frac{V_{BB} - 0.7}{R_E + R_B/\beta_{dc}} = \frac{8 - 0.7}{10{,}000 + 5000/100}$$
$$= \frac{7.3}{10{,}000 + 50} \cong 0.73 \text{ mA}$$

Note how the second term in the denominator is much smaller than the first; the next section tells you why this is important.

7-5.   PRACTICAL VOLTAGE-DIVIDER BIAS

When you get into mass production of linear transistor circuits, one of the main problems is the variation in β_{dc}. It varies from one transistor to another. A 2N3904, for

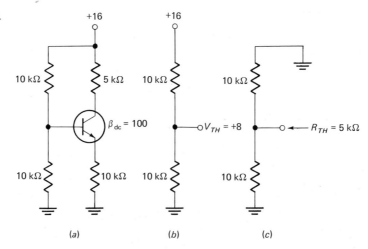

Figure 7-9. Example 7-8.

instance, has a minimum β_{dc} of 100 and a maximum β_{dc} of 300 for an I_C of 10 mA and a temperature of 25°C. This means that when working with thousands of 2N3904s, we can get as much as a 3:1 variation in β_{dc}.

On top of this 3:1 variation is a temperature-produced variation. For a 2N3904, the β_{dc} has an additional variation of about 3:1 for a temperature from −55 to 125°C. A mass-produced circuit using a 2N3904 would therefore have to contend with a possible 9:1 variation in β_{dc} if it had to operate from −55 to 125°C.

I_E independent of β_{dc}

For other transistors and smaller temperature ranges, the variation in β_{dc} may not be as great, but it will be substantial. Therefore, we cannot get a stable value of I_E unless I_E is almost *independent* of β_{dc}.

Here is why voltage-divider bias is the best base-biased circuit when we need a stable I_E value. When we substitute $V_{BB} = V_{TH}$ and $R_B = R_1 || R_2$ into prototype Eq. (7-1),

$$I_E = \frac{V_{TH} - 0.7}{R_E + R_1 || R_2/\beta_{dc}} \tag{7-5}$$

R_E reduces the effect of $R_1 || R_2/\beta_{dc}$. In fact, if

$$R_E \gg \frac{R_1 || R_2}{\beta_{dc}} \tag{7-5a}$$

we can neglect the second denominator term to get

$$I_E \cong \frac{V_{TH} - 0.7}{R_E} \quad \text{(second)} \tag{7-5b} ***$$

Now, β_{dc} can have a 9:1 variation without significantly affecting the value of I_E.

Error in neglecting β_{dc}

When R_E is at least 20 times greater than $R_1 || R_2/\beta_{dc}$, the error is less than 5 percent when using Eq. (7-5b). Symbolically, if

$$R_E \geqslant 20 \frac{R_1 || R_2}{\beta_{dc}}$$

or by rearranging, if

$$R_1 || R_2 \leqslant \frac{\beta_{dc}}{20} R_E \tag{7-5c}$$

we get less than 5 percent error in using Eq. (7-5b).

For instance, if $\beta_{dc} = 100$ and $R_E = 1$ kΩ, then an $R_1 || R_2$ less than 5 kΩ ensures less than 5 percent error when using Eq. (7-5b). Most practical voltage-divider biasing circuits satisfy condition (7-5c).

Summary

Whenever you see a circuit like Fig. 7-10, you should automatically react as follows:

1. Find V_{TH} with Eq. (7-3).
2. Subtract 0.7 V to get $V_{TH} - 0.7$; this is the voltage across R_E in Fig. 7-10. (If you feel the 0.7 V is negligible compared to V_{TH}, skip the subtraction and visualize a voltage of approximately V_{TH} across R_E.)
3. Divide $V_{TH} - 0.7$ by R_E to get the value of I_E.

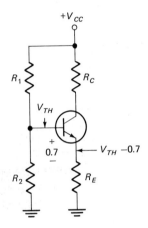

Figure 7-10. Second approximation of voltage-divider bias.

For those rare cases where condition (7-5a) is not satisfied, or if there is any doubt in your mind, you can use Eq. (7-5).

EXAMPLE 7-9.
Calculate I_E in Fig. 7-11a.

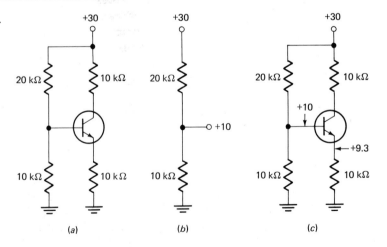

Figure 7-11. Example 7-9.

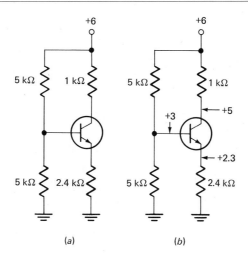

Figure 7-12. Example 7-10.

SOLUTION.
We recognize voltage-divider bias. First, visualize Fig. 7-11b to get $V_{TH} = 10$ V. Second, subtract 0.7 V to get 9.3 V across the 10-kΩ resistor as shown in Fig. 7-11c. Third, divide 9.3 V by 10 kΩ to get an I_E of 0.93 mA.

EXAMPLE 7-10.
Calculate I_E and V_{CE} in Fig. 7-12a.

SOLUTION.
The voltage divider produces a Thevenin voltage of 3 V. After subtracting 0.7 V, we get 2.3 V across the emitter resistor. Therefore, the emitter current equals

$$I_E \cong \frac{2.3}{2.4(10^3)} \cong 1 \text{ mA}$$

This is the approximate value of I_C. When I_C flows through the R_C of 1 kΩ, we get an $I_C R_C$ drop of 1 V. Subtracting this from the supply voltage gives a V_C of 5 V.

Figure 7-12b shows the circuit with its voltages. All these voltages are with respect to ground. Therefore,

$$V_{CE} = V_C - V_E = 5 - 2.3 = 2.7 \text{ V}$$

7-6.   EMITTER-BIAS FORMULA

Base-biased circuits are used when a single power supply is available. When two supplies are available, one positive and the other negative, emitter-biased circuits may be used. Figure 7-13a shows the prototype for all emitter-biased circuits. A negative supply forward-biases the emitter diode through resistance R_E. Base resistor R_B is now

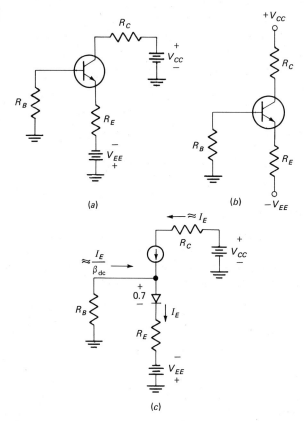

Figure 7-13. Emitter bias. (a) Circuit. (b) Simplified. (c) Bias-model circuit.

grounded. Figure 7-13b shows the simplified way to draw the emitter-biased proto-type.

Deriving the formula for I_E

First, replace the transistor by its bias model as shown in Fig. 7-13c. Second, sum voltages around the base and emitter circuits in a clockwise direction starting with R_B. This gives

$$\frac{I_E}{\beta_{\text{dc}}} R_B + 0.7 + I_E R_E - V_{EE} \cong 0$$

After factoring and solving for I_E,

$$I_E \cong \frac{V_{EE} - 0.7}{R_E + R_B/\beta_{\text{dc}}} \tag{7-6}$$

This prototype equation is almost identical to prototype Eq. (7-1). Instead of V_{BB}, we have V_{EE}. This makes it easy to remember the two prototype formulas, Eqs. (7-1) and (7-6). With these two equations, we can analyze almost all practical biasing circuits.

As before, when 0.7 V is negligible compared to V_{EE}, we get

$$I_E \cong \frac{V_{EE}}{R_E + R_B/\beta_{dc}} \qquad (7\text{-}6a)$$

Or, if we have an accurate value for V_{BE}, we can use

$$I_E \cong \frac{V_{EE} - V_{BE}}{R_E + R_B/\beta_{dc}} \qquad (7\text{-}6b)$$

Making β_{dc} negligible

Whenever you can reduce a circuit to the basic form shown in Fig. 7-13a, you can use prototype Eq. (7-6) to calculate I_E. Most of the time, you will see emitter-biased circuits exactly as shown in Fig. 7-13a or b. Furthermore, to make circuits practical, you can eliminate the effects of β_{dc} by ensuring

$$R_E \gg \frac{R_B}{\beta_{dc}} \qquad (7\text{-}7)$$

For this condition, Eq. (7-6) simplifies to

$$I_E \cong \frac{V_{EE} - 0.7}{R_E} \qquad (\text{second}) \qquad (7\text{-}7a)$$

Similar to the reasoning used for voltage-divider bias, you get less than 5 percent error when using Eq. (7-7a) if

$$R_B \le \frac{\beta_{dc}}{20} R_E \qquad (7\text{-}7b)$$

For instance, if $\beta_{dc} = 100$, and $R_E = 1$ kΩ, then an R_B less than 5 kΩ ensures less than 5 percent error when using Eq. (7-7a).

Ideal value of I_E

Most people find emitter current with the ideal approximation. In Fig. 7-14a, the voltage from base to ground equals

$$V_B = -I_B R_B \qquad (7\text{-}8a)$$

The emitter-to-ground voltage V_E differs from this by the 0.7 V across the emitter diode; therefore,

$$V_E = -(0.7 + I_B R_B) \qquad (7\text{-}8b)$$

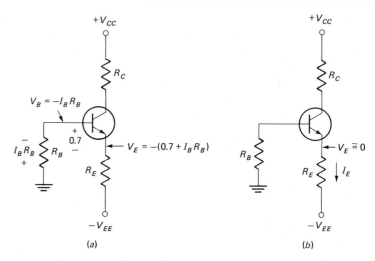

Figure 7-14. (a) *Emitter voltage.* (b) *Ideal emitter voltage.*

(a)

(b)

When condition (7-7b) is satisfied, the R_B is small enough to make $I_B R_B$ a few-tenths volt. For this reason, in a practical emitter-biased circuit, V_E is in the vicinity of -1 V. This is why a troubleshooter or anyone who wants a quick estimate of emitter current will visualize the circuit as shown in Fig. 7-14b. That is, for a fast answer, we think of V_E as approximately 0 V or at ground potential. Because of this, almost all of V_{EE} appears across R_E. Then,

$$I_E \cong \frac{V_{EE}}{R_E} \qquad \text{(ideal)} \qquad\qquad (7\text{-}9)\text{***}$$

Equation (7-9) is an ideal formula, Eq. (7-7a) is a second approximation, and Eq. (7-6) is a third approximation. Usually, Eqs. (7-9) and (7-7a) are adequate for most of our needs.

The emitter-biased circuit of Fig. 7-14b is so fundamental that we will refer to it as *emitter bias.* It is the best two-supply biasing circuit when we want a stable value of I_E.

EXAMPLE 7-11.
Calculate I_E in Fig. 7-15a.

SOLUTION.
Ideally, all of the 10 V from the emitter supply appears across R_E. So,

$$I_E \cong \frac{V_{EE}}{R_E} = \frac{10}{10,000} = 1 \text{ mA}$$

Or, with the second approximation,

$$I_E \cong \frac{V_{EE} - 0.7}{R_E} = \frac{10 - 0.7}{10,000} = 0.93 \text{ mA}$$

And, with the third approximation,

$$I_E \cong \frac{V_{EE} - 0.7}{R_E + R_B/\beta_{dc}} = \frac{10 - 0.7}{10,100} = 0.92 \text{ mA}$$

In this last calculation, you can see how little effect the R_B/β_{dc} term has in a well-designed emitter-bias circuit (one that satisfies $R_E \gg R_B/\beta_{dc}$).

EXAMPLE 7-12.
Calculate I_E in Fig. 7-15b.

SOLUTION.
First, Thevenize the emitter circuit. Break the emitter lead at the point indicated. Then, you get a V_{TH} of -10 V. Next, reduce the supply voltage to zero, which effectively

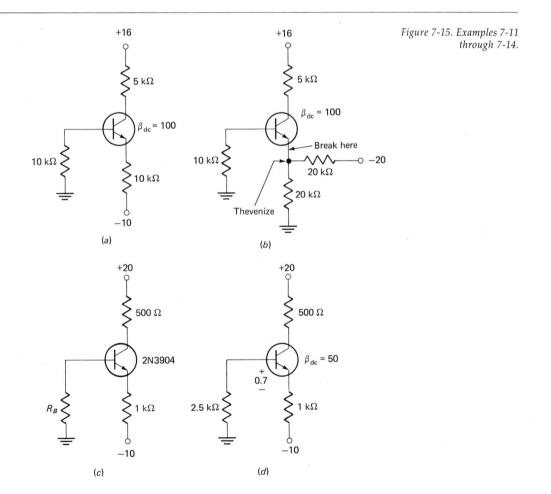

Figure 7-15. Examples 7-11 through 7-14.

grounds the right end of the 20-kΩ resistor. This means R_{TH} is the parallel of two 20-kΩ resistances, or $R_{TH} = 10$ kΩ.

Figure 7-15a is the Thevenized version of Fig. 7-15b. Therefore, the I_E in Fig. 7-15b is 1 mA ideally or 0.93 mA with the second approximation.

EXAMPLE 7-13.

A 2N3904 has a minimum β_{dc} of 100 and a maximum β_{dc} of 300 at $I_C = 10$ mA and $T = 25°C$. At $-55°C$, β_{dc} drops to half of its room-temperature value, but at 125°C it rises to 1.5 times its room-temperature value. What value of R_B satisfies condition (7-7b) in Fig. 7-15c?

SOLUTION.
Condition (7-7b) says

$$R_B \leq \frac{\beta_{dc}}{20} R_E$$

To satisfy this condition, we have to use *worst-case* values. If we use the highest β_{dc} to calculate R_B, the resulting R_B will not satisfy condition (7-7b) when β_{dc} is low. Therefore, the worst-case β_{dc} is the *lowest* value that can occur.

In Fig. 7-15c the emitter current ideally equals 10 mA because ideally all of the 10 V appears across the 1-kΩ emitter resistor. Therefore, a 2N3904 will have a β_{dc} between 100 and 300. If the circuit is to operate always at or near room temperature, the worst-case β_{dc} is 100 and

$$R_B \leq \frac{100}{20} \, 1 \text{ k}\Omega$$

or

$$R_B \leq 5 \text{ k}\Omega$$

On other hand, if the circuit must operate from -55 to 125°C, the worst-case β_{dc} is half of 100, or $\beta_{dc} = 50$. Then,

$$R_B \leq \frac{50}{20} \, 1 \text{ k}\Omega$$

or

$$R_B \leq 2.5 \text{ k}\Omega$$

EXAMPLE 7-14.
For the circuit of Fig. 7-15d, calculate the values of V_B and V_E.

SOLUTION.
First, the ideal value of emitter current is

$$I_E \cong \frac{V_{EE}}{R_E} = \frac{10}{1000} = 10 \text{ mA}$$

and the corresponding base current is

$$I_B \cong \frac{I_E}{\beta_{\text{dc}}} = \frac{10 \text{ mA}}{50} = 0.2 \text{ mA}$$

This flows through R_B, producing a base-to-ground voltage of

$$V_B \cong -I_B R_B = -0.2(10^{-3})2.5(10^3) = -0.5 \text{ V}$$

Second, we can add -0.7 V to this to get

$$V_E \cong -1.2 \text{ V}$$

The answers for V_B and V_E are approximations because we used the ideal value of I_E. If you need more accurate answers, you start by calculating a more accurate value of I_E using second approximation (7-7a) or third approximation (7-6). Then you can find I_B, V_B, and V_E.

7-7.   COLLECTOR-FEEDBACK BIAS

We have discussed the two best ways to bias a transistor when a stable I_E is needed. Besides voltage-divider bias and emitter bias, there is a compromise: *collector-feedback bias*. It uses fewer resistors than voltage-divider bias or emitter bias but is partially dependent on the value of β_{dc}.

Figure 7-16a shows a collector-feedback-biased circuit, and Fig. 7-16b shows a

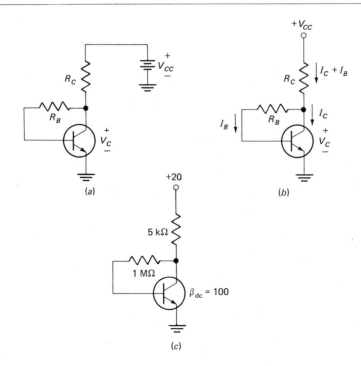

Figure 7-16. Collector-feedback bias.

simpler way of drawing the same circuit. Collector-feedback bias is a special case of base biasing. The collector feeds back a voltage to the base through resistor R_B.

The feedback concept

Instead of a fixed biasing voltage as in Fig. 7-5c, the biasing voltage in Fig. 7-16a comes from the collector whose voltage is *not* fixed. Because of this, changes in collector voltage affect the emitter current. This helps reduce the effects of a changing β_{dc}.

For instance, suppose the temperature increases, causing β_{dc} to increase. This causes an increase in collector current. The larger collector current lowers collector voltage V_C. The smaller V_C reduces the base voltage and consequently the base current. The lower base current tends to reduce the original increase in collector current. In symbols, the changes look like this:

$$\uparrow \quad \uparrow$$
$$I_C = \beta_{dc} I_B$$
$$\downarrow$$

In this equation, the increase in β_{dc} is accompanied by a decrease in I_B; the result is a smaller increase in I_C; therefore, changes in I_C are not as great as changes in β_{dc}.

Trial and error

Since collector-feedback bias is a special case of base biasing, we can use prototype formula (7-1) to find the emitter current. With $V_{BB} = V_C$ and $R_E = 0$, we get

$$I_E = \frac{V_C - 0.7}{R_B / \beta_{dc}} \tag{7-10a}$$

Next, in Fig. 7-16b

$$V_C = V_{CC} - (I_C + I_B) R_C$$

or
$$V_C = V_{CC} - I_E R_C \tag{7-10b}$$

Trial and error is one way to solve problems. To solve Eqs. (7-10a) and (7-10b), we can use trial and error as follows:

1. Guess a value of V_C. Use this value in Eq. (7-10a) to calculate I_E.
2. With this I_E, use Eq. (7-10b) to calculate V_C.
3. When the guessed and calculated V_C's are close in value, the true V_C is halfway between them. If the guessed and calculated V_C's are not close in value, repeat the trial and error.[1]

---

[1] There is no hard-and-fast rule here about how close the guessed and calculated values need to be. Usually, if the two V_C's are within 10 percent of each other, this is close enough.

As an example, here is how to get the approximate value of V_C in Fig. 7-16c. With a supply voltage of 20 V, start by guessing

$$V_C = 10 \text{ V} \qquad \text{(guess)}$$

With Eq. (7-10a),

$$I_E \cong \frac{10 - 0.7}{10^6/100} \cong \frac{10}{10^4} = 1 \text{ mA}$$

Next, calculate the corresponding V_C with Eq. (7-10b):

$$V_C = 20 - 0.001(5000) = 15 \text{ V} \qquad \text{(calculated)}$$

The guessed V_C is 10 V, and the calculated V_C is 15 V; they are too far apart in value. For a second trial,

$$V_C = 13 \text{ V} \qquad \text{(guess)}$$

With Eq. (7-10a),

$$I_E \cong \frac{13 - 0.7}{10^4} \cong 1.3 \text{ mA}$$

With Eq. (7-10b),

$$V_C = 20 - 0.0013(5000) = 13.5 \text{ V} \qquad \text{(calculated)}$$

Now we are close to the true V_C; it must be between 13 and 13.5 V. If we need high accuracy, we could guess again and include the 0.7 V in our calculation of I_E (in the first two trials we neglected the 0.7 V).

Formula for emitter current

Another way to get the true I_E and V_C is by solving Eqs. (7-10a) and (7-10b) simultaneously. By substituting the right-hand member of Eq. (7-10b) into Eq. (7-10a) and solving for I_E,

$$I_E \cong \frac{V_{CC} - 0.7}{R_C + R_B/\beta_{dc}} \qquad \text{(second)} \qquad\qquad (7\text{-}11)^{***}$$

This is a second approximation because it includes the 0.7 V. Neglecting the 0.7 V, the ideal formula is

$$I_E \cong \frac{V_{CC}}{R_C + R_B/\beta_{dc}} \qquad \text{(ideal)} \qquad\qquad (7\text{-}11a)$$

If you want highly accurate answers and have the true value of V_{BE}, use the third approximation:

$$I_E \cong \frac{V_{CC} - V_{BE}}{R_C + R_B/\beta_{dc}} \qquad \text{(third)} \qquad\qquad (7\text{-}11b)$$

As an example, substitute the circuit values of Fig. 7-16c into Eq. (7-11):

$$I_E \cong \frac{20 - 0.7}{5(10^3) + 10^6/100} = 1.29 \text{ mA}$$

and with Eq. (7-10b),

$$V_C \cong 20 - 1.29(10^{-3})5(10^3) = 13.5 \text{ V}$$

Setting up midpoint bias

How can we set up *midpoint bias*, V_C equal to *half* of V_{CC}? For instance, if $V_{CC} = 20$ V, how can we make V_C equal 10 V?

First, we need to find the current for midpoint bias. Starting with Eq. (7-10b),

$$V_C = V_{CC} - I_E R_C$$

For midpoint bias, $V_C = 0.5V_{CC}$; therefore,

$$0.5V_{CC} = V_{CC} - I_E R_C$$

or
$$I_E = \frac{V_{CC}}{2R_C} \qquad \text{(midpoint current)} \qquad (7\text{-}12)$$

Second, the ideal emitter current is given by Eq. (7-11a):

$$I_E \cong \frac{V_{CC}}{R_C + R_B/\beta_{dc}}$$

By substituting the midpoint current into this general expression, we get

$$\frac{V_{CC}}{2R_C} \cong \frac{V_{CC}}{R_C + R_B/\beta_{dc}}$$

or
$$2R_C \cong R_C + \frac{R_B}{\beta_{dc}}$$

Solving for R_B gives

$$R_B \cong \beta_{dc}R_C \qquad \text{(midpoint bias)} \qquad (7\text{-}13)$$

Here is the meaning. When R_B equals $\beta_{dc}R_C$, V_C approximately equals half of V_{CC}. As an example, in Fig. 7-17a we can set up midpoint bias by making

$$R_B \cong \beta_{dc}R_C = 100(10^4) = 1 \text{ M}\Omega$$

With this value of R_B, the circuit looks like Fig. 7-17b and has a V_C of approximately half V_{CC}.

Average value of β_{dc}

In Eq. (7-13) what value of β_{dc} do you use when β_{dc} varies over a large range? You could use the *arithmetic average*

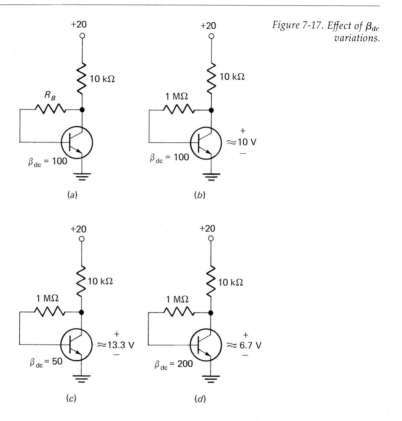

Figure 7-17. Effect of β_{dc} variations.

$$\beta_{(\text{arith av})} = \frac{\beta_{\min} + \beta_{\max}}{2} \qquad (7\text{-}14a)$$

When the spread between the minimum and maximum β_{dc} is not too large, the arithmetic average is useful.

Or, like some designers, you can use the *geometric average* given by

$$\beta_{(\text{geo av})} = \sqrt{\beta_{\min}\beta_{\max}} \qquad (7\text{-}14b)$$

For instance, suppose β_{dc} can be as low as 50 and as high as 200. Then, the arithmetic average is

$$\beta_{(\text{arith av})} = \frac{50 + 200}{2} = 125$$

and the geometric average is

$$\beta_{(\text{geo av})} = \sqrt{50(200)} = 100$$

Geometric average better for large variations

When very large variations in β_{dc} are involved, using the geometric average centers the bias point better because the plus-minus percent changes in V_C will be the same. For a β_{dc} from 50 to 200, the geometric average is 100. Figure 7-17b represents a circuit whose R_B is based on this geometric average.

When the β_{dc} drops to 50, the same circuit looks like Fig. 7-17c. If you work out the V_C value, you will get approximately 13.3 V, an increase of 33 percent from the centered value of 10 V. On the other hand, when β_{dc} is as high as 200, the same circuit has a V_C of 6.7 V, a decrease of approximately 33 percent from the center value of 10 V (see Fig. 7-17d). Therefore, using the geometric average of β_{dc} in Eq. (7-13) gives us a value of R_B that produces equal percent changes from design center.

Collector-feedback bias often used

We have seen a variation in β_{dc} from 50 to 200 produce a change in V_C of ± 33 percent when the transistor is midpoint-biased. This is not bad. In many applications, changes like these would be acceptable. Therefore, collector-feedback bias is a worthwhile biasing circuit. Its main advantage at the moment is simplicity; it uses only two resistors and one power supply. (There is an additional advantage of greater importance; we discuss it in Sec. 16-9.)

EXAMPLE 7-15.
Calculate the I_E and V_C of the collector-feedback bias circuit shown in Fig. 7-18a.

SOLUTION.
You could work this with trial and error. Or, with Eq. (7-11),

$$I_E = \frac{25 - 0.7}{10^3 + 10^5/200} = 16.2 \text{ mA}$$

Figure 7-18. Examples 7-15 and 7-16.

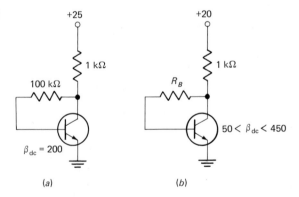

(a)   (b)

And the collector voltage equals

$$V_C = V_{CC} - I_E R_C = 25 - 16.2(10^{-3})10^3$$
$$\cong 9 \text{ V}$$

EXAMPLE 7-16

The β_{dc} of Fig. 7-18b varies over the 9:1 range shown. Select a value of R_B to set up midpoint bias.

SOLUTION.

With a large range in β_{dc}, the geometric average is best to use in calculating R_B.

$$\beta_{dc} = \sqrt{\beta_{min}\beta_{max}} = \sqrt{50(450)} = 150$$

And,
$$R_B = \beta_{dc}R_C = 150(1000) = 150 \text{ k}\Omega$$

7-8. *pnp* BIASING CIRCUITS

Figure 7-19 shows a *pnp* transistor. Since the emitter and collector diodes point in directions opposite to those of an *npn* transistor, all currents and voltages are reversed when the *pnp* transistor is FR-biased. In other words, to forward-bias the emitter diode of a *pnp* transistor, V_{BE} has the minus-plus polarity shown in Fig. 7-19. To reverse-bias the collector diode, V_{CB} must have the minus-plus polarity indicated. From this, it follows that V_{CE} is minus-plus as shown. Since the emitter diode points *in*, conventional *emitter current* flows *into* the *pnp* transistor; base and collector currents flow out.

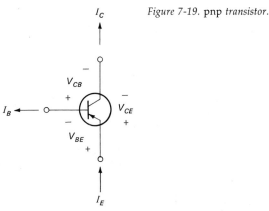

Figure 7-19. pnp *transistor.*

Complementary circuits

The *pnp* transistor is called the *complement* of the *npn* transistor; the word *complement* signifies all voltages and currents are opposite those of the *npn* transistor.

Every *npn* circuit has a complementary *pnp* circuit. To find the complementary *pnp* circuit, all you do is

1. Replace the *npn* transistor by a *pnp* transistor.
2. Complement or reverse all voltages and currents.

As an example, Fig. 7-20*a* shows collector-feedback bias using an *npn* transistor. Emitter current flows down and collector voltage is positive with respect to ground. Figure 7-20*b* shows the complementary *pnp* transistor circuit; all we have done is complement voltages and currents, and replace the *npn* by a *pnp* transistor. Now, emitter current flows up and the collector voltage is negative.

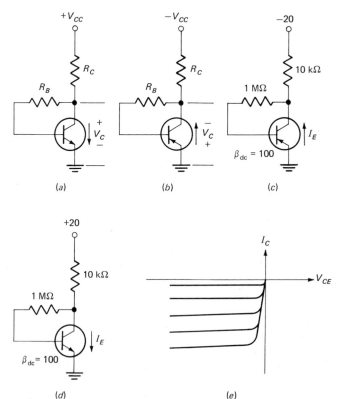

Figure 7-20. Complementary circuits.

Two methods of analysis

When analyzing *pnp* circuits, two approaches are useful:

1. Convert the *pnp* circuit to its complementary *npn* circuit. Analyze this *npn* circuit to get whatever answers you are after. Then apply these answers to the original *pnp* circuit.
2. Analyze the *pnp* circuit directly. If you use *magnitudes,* all the formulas derived for *npn* circuits apply to *pnp* circuits. The true polarity of voltages and the true direction of currents in the *pnp* circuit are found by looking directly at the *pnp* circuit.

As an example of method 1, suppose we want the value of I_E in Fig. 7-20c. With method 1, we would first draw or visualize the complementary *npn* circuit (Fig. 7-20d). Then, with Eq. (7-11),

$$I_E \cong \frac{20 - 0.7}{10^4 + 10^6/100} = 1 \text{ mA}$$

Next, we return to the *pnp* circuit of Fig. 7-20c and conclude that an emitter current of 1 mA flows upward as shown.

With method 2, we stay with Fig. 7-20c throughout the analysis. We use Eq. (7-11) as before to calculate the *magnitude* of I_E which equals 1 mA. We get the true direction of I_E by looking at Fig. 7-20c.

pnp quantities are negative compared to *npn*

We can define the *positive* directions of voltage and current to be those of an FR-biased *npn* transistor; therefore, the voltages and currents in an FR-biased *pnp* transistor are *negative* with respect to the *npn* directions. This is the reason some data sheets give negative values of current and voltage for *pnp* transistors. Also, a transistor curve tracer uses this convention because it simplifies the design of the curve tracer; this is why a curve tracer shows the collector curves of a *pnp* transistor in the third quadrant (Fig. 7-20e) rather than the first quadrant.

7-9. MOVING GROUND AROUND

Don't get the idea the circuits we discuss work only with the ground locations shown. Ground is a reference point you can move around as you please. For instance, Fig. 7-21a shows collector-feedback bias. Figure 7-21b shows the same circuit where we do not rely on ground to conduct current. In a circuit like this, we can remove the ground to get the *floating circuit* of Fig. 7-21c. In this floating circuit, all transistor currents and voltages have the same values as before; the same emitter current is flowing and the same collector-emitter voltage is set up.

Following this line of thinking, we can ground the positive terminal of the supply

Figure 7-21. Moving ground.

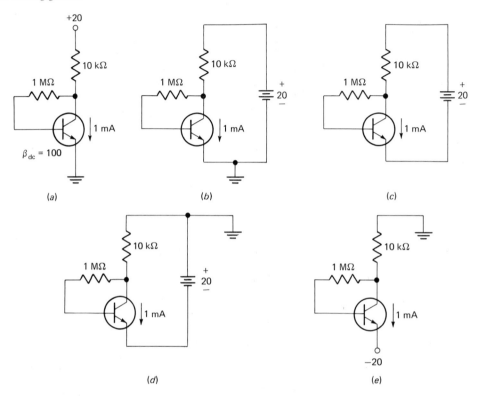

as indicated in Fig. 7-21*d*. And finally, we can draw the circuit as shown in Fig. 7-21*e*. The voltages with respect to ground differ in Fig. 7-21*a* and *e*. For instance, $V_E = 0$ in Fig. 7-21*a*, but $V_E = -20$ V in Fig. 7-21*e*.

What is really important is that the three transistor currents (I_E, I_C, I_B) and the three transistor voltages (V_{BE}, V_{CE}, V_{CB}) have exactly the same values in Fig. 7-21*a* through *e*.

Most of the time, we show one-supply *npn* biasing circuits with the negative supply terminal grounded. Bear in mind, however, that the voltage-divider bias circuit of Fig. 7-22*a* will produce exactly the same transistor currents and voltages if the positive end of the supply is grounded as shown in Fig. 7-22*b*. Likewise, the base-biased circuit of Fig. 7-22*c* sets up exactly the same transistor currents and voltages if the positive supply terminal is grounded as shown in Fig. 7-22*d*.

7-10. THE UPSIDE-DOWN CONVENTION FOR *pnp* TRANSISTORS

More and more schematic diagrams are showing *pnp* transistors drawn upside down. Once you get used to this convention, it simplifies analysis and design because *all emitter currents flow down.*

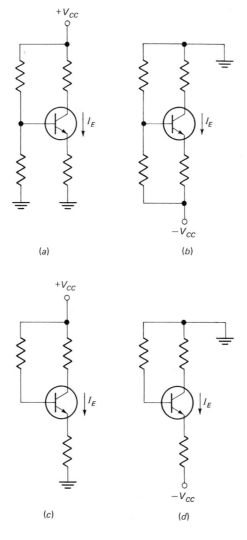

Figure 7-22. Examples of moving ground.

(a)

(b)

(c)

(d)

For instance, Fig. 7-23a shows a *pnp* collector-feedback biasing circuit. If we float the supply, we get the circuit of Fig. 7-23b. And, grounding the negative terminal of the supply, we get Fig. 7-23c. It does not matter how a transistor is oriented in space; therefore, it works just as well upside down as shown in Fig. 7-23d. The emitter current in this circuit has exactly the same value as the emitter current of Fig. 7-23a, b, and c.

Besides unifying analysis and design, drawing *pnp* transistors upside down has another advantage. Often, both *npn* and *pnp* transistors are used in the same circuit. By drawing the *pnp* transistors upside down, the draftsman can produce a simpler schematic diagram because fewer lines have to be crossed. Figure 7-24 is an example of this

Figure 7-23. Drawing pnp
transistors upside down.

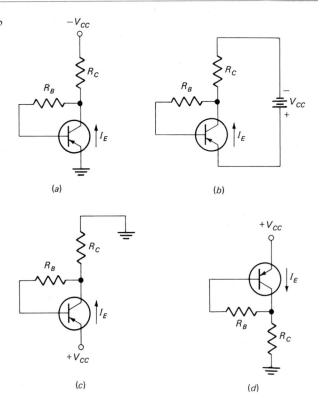

(a)   (b)

(c)   (d)

Figure 7-24. Circuit with npn *and*
pnp *transistors.*

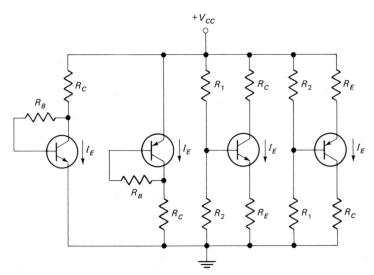

simplification. Note all emitter currents flow down; this definitely helps when calculating voltages.

EXAMPLE 7-17.
Calculate I_E in Fig. 7-25a.

SOLUTION.
Here is the direct approach. About 10 V is across the 20-kΩ resistor in the base circuit. Since the emitter diode is forward-biased, the voltage across the 10-kΩ emitter resistor equals 10 V ideally, or 9.3 V to a second approximation. The ideal emitter current therefore equals 1 mA.

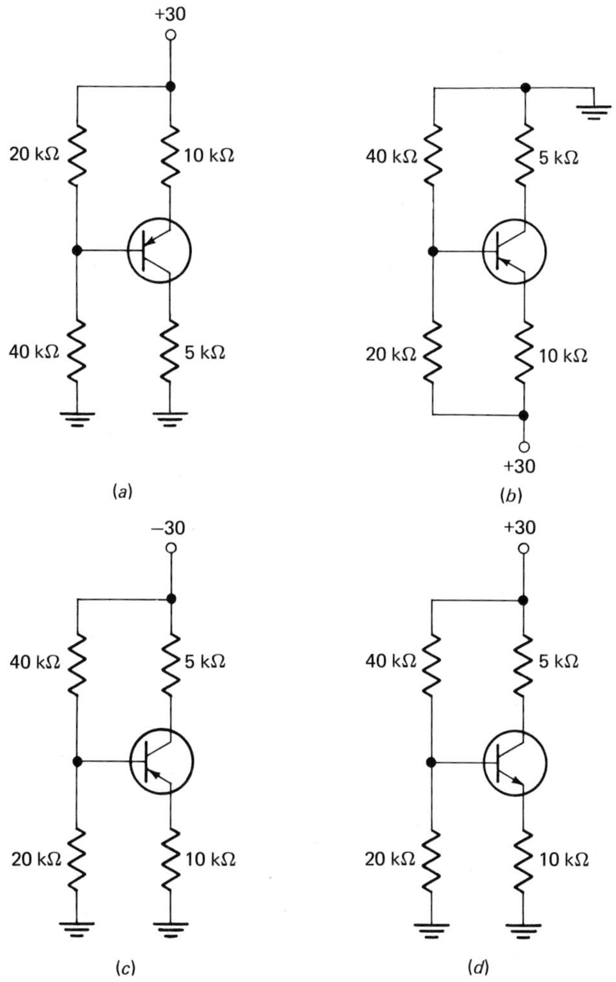

Figure 7-25. Example 7-17.

(a)

(b)

(c)

(d)

If the foregoing steps are not clear, you may prefer the indirect method. First, invert the original circuit to get Fig. 7-25b. Next, move the ground to the opposite end of the power supply to get Fig. 7-25c. Now, draw the complementary *npn* circuit (Fig. 7-25d). The emitter current in this complementary *npn* circuit is exactly the same as in the original circuit. Since Fig. 7-25d is familiar by now, you immediately recognize voltage-divider bias. The V_{TH} in the base circuit is 10 V, almost all of which appears across the 10-kΩ emitter resistor. This produces an I_E of 1 mA.

If necessary, use the indirect approach until familiar with upside-down *pnp* circuits. Eventually, you should use the direct approach because it saves time and keeps you in closer touch with the actual circuit.

If you are having any trouble with *pnp* transistors, practice the following:

1. Draw the complementary *pnp* circuit for each of these: Fig. 7-4b, 7-7, 7-9a, 7-12a, and 7-16a.
2. Draw each complementary *pnp* circuit upside down and move the ground from the positive to the negative supply terminal.

7-11. COLLECTOR CUTOFF CURRENT

All our biasing formulas neglect thermally produced current and surface-leakage current. This is reasonable, *provided you satisfy one condition*. This section tells you what the condition is.

Almost any data sheet lists the value of I_{CBO} at 25°C; this is the current from collector to base with an open emitter (see Fig. 7-26a). I_{CBO} is the reverse current in the

Figure 7-26. (a) *Meaning of* I_{CBO}. (b) *Equivalent circuit.*

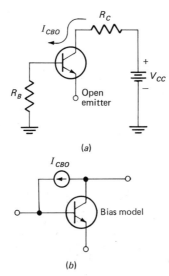

collector diode, partly caused by thermally produced carriers and partly by leakage along the surface. Often, data sheets give values of I_{CBO} at several temperatures besides 25°C. If not, you can estimate I_{CBO} at other temperatures as follows: I_{CBO} roughly doubles for every 10°C rise.

Effect of I_{CBO} on base-biased circuits

Even though I_{CBO} is measured with the emitter open (Fig. 7-26a), it still has an effect when the emitter is not open, that is, when the transistor is FR-biased. To find the effect it has on an FR-biased transistor, we first need a model or equivalent circuit that accounts for I_{CBO}. Figure 7-26b shows this model.[2] All we do is visualize an I_{CBO} current source in shunt with the bias model of the transistor. The bias model produces the desired component of collector current. The I_{CBO} current source, on the other hand, produces an undesired component of collector current.

When we FR-bias a transistor, the I_{CBO} source is still in shunt with the bias model (see Fig. 7-27a). To determine the effect of I_{CBO}, it helps if we draw the equivalent circuit of Fig. 7-27b. The two I_{CBO} sources produce exactly the same effect as the single source of Fig. 7-27a. Why? Because a current of I_{CBO} still flows into node A, and a current of I_{CBO} still flows out of node B. As a result, the voltages and currents of Fig. 7-27a and b are the same.

We can now put the current source on the left into the base circuit and the current source on the right into the collector circuit as shown in Fig. 7-27c. The reason for doing this is to allow us to Thevenize the base circuit. Visualize the base lead broken; this gives a Thevenin base voltage of

$$V_{TH} = V_{BB} + I_{CBO}R_B$$

Next, we reduce all sources to zero; this means shorting the voltage source V_{BB} and opening the current source I_{CBO}. Under this condition, we get a Thevenin resistance of

$$R_{TH} = R_B$$

(You may recall Example 1-5 where we Thevenized the same circuit.)

Figure 7-28 shows the Thevenized base circuit. Now we can see exactly what effect I_{CBO} will have on any base-biased circuit. The I_{CBO} of a transistor makes it appear as though the transistor is driven by a $V_{BB} + I_{CBO}R_B$ voltage source instead of a V_{BB} source. Because of this, all we have to do to prototype Eq. (7-1) is add $I_{CBO}R_B$ to the numerator to get

$$I_E \cong \frac{V_{BB} - 0.7 + I_{CBO}R_B}{R_E + R_B/\beta_{dc}} \tag{7-15}$$

---

[2] This is an approximation because I_{CBO} consists of surface-leakage current as well as thermally produced current. The former is voltage-dependent; because of this, using an I_{CBO} current source is not exact.

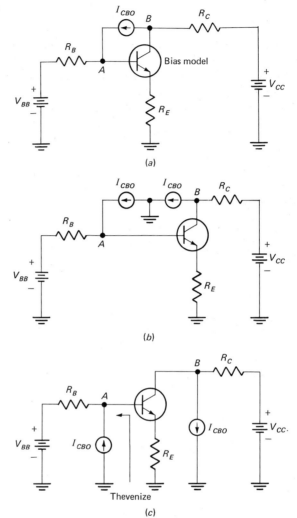

Figure 7-27. *Deriving the effect of* I_{CBO} *on bias.*

(a)

(b)

(c)

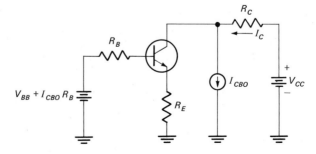

Figure 7-28. I_{CBO} *effects in base and collector circuits.*

Data sheets list the worst-case value of I_{CBO} (the maximum value) for the transistor type. Any transistor of the same type may have an I_{CBO} less than or equal to this maximum value. Since the I_{CBO} varies from one transistor to another of the same type, and also because it is temperature-dependent, the only satisfactory way to handle I_{CBO} is to make it negligible. In Eq. (7-15) we can do this by satisfying the condition:

$$V_{BB} - 0.7 \gg I_{CBO}R_B \qquad (7\text{-}16a)^{***}$$

for I_{CBO} at the highest temperature expected.

Other biasing circuits

By applying a similar derivation to emitter-biased circuits, we find $I_{CBO}R_B$ has to be added to the numerator of prototype formula (7-6). Therefore, the satisfactory solution for emitter-biased circuits is to make

$$V_{EE} - 0.7 \gg I_{CBO}R_B \qquad (7\text{-}16b)$$

For the special case of collector-feedback bias, a similar derivation shows $I_{CBO}R_B$ must be added to Eq. (7-11). Therefore, the condition to satisfy is

$$V_{CC} - 0.7 \gg I_{CBO}R_B \qquad (7\text{-}16c)$$

where I_{CBO} is at the highest expected temperature.

EXAMPLE 7-18.
How small should I_{CBO} be in Fig. 7-29a?

SOLUTION.
We have to satisfy condition (7-16c), which says

$$V_{CC} - 0.7 \gg I_{CBO}R_B$$

Since $V_{CC} = 10$ V, the condition becomes

$$9.3 \text{ V} \gg I_{CBO}R_B$$

The exact value of I_{CBO} is not important; so we can approximate by using

$$10 \text{ V} \gg I_{CBO}R_B$$

How much greater should the 10 V be? At least a factor of 20; more if possible. A factor of 20 means $I_{CBO}R_B$ increases the emitter current by only 5 percent. Assuming $I_{CBO}R_B$ is 1/20 of 10 V,

$$I_{CBO}R_B = \frac{10}{20} = 0.5 \text{ V}$$

With $R_B = 1$ MΩ in Fig. 7-29a,

Figure 7-29. Examples 7-18 and 7-19.

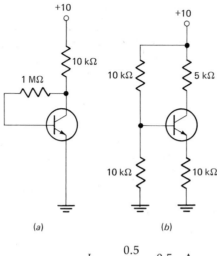

(a)   (b)

$$I_{CBO} = \frac{0.5}{10^6} = 0.5 \ \mu A$$

In selecting a transistor for Fig. 7-29a, therefore, we must satisfy this requirement:

$$I_{CBO} \leqslant 0.5 \ \mu A$$

at the highest expected temperature. With silicon transistors, you can easily satisfy this condition.

EXAMPLE 7-19.
Using the 20-to-1 rule, what is the highest acceptable I_{CBO} in Fig. 7-29b?

SOLUTION.
In this case, condition (7-16a) applies:

$$V_{BB} - 0.7 \gg I_{CBO}R_B$$

Since V_{BB} equals 5 V in Fig. 7-29b, the condition becomes

$$4.3 \ V \gg I_{CBO}R_B$$

With $R_B = 5 \ k\Omega$, a factor of 20 means

$$I_{CBO} = \frac{4.3}{20(5000)} = 43 \ \mu A$$

In other words, a circuit like Fig. 7-29b can use a transistor whose I_{CBO} is as high as 43 μA; the resulting emitter current would be only 5 percent higher than we would calculate by neglecting I_{CBO}.

A final point. The voltage-divider bias circuit of Fig. 7-29b can take an I_{CBO} as high as 43 μA, but the collector-feedback bias circuit of Fig. 7-29a can only tolerate an I_{CBO} of

0.5 μA. Earlier, we said voltage-divider bias and emitter bias are the two best biasing circuits when a stable I_E is needed. In other words, these circuits showed the least dependence on β_{dc}. The same idea holds true for the effects of I_{CBO}; of all the biasing circuits, voltage-divider bias and emitter bias show the least dependence on I_{CBO}. Therefore, if you ever have an application where I_{CBO} is a problem, the best biasing circuits to use are voltage-divider bias and emitter bias.

7-12. SUMMARY

As mentioned at the beginning of this chapter, in linear transistor circuits the transistor must remain FR-biased throughout the ac cycle. We intend to drive the transistor with an ac voltage to produce changes in transistor currents and voltages. To prevent this ac voltage from reverse-biasing the emitter diode or forward-biasing the collector diode, we first set up dc currents and voltages. Then, if the ac signal is not too large, the transistor remains FR-biased throughout the cycle.

This chapter has shown us how to set up the dc currents and voltages. The current

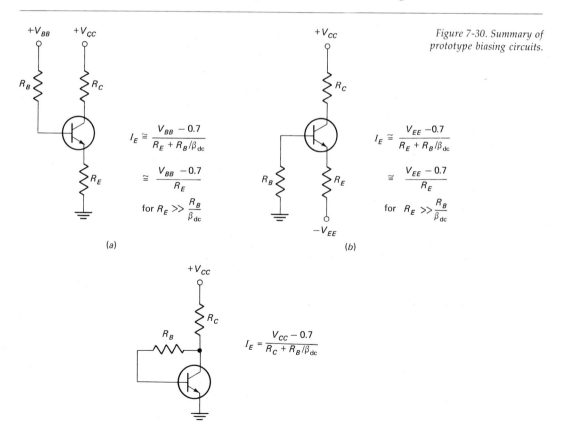

Figure 7-30. Summary of prototype biasing circuits.

$$I_E \cong \frac{V_{BB} - 0.7}{R_E + R_B/\beta_{dc}}$$

$$\cong \frac{V_{BB} - 0.7}{R_E}$$

for $R_E \gg \dfrac{R_B}{\beta_{dc}}$

(a)

$$I_E \cong \frac{V_{EE} - 0.7}{R_E + R_B/\beta_{dc}}$$

$$\cong \frac{V_{EE} - 0.7}{R_E}$$

for $R_E \gg \dfrac{R_B}{\beta_{dc}}$

(b)

$$I_E = \frac{V_{CC} - 0.7}{R_C + R_B/\beta_{dc}}$$

(c)

playing the central role is the emitter current; once we have it, we can easily calculate the other currents and voltages. Figure 7-30 summarizes the prototype biasing circuits and their formulas. Voltage-divider bias is a special case of Fig. 7-30a with $V_{BB} = V_{TH}$ and $R_B = R_{TH}$ found by the methods discussed earlier. Figure 7-30c shows the special case of collector-feedback bias. By referring to Fig. 7-30 throughout this book, we can easily calculate the dc emitter current in the many different biasing circuits we discuss.

Problems

7-1.　Calculate the value of I_E in Fig. 7-31a.

7-2.　What values do the three transistor currents have in Fig. 7-31b?

Figure 7-31.

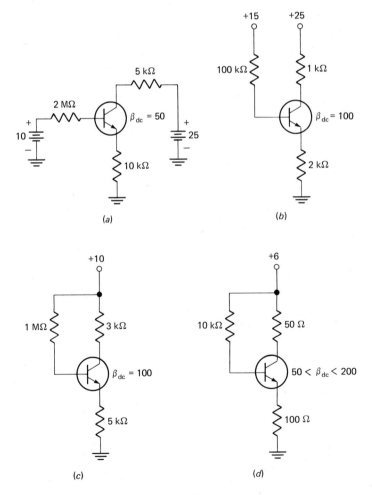

(a)

(b)

(c)

(d)

7-3. Find the emitter current in Fig. 7-31c.

7-4. Calculate the minimum and maximum values of I_E for the transistor of Fig. 7-31d.

7-5. What value does V_C have in Fig. 7-31c? V_E? And V_{CE}?

7-6. Find the minimum and maximum values of V_{CE} in Fig. 7-31d.

7-7. Calculate the emitter current in Fig. 7-32a.

7-8. What is the value of emitter current for the circuit shown in Fig. 7-32b?

7-9. Calculate all three transistor currents in Fig. 7-32c.

7-10. How much emitter current is there in Fig. 7-32d?

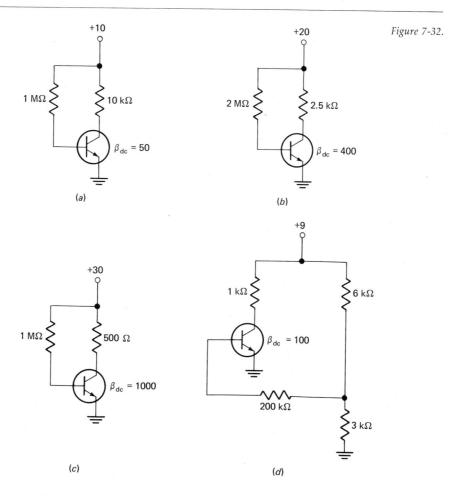

Figure 7-32.

(a)

(b)

(c)

(d)

7-11. What value does V_{CE} have in Fig. 7-32b? If β_{dc} increases to 800 because of a temperature change, what will V_{CE} equal? Why won't the transistor work normally in the second case?

7-12. If a transistor with a β_{dc} from 50 to 200 is used in Fig. 7-32a, what is the minimum and maximum value of V_{CE}?

7-13. Calculate the emitter current in the voltage-divider bias circuit of Fig. 7-33a. What value does V_C have? V_{CE}?

Figure 7-33.

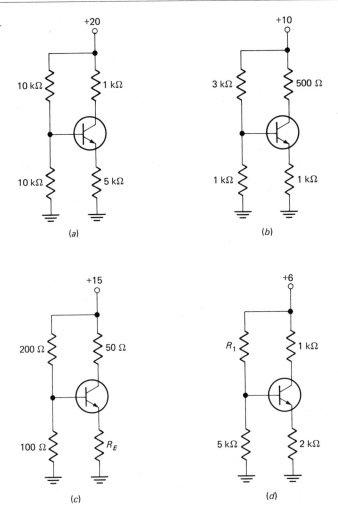

(a)

(b)

(c)

(d)

7-14. Work out the value of I_E, V_C, and V_{CE} in Fig. 7-33b.

7-15. To set up 100 mA of emitter current, what value should R_E have in Fig. 7-33c?

7-16. To have 1 mA of emitter current in Fig. 7-33d, what value should R_1 have?

7-17. If the transistor of Fig. 7-33a has a β_{dc} of 50, what value do you get for I_E when you use Eq. (7-5)?

7-18. Calculate the emitter current I_E and the collector voltage V_C in Fig. 7-34a.

7-19. To have an emitter current of 50 mA in Fig. 7-34b, what value should R_E have? What will V_C equal in this case?

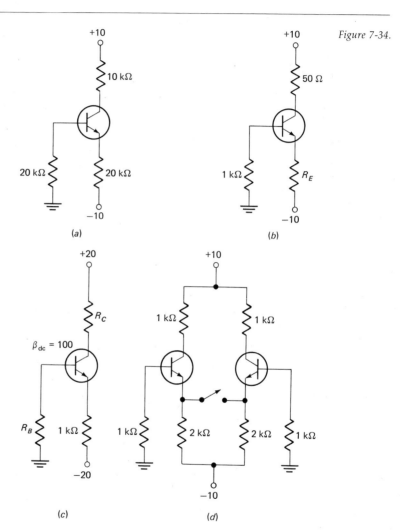

Figure 7-34.

7-20. To have a V_C of 10 V in Fig. 7-34c, what value should R_C have? To satisfy condition (7-7b), what is the maximum value of R_B?

7-21. The two transistors of Fig. 7-34d are identical. How much emitter current is there in each transistor? If you close the switch between the emitters, how much current will flow through this switch? What is the collector voltage V_C of each transistor? What voltage will there be from one collector to the other?

7-22.   How much emitter current is there in Fig. 7-35a? What value does V_C have?

7-23.   Calculate the minimum and maximum emitter current in the circuit of Fig. 7-35b.

Figure 7-35.

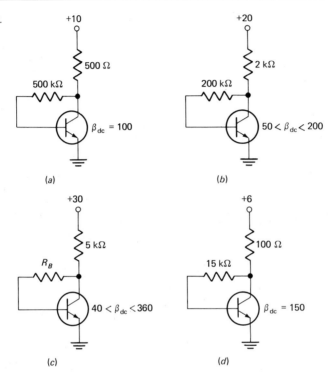

(a)   (b)

(c)   (d)

7-24.   Select a value of R_B to set up midpoint bias in Fig. 7-35c. Use the geometric average of β_{dc}.

7-25.   Calculate the three transistor currents and the three transistor voltages (V_{BE}, V_{CE}, and V_{CB}) in Fig. 7-35d.

7-26.   Draw the complementary *pnp* circuit for each voltage-divider bias circuit of Fig. 7-33.

7-27.   For each emitter-biased circuit in Fig. 7-34, show its complementary *pnp* circuit.

7-28.   Show the complementary *pnp* circuit for each collector-feedback bias circuit in Fig. 7-35.

7-29.   Redraw each circuit of Fig. 7-33 with the positive end of the power supply grounded instead of the negative end.

7-30.   Show each collector-feedback bias circuit of Fig. 7-35 with the positive supply terminal grounded instead of the negative.

7-31.   For each voltage-divider bias circuit of Fig. 7-33, do the following: draw the complementary *pnp* circuit, turn the resulting circuit upside down, and move the ground from the positive end of the supply to the negative end.

7-32. Repeat the preceding problem, for each circuit of Fig. 7-34.

7-33. Calculate I_E in Fig. 7-36a. Also, what are values of V_E and V_C?

7-34. Calculate I_E, V_E, and V_C in Fig. 7-36b.

7-35. Work out the emitter current and collector-to-ground voltage in Fig. 7-36c.

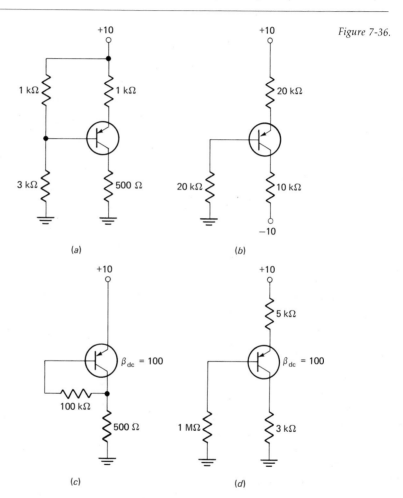

Figure 7-36.

(a)   (b)

(c)   (d)

7-36. Calculate the three transistor currents in Fig. 7-36d. What is the value of V_C?

7-37. If a transistor has an I_{CBO} of 1 μA, what will $I_{CBO}R_B$ equal in each of the following:

1. Fig. 7-32c   3. Fig. 7-34b
2. Fig. 7-33b   4. Fig. 7-35a

7-38. For each figure listed in the preceding problem, work out the largest permitted I_{CBO} assuming a 20:1 factor as done in Example 7-19.

8. AC Equivalent Circuits

We can apply an ac voltage across the emitter diode; this forces transistor currents and voltages to have ac variations of the same frequency. By proper design, we can *amplify* the ac signal, that is, increase its peak-to-peak value. To analyze amplifier circuits, we first must discuss *ac equivalent circuits*.

8-1. COUPLING AND BYPASS CAPACITORS

Most capacitors in transistor circuits are either *coupling* or *bypass* capacitors. A coupling capacitor passes an ac signal from one ungrounded point to another ungrounded point. For instance, in Fig. 8-1a the ac voltage at point A also appears at point B. For this to happen, the capacitive reactance X_C must be very small compared with the resistances.

In Fig. 8-1a the circuit to the left of point A may be a single ac source and resistor, or it may be the Thevenin equivalent circuit of something more complicated. Likewise, resistance R_L may be a single load resistor, or it may be the equivalent resistance of a more complex network. It doesn't matter what the actual circuits are on either side of the capacitor; as long as we can reduce the circuit to a single loop as shown, the ac current flows through a total resistance of $R_{TH} + R_L$.

As you recall from basic circuit theory, the ac current in a one-loop RC circuit equals

$$I = \frac{V}{\sqrt{R^2 + X_C^2}} \tag{8-1}$$

where R is the total resistance of the loop. In Fig. 8-1a, $R = R_{TH} + R_L$. The maximum ac

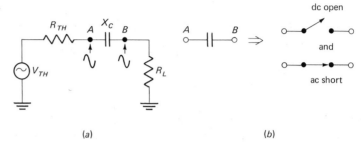

Figure 8-1. Coupling capacitor.
(a) *Circuit.* (b) *Equivalent.*

(a)

(b)

current flows when X_C is much smaller than R. In other words, the capacitor couples the signal properly from A to B when $X_C \ll R$.

Capacitor size

The size of the coupling capacitor depends on the lowest frequency you are trying to couple because X_C increases when frequency decreases. We will use this rule:

$$T = RC \qquad \text{(at lowest frequency)} \qquad (8\text{-}2)\,{}^{***}$$

where T is the period, that is, $1/f$. In other words, we will use a capacitor large enough to satisfy Eq. (8-2) at the lowest frequency to be coupled; then, all higher frequencies will be well coupled.[1]

For instance, suppose we are trying to couple frequencies from 20 Hz to 50 kHz into an amplifier. Then, the lowest frequency 20 Hz has a period of $T = 1/f = 1/20 = 0.05$ s. If the total resistance in the one-loop circuit (Fig. 8-1a) is 10 kΩ, the coupling capacitor must satisfy

$$T = RC$$

$$0.05 = 10^4 C$$

or

$$C = \frac{0.05}{10^4} = 5 \ \mu\text{F}$$

This size capacitance will do an excellent job of coupling all frequencies above 20 Hz. (Another widely used rule is to keep X_C less than one-tenth R; we prefer $T = RC$ because it is easier to work with.)

Ideally, a capacitor looks open to dc current. For this reason, we can think of a coupling capacitor as shown in Fig. 8-1b. It acts like a switch that is open to dc current

---

[1] By substitution into Eq. (8-1), you can prove the ac current is down only 1 percent from the maximum value when $T = RC$.

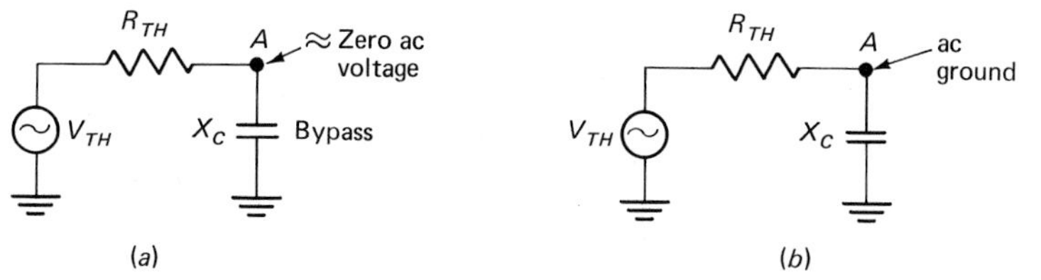

Figure 8-2. Bypass capacitor. (a)
Circuit. (b) Ac ground.

but shorted to ac current. This discriminating action allows us to get an ac signal from one circuit into another without disturbing the dc biasing of each circuit.

AC ground

A bypass capacitor is similar to a coupling capacitor except it couples an un-grounded point to a *grounded point* as shown in Fig. 8-2a. Again, V_{TH} and R_{TH} may be a single source and resistor as shown, or may be a Thevenin circuit. To the capacitor it makes no difference. It sees a total resistance of R_{TH}. Equations (8-1) and (8-2) still apply because we have a one-loop RC circuit. The only difference is $R = R_{TH}$.

Bypass capacitors bring a new idea with them. In Fig. 8-2b, the capacitor ideally looks like a short to an ac signal. Because of this, point A is shorted to ground as far as the ac signal is concerned. This is why we have labeled point A as *ac ground*. A bypass capacitor will not disturb the dc voltage at point A because it looks open to dc current. However, a bypass capacitor automatically makes point A an ac ground point.

In the normal frequency range of an amplifier, all coupling and bypass capacitors look like ac shorts. For this reason, we will approximate all capacitors as dc opens and ac shorts, unless otherwise indicated.

EXAMPLE 8-1.
The input signal of Fig. 8-3a can have a frequency between 10 Hz and 50 kHz. For the coupling capacitor to work properly, what size should it be?

Figure 8-3. Example 8-1.

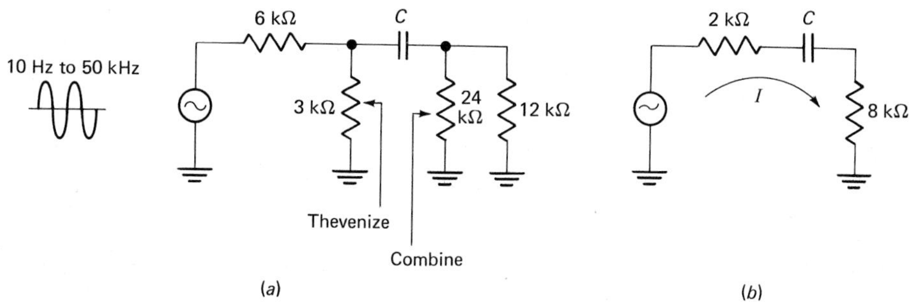

SOLUTION.

The first step is to reduce the circuit to a single loop. When you Thevenize the circuit to the left of capacitor, you get an R_{TH} of 2 kΩ. When you combine the two load resistors, you get an equivalent resistance of 8 kΩ.

Figure 8-3b shows the original circuit reduced to a single loop. The total resistance in the loop is 10 kΩ. The lowest frequency, 10 Hz, has a period of $T = 1/f = 1/10 = 0.1$ s. Therefore, the coupling capacitor must satisfy

$$T = RC$$

$$0.1 = 10^4 C$$

or
$$C = \frac{0.1}{10^4} = 10 \ \mu F$$

EXAMPLE 8-2.

We want an ac ground on point A in Fig. 8-4a. What size should the bypass capacitor be?

SOLUTION.

Again, we must reduce the circuit to a single loop. Looking back into point A, we see a Thevenin resistance of

$$R_{TH} = 125 || 10,000 \cong 125 \ \Omega$$

Figure 8-4b shows the circuit in one-loop form. The lowest frequency is 10 Hz. So,

$$T = RC$$

$$0.1 = 125C$$

or
$$C = \frac{0.1}{125} = 800 \ \mu F$$

Therefore, to bypass point A to ground we need at least 800 μF.

Figure 8-4. Example 8-2.

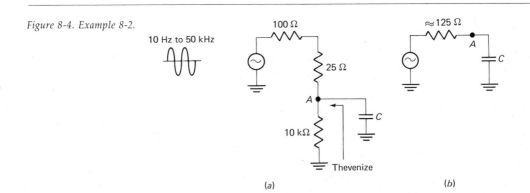

(a) (b)

8-2. THE SUPERPOSITION THEOREM FOR AC-DC CIRCUITS

In a transistor amplifier, the dc sources set up dc currents and voltages. The ac source produces fluctuations in the transistor currents and voltages. The simplest way to analyze the action of transistor circuits is to split the analysis into two parts: a dc analysis and an ac analysis. In other words, we can analyze transistor circuits by applying the superposition theorem in a special way. Instead of taking one source at a time, we take all dc sources at the same time and work out the dc currents and voltages using the methods of the preceding chapter. Next, we take all ac sources at the same time and calculate the ac currents and voltages. By adding the dc and ac currents and voltages, we get the total currents and voltages.

Ac and dc equivalent circuits

Here are the steps in applying superposition to transistor circuits:

1. Reduce all ac sources to zero; open all capacitors. The circuit that remains is all that matters as far as dc currents and voltages are concerned. Because of this, we call this circuit the *dc equivalent circuit*. Using this circuit, we calculate whatever dc currents and voltages we are interested in.
2. Reduce all dc sources to zero; short all coupling and bypass capacitors. The circuit that remains is all that matters as far as ac currents and voltages are concerned. We call this circuit the *ac equivalent circuit*. This is the circuit to use in calculating ac currents and voltages.
3. The total current in any branch is the sum of the dc current and ac current through that branch. And, the total voltage across any branch is the sum of the dc voltage and ac voltage across that branch.

Here is how we apply the superposition theorem to the transistor amplifier of Fig. 8-5a. First, reduce all ac sources to zero and open all capacitors. All that remains is the circuit of Fig. 8-5b; this is the dc equivalent circuit. This is all that matters as far as dc currents and voltages are concerned. With this circuit, we can find any dc current or voltage of interest.

Second, reduce all dc sources to zero and short all coupling and bypass capacitors. What remains is the ac equivalent circuit shown in Fig. 8-5c. Especially important, reducing a voltage source to zero is the same as shorting it. This is why R_B and R_C are shorted to ac ground through the V_{CC} source. Also, the bypass capacitor places the emitter at ac ground. With the ac equivalent circuit of Fig. 8-5c, we will be able to calculate any ac currents and voltages of interest.

Notation

To avoid confusing ac and dc currents and voltages, we will use capital letters and subscripts for dc quantities. That is, we will use

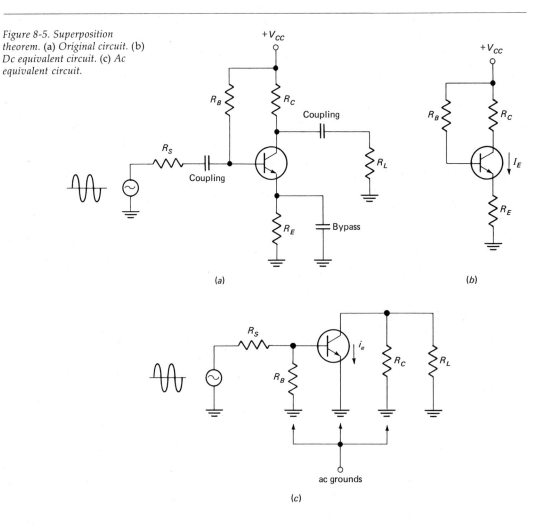

Figure 8-5. Superposition theorem. (a) Original circuit. (b) Dc equivalent circuit. (c) Ac equivalent circuit.

I_E, I_C, I_B for the dc currents
V_E, V_C, V_B for the dc voltages to ground
V_{BE}, V_{CE}, V_{CB} for the transistor dc voltages

For the ac currents and voltages we will use lowercase letters and subscripts as follows:

i_e, i_c, i_b for the ac currents
v_e, v_c, v_b for the ac voltages to ground
v_{be}, v_{ce}, v_{cb} for the transistor ac voltages

These are standard notations used by most people; you should become familiar with them because we use them extensively in the dc and ac analysis of transistor circuits.

EXAMPLE 8-3.

Show the dc and ac equivalent circuits for the three-transistor amplifier of Fig. 8-6a.

SOLUTION.

Working from left to right, we open each capacitor. All that remains when we are finished is the dc equivalent circuit of Fig. 8-6b. Here we recognize the first *stage* (a transistor and its associated circuitry) as an *npn* collector-feedback bias circuit. The second stage is an *npn* voltage-divider circuit. And the third stage is an upside-down *pnp* voltage-divider circuit. With the methods of Chap. 7, we can calculate any dc current or voltage of interest.

Next, reduce V_{CC} to zero in Fig. 8-6a (equivalent to grounding the V_{CC} line) and short all coupling and bypass capacitors. What remains is the ac equivalent circuit of Fig. 8-6c. By combining parallel resistances, we get the simple circuit of Fig. 8-6d; later chapters show you how to analyze this circuit.

8-3.   TRANSISTOR AC EQUIVALENT CIRCUITS

Remember what happened to a piece of wire when we took everything into account (Chap. 1)?

Figure 8-7 shows the three doped regions and the two depletion layers of a transistor. Each of these affects ac current differently from dc current. Why? For one thing, there are diode capacitances in the transistor; these look open to dc currents but may affect the ac currents.

The exact ac model

By taking as many effects as possible into account, we get the ac equivalent circuit of Fig. 8-8. Starting at the left, the ac emitter current i_e must flow through the inductance of the emitter lead. After this current enters the emitter region, it must flow through the bulk resistance $r_{e(\text{bulk})}$ of the emitter. When the current reaches the emitter depletion layer, part of it may flow through capacitance C'_e and the rest of it through r'_e. Next, the ac emitter current i_e flows through a small voltage source $\mu v'_c$.

On reaching the central node, i_e splits into i_c and i_b. The much smaller i_b flows down through r'_b shunted by C'_b and out the transistor through the inductance of the base lead. The ac collector current i_c flows through the parallel network of C'_c, r'_c, and the αi_e current source. Then, i_c flows through the bulk resistance of the collector and finally through the inductance of the collector lead.

A circuit like Fig. 8-8 is fit only for a computer to work with. We have no intention of using it except as a reminder of some of the smaller effects in a transistor. What we

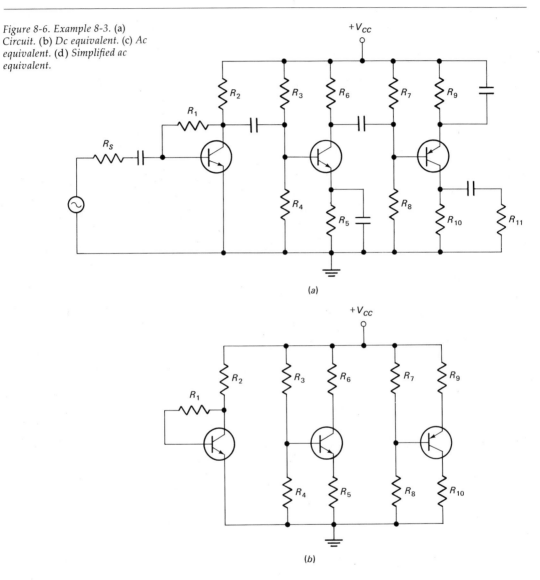

Figure 8-6. Example 8-3. (a) Circuit. (b) Dc equivalent. (c) Ac equivalent. (d) Simplified ac equivalent.

(a)

(b)

have to do is eliminate as much as possible from Fig. 8-8 until we have reasonable equivalent circuits.

The low-frequency model

To begin with, we can usually neglect the lead inductances when the frequency is less than 100 MHz. Also, because the emitter and collector are larger and more heavily

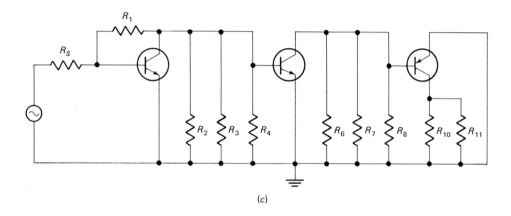

(c)

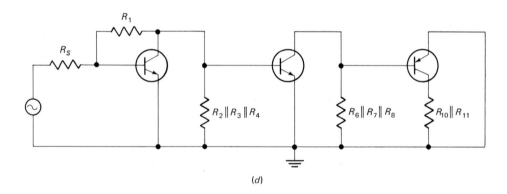

(d)

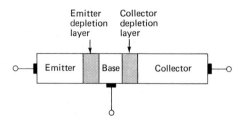

Emitter
depletion
layer

Collector
depletion
layer

Figure 8-7. Regions of a transistor.

Figure 8-8. Exact ac equivalent circuit.

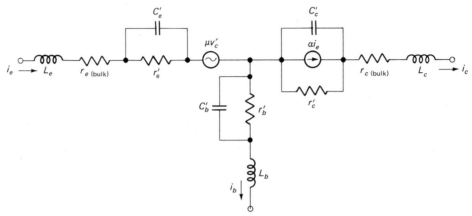

doped than the base, $r_{e(\text{bulk})}$ and $r_{c(\text{bulk})}$ are small enough to neglect. Capacitance C_b' is so small that it too drops out. What remains is the equivalent circuit of Fig. 8-9.

But even Fig. 8-9 is too complicated for everyday work. The internal capacitances C_e' and C_c' shunt ac currents away from desirable paths. These capacitances are important only at higher frequencies called the *cutoff frequencies* (Chap. 16). As long as we operate a transistor *below* its cutoff frequencies, we can neglect C_e' and C_c'.

Figure 8-10 shows the low-frequency model of a transistor. Voltage v_c' is the difference of potential across the collector depletion layer. When the collector voltage changes, the width of the depletion layer changes. As you recall, this affects the base current slightly (Fig. 6-15b). To account for this effect, the $\mu v_c'$ generator is included in the emitter circuit. We will call Fig. 8-10 the *third approximation*. This or something like it is what we will use for accuracies better than 1 percent. Section 15-7 discusses the *h*-parameter method, equivalent to working with Fig. 8-10.

Figure 8-10 is not good enough. If we are trying to analyze and design new circuits, an equivalent circuit like Fig. 8-10 will so confuse and cloud our thinking that we

Figure 8-9. Simplified model with internal capacitances.

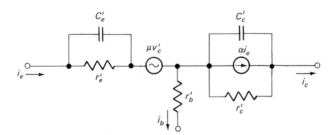

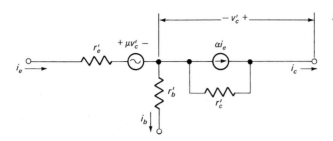

Figure 8-10. Third approximation.

will not see important relations. What we need is an equivalent circuit that gets to the bones of transistor action.

8-4. THE IDEAL-TRANSISTOR APPROXIMATION

Accuracy may be important near the end of an analysis or design, but during the early stages it loses its value. If you try for high accuracy from the start, you often waste time. The reason is simple enough: when you try to analyze or design a new circuit, you make many false starts until you find the right approach. If you are trying to find the right approach with an equivalent circuit like Fig. 8-10, you are sure to get lost.

We can prune Fig. 8-10 further. The $\mu v_c'$ generator has values in the millivolt range and can be neglected in a first approximation. The quantity α is the ac alpha, the ratio of ac collector current to ac emitter current, i_c/i_e. For almost any transistor, α is greater than 0.95; therefore, a reasonable approximation is $\alpha \cong 1$. Resistance r_c' accounts for the slight increase in collector current when the collector depletion layer penetrates deeper into the base. For most transistors, r_c' is in megohms and is high enough to neglect in a first approximation.

Finally, r_b' may be negligible. It depends on how much dc collector current is flowing. For I_C less than about 10 mA, the ac voltage across r_b' is almost always small. In other words, for I_C less than 10 mA or thereabouts, the $i_b r_b'$ voltage drop is usually small enough to neglect in a first approximation (Chap. 10 deals with I_C greater than 10 mA).

With the foregoing approximations, Fig. 8-10 reduces to Fig. 8-11. Now we have arrived. This is the *ideal-transistor* approximation. This model, simple as it is, hangs onto the main ideas of transistor action. It is satisfactory for much transistor analysis and design. Using Fig. 8-11, a creative mind can sail through all those false starts that

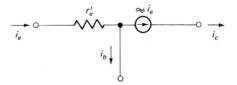

Figure 8-11. Ideal transistor.

are part of preliminary analysis and design. Then, if necessary, the ideal answers can be improved by using second and third approximations to be discussed later.

Figure 8-11 gets to the point. As far as the ac signal is concerned, the emitter diode acts like a *resistance* of r'_e and the collector diode like a current source of approximately i_c. We will use the ideal transistor as the foundation for understanding ac action of transistors. The ideal transistor is not far from the truth; actual transistors would not be as useful as they are unless the higher-order effects were minor.

What we are doing with the transistor is similar to what everybody does with a piece of wire, a resistor, a capacitor, or an inductor. We are neglecting higher-order effects with the understanding that they are important in special cases. For many people, the ideal-transistor approximation is adequate for more than 90 percent of the analysis and design they do. And for the remaining 10 percent, the second and third approximations can be used.

8-5. EMITTER-DIODE AC RESISTANCE

To use the ideal-transistor approximation, we need to know more about r'_e. Figure 8-12 shows the typical diode curve relating I_E and V_{BE}. Point Q is the *quiescent point*; the coordinates of this point are the dc emitter current and dc base-emitter voltage. In the ideal-transistor approximation, we neglect the effects of r'_b; this means that V_{BE} is the total voltage appearing across the emitter diode.

Small signal required

When an ac signal drives an FR-biased transistor, it forces the emitter current and voltage to change. If the signal is small, the operating point swings from Q to A, back to Q, down to B, and back to Q. This action repeats for the next cycle. Since A and B are close to Q, only a small arc of the diode curve is used. Whenever a small piece of a curve is used like this, the operation is approximately *linear*. All this means is that the

Figure 8-12. Graphical meaning of r'_e.

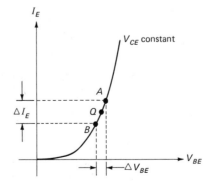

arc from A to B is essentially a straight line. Because of this, the changes in voltage and current are directly proportional. In symbols,

$$\Delta V_{BE} = K \Delta I_E$$

where Δ stands for "the change in" and where K is a constant of proportionality.

If we reduce the ac signal, the changes in voltage and current are still directly proportional with the same value of K as before. In fact, changing the peak value of any small ac signal does not change the value of K.

This is what r'_e is all about. It is nothing more than the constant of proportionality K, that is,

$$r'_e = K = \frac{\Delta V_{BE}}{\Delta I_E} \tag{8-3}$$

Figure 8-12 illustrates these changes between points A and B. Since the arc between A and B is essentially linear, you can use any two points between A and B to measure the changes. For instance, if desired, you could measure the change in voltage and current between Q and A. The ratio of these changes still gives the same value of r'_e.

Getting the same value of r'_e for all points between A and B requires a small ac signal. How small? Theoretically, it should be infinitesimally small. As a practical guide, however, we will accept a signal as small when the peak-to-peak change in emitter current is less than 10 percent of the quiescent current. For example, if the dc current is 10 mA, we will consider an ac signal small if the peak-to-peak change in emitter current is less than 1 mA. This is arbitrary, but we need some basic rule to work with until we discuss *harmonic distortion* (Chap. 20).

Ac signal identical to changes

Since the ac signal causes the operating point to swing from Q to A on one half cycle and from Q to B on the other half cycle, the changes in current and voltage are the same as the ac current and voltage. In symbols,

$$i_e = \Delta I_E$$

and
$$v_{be} = \Delta V_{BE}$$

If the changes are from points B to A, i_e and v_{be} are peak-to-peak values. On the other hand, if the changes are from Q to A, i_e and v_{be} are peak values. Unless otherwise indicated, we will use peak-to-peak values in our discussions. That is, i_e and v_{be} represent the peak-to-peak ac emitter current and voltage.

Because ac current and voltage are identical to the changes in total current and voltage, we can rewrite Eq. (8-3) as

$$r'_e = \frac{v_{be}}{i_e} \tag{8-4}$$

For instance, if $v_{be} = 25$ mV and $i_e = 1$ mA, r'_e will equal 25 Ω. Once we have the value of r'_e, we can use it for other voltages and currents within the small arc.

The formula for r'_e

Look at Fig. 8-12. Since r'_e is the ratio of the change in V_{BE} to the change in I_E, its value depends on the location of Q. The higher Q is up the curve, the smaller r'_e becomes because the same change in voltage produces a larger change in current. In other words, the slope of the diode curve at point Q determines the value of r'_e.

By using calculus to find this slope, we can prove that

$$r'_e = \frac{25 \text{ mV}}{I_E} \tag{8-5} ***$$

For instance, if the Q point has an I_E of 0.1 mA

$$r'_e = \frac{25 \text{ mV}}{0.1 \text{ mA}} = 250 \text{ Ω}$$

Or, for a higher Q point with $I_E = 1$ mA, r'_e decreases to

$$r'_e = \frac{25 \text{ mV}}{1 \text{ mA}} = 25 \text{ Ω}$$

We already know how to calculate I_E using the biasing formulas of the preceding chapter. Therefore, once we have I_E, we can find the corresponding value of r'_e.

The Appendix derives Eq. (8-5) and describes some of the conditions attached to it. For instance, Eq. (8-5) is a room-temperature formula; it applies to temperatures near 65°F or 18°C. Also, Eq. (8-5) is based on a rectangular pn junction like the one shown in Fig. 8-7. Because the shape of a diode curve changes slightly with nonrectangular junctions, the value of r'_e may differ slightly from Eq. (8-5). Nevertheless, 25 mV/I_E is an excellent approximation for any transistor, germanium or silicon, no matter what the actual shape of its base-emitter junction. A final condition worth mentioning is that I_E must be greater than zero, always satisfied when a transistor is FR-biased.

At any temperature

Occasionally, we need the value of r'_e for temperatures above and below room temperature. The formula is

$$r'_e = \frac{C + 273}{291} \frac{25 \text{ mV}}{I_E} \tag{8-6}$$

where C is the junction temperature in Celsius degrees. For instance, at 100°C,

$$r'_e = \frac{100 + 273}{291} \frac{25 \text{ mV}}{I_E}$$

$$= \frac{32 \text{ mV}}{I_E}$$

Therefore, at 100°C an I_E of 1 mA means r'_e equals 32 Ω.

8-6. AC BETA

Figure 8-13 shows a typical graph of I_C versus I_B. β_{dc} is the ratio of total collector current I_C to total base current I_B. Since the graph is nonlinear, β_{dc} depends on where the Q point is located. This is why data sheets specify β_{dc} for a particular value of I_C.

The ac beta (designated β_{ac} or simply β) is a small-signal quantity and depends on the location of point Q. In terms of Fig. 8-13, ac beta is defined as

$$\beta = \frac{\Delta I_C}{\Delta I_B} \tag{8-7}$$

or since ac currents are the same as the changes in total currents,

$$\beta = \frac{i_c}{i_b} \tag{8-8 ***}$$

Graphically, ac beta is the slope of the curve at point Q. For this reason, it has different values for different Q locations.

Data sheets sometimes list the value of h_{fe}, which is the same as β. Especially note the subscripts on h_{fe} are lowercase letters, whereas the subscripts on h_{FE} are capital letters. Therefore, when reading data sheets, do not confuse these current gains. Quantity h_{FE} is the dc current gain and is equivalent to β_{dc}. Quantity h_{fe} is the small-signal current gain, equivalent to β. You will find h_{FE} is almost always listed on data sheets, but h_{fe} is not shown as often. If a data sheet does not give the value of h_{fe}, you can use the dc current gain h_{FE} as an approximation for h_{fe}. The two quantities are usually close enough in value to justify this as a first approximation.

A minor point. Notice in Fig. 8-13 the curve is labeled V_{CE} *constant*. This is necessary to avoid the effect of a changing collector depletion layer. As you recall, if you let the collector voltage increase, the collector depletion layer reaches deeper into the base and captures a few additional electrons. To avoid this effect while measuring or calculating β, we must hold V_{CE} constant. As a practical matter, however, the effect of a changing collector depletion layer is so small that only a small error in the β value occurs if we let V_{CE} change while measuring changes in I_C and I_B.

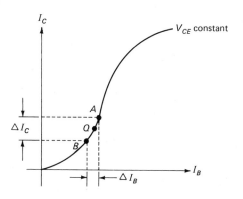

Figure 8-13. Graphical meaning of β.

Figure 8-14. Ideal model.

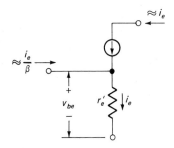

8-7. THE IDEAL MODEL

For almost any transistor, α is greater than 0.95 and β is greater than 20. Therefore, we can list the following approximations for ac analysis:

$$\alpha > 0.95 \Longrightarrow \alpha \cong 1 \Longrightarrow i_c \cong i_e$$

$$i_c \cong i_e \Longrightarrow \frac{i_c}{i_b} \cong \frac{i_e}{i_b} \cong \beta$$

$$\beta > 20 \Longrightarrow \beta \gg 1 \Longrightarrow \beta + 1 \cong \beta$$

$$\beta \gg 1 \Longrightarrow i_c + i_b \cong i_c$$

These approximations are the same as those used in Sec. 7-2, except that we use ac quantities instead of dc quantities.

To keep the ac analysis similar to the dc analysis of the preceding chapter, we can redraw the ideal transistor as shown in Fig. 8-14. During the positive half cycle of the ac signal, a voltage of v_{be} appears across the emitter diode with the plus-minus polarity

Figure 8-15. Examples 8-4 and 8-5.

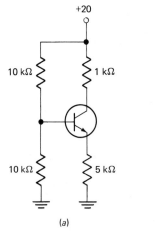

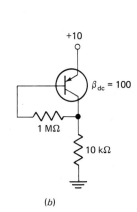

(a)   (b)

shown. This sets up an ac emitter current of i_e. Because of this, the ac collector current approximately equals i_e and the ac base current approximately equals i_e/β.

Since Fig. 8-14 is so important in our analysis, we will call it the *ideal model*. As you can see, it resembles the *bias model* used in Chap. 7. Just as the bias model was the key to dc analysis of transistor circuits, so too is the ideal model the key to ac analysis.

EXAMPLE 8-4.

What is the value of r'_e in Fig. 8-15a?

SOLUTION.

We recognize voltage-divider bias. The dc voltage across the emitter resistor is about 10 V. Therefore,

$$I_E \cong \frac{10}{5000} = 2 \text{ mA}$$

With Eq. (8-5),

$$r'_e = \frac{25 \text{ mV}}{I_E} = \frac{25 \text{ mV}}{2 \text{ mA}} \cong 12.5 \ \Omega$$

EXAMPLE 8-5.

What value does r'_e have in Fig. 8-15b?

SOLUTION.

This is an upside-down *pnp* collector-feedback bias circuit. Therefore, the dc emitter current equals

$$I_E \cong \frac{V_{CC} - 0.7}{R_C + R_B/\beta_{dc}} \cong \frac{10}{10^4 + 10^6/100} = 0.5 \text{ mA}$$

and the ac emitter resistance equals

$$r'_e = \frac{25 \text{ mV}}{I_E} = \frac{25 \text{ mV}}{0.5 \text{ mA}} = 50 \ \Omega$$

Problems

8-1. The ac source of Fig. 8-16a can have a frequency between 100 Hz and 200 kHz. To couple the signal properly over this range, what size should the coupling capacitor be?

8-2. We want the coupling capacitor of Fig. 8-16b to couple all frequencies from 500 Hz to 1 MHz. What size should it have?

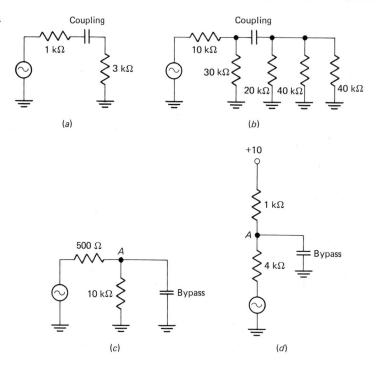

Figure 8-16.

(a)

(b)

(c)

(d)

8-3. To bypass point A to ground in Fig. 8-16c for all frequencies from 20 Hz up, what size should the bypass capacitor have?

8-4. We want point A in Fig. 8-16d to look like ac ground from 10 Hz to 200 kHz. What size should the bypass capacitor be?

8-5. In Fig. 8-3a, if the lowest frequency is changed from 10 to 1 Hz, what size coupling capacitor do we need?

8-6. Suppose we change the lowest frequency in Fig. 8-4a to 2 Hz. What size does the bypass capacitor need to be? Is this large size commercially available?

8-7. Draw the dc equivalent circuit for the amplifier of Fig. 8-17a; label the three currents with standard dc notation. Next, draw the ac equivalent circuit.

8-8. Draw the dc and ac equivalent circuits for Fig. 8-17b. What is the approximate value of dc emitter current?

8-9. Show the dc and ac equivalent circuit for the upside-down *pnp* amplifier of Fig. 8-17c.

8-10. Approximately how much dc emitter current is there in the amplifier of Fig. 8-17d? Draw the ac equivalent circuit.

Figure 8-17.

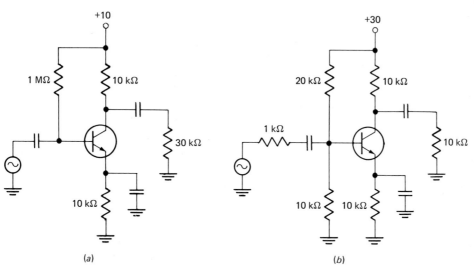

(a) (b)

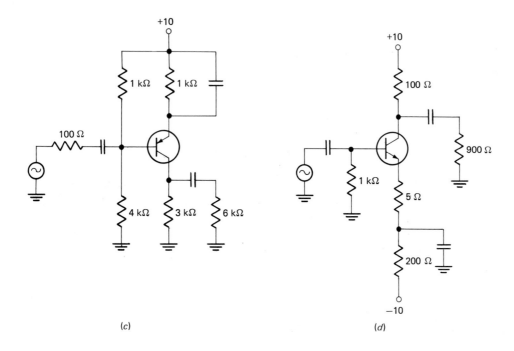

(c) (d)

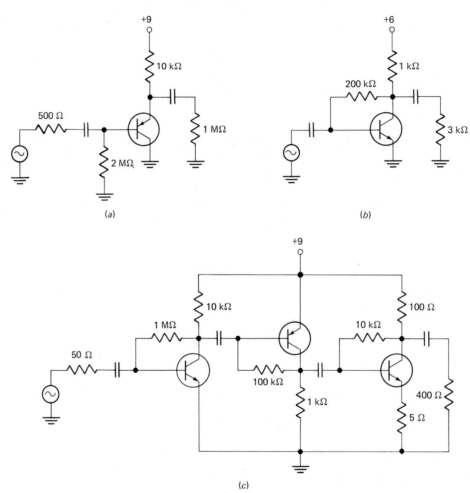

Figure 8-18.

(a)

(b)

(c)

8-11. Draw the dc and ac equivalent circuits for the *pnp* amplifier shown in Fig. 8-18*a*.

8-12. For the collector-feedback-biased amplifier of Fig. 8-18*b*, show the ac equivalent circuit.

8-13. Draw the dc and ac equivalent circuits for the three-transistor amplifier of Fig. 8-18*c*.

8-14. In the voltage-divider-biased amplifier of Fig. 8-17*b*, to have small-signal operation what is the largest permitted peak-to-peak change in total emitter current?

8-15. To have small-signal operation in the emitter-biased amplifier of Fig. 8-17*d*, what is the largest permitted peak-to-peak emitter current?

8-16. In measuring ac currents and voltages, suppose you find that $v_{be} = 10 \ \mu V$ and $i_e = 1 \ \mu A$. What is the value of r'_e?

8-17. If the peak-to-peak change in base-emitter voltage is 5 mV and the peak-to-peak change in emitter current is 0.01 mA, what is the value of r'_e?

8-18. With Eq. (8-5), calculate the value of r'_e for each of these dc emitter currents: 0.01 mA, 0.05 mA, 0.1 mA, 0.5 mA, 1 mA, 5 mA, and 10 mA.

8-19. What is the value of r'_e in the amplifier of Fig. 8-17b?

8-20. In the emitter-biased amplifier of Fig. 8-17d, what value does r'_e have?

8-21. If the transistor of Fig. 8-18b has a β_{dc} of 100, what value does r'_e have?

8-22. A transistor has a dc emitter current of 1 mA. What is the value of r'_e at $-50°$C? At $0°$C? At $150°$C?

8-23. The transistor amplifier of Fig. 8-17b operates in outer space and has a junction temperature of $-30°$C. What is the value of r'_e?

8-24. In Fig. 8-13, suppose the changes between points B and A are as follows: $\Delta I_C = 0.1$ mA and $\Delta I_B = 0.002$ mA. What is the value of β? And h_{fe}?

8-25. Suppose the Q point of Fig. 8-13 has coordinates of $I_C = 10$ mA and $I_B = 0.05$ mA. What value does β_{dc} have? If point A has coordinates of $I_C = 10.3$ mA and $I_B = 0.051$ mA, what is the value of β?

8-26. A transistor amplifier has an $I_E = 1$ mA. Figure 8-19 shows the ideal model of the transistor. If $i_e = 0.05$ mA, what value does v_{be} have? For $\beta = 100$, what is the value of i_b and i_c?

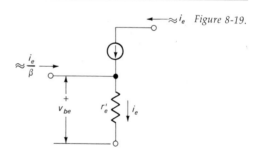

$\leftarrow \approx i_e$ Figure 8-19.

8-27. Suppose the r'_e in Fig. 8-19 equals 10 Ω. If v_{be} equals 1 mV, what is the value of i_e? If you increase v_{be} to 5 mV, what is the new value of i_e?

9. Small-Signal Amplifiers

Chapter 7 worked out an almost universal method of dc transistor analysis by deriving formulas for two prototype circuits: base-biased and emitter-biased. Then, it showed how to Thevenize practical biasing circuits to reduce them to prototype form.

This chapter does the same thing for ac analysis. We derive formulas for two ac prototypes. Then, we show how to Thevenize amplifier circuits to reduce them to prototype form. In this way, we are relieved of trying to remember many formulas for special cases. Instead, with a few key prototype formulas and Thevenin's theorem, we will be able to analyze many different amplifiers.

9-1. BASE DRIVE AND EMITTER DRIVE

You will encounter many different amplifier circuits. With Thevenin's theorem, you can reduce most ac equivalent circuits to either of two basic ac forms: *base-driven* or *emitter-driven.*

Figure 9-1a shows a base-driven circuit, so called because source v_{bb} drives the base through resistance r_B. Source v_{bb} and resistance r_B may be a single source and resistor as shown, or they may represent the Thevenin equivalent circuit for whatever is driving the base. Resistor r_E may be a single resistor or may represent the combined resistance of several resistors. Similarly, r_C may be a single resistor as shown or may be a combined resistance. Regardless of what the actual ac equivalent circuit is, if you can reduce it to the form of Fig. 9-1a you will have the circuit in base-driven prototype form.

Sometimes, Thevenin's and other reducing theorems result in a circuit like Fig.

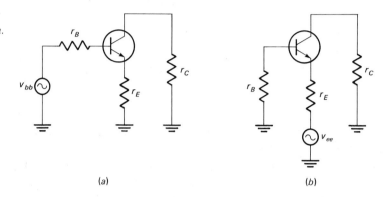

Figure 9-1. Ac prototypes. (a) Base-driven. (b) Emitter-driven.

9-1*b*. This is an emitter-driven prototype because the source v_{ee} drives the emitter through resistance r_E.

Here is our plan for the remainder of this chapter. We will derive a few formulas for the base-driven and emitter-driven prototypes. Then we will practice reducing amplifiers to base-driven or emitter-driven form. By applying the prototype formulas, we will know how the original amplifiers work.

9-2. BASE-DRIVEN FORMULAS

Again, *emitter current* plays the *central role* in transistor action. Once we have found the value of i_e, we can get i_c (approximately equal to i_e) or we can get i_b (divide by β). Because of this, we will concentrate on emitter current and its effects.

Formula for ac emitter current

Figure 9-2*a* shows the base-driven prototype with the three ac currents that flow in the transistor. It is immaterial whether we analyze the circuit for the positive half cycle of input voltage or the negative half cycle. The results for either half cycle apply to the other half when we complement all currents and voltages. For this reason, we can analyze Fig. 9-2*a* for the positive half cycle of input voltage; this is why v_{bb} has the plus-minus polarity shown. During this positive half cycle of input voltage, base and collector currents flow into the transistor and emitter current flows out.

Visualize the transistor as the ideal model discussed in Chap. 8. Figure 9-2*b* shows the resulting circuit. Starting with the v_{bb} source, we can sum voltages in a clockwise direction to get

$$-v_{bb} + \frac{i_e}{\beta} r_B + i_e(r_e' + r_E) \cong 0$$

After factoring and solving for i_e, we have

$$i_e \cong \frac{v_{bb}}{r_E + r_e' + r_B/\beta} \qquad\qquad (9\text{-}1)^{***}$$

Equation (9-1) is important. Because many amplifiers can be reduced to the base-driven prototype, we can use Eq. (9-1) to calculate the ac emitter current in these amplifiers. Once we have i_e, all other ac calculations are easy.

Voltage formulas

When you build a transistor amplifier, you often measure transistor voltages to ground with an oscilloscope or voltmeter. For this reason, you should know approximately what ac voltages exist in an amplifier. The three basic ac voltages in any transistor amplifier are the collector-to-ground voltage v_c, the emitter-to-ground voltage v_e, and the base-to-ground voltage v_b. By referring to Fig. 9-2b, we can work out a formula for each of these in terms of i_e.

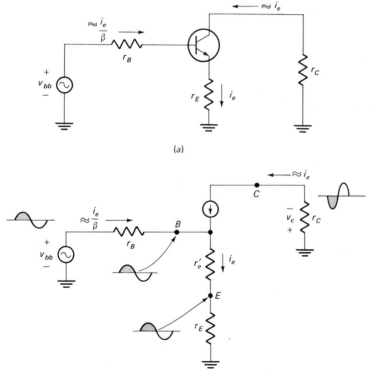

Figure 9-2. Deriving current and voltage formulas.

(a)

(b)

Start with v_c. The ac voltage from collector to ground must equal the voltage across r_C. Since the current flows up through r_C during the positive half cycle of input voltage, v_c has the minus-plus polarity shown in Fig. 9-2b. The magnitude of v_c is

$$v_c \cong i_e r_C \qquad (9\text{-}1a)$$

When the source voltage reverses polarity, that is, during the negative half cycle of source voltage, the voltage across r_C reverses polarity, making the collector positive with respect to ac ground. Therefore, the ac collector voltage is 180° out of phase with the source voltage. This *phase inversion* between the source and collector voltages happens in all base-driven amplifiers.

Next, by looking at Fig. 9-2b we can see that i_e flows down through r_E during the positive half cycle of input voltage. Therefore, the ac emitter-to-ground voltage is

$$v_e = i_e r_E \qquad (9\text{-}1b)$$

Since the voltage is plus-minus across r_E, the phase of the emitter voltage is the same as the phase of the ac source voltage.

The third voltage of interest is the base-to-ground voltage v_b. In Fig. 9-2b, v_b is identical to the ac voltage across $r'_e + r_E$. Therefore,

$$v_b = i_e(r'_e + r_E) \qquad (9\text{-}1c)$$

This voltage is in phase with the ac source voltage v_{bb}.

In summary, if we can reduce the ac equivalent circuit of an amplifier to the base-driven prototype, we can use Eq. (9-1) to calculate the ac emitter current. Once we have i_e, the rest is easy. We can find the three transistor voltages by using Eqs. (9-1a) through (9-1c). The phase relations are always the same: v_c is inverted with respect to v_b, and v_e *follows* or is in phase with v_b.

Voltage gain

The whole point of an amplifier is to increase the peak-to-peak signal. In this chapter, we are interested in *voltage amplifiers,* those that increase the signal voltage. *Voltage gain* from point x to point y is defined as

$$A = \frac{v_y}{v_x} \qquad (9\text{-}2)$$

where v_y is the ac voltage from point y to ground and v_x is the ac voltage from point x to ground. For instance, if $v_y = 1$ V and $v_x = 0.1$ V, we would have a voltage gain of

$$A = \frac{1\text{ V}}{0.1\text{ V}} = 10$$

This tells us the voltage at point y is 10 times greater than the voltage at point x.

One of the most important voltage gains in an amplifier is the voltage gain from base to collector, that is, the ratio of ac collector voltage to ac base voltage. In symbols, the voltage gain from base to collector equals

$$A = \frac{v_c}{v_b}$$

When you build transistor amplifiers and measure v_c and v_b, you find that v_c is normally much greater than v_b. For example, you may find $v_c = 10$ V and $v_b = 0.05$ V. In this case, the amplifier has a voltage gain from base to collector of

$$A = \frac{10 \text{ V}}{0.05 \text{ V}} = 200$$

Because the voltage gain from base to collector is so important, we often need a formula for it. With Eqs. (9-1a) and (9-1c), we have

$$A = \frac{v_c}{v_e} \cong \frac{i_e r_C}{i_e(r'_e + r_E)}$$

or
$$A \cong \frac{r_C}{r_E + r'_e} \qquad\qquad (9\text{-}3)\,{***}$$

Therefore, to get voltage gain from base to collector, all we need is r_C larger than $r_E + r'_e$.

EXAMPLE 9-1.
Work out the value of r'_e in Fig. 9-3a. Then reduce the amplifier to base-driven prototype form.

SOLUTION.
To get r'_e, we need the dc emitter current I_E. Visualize all coupling and bypass capacitors as open. What remains is the dc equivalent circuit of Fig. 9-3b. At this point, you should be able to see at a glance that the dc voltage from base to ground is 10 V; therefore, the dc voltage from emitter to ground is almost 10 V. And finally, the dc emitter current I_E equals approximately 1 mA. As a result,

$$r'_e = \frac{25 \text{ mV}}{I_E} = \frac{25 \text{ mV}}{1 \text{ mA}} = 25 \text{ }\Omega$$

Return to Fig. 9-3a. The first step in reducing this amplifier to its base-driven prototype form is this: visualize all capacitors shorted, and reduce the V_{CC} supply voltage to zero. What remains is the ac equivalent circuit of Fig. 9-3c. To reduce this ac equivalent circuit as much as possible, we Thevenize the base and collector circuits. In the base circuit, the Thevenin voltage is v_{bb} and the Thevenin resistance is zero. In the collector circuit, the Thevenin voltage is zero, and the Thevenin resistance is 5 kΩ.

Figure 9-3d shows the amplifier after Thevenizing the base and collector circuits. We recognize this as a base-driven prototype with $r_B = 0$, $r_E = 100$ Ω, and $r_C = 5$ kΩ.

Figure 9-3. Examples 9-1 through 9-3.

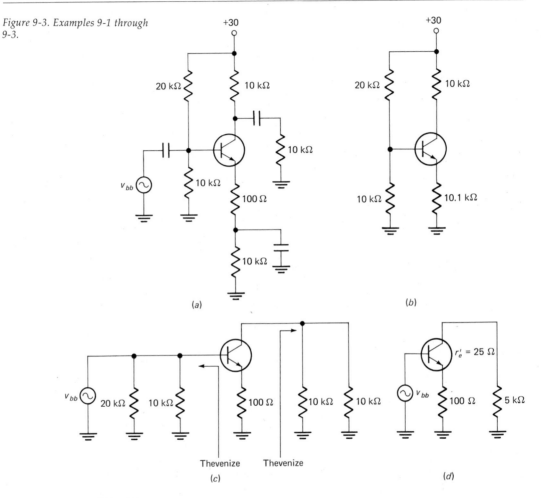

(a)

(b)

Thevenize   Thevenize

(c)

(d)

EXAMPLE 9-2.

Calculate ac emitter current i_e for the amplifier of Fig. 9-3d. After you have it, work out the three ac voltages v_c, v_e, and v_b. Use a v_{bb} of 1-mV peak.

SOLUTION.

We start by using prototype Eq. (9-1).

$$i_e = \frac{v_{bb}}{r_E + r_e' + r_B/\beta} = \frac{0.001}{100 + 25 + 0}$$

$$= 8\ \mu A$$

With this emitter current, we can work out each ac voltage as follows:

$$v_c \cong i_e r_C = 8(10^{-6})5(10^3) = 40 \text{ mV}$$

$$v_e = i_e r_E \cong 8(10^{-6})100 = 0.8 \text{ mV}$$

$$v_b = i_e(r'_e + r_E) = 8(10^{-6})(25 + 100) = 1 \text{ mV}$$

The last calculation (for v_b) is unnecessary in this particular example because $r_B = 0$. Whenever $r_B = 0$, the base voltage must equal the source voltage, that is, $v_b = v_{bb}$.

Also note that all answers are peak values because we used the peak value of v_{bb} when we calculated i_e. If v_{bb} were a sine wave, you could convert all answers to rms values by multiplying each peak value by 0.707. You do this only if you need rms answers. (When working with an oscilloscope, peak values or peak-to-peak values are more convenient. But with ac voltmeters, rms values are usually easier to work with.)

EXAMPLE 9-3.
Calculate the voltage gain from base to collector for the amplifier of Fig. 9-3a.

SOLUTION.
We will do this in two different ways. First, working directly with the v_c and v_b found in the preceding example, we get

$$A = \frac{v_c}{v_b} = \frac{40 \text{ mV}}{1 \text{ mV}} = 40$$

This is the way you would calculate voltage gain after you have measured v_c and v_b in a transistor amplifier.

The second approach is the formula approach using Eq. (9-3).

$$A = \frac{r_C}{r_E + r'_e} = \frac{5000}{100 + 25} = 40$$

This is the way you calculate voltage gain when analyzing or designing an amplifier on paper. Also, when testing amplifiers you can calculate voltage gain from the schematic using Eq. (9-3); then, you can measure the actual amplifier voltages and calculate voltage gain with measured values. If the two voltage gains agree, the amplifier is working correctly.

9-3.   SPECIAL CASE: $r_E = 0$

A special case of the base-driven prototype is the case of $r_E = 0$. When $r_E = 0$, there is no ac resistance between the emitter and ground. In other words, the emitter is at ac ground. This special case is called the *grounded-emitter* or *common-emitter* (CE) amplifier.

Figure 9-4a shows the base-driven prototype of a CE amplifier. During the positive half cycle of source voltage, the base and collector currents flow into the transistor, and

Figure 9-4. *Common-emitter amplifier.*

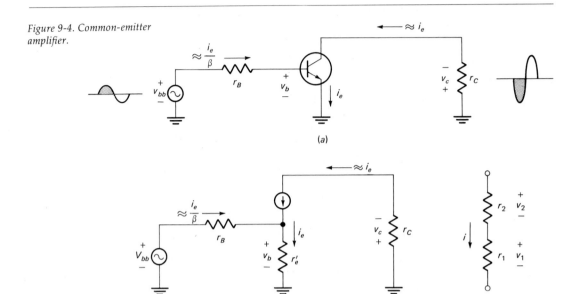

(a)

(b)   (c)

the emitter current flows out. As in any base-driven amplifier, the ac collector voltage is 180° out of phase with ac base voltage.

What is the voltage gain from base to collector in a CE amplifier? The easy way to see this is to replace the transistor by its ideal model shown in Fig. 9-4b. The magnitude of the collector voltage is

$$v_c \cong i_e r_C$$

and the base voltage is

$$v_b = i_e r'_e$$

because v_b appears directly across r'_e in Fig. 9-4b. By taking the ratio of v_c to v_b, we get the formula for the voltage gain from base to collector.

$$A = \frac{v_c}{v_b} \cong \frac{i_e r_C}{i_e r'_e}$$

or

$$A \cong \frac{r_C}{r'_e} \qquad\qquad (9\text{-}3a) ***$$

You can also get the same result by substituting $r_E = 0$ into the gain formula (9-3). Equation (9-3) applies to any base-driven amplifier; Eq. (9-3a) is a special case that applies only to a CE amplifier, one where $r_E = 0$.

As an example, suppose we short the 100-Ω resistor in the emitter circuit of Fig.

9-3a. Then, in the final prototype form of Fig. 9-3d the emitter is at ac ground, that is, r_E = 0. In this case, the voltage gain from base to collector equals

$$A \cong \frac{r_C}{r'_e} = \frac{5000}{25} = 200$$

With this kind of gain, a 1-mV base voltage will produce a 200-mV collector voltage.

Equation (9-3a) makes sense. With the same current (approximately) flowing through r_C and r'_e, the *ratio of voltages* has to equal the *ratio of resistances.* In other words, look at Fig. 9-4c. The same current i flows through r_1 and r_2; this sets up voltages v_1 and v_2. Since voltage is directly proportional to resistance, the ratio v_2/v_1 equals the ratio r_2/r_1. In any CE amplifier, i_c equals i_e to a close approximation. Therefore, essentially the same current flows through r_C and r'_e. Because of this, the voltage ratio v_c/v_b must equal the resistance ratio r_C/r'_e.

Since the CE amplifier is used so often, you should remember that its approximate voltage gain from base to collector equals r_C/r'_e. This is an ideal answer; sometimes we have to use the second or third approximation discussed later.

EXAMPLE 9-4.
What is the voltage gain from base to collector for each CE amplifier shown in Fig. 9-5?

SOLUTION.
By using the methods of dc analysis discussed in Chap. 7 and summarized by Fig. 7-30, you can calculate the dc emitter current I_E in Figs. 9-5a through c. If you do this, you will find $I_E \cong 1$ mA in each amplifier. Therefore, $r'_e = 25$ Ω for each amplifier.

If you now short all capacitors and ground the V_{CC} supply, you get the ac equivalent circuit of each amplifier. After Thevenizing the base and collector circuits (similar to Example 9-1), you get the same prototype form for each amplifier; this prototype is shown in Fig. 9-5d.

The voltage gain from base to collector equals

$$A \cong \frac{r_C}{r'_e} = \frac{5000}{25} = 200$$

EXAMPLE 9-5.
Figure 9-6a shows a *pnp* CE amplifier. What is the voltage gain from base to collector?

SOLUTION.
As far as the ac equivalent circuit is concerned, a *pnp* transistor acts exactly the same as an *npn* transistor, provided it remains FR-biased through the ac cycle. During the positive half cycle of ac source voltage, ac base and collector currents flow into the transistor, and ac emitter current flows out of the transistor whether the transistor is *npn* or *pnp*. In other words, the ideal model is the same for either an *npn* or a *pnp* transistor; for this reason, all ac formulas we derive for *npn* circuits apply to *pnp* circuits.

Figure 9-5. Example 9-4.

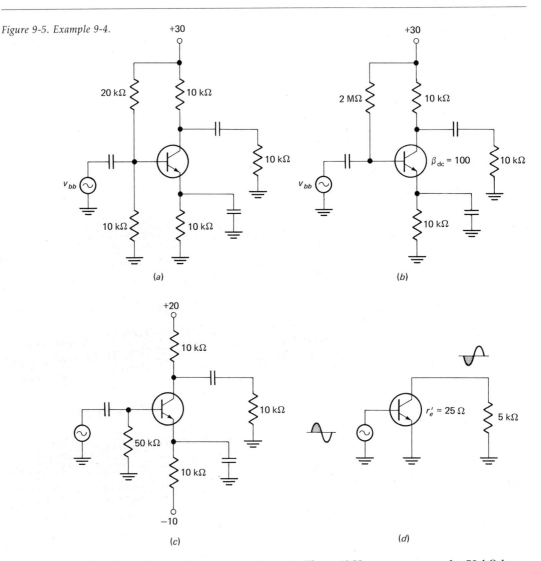

(a)

(b)

(c)

(d)

Visualize all capacitors open in Fig. 9-6a. Then, 10 V appears across the 50-kΩ base resistor, and almost all of this 10 V appears across the 20-kΩ emitter resistor. Therefore, the dc emitter current approximately equals 0.5 mA. And $r_e' = 50$ Ω.

Next, visualize all capacitors shorted and V_{CC} reduced to zero in Fig. 9-6a. By inverting the transistor, we get the ac equivalent circuit of Fig. 9-6b. When we Thevenize the base circuit, the 50-kΩ and 100-kΩ resistors are so much larger than the 1-kΩ resistor that the Thevenin voltage is still approximately v_{bb} and the Thevenin resistance is still close to 1 kΩ. When we Thevenize the collector circuit, we get a V_{TH} of zero and an R_{TH} of

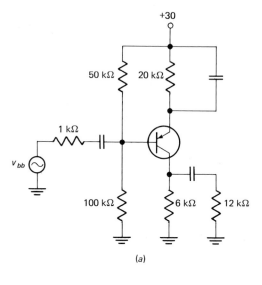

Figure 9-6. Example 9-5.

(a)

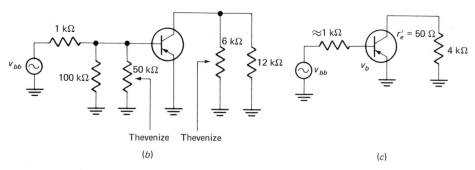

(b)

(c)

$$R_{TH} = 6000||12{,}000 = 4 \text{ k}\Omega$$

Figure 9-6c shows the amplifier reduced to prototype form. Since the *pnp* transistor has the same ideal model as an *npn* transistor, the voltage gain from base to collector equals r_c/r'_e. In this particular case,

$$A = \frac{r_C}{r'_e} = \frac{4000}{50} = 80$$

EXAMPLE 9-6.

Explain the voltage waveforms shown in Fig. 9-7a.

SOLUTION.

Section 8-2 described the superposition theorem applied to transistor-circuit analysis. Step 1 of the method deals with dc currents and voltages; Chap. 7 gave the methods for calculating these dc currents and voltages. Step 2 in the superposition theorem deals with ac currents and voltages. Most of this chapter shows you how to calculate ac currents and voltages. Step 3 of the superposition theorem is to add the dc and ac currents or voltages, whichever you are interested in, to get the total current or voltage for any branch.

When you look at voltage waveforms with a dc-coupled oscilloscope, you will see *total* voltages, that is, the sum of dc and ac voltage. Here is why the following waveforms appear in the amplifier of Fig. 9-7a. At point *I* (the input), an oscilloscope will show whatever the ac source voltage is; in this case, we are assuming a sine wave with a positive peak of 10 mV and a negative peak of −10 mV.

At point *B* (the base) the oscilloscope will show the source sine wave shifted upward by 10 V; the dc voltage from base to ground is

$$V_B = 10 \text{ V}$$

and the ac voltage from base to ground equals

$$v_b = 10 \text{ mV peak} = 0.01 \text{ V peak}$$

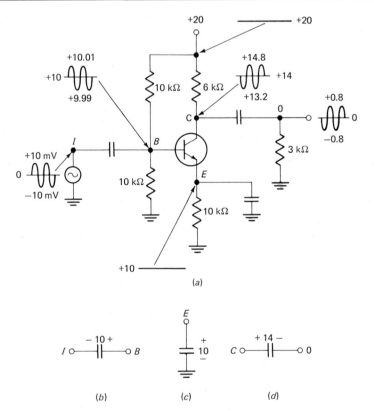

Figure 9-7. Example 9-6.

When we add these dc and ac voltages, we get the total base voltage shown; the maximum voltage is 10.01 V, and the minimum is 9.99 V.

At point E (the emitter) we get a straight line on an oscilloscope at $+10$ V ideally (9.3 V for the second approximation). There is no ac voltage at point E because the bypass capacitor ac grounds the emitter.

The 1 mA of dc emitter current produces a 6-V drop across the 6-kΩ collector resistor. This leaves a dc voltage of 14 V from collector to ground. Since $r'_e = 25$ Ω and $r_C = 2$ kΩ, the voltage gain is

$$A = \frac{r_C}{r'_e} = \frac{2000}{25} = 80$$

Therefore, the ac collector voltage equals

$$v_c = Av_b = 80(0.01) = 0.8 \text{ V peak}$$

After we add the dc and ac collector voltage, we get the inverted sine wave shown. It has a maximum value of $+14.8$ V and a minimum of $+13.2$ V.

At point O (the final output), all we see is a sine wave with a positive peak of 0.8 V and a negative peak of -0.8 V. The coupling capacitor has blocked the dc collector voltage but passed the ac collector voltage.

Finally, the V_{CC} point produces a straight line at $+20$ V on an oscilloscope, assuming the power-supply ripple is too small to see.

Incidentally, the capacitor voltage ratings must be greater than the dc voltages across the capacitors. Figure 9-7b through d shows each capacitor and its dc voltage. The voltage rating of each capacitor must be greater than the dc voltage across it. Furthermore, if you use electrolytic capacitors, you must connect them with the polarities shown.

9-4. SWAMPING THE EMITTER DIODE

The voltage gain from base to collector in a CE amplifier equals r_C/r'_e. The quantity r'_e equals 25 mV/I_E only at room temperature and for a rectangular base-emitter junction. As indicated earlier, even though the junction is not rectangular in most transistors, 25 mV/I_E is still a good approximation for r'_e. Most data sheets do not list h_{fe} and h_{ie}; but if you see these quantities on a data sheet, you can use

$$r'_e \cong \frac{h_{ie}}{h_{fe}} \tag{9-4}$$

to get an accurate value of r'_e. We will discuss the quantities h_{fe} and h_{ie} in detail in Sec. 15-7.

Besides the differences from one transistor type to another r'_e has a temperature dependence given by Eq. (8-6). Any change in the value of r'_e will change the voltage gain in a CE amplifier. In some applications, a change in voltage gain is acceptable. For instance, in a transistor radio, you can offset changes in voltage gain by adjusting the volume control. But there are many applications where you need as stable a voltage gain as possible.

What swamping means

Many people *swamp* the emitter diode to reduce the effects of r_e'. Figure 9-8a shows the base-driven prototype and Fig. 9-8b shows the same circuit with the ideal model for the transistor. Earlier, we proved the voltage gain from base to collector equals

$$\frac{v_c}{v_b} = \frac{r_C}{r_E + r_e'}$$

(Again, the way to remember this is by recognizing the same approximate current flows through collector and emitter circuits; therefore, the voltage ratio must equal the resistance ratio.) Swamping the emitter diode means making r_E much greater than r_e'. In this case, the voltage gain becomes

Figure 9-8. The swamped amplifier.

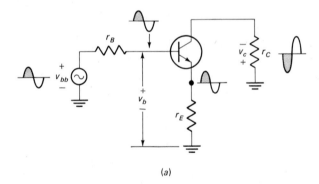

(a)

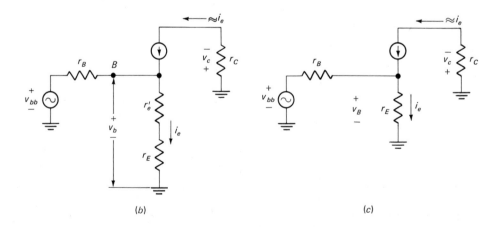

(b)   (c)

$$\frac{v_c}{v_b} \cong \frac{r_C}{r_E} \qquad \text{for } r_E \gg r_e' \qquad\qquad (9\text{-}5)\,***$$

Figure 9-8c illustrates a swamped amplifier. In a swamped amplifier, r_e' is so small compared to r_E that it produces a negligible effect on emitter current. For this reason, you can visualize the emitter branch with only r_E in it. With the same approximate current in the collector and emitter circuits, the voltage ratio v_c/v_b equals the resistance ratio r_C/r_E. This is the way to remember the voltage gain of a swamped amplifier.

If you swamp too heavily, that is, use too large an r_E, you will get a small voltage gain. For instance, if $r_e' = 25 \ \Omega$, $r_C = 5 \ k\Omega$, and $r_E = 1 \ k\Omega$, the voltage gain equals

$$\frac{v_c}{v_b} \cong \frac{r_C}{r_E + r_e'} = \frac{5000}{1000 + 25} \cong 5$$

This is heavy swamping; r_E is 40 times greater than r_e'. Since r_e' adds only 25 Ω to the denominator, any changes in r_e' cause only minor changes in the denominator. The circuit is almost insensitive to changes in r_e', but the voltage gain is only 5.

The amount of swamping to use depends on the temperature range and other factors affecting the value of r_e'. If r_E is 10 times greater than r_e', the effect of changes in r_e' is reduced by a factor of 10. If r_E is 20 times greater than r_e', the effect of changes in r_e' is reduced by a factor of 20. The final choice of how much larger r_E should be than r_e' is a decision the designer makes in a particular application. The point is that by swamping the emitter diode, we can get a voltage gain from base to collector hardly affected by r_e'.

Emitter follows base

A final point about a swamped amplifier. In Fig. 9-8a, the ac emitter-to-ground voltage equals

$$v_e = i_e r_E \qquad\qquad (9\text{-}6a)$$

As previously described, this ac voltage follows the base voltage, that is, v_e is in phase with v_b. In Fig. 9-8b, the base voltage equals

$$v_b = i_e(r_e' + r_E)$$

or $\qquad\qquad\qquad\qquad v_b \cong i_e r_E \qquad \text{when } r_E \gg r_e' \qquad\qquad (9\text{-}6b)$

Compare Eqs. (9-6a) and (9-6b). This comparison shows

$$v_e \cong v_b \qquad \text{when } r_E \gg r_e' \qquad\qquad (9\text{-}7)\,***$$

This says the ac emitter voltage approximately equals the ac base voltage in a swamped amplifier. This is important to know for troubleshooting if nothing else. You can visualize this result by looking at Fig. 9-8b. When r_e' is much smaller than r_E, almost all of v_b appears across r_E, the ac resistance from emitter to ground.

EXAMPLE 9-7.

What is the approximate ac voltage across the output 3-kΩ resistor of Fig. 9-9?

SOLUTION.

First, its voltage-divider bias is about 10 V dc from base to ground. Therefore, the dc emitter current I_E approximately equals 10 mA. This means an r'_e of about 2.5 Ω.

In the ac equivalent circuit, the bottom of the 50-Ω emitter resistor is at ac ground. Therefore, $r_E = 50$ Ω, more than enough to swamp an r'_e of 2.5 Ω (a factor of 20). So, the swamped voltage gain is

$$A = \frac{v_c}{v_b} \cong \frac{r_C}{r_E} = \frac{1000||3000}{50} = 15$$

The ac voltage across the output 3-kΩ resistor is

$$v_c = Av_b = 15(10 \text{ mV}) = 150 \text{ mV}$$

The actual waveform will be a sine wave with a peak of 150 mV, and 180° out of phase with the base voltage.

Incidentally, since the emitter diode is swamped, the ac voltage from emitter to ground will be a sine wave of approximately 10-mV peak, the same as the base voltage.

EXAMPLE 9-8.

Suppose in wiring a circuit like Fig. 9-9 you accidentally leave out the bypass capacitor. What happens to the voltage gain?

Figure 9-9. Examples 9-7 and 9-8.

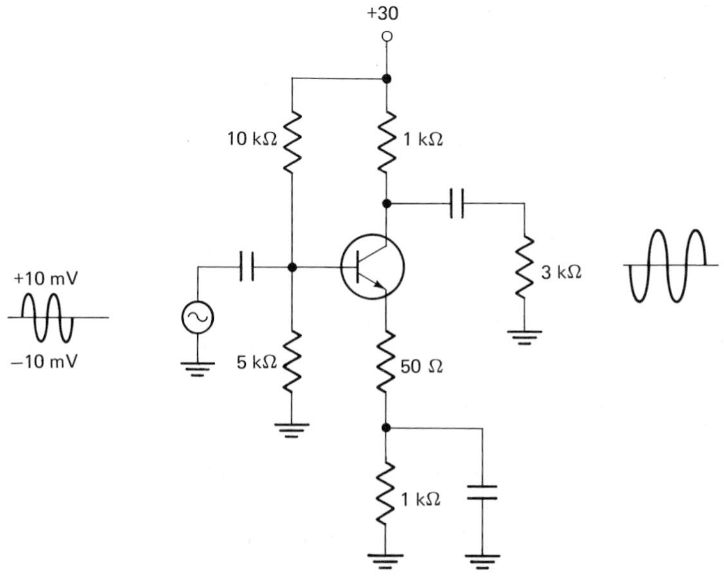

SOLUTION.

Leaving out the emitter bypass capacitor means

$$r_E = 50 + 1000 \cong 1000 \ \Omega$$

because the bottom of the 50-Ω resistor is no longer at ac ground. In a case like this, the swamping becomes enormously heavy, and the voltage gain drops to

$$\frac{v_c}{v_b} \cong \frac{1000 || 3000}{1000} = 0.75$$

This tells us the ac collector voltage is less than the ac base voltage. In other words, we do not get voltage gain, but instead get *attenuation*, a loss of voltage.

An open bypass capacitor in a CE amplifier like Fig. 9-7a produces the same result: a loss of voltage. Remember the open bypass capacitor; it is a common trouble in discrete amplifiers.

9-5.   INPUT IMPEDANCE

The ac source driving an amplifier must supply ac current to the amplifier. Usually, the less current the amplifier draws from the source, the better. The *input impedance* of an amplifier determines how much current the amplifier takes from the source. This section defines ac input impedance and works out a few basic formulas for it.

Definition

In the normal frequency range of an amplifier, that is, where coupling and bypass capacitors look like ac shorts, and all other reactances are negligible, the ac input impedance is defined as

$$z_{in} = \frac{v_{in}}{i_{in}} \qquad (9\text{-}8)$$

where v_{in} and i_{in} are peak values, peak-to-peak values, rms values, or any consistent pair of values.

Figure 9-10a illustrates the idea of input impedance. An amplifier is inside the box. For the amplifier to work, it must have an ac voltage v_{in} across the input terminals. The ac source (not shown) delivers an ac current i_{in} to the amplifier. The ratio of v_{in} to i_{in} is the input impedance of the amplifier. For a given v_{in}, the less current the amplifier draws, the higher its input impedance.

As an example, suppose an amplifier draws 2 μA when the input voltage is 10 mV (Fig. 9-10b). Then, the ac input impedance equals

$$z_{in} = \frac{10 \ mV}{2 \ \mu A} = 5 \ k\Omega$$

Figure 9-10. Input-impedance concept.

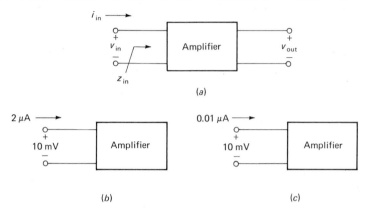

(a)

(b)   (c)

If another amplifier draws only 0.01 μA when 10 mV is across its input terminals (Fig. 9-10c), this second amplifier has an input impedance of

$$z_{in} = \frac{10 \text{ mV}}{0.01 \ \mu A} = 1 \text{ M}\Omega$$

If all other characteristics are identical, the amplifier with a 1-MΩ input impedance is more desirable because it *loads* the source more lightly, that is, draws less current and absorbs less power.

Two components of input impedance

In a base-driven amplifier, the ac source has to supply ac current to the base; in addition, the source has to deliver current to the biasing resistors in the base circuit. For instance, Fig. 9-11a shows a base-driven amplifier, and Fig. 9-11b the ac equivalent circuit. R_B is the combined sum of R_1 and R_2, that is,

$$R_B = R_1 || R_2$$

On the positive half cycle of source voltage, the input current i_{in} flows into the amplifier and splits into two components at node B. Component i_B flows down through R_B, and component i_b flows into the base (Fig. 9-11b).

Because the current splits into two parts, the input impedance of the amplifier is made up of two parts. The first is R_B and the second is $z_{in(base)}$ shown in Fig. 9-11b. Since these two impedances are in parallel, the input impedance of the amplifier is

$$z_{in} = R_B || z_{in(base)} \tag{9-9}$$

To find the input impedance of an amplifier, we need the value of R_B and $z_{in(base)}$.

How do we calculate R_B? For many amplifiers, the biasing resistors in the base are in parallel when we visualize the ac equivalent circuit. For instance, in Fig. 9-11a if R_1

and R_2 are 10 kΩ each, they appear as two parallel 10-kΩ resistors in the ac equivalent circuit. Therefore, in Fig. 9-11b,

$$R_B = 10^4 || 10^4 = 5 \text{ k}\Omega$$

So whatever the amplifier, if the biasing resistors appear in parallel in the ac equivalent circuit, you combine these into a single resistance R_B as shown in Fig. 9-11b.

Collector-feedback bias is an exceptional case. When you visualize the ac equivalent circuit, the biasing resistor in the base does *not* appear across the input terminals; it still appears from collector to base as shown in Fig. 9-11c. The feedback effect

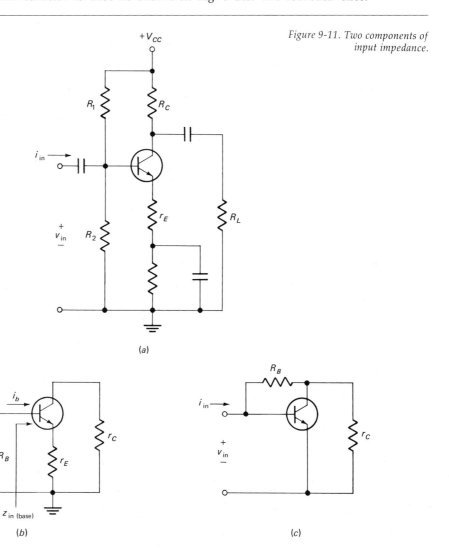

Figure 9-11. *Two components of input impedance.*

(a)

(b)

(c)

described in Chap. 7 complicates the analysis slightly. For this reason, we will post-pone the discussion of voltage gain and input impedance for this special circuit until Chap. 15.

For most amplifiers, R_B appears across the input terminals as shown in Fig. 9-11b; the only remaining question is how do you find $z_{\text{in(base)}}$.

Impedance looking into the base

Figure 9-12a shows the base-driven prototype to the right of R_B. The input imped-ance of the base is

$$z_{\text{in(base)}} = \frac{v_b}{i_b} \qquad (9\text{-}10a)$$

That is, you divide the ac base voltage by the ac base current to get the input imped-ance of the base. If v_b is 1 mV and i_b is 1 μA, the input impedance of the base is

$$z_{\text{in(base)}} = \frac{1 \text{ mV}}{1 \text{ } \mu\text{A}} = 1 \text{ k}\Omega$$

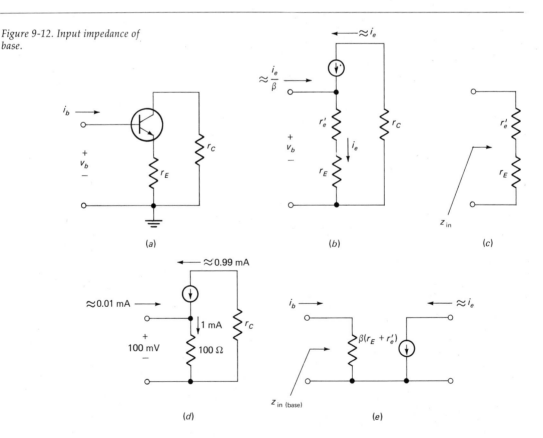

Figure 9-12. Input impedance of base.

Equation (9-10a) is useful when you measure v_b and i_b. Often, you would prefer a formula for $z_{in(base)}$ in terms of schematic values. To get this formula, examine Fig. 9-12b, which shows the ideal model for the transistor. Since v_b appears across r'_e and r_E, we may write

$$z_{in(base)} = \frac{v_b}{i_b} \cong \frac{i_e(r'_e + r_E)}{i_e/\beta}$$

or
$$z_{in(base)} = \beta(r_E + r'_e) \qquad\qquad (9\text{-}10b)\,{***}$$

This says the input impedance of the base is β times the sum of r_E and r'_e. For instance, if $\beta = 100$ and $r_E + r'_e = 100\ \Omega$, the input impedance is 10 kΩ.

This result should strike you as unusual. In fact, if you have not seen it before, it should disturb you. In Fig. 9-12b, when you look into the base, you see $r_E + r'_e$. Why isn't this the impedance? To begin with, Fig. 9-12b is *not* a one-branch circuit. Only when you have a one-branch circuit like Fig. 9-12c is the impedance equal to what you see. If the collector diode of a transistor were burned out, Fig. 9-12c would apply and the input impedance would equal $r_E + r'_e$. But when the transistor operates normally, the collector diode acts like a current source and the T circuit of Fig. 9-12b applies. In a T circuit like this, the input impedance is approximately β times the sum of $r_E + r'_e$.

To clear up any difficulties, look at Fig. 9-12d. The emitter branch has a 100-Ω resistor; this is not the input impedance. The ac emitter current is 1 mA; therefore, v_b equals

$$v_b = i_e(r'_e + r_E) = 0.001(100) = 100\ \text{mV}$$

The ac source does not have to supply the 1 mA of ac emitter current; all it has to supply is a much smaller ac base current. For a β of 100, the base current approximately equals 0.01 mA. The input impedance of the base is

$$z_{in(base)} = \frac{v_b}{i_b} \cong \frac{100\ \text{mV}}{0.01\ \text{mA}} = 10\ \text{k}\Omega$$

which agrees with the result you get by using

$$z_{in(base)} \cong \beta(r_E + r'_e) = 100(100) = 10\ \text{k}\Omega$$

In summary, as far as the ac source is concerned, the impedance in the emitter circuit is increased by a factor of β. For this reason, the equivalent circuit of Fig. 9-12e looks exactly the same to the source as the actual circuit of Fig. 9-12b. In terms of the current the source must supply, the input impedance of the base is $\beta(r_E + r'_e)$.

Special cases

Equation (9-10b) applies to all base-driven amplifiers. For the special case of a CE amplifier, $r_E = 0$; therefore, in a CE amplifier,

$$z_{\text{in(base)}} \cong \beta r'_e \qquad \text{for } r_E = 0 \tag{9-11a}$$

For example, if $\beta = 100$ and $r'_e = 25\ \Omega$, $z_{\text{in(base)}}$ will equal 2.5 kΩ.

On the other hand, for the special case of a swamped amplifier, r_E is much greater than r'_e. So,

$$z_{\text{in(base)}} \cong \beta r_E \qquad \text{for } r_E \gg r'_e \tag{9-11b}$$

You can see right away a swamped amplifier has a larger input impedance than a CE amplifier. This is another advantage of the swamped amplifier.

EXAMPLE 9-9.
What is the input impedance of the amplifier shown in Fig. 9-13a?

SOLUTION.
The dc emitter current is approximately 1 mA; therefore, $r'_e = 25\ \Omega$.

When we visualize the ac equivalent circuit and combine all resistances, we get Fig. 9-13b. In this circuit,

$$R_B = 20\ \text{k}\Omega$$

and
$$z_{\text{in(base)}} = \beta r'_e = 200(25) = 5\ \text{k}\Omega$$

To the source, the input impedance is

$$z_{\text{in}} = R_B || z_{\text{in(base)}} = 20(10^3) || 5(10^3) = 4\ \text{k}\Omega$$

EXAMPLE 9-10.
What is the approximate value of input impedance in Fig. 9-13c?

SOLUTION.
Visualize the dc equivalent circuit. In this upside-down pnp voltage-divider-biased circuit, approximately 6 V appears across the 2-kΩ emitter resistor. Therefore, the dc emitter current is approximately 3 mA and $r'_e \cong 8\ \Omega$. (You can get a more accurate value by subtracting 0.7 V from 6 V to get 5.3 V across the emitter resistor.)

Next, visualize the ac equivalent circuit (Fig. 9-13d). The two base resistors combine into a single 8-kΩ resistor. The two collector resistors combine into a single 2-kΩ resistor. In the emitter circuit, the 2-kΩ resistor is shunted by the 100-Ω resistor. Because of this,

$$r_E = 2000 || 100 \cong 100\ \Omega$$

and
$$z_{\text{in(base)}} \cong \beta(r_E + r'_e) = 50(100 + 8) = 5.4\ \text{k}\Omega$$

The source cares only about the current it has to supply. As far as it is concerned, it sees

$$z_{\text{in}} = R_B || z_{\text{in(base)}} = 8000 || 5400 \cong 3200\ \Omega$$

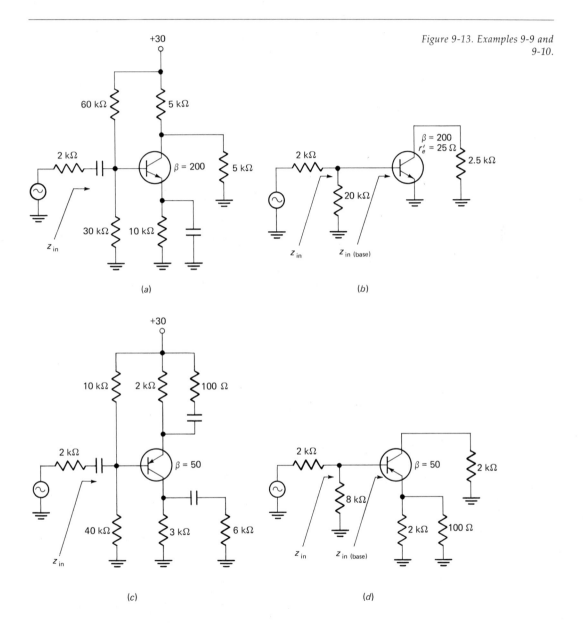

Figure 9-13. Examples 9-9 and 9-10.

9-6. SOURCE IMPEDANCE

Every ac source has some internal impedance. This impedance often is small enough to neglect, but there are many cases where we cannot neglect it. This section tells you how to deal with source impedance.

The interface

Figure 9-14a shows an ac source driving an amplifier through source impedance R_S. Since the amplifier has input impedance z_{in}, we can visualize the input circuit as a voltage divider (Fig. 9-14b). Therefore, the ac voltage appearing across the input terminals of the amplifier is

$$v_{in} = \frac{z_{in}}{R_S + z_{in}} v_S \qquad (9\text{-}12)\,***$$

As an example, if $z_{in} = 4$ kΩ and $R_S = 1$ kΩ,

$$v_{in} = \frac{4000}{1000 + 4000} v_S = 0.8v_S$$

which tells us 80 percent of the ac source voltage appears across the input terminals of the amplifier. This is the voltage the amplifier responds to, that is, amplifies by a factor of $r_C/(r_E + r'_e)$.

Now is a good time to mention the idea of the *interface*. Figure 9-15a shows one amplifier driving another; the output voltage of the first amplifier is the input voltage to the second amplifier. In Fig. 9-15a you can think of an imaginary plane between the two amplifiers. This imaginary plane is called the *interface*.

When the first amplifier looks through the interface at the second amplifier, it sees input impedance z_{in} (Fig. 9-15b). On the other hand, when the second amplifier looks back at the first amplifier through the interface, it sees a Thevenin equivalent circuit with v_S and R_S (Fig. 9-15c).

The interface is helpful whenever we connect two circuits together in *cascade*, that is, the output of one circuit used as the input to the other. By visualizing what each circuit sees through the interface, we can determine how the circuits interact. For the simple case of one amplifier driving another, the interaction is summarized by the voltage divider of Fig. 9-15d.

Figure 9-14. Effect of source impedance.

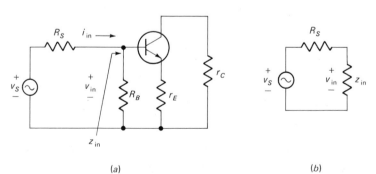

(a) (b)

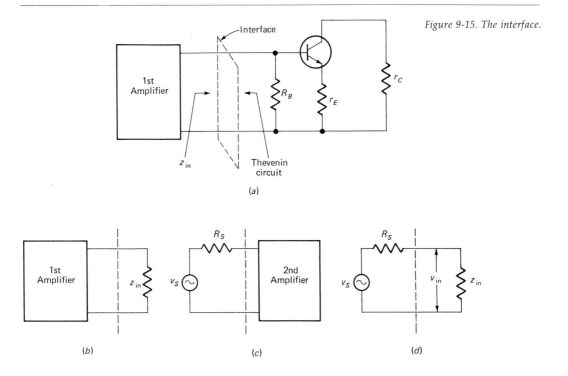

Figure 9-15. The interface.

Special cases

With Eq. (9-12) you can calculate the input voltage to any base-driven amplifier. There are two special cases worth mentioning. First, the input impedance z_{in} may be much larger than the source impedance R_S. In this case, Eq. (9-12) simplifies to

$$v_{in} \cong v_S \qquad \text{when } z_{in} \gg R_S \qquad (9\text{-}12a)$$

This says almost all the source voltage appears across the input terminals of the amplifier.

Figure 9-16a summarizes the special case of z_{in} much greater than R_S. For this case, the ac source acts like an ideal voltage source. That is, we can approximate R_S as zero because it has negligible effect on the current flowing in the circuit.

The second special case to know about is the *impedance-matched* case, $z_{in} = R_S$. For this situation, Eq. (9-12) reduces to

$$v_{in} = 0.5v_S \qquad \text{when } z_{in} = R_S \qquad (9\text{-}12b)$$

Figure 9-16b shows this important case. Half the source voltage appears across the input terminals of the amplifier. As you recall from basic circuit theory, when $z_{in} = R_S$, maximum power is delivered to z_{in}. Later chapters discuss this case further.

Figure 9-16. (a) Very small source impedance. (b) Impedance match.

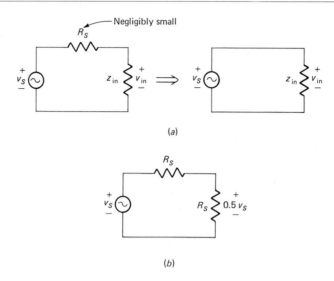

(a)

(b)

Voltage gain from source to collector

When R_S is not small enough to neglect, v_{in} is less than v_S in Fig. 9-17. Because of this, the ratio of ac collector voltage to ac source voltage is less than the ratio of v_c/v_b. In other words, the voltage gain from the source to the collector is less than the voltage gain from the base to the collector.

What is the formula for source-to-collector voltage gain? First, we rearrange Eq. (9-12) to get

$$\frac{v_{in}}{v_S} = \frac{z_{in}}{R_S + z_{in}}$$

This is the voltage gain from source to input. Even though we lose voltage going from source to input, v_{in}/v_S will be called a voltage gain because it is still a voltage ratio. Second, $v_{in} = v_b$ in Fig. 9-17; therefore,

Figure 9-17. Voltage gain with source impedance.

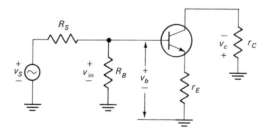

$$\frac{v_c}{v_{in}} = \frac{v_c}{v_b} \cong \frac{r_C}{r_E + r_e'}$$

The product of v_{in}/v_S and v_c/v_{in} is

$$\frac{v_c}{v_S} = \frac{v_{in}}{v_S}\frac{v_c}{v_{in}} = \frac{z_{in}}{R_S + z_{in}}\frac{r_C}{r_E + r_e'} \qquad (9\text{-}13)$$

In Eq. (9-13), the first factor is the voltage gain from source to input; the second factor is the voltage gain from input to collector. Because the first factor is less than unity (except for $R_S = 0$), the source-to-collector voltage gain is less than the base-to-collector gain.

In fact, if the signal comes out of a very high-impedance source, you may get less than unity voltage gain from source to collector. For instance, if $R_S = 1$ MΩ, $z_{in} = 1$ kΩ, and $v_c/v_b = 200$,

$$\frac{v_c}{v_S} = \frac{10^3}{10^6 + 10^3}\,200 \cong 0.2$$

In this case, we get attenuation. The point is clear enough: you may not get source-to-collector gain if $R_S \gg z_{in}$.

In most practical cases, R_S is not large enough to offset the voltage gain from base to collector. For the special case of R_S much smaller than z_{in},

$$\frac{v_c}{v_S} \cong \frac{v_c}{v_b} \qquad \text{for } z_{in} \gg R_S \qquad (9\text{-}13a)$$

For the special case of impedance-matching z_{in} to R_S, we have

$$\frac{v_c}{v_S} = 0.5\,\frac{v_c}{v_b} \qquad \text{for } z_{in} = R_S \qquad (9\text{-}13b)$$

EXAMPLE 9-11.
The amplifier of Fig. 9-13a has an input impedance of 4 kΩ (found in Example 9-9). Work out the voltage gain from source to input. Also, calculate the voltage gain from source to collector.

SOLUTION.
First,

$$\frac{v_{in}}{v_S} \cong \frac{4000}{2000 + 4000} = 0.667$$

So, about two-thirds of the source voltage appears across the input of the amplifier.
Next,

$$\frac{v_c}{v_S} = \frac{v_{in}}{v_S}\frac{v_c}{v_{in}} = 0.667\,\frac{5000||5000}{25} = 66.7$$

As you can see, the source-to-collector voltage gain is 66.7, which is less than the base-to-collector voltage gain of 100. (The gain of 100 is evident when you look at Fig. 9-13b.)

9-7. THE EMITTER FOLLOWER

To avoid $R_S \gg z_{in}$, you sometimes have to step up the impedance level. For instance, Fig. 9-18a shows a heavily loaded source, one where R_L is smaller than R_S. In a case like this, most of the voltage is lost across the internal source impedance. One way to get around this is with a transformer as shown in Fig. 9-18b. Now, the input impedance seen by the source is

$$z_{in} = n^2 R_L = 10^2 (100) = 10 \text{ k}\Omega$$

Alternatively, you can use an *emitter follower* to step up the impedance.

Input impedance

Figure 9-19a shows the *ac equivalent circuit* of an emitter follower. The ac source drives the base, and the output signal is taken from the emitter. Since the collector is at ac ground, the circuit is sometimes called a *grounded-collector* or *common-collector* (CC) amplifier. As proved earlier, the input impedance of the base is

$$z_{in(base)} \cong \beta (r_E + r_e') \qquad \text{(9-14a)}$$

or
$$z_{in(base)} \cong \beta r_E \qquad \text{for } r_E \gg r_e' \qquad \text{(9-14b)}$$

This means the emitter follower steps up the impedance by a factor of β.

As an example, Fig. 9-19c shows an emitter follower loaded by an R_L of 100 Ω. About 10 mA of dc emitter current flows so that r_e' is around 2.5 Ω. Figure 9-19b shows the ac equivalent circuit. With a β of 100,

$$z_{in(base)} \cong 100(90 + 2.5) \cong 9 \text{ k}\Omega$$

and
$$z_{in} = 10^5 || 9(10^3) \cong 8.3 \text{ k}\Omega$$

Since z_{in} is greater than the R_S of 1 kΩ, most of the ac source voltage appears from base to ground. If we did not use the emitter follower to step up the impedance, we would lose most of the voltage across the source impedance (Fig. 9-18a).

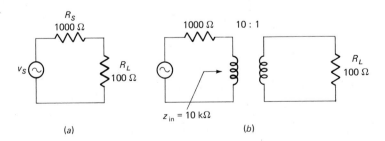

Figure 9-18. (a) *Large source impedance.* (b) *Transformer step-up in impedance level.*

(a)

(b)

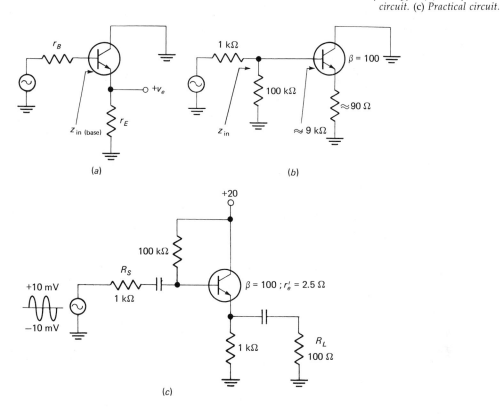

Figure 9-19. Emitter follower. (a) Ac prototype. (b) Ac equivalent circuit. (c) Practical circuit.

Stepping up an impedance is the main reason for using an emitter follower. Not only is it more convenient than a transformer, an emitter follower has a much better frequency response, that is, works over a larger frequency range.

Voltage gain

With an emitter follower, the voltage gain is less than unity. We can prove this as follows. We found earlier

$$v_e = i_e r_E$$

and

$$v_b \cong i_e(r_E + r'_e)$$

Therefore, the voltage gain from base to emitter is

$$\frac{v_e}{v_b} \cong \frac{r_E}{r_E + r'_e} \qquad\qquad (9\text{-}15a)^{***}$$

The denominator is greater than the numerator; so, the ratio is less than unity. For the case of r_E much greater than r_e', we get

$$\frac{v_e}{v_b} \cong 1 \qquad \text{for } r_E \gg r_e' \tag{9-15b}$$

In Fig. 9-19c, r_E is much greater than r_e', and the voltage gain is near unity. Because of this, the final output voltage is a sine wave with a peak of about 10 mV. (A closer answer is 9 mV because we lose about 1 mV across R_S.)

Power gain

We do not get voltage gain with an emitter follower, but we do get *power gain*. The output power in Fig. 9-19a is

$$p_e = i_e^2 r_E$$

and the input power to the base is

$$p_b = i_b^2 z_{\text{in(base)}} \cong i_b^2 \beta (r_E + r_e')$$

The power gain or ratio is

$$\frac{p_e}{p_b} \cong \frac{i_e^2 r_E}{i_b^2 \beta (r_E + r_e')}$$

or since $i_e/i_b \cong \beta$,

$$\frac{p_e}{p_b} \cong \beta \frac{r_E}{r_E + r_e'} \tag{9-16a}$$

The first factor is the current gain; the second factor is the voltage gain. The product of the two gains is the power gain.

For the usual case of r_E much greater than r_e', we get

$$\frac{p_e}{p_b} \cong \beta \qquad \text{for } r_E \gg r_e' \tag{9-16b}$$

This tells us the power gain of an emitter follower is approximately the same as the current gain. So, if we use a transistor with a β of 100, we get a power gain of 100. Put 1 μW into the base and 100 μW comes out the emitter. Conversely, if 500 μW comes out the emitter, only 5 μW goes into the base.

Because it is easier to work with, has a better frequency response, and has power gain, the emitter follower is almost always preferred to a transformer.

EXAMPLE 9-12.
Calculate $z_{\text{in(base)}}$, voltage gain, and power gain for the emitter follower of Fig. 9-20.

SOLUTION.
Visualize the ac equivalent circuit. Then, you can see $r_E = 10,000 || 500$, which is approximately 500 Ω. With a β of 200,

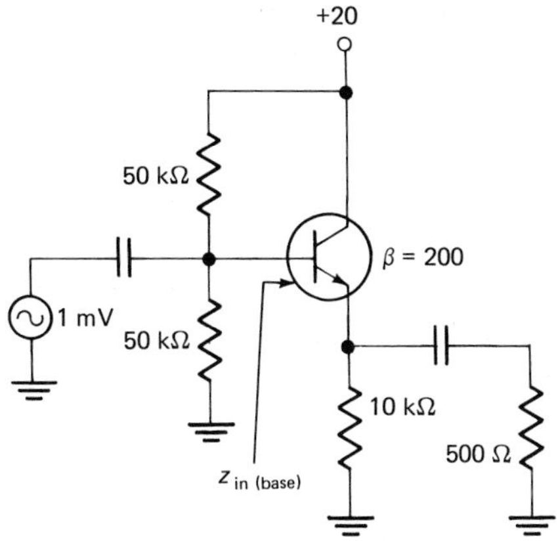

Figure 9-20. Example 9-12.

$$z_{\text{in(base)}} \cong 200(500) = 100 \text{ k}\Omega$$

(The r_e' is negligible since it is only 25 Ω, apparent in Fig. 9-20 because 1 mA of dc emitter current flows.)

The voltage gain is

$$\frac{v_e}{v_b} \cong 1 \qquad \text{because } r_E \gg r_e'$$

and the power gain is

$$\frac{p_e}{p_b} \cong \beta = 200$$

9-8. THE DARLINGTON PAIR

The higher the β, the higher the input impedance of the base. Many transistors have β's up to 300; some as high as 1000. With a *Darlington pair*, we can get even higher β's.

Figure 9-21*a* shows the Darlington pair. Assuming external circuits FR-bias the

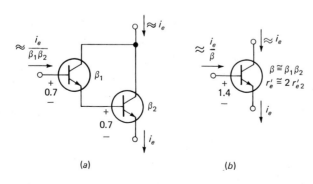

Figure 9-21. (a) npn Darlington pair. (b) Single-transistor equivalent of Darlington pair.

(a) (b)

transistors, about 0.7 V appears across each emitter diode. The emitter current out of the first transistor drives the base of the second transistor. Because of this, the overall β is the product of each β. For example, if $\beta_1 = 50$ and $\beta_2 = 100$, the current gain from the first base to the second emitter is approximately 5000.

We can visualize a Darlington pair as a single transistor (Fig. 9-21b) whose $\beta = \beta_1\beta_2$. The Appendix proves that the r_e' of this equivalent transistor is

$$r_e' = r_{e2}' + \frac{r_{e1}'}{\beta_2} \tag{9-17a}$$

For similar transistors, this closely equals

$$r_e' \cong 2r_{e2}' \tag{9-17b}$$

where r_{e2}' is the ac emitter resistance of the second transistor. Note also the total base-emitter voltage across the equivalent transistor is 1.4 V as shown.

By complementing and inverting the Darlington pair of Fig. 9-21a, we get the complementary *pnp* Darlington pair shown in Fig. 9-22a. This connection also acts like a single transistor whose β equals the product of individual β's and whose r_e' equals $2r_{e2}'$ as shown in Fig. 9-22b.

Because the Darlington pair acts like a single transistor, you can use it in any of the circuits discussed earlier. The formulas derived earlier for voltage gain and input impedance still apply; simply use $\beta_1\beta_2$ for β, and use the r_e' given by Eq. (9-17a) or (9-17b). Equation (9-17a) is more accurate than (9-17b), but we use (9-17b) for all preliminary analysis. As an example, the Darlington CE amplifier of Fig. 9-23a has

$$z_{\text{in(base)}} \cong \beta r_e' \cong 2\beta_1\beta_2 r_{e2}'$$

and
$$\frac{v_c}{v_b} \cong \frac{r_C}{r_e'} \cong \frac{r_C}{2r_{e2}'}$$

As another example, the Darlington emitter follower of Fig. 9-24 ideally has

$$z_{\text{in(base)}} \cong \beta r_E \cong \beta_1\beta_2 r_E$$

Figure 9-22. (a) pnp Darlington pair. (b) Single-transistor equivalent.

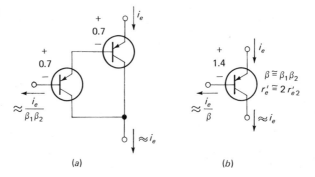

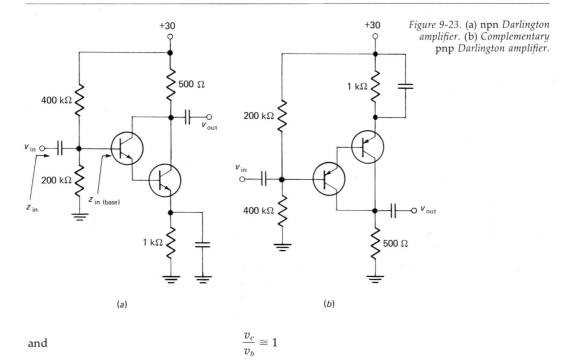

Figure 9-23. (a) npn *Darlington amplifier.* (b) *Complementary* pnp *Darlington amplifier.*

and

$$\frac{v_c}{v_b} \cong 1$$

The increase in β of a Darlington pair holds for dc betas as well as ac betas. That is, you multiply the individual dc betas to get the overall dc beta.

Occasionally, the high β of a Darlington emitter follower produces a $z_{\text{in(base)}}$ over a megohm. In a case like this, you can no longer neglect r_c' in Fig. 8-10 (the third approximation). The input impedance of the base becomes

$$z_{\text{in(base)}} \cong \beta r_E || r_c' \cong r_c' \qquad \text{when } \beta r_E \gg r_c'$$

As discussed earlier, r_c' is typically in megohms. For this reason, the r_c' of the first transistor represents the limit on how high an input impedance can become. Typically, the highest input impedance you can get with bipolar transistors is a couple of megohms.

When we need input impedances much greater than a megohm, we turn to other devices such as the *field-effect transistor* (FET), which is discussed in Chaps. 13 and 14.

EXAMPLE 9-13.

What is the voltage gain and input impedance for the Darlington CE amplifier of Fig. 9-23a? Each β equals 100.

SOLUTION.

There's 10 V dc on the first base. Subtracting 1.4 V gives us 8.6 V dc across the 1-kΩ emitter resistor. Therefore,

$$I_E = \frac{8.6}{1000} = 8.6 \text{ mA}$$

and
$$r'_{e2} = \frac{25 \text{ mV}}{8.6 \text{ mA}} = 2.9 \text{ }\Omega$$

The voltage gain equals

$$\frac{v_c}{v_b} = \frac{r_C}{r'_e} = \frac{r_C}{2r'_{e2}} = \frac{500}{5.8} = 86.2$$

With each β equal to 100, the overall β equals

$$\beta = 100(100) = 10{,}000$$

and the input impedance of the first base is

$$z_{\text{in(base)}} \cong \beta r'_e = 10{,}000(5.8) = 58 \text{ k}\Omega$$

And the input impedance of the stage is

$$z_{\text{in}} = R_1 || R_2 || z_{\text{in(base)}} = 400(10^3) || 200(10^3) || 58(10^3)$$
$$= 40.4 \text{ k}\Omega$$

Figure 9-23b is the upside-down *pnp* version of Fig. 9-23a. For this reason, it has the same voltage gain and input impedance.

EXAMPLE 9-14.
The β of each transistor in Fig. 9-24 is 100. What is the value of input impedance and power gain for the Darlington emitter follower?

SOLUTION.
The circuit is biased the same as the Darlington amplifier discussed in Example 9-13. Therefore,

$$z_{\text{in(base)}} \cong \beta(r_E + r'_e)$$
$$= 10{,}000(91 + 5.8) \cong 1 \text{ M}\Omega$$

and
$$z_{\text{in}} = R_1 || R_2 || z_{\text{in(base)}}$$
$$= 400(10^3) || 200(10^3) || 10^6 = 118 \text{ k}\Omega$$

The power gain equals

$$\frac{p_e}{p_b} \cong \beta = 10{,}000$$

Remember that p_e is the ac power coming out of the emitter; the 1-kΩ and 100-Ω resistors share this p_e; the 100-Ω resistor gets 10 times as much power as the 1-kΩ resistor.

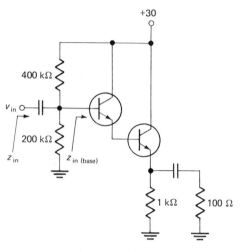

Figure 9-24. Darlington emitter follower.

9-9. THE SECOND APPROXIMATION OF A TRANSISTOR

For our second approximation of the transistor, we include the base-spreading resistance r_b' shown in Fig. 9-25a. The r_b' resistance drops some signal voltage before it reaches the emitter depletion layer. Because of this, we get slightly less voltage gain than with the ideal model.

Ac emitter current

Since the CE amplifier, swamped amplifier, emitter follower, or any base-driven amplifier can be reduced to the base-driven prototype of Fig. 9-25b, all we need to do is to determine what change r_b' causes in the ac operation of Fig. 9-25b. Visualize the transistor replaced by the second approximation shown in Fig. 9-25c. Start with v_{bb} and sum voltages in a clockwise direction to get

$$-v_{bb} + \frac{i_e}{\beta}\,(r_B + r_b') + i_e(r_e' + r_E) \cong 0$$

After solving for i_e, we have

$$i_e \cong \frac{v_{bb}}{r_E + r_e' + (r_B + r_b')/\beta} \tag{9-18}$$

With this key formula, we can rederive any formula of interest. The rederivations are straightforward and will not be reproduced here. After you have rederived a few formulas using the second approximation, you will notice the following: in any ideal formula containing r_e', if you add r_b'/β to r_e', you will get the second-approximation formula. Symbolically,

Figure 9-25. Including the effects of r_b'.

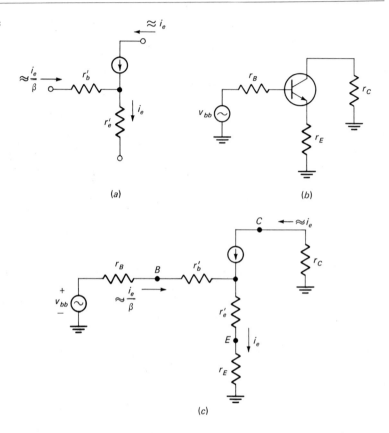

(a)

(b)

(c)

Ideal formula $\longrightarrow$ second-approximation formula

$$r_e' \longrightarrow r_e' + \frac{r_b'}{\beta}$$

In other words, given any ideal formula derived earlier, replace r_e' by $r_e' + r_b'/\beta$; the resulting formula is the second-approximation formula.

For instance, in a CE amplifier the ideal formulas for voltage gain and input impedance are

$$\frac{v_c}{v_b} \cong \frac{r_C}{r_e'} \qquad \text{(ideal)}$$

and
$$z_{\text{in(base)}} = \beta r_e' \qquad \text{(ideal)}$$

Replacing r_e' by $r_e' + r_b'/\beta$ gives the formulas for the second approximation:

$$\frac{v_c}{v_b} \cong \frac{r_C}{r_e' + r_b'/\beta} \qquad \text{(second)}$$

and

$$z_{in(base)} = \beta(r_e' + r_b'/\beta)$$
$$= \beta r_e' + r_b' \quad \text{(second)}$$

As another example, if you want the voltage gain in a swamped amplifier, you transform

$$\frac{v_c}{v_b} = \frac{r_C}{r_E + r_e'} \quad \text{(ideal)}$$

to

$$\frac{v_c}{v_b} = \frac{r_C}{r_E + r_e' + r_b'/\beta} \quad \text{(second)}$$

This is a sensible approach. Rather than clutter our minds with new formulas for the second approximation, we will continue to concentrate on ideal formulas. Whenever we want to include the effects of r_b', we can modify the ideal formulas by replacing r_e by $r_e' + r_b'/\beta$.

EXAMPLE 9-15.
A CE amplifier has $r_C = 5$ kΩ, $r_e' = 50$ Ω, $r_b' = 150$ Ω, and $\beta = 50$. Calculate the ideal voltage gain from base to collector, and also the second-approximation voltage gain.

SOLUTION.
The ideal gain is

$$\frac{v_c}{v_b} \cong \frac{r_C}{r_e'} = \frac{5000}{50} = 100$$

and the second-approximation gain is

$$\frac{v_c}{v_b} \cong \frac{r_C}{r_e' + r_b'/\beta} = \frac{5000}{50 + 150/50} = \frac{5000}{53} = 94.3$$

9-10. EMITTER-DRIVEN FORMULAS

The emitter-driven amplifier is much less important than the base-driven amplifier except in a few applications. For this reason, we devote less time to it. Later, in the special cases where it is important, we will discuss the emitter-driven amplifier again. For now, we will work out a few basic formulas for an emitter-driven prototype.

Ac emitter current

Figure 9-26a shows the emitter-driven prototype, and Fig. 9-26b is the same ac circuit with an ideal model for the transistor. Starting at r_B, we can sum voltages in a clockwise direction to get

$$\frac{i_e}{\beta} r_B + i_e(r_e' + r_E) - v_{ee} \cong 0$$

Figure 9-26. Emitter-driven amplifier analysis.

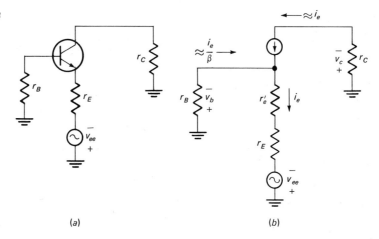

(a)   (b)

After solving for i_e, we have

$$i_e = \frac{v_{ee}}{r_E + r'_e + r_B/\beta}$$

(9-19)

This formula for the emitter-driven prototype is easy to remember; it differs only slightly from the formula for the base-driven prototype; instead of v_{bb} in the numerator, we have v_{ee}.

CB amplifier

The most important special case of an emitter-driven prototype (Fig. 9-26a) is for r_B = 0. When r_B is zero, the base is at ac ground. Such a circuit is called a *grounded-base* or *common-base* (CB) amplifier.

Usually, the CB amplifier is drawn as shown in Fig. 9-27a. In this way, the ac source voltage drives the input from the left side. Voltage v_e is the ac voltage reaching the emitter diode, and v_c is the ac output voltage across r_C. Figure 9-27a shows voltage polarities and current directions for the negative half cycle of ac source voltage. As a convenience, we can complement all voltages and currents to get Fig. 9-27b, which applies for the positive half cycle of ac source. Note that v_c has the same phase as v_e. In other words, the output voltage in a CB amplifier is *in phase* with the input voltage.

CB voltage gain

Figure 9-27c shows the emitter-driven prototype with the ideal model of the transistor. In this circuit, the ac collector voltage equals

$$v_c \cong i_e r_C$$

and the ac emitter voltage is

$$v_e = i_e r'_e$$

because all of v_e appears across r'_e when the base is grounded. The voltage gain from emitter to collector therefore equals

$$\frac{v_c}{v_e} \cong \frac{i_e r_C}{i_e r'_e}$$

or

$$\frac{v_c}{v_e} \cong \frac{r_C}{r'_e} \qquad\qquad (9\text{-}20a)$$

This is identical to the voltage gain from base to collector in a CE amplifier. In other words, both the CB and CE amplifiers have the same ideal voltage gain from the input terminal to the collector.

Also important in a CB amplifier is the voltage gain from source to collector. The ac source voltage appears across the series circuit of r_E and r'_e (Fig. 9-27c). Therefore, the ac source voltage equals

$$v_{ee} = i_e(r_E + r'_e)$$

The voltage gain from source to collector equals

$$\frac{v_c}{v_{ee}} \cong \frac{i_e r_C}{i_e(r_E + r'_e)}$$

or

$$\frac{v_c}{v_{ee}} \cong \frac{r_C}{r_E + r'_e} \qquad\qquad (9\text{-}20b)$$

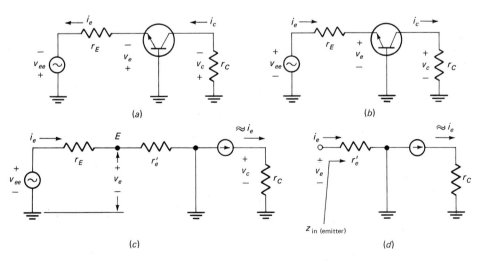

Figure 9-27. Common-base amplifier.

The easy way to understand and remember the two voltage gains is by remembering an idea described earlier. As you recall, when the same current flows through resistances, the voltage ratio equals the resistance ratio. In Fig. 9-27c, i_e flows through r_E and r_e'. An ac collector current of i_e (approximately) flows through r_C. Therefore, the voltage ratio v_c/v_e equals the resistance ratio r_C/r_e', and the voltage ratio v_c/v_{ee} equals the resistance ratio $r_C/(r_E + r_e')$.

Input impedance

One reason for the limited use of a CB amplifier is its *low* input impedance. Looking into the emitter of Fig. 9-27d, we have an input impedance of

$$z_{\text{in(emitter)}} \cong \frac{v_e}{i_e} = \frac{i_e r_e'}{i_e}$$

or
$$z_{\text{in(emitter)}} \cong r_e' \qquad\qquad (9\text{-}21)***$$

This means a CB amplifier has an input impedance of only 25 Ω when I_E is 1 mA.

On the other hand, the input impedance of a CE amplifier is

$$z_{\text{in(base)}} \cong \beta r_e'$$

Because of this, the CE amplifier draws less current from an ac source. Since the CE amplifier has ideally the same voltage gain as a CB amplifier, the CE amplifier is always preferred to the CB amplifier except for a few special cases to be discussed later.

EXAMPLE 9-16.
Calculate the voltage gains and input impedance of the CB amplifier of Fig. 9-28a.

SOLUTION.
Visualize the dc equivalent circuit and you will see the familiar voltage-divider bias circuit analyzed many times in earlier examples. At this point, you should have no trouble seeing that $I_E \cong 1$ mA and $r_e' = 25$ Ω.

Next, get the ac equivalent circuit by shorting all capacitors and connecting V_{CC} to ground. When you do this, you get Fig. 9-28b. Since the 20-kΩ and 10-kΩ base resistors have grounds on each end, they are irrelevant in the ac equivalent circuit.

By drawing the CB amplifier in the conventional way, the circuit looks like Fig. 9-28c. After Thevenizing the input and output circuits, we get the CB prototype of Fig. 9-28d. The rest is easy.

$$\frac{v_c}{v_e} \cong \frac{r_C}{r_e'} = \frac{5000}{25} = 200$$

$$\frac{v_c}{v_{ee}} \cong \frac{r_C}{r_E + r_e'} = \frac{5000}{100 + 25} = 40$$

and
$$z_{\text{in(emitter)}} \cong r_e' = 25 \ \Omega$$

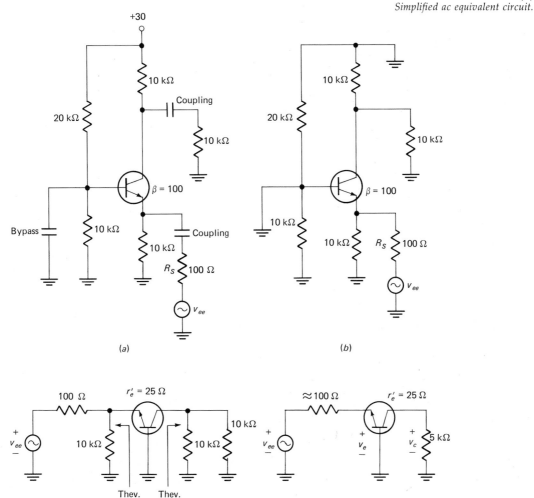

Figure 9-28. CB amplifier. (a) Complete circuit. (b) Ac equivalent circuit. (c) Conventional orientation. (d) Simplified ac equivalent circuit.

The gain from source to collector is lower because the 100-Ω source impedance drops most of the ac source voltage. If we use a CE amplifier with the same value of I_E, the input impedance of the base will be 2500 Ω for a β of 100. With a source impedance of 100 Ω, almost all the ac source voltage will appear across the input to the CE amplifier. Because of this, we would get the full voltage gain of 200 using the CE amplifier.

Problems

9-1. Calculate the ac emitter current in the emitter-driven prototype of Fig. 9-29a.

9-2. In Fig. 9-29a, work out these ac voltages with respect to ground: v_e, v_b, and v_c. Calculate the voltage ratios v_c/v_b and v_c/v_e.

Figure 9-29.

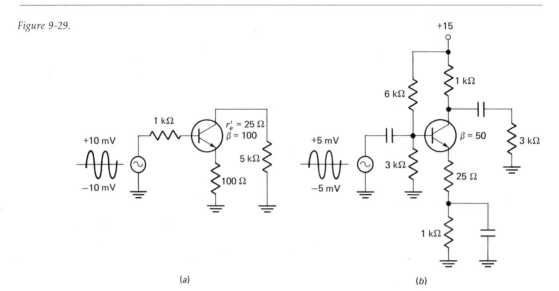

(a)

(b)

9-3. The amplifier of Fig. 9-29b has an r_e' of approximately 5 Ω. Calculate the ac emitter current. What is the value of v_e? And the voltage gain from base to collector?

9-4. Work out the voltage gain from base to collector for the CE amplifier of Fig. 9-30a.

9-5. In the *pnp* amplifier of Fig. 9-30b, how much ac collector voltage do you get if the ac source voltage is a 5-mV peak sine wave?

9-6. Calculate the voltage gain for the amplifier of Fig. 9-30c.

9-7. What is the voltage gain in Fig. 9-30d?

9-8. If you connect a dc-coupled oscilloscope from the collector to ground in Fig. 9-30a, what will the total waveform be if the ac source is a 1-mV-peak sine wave?

9-9. In Fig. 9-30b, the ac source puts out a sine wave with a peak of 5 mV. Describe the total voltage waveforms you would see with a dc-coupled oscilloscope at each of these points: the base, the emitter, and the collector (all with respect to ground).

9-10. What are the minimum dc voltage ratings for the capacitors in Fig. 9-30a?

9-11. Find the voltage gain of the swamped amplifier in Fig. 9-31a.

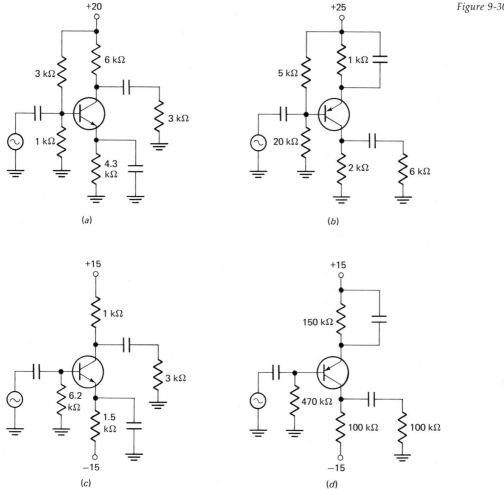

Figure 9-30.

9-12. Calculate the voltage gain from base to collector for the swamped amplifier shown in Fig. 9-31b. If the ac source signal is a sine wave with a 0.1-V peak, what does v_c equal? And v_e?

9-13. In Fig. 9-31c, what is the value of r_E? And the voltage gain from base to collector?

9-14. How much ac collector voltage v_c is there in Fig. 9-31d? And how much v_e?

9-15. Suppose the emitter bypass capacitor of Fig. 9-31a opens. What will the voltage gain equal?

9-16. If the collector coupling capacitor in Fig. 9-31b opens, what is the new value of

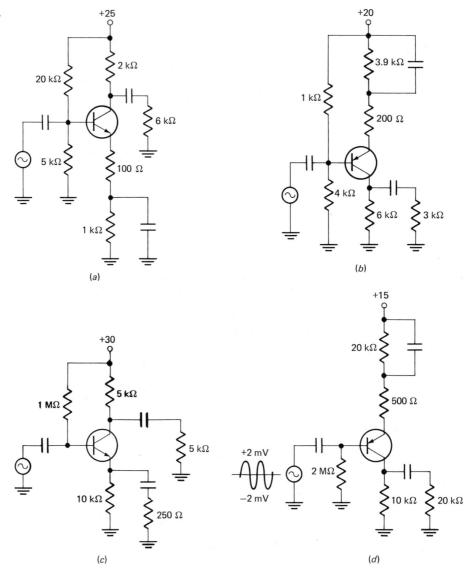

Figure 9-31.

(a)

(b)

(c)

(d)

voltage gain from base to collector? Suppose the capacitor shorts instead of opening; what effect will this have on the dc operation and the ac operation?

9-17.   If the base coupling capacitor of Fig. 9-31c shorts, what will this do to the dc operation?

9-18.   Ac current is difficult to measure. For this reason, rather than try to measure ac input current in Fig. 9-32, we can insert a *test resistor* as shown. By measuring

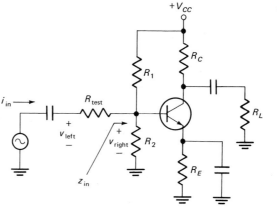

Figure 9-32.

the ac voltage on the left and right ends of this resistor, we can calculate the ac input current. Suppose the values are $v_{\text{left}} = 15$ mV, $v_{\text{right}} = 3$ mV, and $R_{\text{test}} = 1$ kΩ. Calculate the ac input current i_{in}; also, work out the input impedance z_{in}.

9-19. If the transistor of Fig. 9-30a has a β of 50, what is the value of $z_{\text{in(base)}}$? Of z_{in}?

9-20. The transistor in Fig. 9-30b has a β equal to 300. Calculate $z_{\text{in(base)}}$ and z_{in}.

9-21. Suppose the β is 100 for each transistor shown in Fig. 9-31. Assume r_e' is negligible in each circuit. What is the value of $z_{\text{in(base)}}$ for each amplifier?

9-22. The ac source voltage v_S of Fig. 9-33 equals 1 mV rms. If β equals 100, what does v_b equal? And v_c?

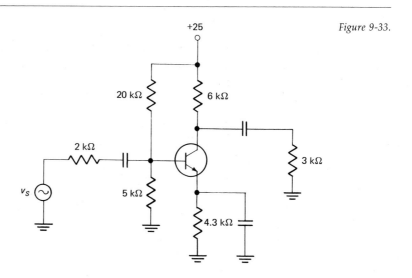

Figure 9-33.

9-23. Suppose the transistor of Fig. 9-33 has a β of 200. What is the voltage gain from source to base? From base to collector? From source to collector?

9-24. What is the value of β in Fig. 9-33 that produces an impedance match between z_{in} and R_S?

9-25. Suppose the R_S of Fig. 9-33 is changed from 2 kΩ to 150 kΩ; what is the voltage gain from source to base if β equals 100? From source to collector?

9-26. In Fig. 9-34a, what is the value of $z_{in(base)}$? And z_{in}? What is the voltage gain from base to emitter? And the power gain?

9-27. In Fig. 9-34a, calculate the voltage gain from source to base. And from source to emitter.

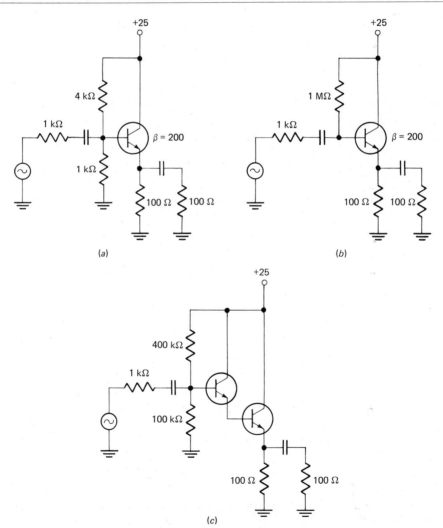

Figure 9-34.

(a)

(b)

(c)

9-28. Work out the z_{in} for the emitter follower of Fig. 9-34b. What is the voltage gain from base to emitter? From source to emitter?

9-29. Calculate z_{in} for the Darlington emitter follower of Fig. 9-34c. (Use betas of 100 for each transistor.) What is the power gain from input base to output emitter?

9-30. The Darlington amplifiers of Fig. 9-35 have betas of 100 and V_{BE}'s of 0.7 V for all transistors. Calculate the z_{in} of each amplifier. And the voltage gain from base to output.

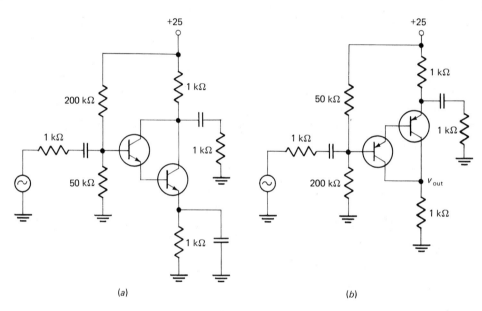

Figure 9-35.

(a)

(b)

9-31. For the CE amplifiers of Fig. 9-30a and b, $r'_b = 200\ \Omega$ and $\beta = 100$. Calculate the second-approximation voltage gain from base to collector.

9-32. In Fig. 9-30c and d, $r'_b = 150\ \Omega$ and $\beta = 50$. Work out the voltage gain from base to collector for each amplifier using the second approximation.

9-33. Suppose r'_b is $100\ \Omega$ and h_{fe} is 100 for the emitter follower of Fig. 9-34a. What value does $z_{in(base)}$ have?

9-34. Figure 9-36a shows a CB amplifier using two supplies to set up the FR bias. Calculate the dc emitter current, the voltage gain from emitter to collector, the input impedance looking into the emitter, and the voltage gain from source to collector.

9-35. The CB amplifier of Fig. 9-36b uses one positive supply to FR-bias the transistor. In the ac equivalent circuit, what are the following values: r_E, r_C, and z_{in} (emitter)? How much voltage gain is there from emitter to collector? From source to collector?

Figure 9-36.

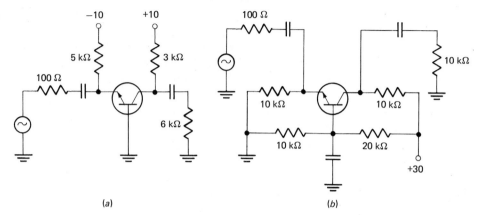

(a)   (b)

9-36. Some electrolytic capacitors may be too leaky (have too much leakage current). As mentioned in Chap. 1, we include a shunt resistance in the second approximation of a capacitor (Fig. 1-4b). In the voltage-divider-biased amplifier of Fig. 9-37,

1. What value does I_E have if the input coupling capacitor has a shunt resistance of 100 MΩ?
2. If the input capacitor has a shunt resistance of only 200 kΩ, what is the value of I_E?

Figure 9-37.

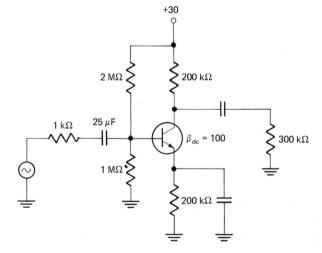

10. Class A Power Amplifiers

The early stages in many systems are small-signal amplifiers. After peak-to-peak voltage reaches its limit, however, we can no longer get an increase in signal voltage. But we can still get an increase in signal power. Near the end of a system, therefore, the stages are large-signal amplifiers where the emphasis is on power gain. Amplifiers of this type are called *power amplifiers*.

The emitter follower is an example of a power amplifier. With it, we get a voltage gain of approximately unity and a power gain of β. In general, a power amplifier may have a voltage gain greater than unity; however, it is called a power amplifier because it is optimized for power gain rather than voltage gain.

The transistors used in small-signal amplifiers are referred to as *small-signal transistors;* those used in power amplifiers are called *power transistors.* As a rule, a small-signal transistor has a power dissipation less than half a watt; a power transistor, more than half a watt.

10-1. THE Q POINT

Over and over we have stressed the following: for linear operation, the transistor must remain FR-biased through the ac cycle. This is why we bias a transistor to set up a dc collector current and voltage; if a small ac signal drives the transistor, the emitter diode remains forward-biased and the collector diode reverse-biased through the ac cycle.

In a power amplifier, the signal is no longer small. Instead, it produces large changes in current and voltage. Unless we are careful, the transistor will come out of FR bias at the positive or negative peak of the signal.

To check that the positive or negative peak of the signal does not force the transistor out of FR bias, we will use the *ac load line,* a graphical method of analysis. The next section discusses the ac load line in detail. In this section all we are interested in is one point on the ac load line: the *quiescent* (*Q*) point.

We get the values of quiescent or no-signal current and voltage from the dc equivalent circuit. For instance, in the voltage-divider-biased circuit of Fig. 10-1*a,* we can see right away that I_E is approximately 1 mA. Therefore, I_C is approximately 1 mA. To get the dc collector-to-emitter voltage, we use

$$V_{CE} = V_C - V_E$$

In Fig. 10-1*a,* the collector-to-ground voltage V_C equals

$$V_C = V_{CC} - I_C R_C$$
$$= 20 - 0.001(3000) = 17 \text{ V}$$

and the emitter-to-ground voltage is

$$V_E = I_E R_E$$
$$= 0.001(10{,}000) = 10 \text{ V}$$

Therefore,

$$V_{CE} = 17 - 10 = 7 \text{ V}$$

Now, we can plot the *Q* point as shown in Fig. 10-1*b.*

Whether you have a base-biased prototype (Fig. 10-1*c*) or an emitter-biased prototype (Fig. 10-1*d*), the procedure for locating the *Q* point is the same: you calculate I_C and V_{CE} using the dc equivalent circuit; then you plot the *Q* point with I_C on the vertical axis and V_{CE} on the horizontal axis as shown in Fig. 10-1*e.* To keep track of the *Q* point when things get complicated, we give it the subscripts shown. That is, I_{CQ} stands for the collector current when no ac signal is present. Similarly, V_{CEQ} represents the collector-to-emitter voltage with no ac signal.

EXAMPLE 10-1.
Plot the *Q* point for the CE amplifier of Fig. 10-2*a.*

SOLUTION.
The circuit is emitter-biased. To a good approximation, the emitter is at dc ground, that is, $V_E \cong 0$. Therefore,

$$I_{CQ} \cong I_E \cong \frac{V_{EE}}{R_E} = \frac{10}{2000} = 5 \text{ mA}$$

[When you need top accuracy, you can use Eq. (7-6) to calculate I_E.]
Next,

$$V_C = V_{CC} - I_C R_C = 15 - 0.005(1000) = 10 \text{ V}$$

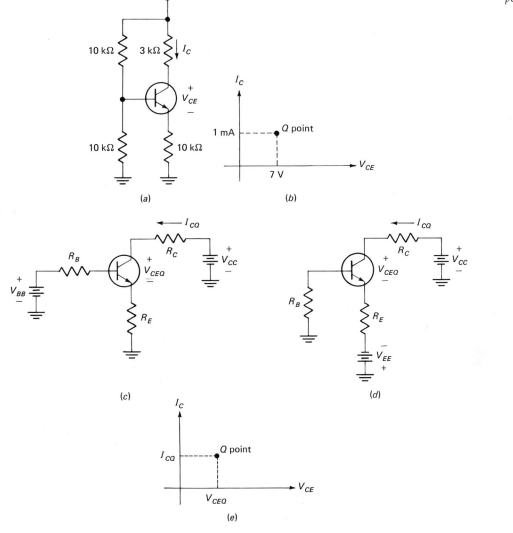

Figure 10-1. Concept of the Q point.

Since $V_E \cong 0$,

$$V_{CEQ} = V_C - V_E \cong 10 \text{ V}$$

Figure 10-2b shows the Q point. When the ac signal comes in, it will force the collector current and voltage to change.

Figure 10-2. Examples 10-1 and 10-2.

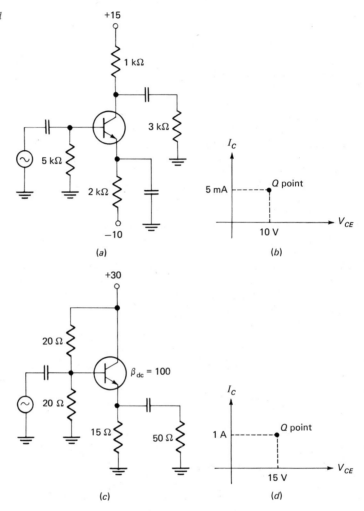

(a)

(b)

(c)

(d)

EXAMPLE 10-2.
Locate the Q point for the emitter follower of Fig. 10-2c.

SOLUTION.
The dc Thevenin base voltage is 15 V. Almost all of this 15 V appears across the 15-Ω emitter resistor. So,

$$I_{CQ} \cong I_E \cong \frac{15}{15} = 1 \text{ A}$$

With zero collector resistance, $V_C = 30$ V. The dc emitter voltage is

$$V_E \cong V_B = V_{TH} = 15 \text{ V}$$

Therefore,

$$V_{CEQ} = V_C - V_E = 30 - 15 = 15 \text{ V}$$

Figure 10-2d shows the Q point.

10-2. THE AC LOAD LINE

When an ac signal drives an amplifier, it causes changes in collector current I_C and voltage V_{CE}. We designate these changes by

$$\Delta I_C = \text{change in collector current}$$
$$= I_C - I_{CQ} \tag{10-1}$$

and

$$\Delta V_{CE} = \text{change in collector-emitter voltage}$$
$$= V_{CE} - V_{CEQ} \tag{10-2}$$

In these equations, I_C and V_{CE} represent *total quantities,* that is, the sum of dc and ac components at any instant in time. For instance, suppose an ac signal forces the I_C and V_{CE} of Fig. 10-2d to change to $I_C = 1.5$ A and $V_{CE} = 7.5$ V. With Eqs. (10-1) and (10-2), we calculate changes of

$$\Delta I_C = I_C - I_{CQ} = 1.5 - 1 = 0.5 \text{ A}$$

and

$$\Delta V_{CE} = V_{CE} - V_{CEQ} = 7.5 - 15 = -7.5 \text{ V}$$

Ac signal identical to changes

Since the ac signal causes the changes in current and voltage, we can use the ac equivalent circuit to calculate these changes. For instance, Fig. 10-3a shows a CE amplifier. The dc equivalent circuit for this amplifier is the same as Fig. 10-1a; therefore, the amplifier has the Q point shown in Fig. 10-3d: $I_{CQ} = 1$ mA and $V_{CEQ} = 7$ V. When the ac signal comes in, it causes *excursions* or changes from the Q point. To find these changes, we visualize the ac equivalent circuit of the CE amplifier; after Thevenizing the base and collector circuits, we get Fig. 10-3b.

As emphasized in Chap. 8, *ac current and voltage are identical to changes in total current and voltage.* In symbols,

$$i_c = \Delta I_C = \text{change in current with respect to } Q$$

and

$$v_{ce} = \Delta V_{CE} = \text{change in voltage with respect to } Q$$

As an example, suppose the ac signal in Fig. 10-3a forces the collector current to change from a quiescent value of 1 mA to a maximum value of 1.5 mA. Then, the peak ac current (maximum current change from Q point) is

$$i_c = \Delta I_C = I_C - I_{CQ} = 1.5 \text{ mA} - 1 \text{ mA} = 0.5 \text{ mA}$$

Peak ac voltage

Once you know the peak ac current, you can find the peak ac voltage. For instance, if the peak ac current is 0.5 mA, you visualize the ac equivalent circuit of Fig. 10-3b as shown in Fig. 10-3c. The 0.5 mA flows up through the 2-kΩ resistor. Therefore, the peak ac voltage equals

$$v_{ce} = \Delta V_{CE} = -\Delta I_C r_C = -0.5(10^{-3})2000$$
$$= -1 \text{ V}$$

The minus sign is consistent with what we said about phase inversion. During the positive half cycle of ac source voltage, collector current increases, causing a decrease in collector-emitter voltage.

Figure 10-3. Peak changes in current and voltage.

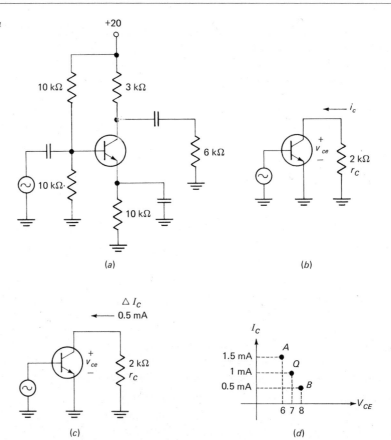

The positive peak point

We have found a change of +0.5 mA in collector current produces a change of −1 V in collector-emitter voltage. To find the total current and voltage, we can rearrange Eqs. (10-1) and (10-2) to get

$$I_C = I_{CQ} + \Delta I_C \qquad\qquad (10\text{-}3)$$
$$= 1 \text{ mA} + 0.5 \text{ mA} = 1.5 \text{ mA}$$

and
$$V_{CE} = V_{CEQ} + \Delta V_{CE} \qquad\qquad (10\text{-}4)$$
$$= 7 + (-1) = 6 \text{ V}$$

These values of I_C and V_{CE} are the coordinates of point A in Fig. 10-3d.

Let's review what we did. In Fig. 10-3a, the quiescent current and voltage equal 1 mA and 7 V. The ac signal causes an increase of 0.5 mA in collector current. This 0.5 mA flows through an r_C of 2 kΩ. Because of this, the change in collector-emitter voltage is −1 V. By plotting the total collector current (1.5 mA) and the total collector voltage (6 V), we get point A of Fig. 10-3d.

Point A occurs at the positive peak of the ac signal. If we increase the ac signal, the current and voltage changes will be greater and point A will be higher.

The negative peak point

The argument for finding the negative peak point is similar to that used for the positive peak point. All we do is complement the changes in collector current and voltage. In Fig. 10-3c this means the change in current flows down through the 2-kΩ resistor, producing a positive change of 1 V. Therefore, at the negative peak of the input signal, the total collector current and the total collector-emitter voltage are 0.5 mA and 8 V, point B of Fig. 10-3d.

Other points lie on a straight line

In Fig. 10-3b, the ac voltage equals

$$v_{ce} = -i_c r_C$$

or since ac quantities are identical to changes,

$$\Delta V_{CE} = -\Delta I_C r_C$$

This tells us the change in collector-emitter voltage is *directly proportional* to the change in collector current. Because of this linear relation, a change in collector current less than 0.5 mA produces a change in collector voltage proportionately less than −1 V. In other words, at any instant during the ac cycle, the coordinates of total collector current and voltage *lie along a straight line* as shown in Fig. 10-4. We call this the *ac load line.*

The ac load line is a visual aid for understanding large-signal operation. During

Figure 10-4. Part of ac load line.

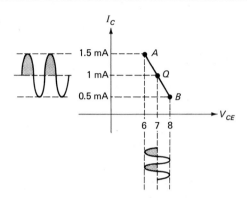

the positive half cycle of ac source voltage, the collector current swings from 1 mA (point Q) to 1.5 mA (point A); during this half cycle, the collector-emitter voltage swings from 7 V to 6 V. On the negative half cycle of ac source voltage, the collector current swings from 1 mA to 0.5 mA, while the collector-emitter voltage swings from 7 to 8 V.

At any instant during the ac cycle, the coordinates of collector current and voltage are on the ac load line. Because of this, we can visualize the *instantaneous operating point* starting at point Q and moving along the ac load line toward point A; after peaking at A, the instantaneous operating point swings from A through Q to B. Then it returns to Q, where the next cycle begins.

A final idea. The ac load line of Fig. 10-4 belongs to the CE amplifier of Fig. 10-3*a*. If we were to increase the ac source signal, the changes in collector current and voltage would increase. In Fig. 10-4 this means the instantaneous operating point would move further away from Q at the positive and negative peaks. How far can it move? The next section answers this question.

10-3. LOAD LINE OF BASE-DRIVEN PROTOTYPE

The CE amplifier, the swamped amplifier, the emitter follower, and many other amplifiers are special cases of the base-driven prototype. This section discusses the load line for the base-driven prototype. Once we have this load line, we will have the key to large-signal operation of many practical amplifiers.

Relation between ac collector current and voltage

Figure 10-5*a* shows the base-driven prototype. Since the ac collector current flows through r_C and the ac emitter current through r_E, the ac collector voltage has to equal a value that satisfies Kirchhoff's voltage law. In other words, the sum of ac voltages around the collector and emitter circuits must be zero. Starting with v_{ce} and summing in a counterclockwise direction, we get

$$v_{ce} + i_e r_E + i_c r_C = 0$$

As we have done so often, we will use $i_c \cong i_e$ to simplify the equation. Substituting i_c for i_e gives

$$v_{ce} + i_c r_E + i_c r_C \cong 0$$

or by rearranging,

$$v_{ce} \cong -i_c(r_C + r_E) \tag{10-5a}$$

This is an important relation. It tells how the ac collector voltage is related to the ac collector current in any base-driven amplifier. Since ac quantities are the same as changes in total quantities, we can rewrite the equation as

$$\Delta V_{CE} \cong -\Delta I_C(r_C + r_E) \tag{10-5b} ***$$

This tells us the change in collector voltage equals minus the change in collector current times the ac resistance of the collector and emitter circuits.

By rearranging Eq. (10-5b), we can get another useful relation:

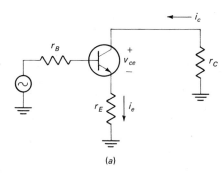

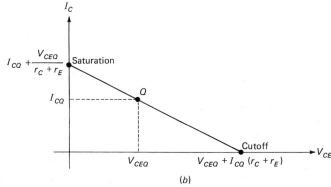

Figure 10-5. (a) Base-driven prototype. (b) Ac load line.

$$\Delta I_C = - \frac{\Delta V_{CE}}{r_C + r_E} \qquad\qquad (10\text{-}5c)$$

This says the change in collector current equals minus the change in collector voltage divided by the ac resistance in the collector and emitter circuits.

Load line must lie in first quadrant

Figure 10-5*b* shows the ac load line for any base-driven amplifier. The coordinates of the *Q* point are I_{CQ} and V_{CEQ}. You calculate these values from the dc equivalent circuit using the methods of Chap. 7.

When an ac signal drives the amplifier, the instantaneous operating point moves away from point *Q*. How far can it move? As shown in Fig. 10-5*b*, the instantaneous operating point can move no higher than a point called *saturation* and no lower than a point known as *cutoff*. In other words, all instantaneous operating points must lie in the first quadrant. Symbolically,

$$I_C > 0 \qquad \text{and} \qquad V_{CE} > 0$$

Why is this true? As we have said many times, linear operation of a transistor requires FR bias throughout the ac cycle. This means the emitter diode must stay forward-biased and the collector diode reverse-biased at all times. When the instantaneous operating point moves toward saturation, V_{CE} is approaching zero. As you recall from Chap. 6, V_{CE} can drop no lower than $V_{CE(\text{sat})}$ because this is the voltage at which the collector diode comes out of reverse bias. $V_{CE(\text{sat})}$ is a few-tenths volt in most cases; therefore, to a first approximation, V_{CE} can decrease almost to zero. This explains why the upper end of the load line stops at the saturation point (Fig. 10-5*b*).

On the other hand, when the instantaneous operating point moves toward cutoff, I_C is approaching zero. This means the forward bias on the emitter diode is decreasing. If I_C reaches zero, the emitter diode has come out of forward bias. When this happens, we lose normal transistor action. This explains why the instantaneous operating point can move no lower than the cutoff point shown in Fig. 10-5*b*.

The saturation current

During the positive half cycle of ac source voltage, the instantaneous operating point moves toward saturation. The maximum current the transistor must handle occurs at saturation. For this reason, we need a formula for the current at the saturation point.

At the saturation point, V_{CE} equals $V_{CE(\text{sat})}$, a few-tenths volt in most cases. To a first approximation, V_{CE} is zero at saturation. In Fig. 10-5*b*, when the instantaneous operating point moves from the *Q* point to the saturation point, the change in collector-emitter voltage is

$$\Delta V_{CE} = V_{CE(\text{sat})} - V_{CEQ}$$
$$\cong -V_{CEQ}$$

This makes sense. In Fig. 10-5b, if $V_{CEQ} = 10$ V, the voltage will change by -10 V when the instantaneous operating point swings from the Q point to the saturation point.

When we substitute $\Delta V_{CE} = -V_{CEQ}$ into Eq. (10-5c), we get the corresponding change in collector current as follows:

$$\Delta I_C \cong -\frac{\Delta V_{CE}}{r_C + r_E} \cong -\frac{-V_{CEQ}}{r_C + r_E}$$

or
$$\Delta I_C \cong \frac{V_{CEQ}}{r_C + r_E} \qquad\qquad (10\text{-}6a)\,{}^{***}$$

So, *to find the maximum possible increase in collector current, all we do is divide the quiescent voltage by the ac collector and emitter resistance.* For instance, if $V_{CEQ} = 10$ V, $r_C = 9$ kΩ, and $r_E = 1$ kΩ, the change in collector current between the Q point and the saturation point is

$$\Delta I_C = \frac{10}{9000 + 1000} = 1 \text{ mA}$$

Look at Fig. 10-5b again. The total collector current at saturation must equal the quiescent current I_{CQ} plus the change in collector current between Q and saturation. In symbols,

$$I_{C(\text{sat})} = I_{CQ} + \Delta I_C$$

With Eq. (10-6a), we can substitute for ΔI_C to get

$$I_{C(\text{sat})} = I_{CQ} + \frac{V_{CEQ}}{r_C + r_E} \qquad\qquad (10\text{-}6b)\,{}^{***}$$

With this formula, you can calculate the maximum possible value of collector current. This maximum current $I_{C(\text{sat})}$ occurs at the positive peak of ac source voltage.

Cutoff voltage

During the negative half cycle of ac source voltage, the instantaneous operating point moves toward cutoff. The maximum voltage the transistor must handle occurs at cutoff. This voltage must be less than the $V_{CE(\text{max})}$ rating given on the transistor data sheet. In other words, $V_{CE(\text{cutoff})}$ must be less than the breakdown voltage BV_{CEO}. For this reason, we need a formula for $V_{CE(\text{cutoff})}$.

At the cutoff point in Fig. 10-5b, the emitter diode is just coming out of forward bias, so that $I_E = 0$. The only current flowing in the collector diode is I_{CBO}. Since I_{CBO} is small compared to normal values of quiescent current, we can neglect I_{CBO} in calculating V_{CE} at cutoff. In other words, we can visualize the cutoff point in Fig. 10-5b on the V_{CE} axis.

When the instantaneous operating point moves from the Q point to the cutoff point, the change in I_C is

$$\Delta I_C = I_C - I_{CQ} \cong 0 - I_{CQ}$$
$$= -I_{CQ}$$

This makes sense when we look at Fig. 10-5b. If $I_{CQ} = 1$ mA, the change in collector current between Q and cutoff must be -1 mA.

When we substitute $\Delta I_C = -I_{CQ}$ into Eq. (10-5b), we get

$$\Delta V_{CE} = I_{CQ}(r_C + r_E) \qquad (10\text{-}7a)^{***}$$

With this formula, *we can find the maximum possible increase in collector-emitter voltage.* For example, if $I_{CQ} = 2$ A, $r_C = 4$ Ω, and $r_E = 1$ Ω, the change in collector-emitter voltage between the Q point and the cutoff point is

$$\Delta V_{CE} = 2(4 + 1) = 10 \text{ V}$$

In Fig. 10-5b, the total collector-emitter voltage at cutoff equals the quiescent voltage plus the change in voltage between Q and cutoff. In symbols,

$$V_{CE(\text{cutoff})} = V_{CEQ} + \Delta V_{CE}$$

With Eq. (10-7a), we can substitute for ΔV_{CE} to get

$$V_{CE(\text{cutoff})} = V_{CEQ} + I_{CQ}(r_C + r_E) \qquad (10\text{-}7b)^{***}$$

With this formula, we can calculate the largest possible collector-emitter voltage. The value we get must be less than the BV_{CEO} of the transistor; otherwise, the transistor breaks down.

Summary

Figure 10-5b summarizes the important ideas for the ac load line of a base-driven amplifier. By referring to Fig. 10-5b, we can calculate the saturation current and cutoff voltage.

Figure 10-5b has great practical value. With it, we can analyze the large-signal operation of all kinds of base-driven amplifiers. The rest of this chapter tells you how.

EXAMPLE 10-3.

The BC107 of Fig. 10-6a has the following maximum ratings: $I_{C(\text{max})} = 100$ mA and $BV_{CEO} = 45$ V. Show that neither of these is exceeded during the ac cycle.

SOLUTION.

Visualize the dc equivalent circuit and you can calculate $I_{CQ} = 10$ mA and $V_{CEQ} = 15$ V. Figure 10-6b shows this Q point.

Next, visualize the ac equivalent circuit. After Thevenizing the base and collector

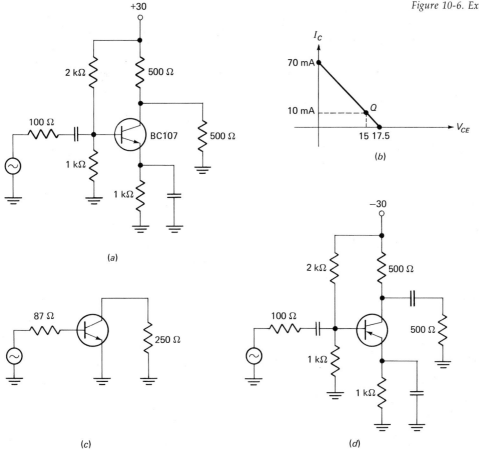

Figure 10-6. Example 10-3.

circuits, we get Fig. 10-6c. As shown, $r_E = 0$ and $r_C = 250$ Ω. Referring to the general ac load line of Fig. 10-5b, we can calculate the end points on the load line as follows:

$$I_{C(\text{sat})} = I_{CQ} + \frac{V_{CEQ}}{r_C + r_E}$$

$$= 10(10^{-3}) + \frac{15}{250} = 70 \text{ mA}$$

and

$$V_{CE(\text{cutoff})} = V_{CEQ} + I_{CQ}(r_C + r_E)$$
$$= 15 + 10^{-2}(250) = 17.5 \text{ V}$$

Figure 10-6b shows the saturation current and the cutoff voltage. Since the BC107 has an $I_{C(\text{max})}$ rating of 100 mA, there is no danger of exceeding this rating because at most we can have 70 mA of collector current. Also, the largest collector voltage occurs at cutoff and equals 17.5 V, which is much less than the BV_{CEO} of 45 V.

If we use magnitudes of current and voltage, the complementary *pnp* amplifier of Fig. 10-6d also has the ac load line shown in Fig. 10-6b.

EXAMPLE 10-4.
Show the ac load line for the emitter follower of Fig. 10-7a.

SOLUTION.
First, get I_{CQ} and V_{CEQ}. With about 10 V dc across the 50-Ω emitter resistor, we get

$$I_{CQ} = \frac{10}{50} = 0.2 \text{ A}$$

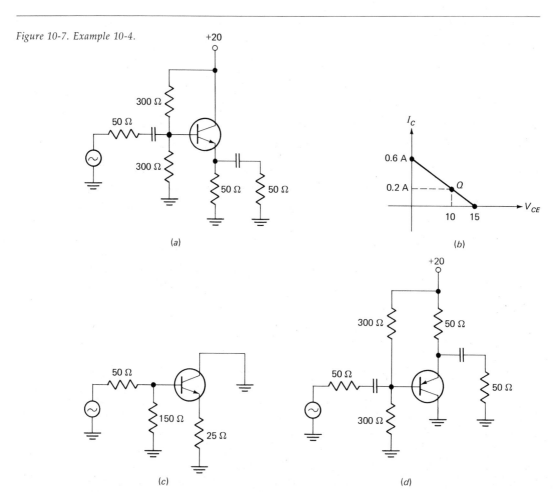

Figure 10-7. Example 10-4.

and $$V_{CEQ} = V_C - V_E = 20 - 10 = 10 \text{ V}$$

Figure 10-7b shows this Q point.

Next, visualize the ac equivalent circuit as shown in Fig. 10-7c. Since $r_C = 0$ and r_E $= 25 \ \Omega$, we calculate the ends of the load line as follows:

$$I_{C(\text{sat})} = I_{CQ} + \frac{V_{CEQ}}{r_C + r_E}$$

$$= 0.2 + \frac{10}{25} = 0.6 \text{ A}$$

and $$V_{CE(\text{cutoff})} = V_{CEQ} + I_{CQ}(r_C + r_E)$$
$$= 10 + 0.2(25) = 15 \text{ V}$$

Figure 10-7b shows the ac load line with these end points.

The upside-down *pnp* emitter follower of Fig. 10-7d also has the ac load line shown in Fig. 10-7b.

10-4. OPTIMUM Q POINT FOR CLASS A

If you *overdrive* an amplifier, the output signal will be clipped on either or both peaks. Figure 10-8a shows clipping on the negative half cycle of source voltage. Since the Q point is closer to cutoff than to saturation, the instantaneous operating point hits the cutoff point before the saturation point. Because of this, we get *cutoff clipping* as shown.

If the Q point is too high, that is, closer to saturation than to cutoff, we get *saturation clipping* as shown in Fig. 10-8b. On the positive half cycle of ac source voltage, the instantaneous operating point drives into the saturation point and stays there temporarily; this results in positive clipping of the collector current.

To get the maximum unclipped signal, we can locate the Q point in the center of the ac load line (Fig. 10-8c). In this way, the instantaneous operating point can swing equally in both directions before clipping occurs. With the right size of input signal, we can get the maximum possible unclipped output signal.

Definition of class A

All along, we have said a transistor should stay FR-biased through the ac cycle. To distinguish this operation from other kinds, we call it *class A operation*. In other words, in a class A amplifier all transistors remain FR-biased during the cycle. In terms of the load line, class A operation means that no clipping occurs at either end of the load line. If we did get clipping, the operation would no longer be called class A operation.

In a class A amplifier the best place to locate the Q point is in the center of the ac load line as shown in Fig. 10-8c. The reason is clear enough. When the Q point is in the middle of the ac load line, we can get the largest possible unclipped output signal.

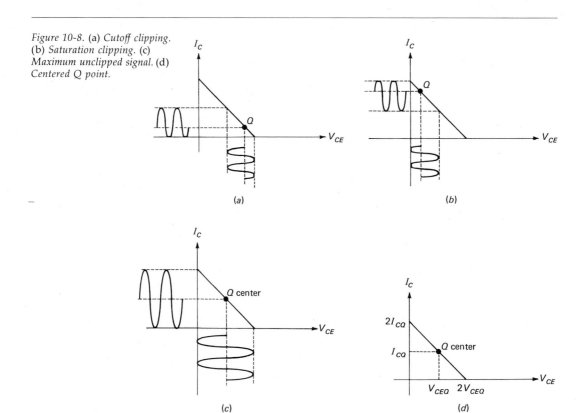

Figure 10-8. (a) *Cutoff clipping.* (b) *Saturation clipping.* (c) *Maximum unclipped signal.* (d) *Centered Q point.*

(a)

(b)

(c)

(d)

Centered Q point

Figure 10-8d shows a centered Q point. I_{CQ} and V_{CEQ} are the quiescent current and voltage. Simple geometry tells us the saturation current must be twice as large as I_{CQ} and the cutoff voltage twice as large as V_{CEQ}; otherwise, the Q point would not be centered. In symbols,

$$I_{C(\text{sat})} = 2I_{CQ} \qquad (\text{centered } Q) \qquad\qquad (10\text{-}8a)\,^{***}$$

and

$$V_{CE(\text{cutoff})} = 2V_{CEQ} \qquad (\text{centered } Q) \qquad\qquad (10\text{-}8b)\,^{***}$$

Since the cutoff voltage of any base-driven amplifier is given by Fig. 10-5b, we can find an important relation between quiescent values and ac resistances as follows. In Fig. 10-5b, the cutoff voltage is

$$V_{CE(\text{cutoff})} = V_{CEQ} + I_{CQ}(r_C + r_E)$$

Because of this, we can substitute the right member of Eq. (10-8b) to get

$$2V_{CEQ} = V_{CEQ} + I_{CQ}(r_C + r_E)$$

By rearranging this, we get

$$r_C + r_E = \frac{V_{CEQ}}{I_{CQ}} \qquad \text{(centered } Q\text{)} \qquad \text{(10-9)}^{***}$$

This final result is one of the most important in this chapter. Here is what it tells you. *To have a centered Q point, the ac resistance of the collector and emitter circuits must equal the ratio of the quiescent collector voltage to the quiescent collector current.* You will find this equation useful in analyzing and designing power amplifiers.

EXAMPLE 10-5.
Earlier, we analyzed Fig. 10-2a and found its Q point. Check if the Q point is in the center of the ac load line.

SOLUTION.
The emitter-biased amplifier of Fig. 10-2a is a CE amplifier because the emitter is bypassed to ground. Therefore, $r_E = 0$. A glance at the collector circuit indicates

$$r_C = 1000 || 3000 = 750 \ \Omega$$

And, Fig. 10-2b shows $I_{CQ} = 5$ mA and $V_{CEQ} = 10$ V. Now, we can check if Eq. (10-9) is satisfied.

$$\frac{V_{CEQ}}{I_{CQ}} = \frac{10 \text{ V}}{5 \text{ mA}} = 2 \text{ k}\Omega$$

which does not equal

$$r_C + r_E = 750 \ \Omega$$

Therefore, the Q point cannot be in the center of the ac load line. As a result, the amplifier of Fig. 10-2a delivers a smaller unclipped signal than it could with a centered Q point.

EXAMPLE 10-6.
Is the Q point centered in Fig. 10-2c?

SOLUTION.
First, note that

$$r_C = 0$$

and

$$r_E = 15 || 50 = 11.5 \ \Omega$$

So, $r_C + r_E = 11.5 \ \Omega$ in Fig. 10-2c.

Second, check the ratio of V_{CEQ} to I_{CQ}. In Fig. 10-2d, the quiescent values are 1 A and 15 V, which gives a ratio of

$$\frac{V_{CEQ}}{I_{CQ}} = \frac{15 \text{ V}}{1 \text{ A}} = 15 \text{ } \Omega$$

which does not equal $r_C + r_E$. Again, the Q point is not in the center of the ac load line.

EXAMPLE 10-7.
Figure 10-9a is a *phase splitter* (sometimes called a phase inverter or a paraphase ampli-fier). What does it do, and is its Q point centered?

SOLUTION.
First, visualize the ac equivalent circuit. After reducing, we have Fig. 10-9b. Since i_c closely approximates i_e, essentially the same peak voltages appear across r_C and r_E. As you recall, the collector signal is 180° out of phase with the base signal, and the emitter signal is in phase with the base signal. Therefore, the output of a phase splitter is a pair of sine waves of equal amplitude but opposite phase.

Next, check if the Q point is centered. The sum of collector and emitter ac resis-tance is

$$r_C + r_E = 50 + 50 = 100 \text{ } \Omega$$

When we visualize the dc equivalent circuit of Fig. 10-9a, we can see a dc Thevenin base voltage of 5 V. Ideally, all of this appears across the 100-Ω emitter resistor. Therefore,

$$I_{CQ} \cong I_E \cong \frac{V_E}{R_E} \cong \frac{5}{100} = 50 \text{ mA}$$

and
$$V_{CEQ} = V_C - V_E = 10 - 5 = 5 \text{ V}$$

Figure 10-9. Phase splitter.

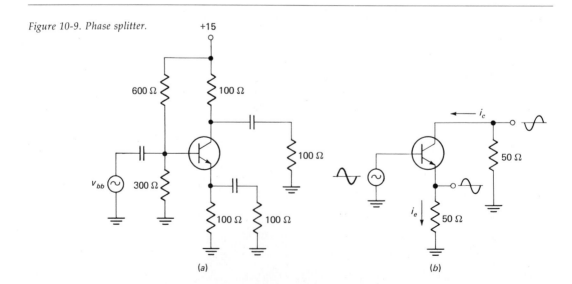

(a)   (b)

The ratio is

$$\frac{V_{CEQ}}{I_{CQ}} = \frac{5}{0.05} = 100 \ \Omega$$

which does equal $r_C + r_E$. So, the phase splitter of Fig. 10-9a has a centered Q point, at least ideally. (You can improve the calculations by taking the 0.7 V into account.)

10-5.   MAXIMUM LOAD POWER FOR CLASS A

What is the maximum power you can get out of a class A amplifier? How much power must a transistor dissipate?

Load power

Figure 10-10a shows a base-driven prototype, and Fig. 10-10b is the ac load line with a centered Q point. When the ac source signal is large enough to produce the maximum unclipped output signal, we get the current and voltage waveforms of Fig. 10-10b.

As you can see, the ac collector current is a sine wave with a peak-to-peak value of $2I_{CQ}$. Therefore, it has a peak value of I_{CQ} or an rms value of $0.707I_{CQ}$. Since this ac current flows through r_C and r_E, the average power delivered to the ac load resistance is

$$P_{\text{load}} = (0.707I_{CQ})^2(r_C + r_E)$$

or
$$P_{\text{load}} = 0.5I_{CQ}^2(r_C + r_E) \tag{10-10a}$$

Similarly, the ac collector-voltage waveform of Fig. 10-10b is a sine wave with a peak-to-peak value of $2V_{CEQ}$, which implies an rms value of $0.707V_{CEQ}$. Since this sine wave appears across a total ac load resistance $r_C + r_E$, an alternative formula for ac load power is

$$P_{\text{load}} = \frac{(0.707V_{CEQ})^2}{r_C + r_E}$$

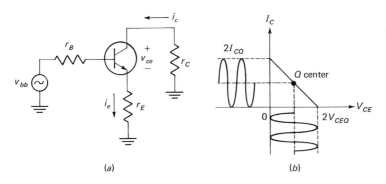

(a)

(b)

Figure 10-10. (a) Base-driven prototype. (b) Centered Q point.

or
$$P_{\text{load}} = 0.5 \, \frac{V_{CEQ}^2}{r_C + r_E} \qquad\qquad (10\text{-}10b)$$

Either Eq. (10-10a) or (10-10b) is useful for calculating the largest possible output power from a class A amplifier. As an example, the phase splitter of Fig. 10-9a has a V_{CEQ} of approximately 5 V. Therefore, the maximum ac load power delivered to r_C and r_E is

$$P_{\text{load}} = 0.5 \, \frac{5^2}{50 + 50} = 0.125 \text{ W}$$

We would get the same result if we used the current formula (10-10a). Use whichever is more convenient in your application.

Power dissipation

In our discussion of the physics behind transistor action, we described how conduction-band electrons fall down the collector energy hill. As they fall, they give up energy, mostly in the form of heat. The transistor must dissipate this heat to its surroundings. During the ac cycle, the collector current and voltage change, but on the average, a fixed amount of heat is produced. In other words, the transistor has to dissipate an *average power* which we designate P_D. Data sheets always include the maximum power dissipation of the transistor type. You cannot exceed this $P_{D(\text{max})}$ without risking damage to the transistor.

The *quiescent power dissipation* of a transistor is

$$P_{DQ} = V_{CEQ} I_{CQ} \qquad\qquad (10\text{-}11)\,***$$

This tells us that when no ac signal is present, the transistor must dissipate power equal to the product of quiescent voltage and current. For example, in Fig. 10-9a, we found $I_{CQ} = 50$ mA and $V_{CEQ} = 5$ V. Therefore,

$$P_{DQ} = 5(0.05) = 0.25 \text{ W}$$

To avoid damage, the transistor used in Fig. 10-9a would need a $P_{D(\text{max})}$ rating greater than 0.25 W.[1]

With an ac signal and a centered Q point, the transistor dissipates less power than it does under quiescent conditions. One way to prove this is with calculus. Another way is as follows. Calculate the power for each point along the ac load line; you will find maximum power occurs at the centered Q point; therefore, with an ac signal, the average power must be less than the quiescent power.

For instance, at the Q point,

$$p = V_{CEQ} I_{CQ}$$

When the operating point is midway between Q and saturation,

---

[1] Sometimes, a *heat sink* (mass of metal) is attached to the transistor case to help dissipate heat. This increases the $P_{D(\text{max})}$ rating.

$$p = 0.5V_{CEQ}(1.5I_{CQ})$$
$$= 0.75V_{CEQ}I_{CQ}$$

At saturation, the ideal power is

$$p = 0(2I_{CQ}) = 0$$

In this way, you can prove that any point corresponds to a power less than or equal to the quiescent power. For this reason, the average power must be less than the quiescent power when an ac signal drives the transistor. This also means quiescent power is the worst-case condition, that is, the transistor is safe from damage if it can dissipate P_{DQ}, the quiescent power dissipation for a centered Q point.

Relation between ac load power and quiescent power

In Fig. 10-10b, the ac collector current is a sine wave with an rms value of $0.707I_{CQ}$, and the ac collector voltage is a sine wave with an rms value of $0.707V_{CEQ}$. The product of rms collector current and rms collector voltage gives us the total ac load power:

$$P_{load} = (0.707V_{CEQ})(0.707I_{CQ}) = 0.5V_{CEQ}I_{CQ}$$

If we take the ratio of this load power to quiescent power, we get

$$\frac{P_{load}}{P_{DQ}} = \frac{0.5V_{CEQ}I_{CQ}}{V_{CEQ}I_{CQ}}$$

or
$$\frac{P_{load}}{P_{DQ}} = 0.5 \qquad \text{(centered } Q\text{)} \qquad\qquad (10\text{-}12)\,{***}$$

This says the maximum ac load power in a class A amplifier equals half the quiescent power. This is the best you can do with a class A amplifier, and you get this only when the Q point is centered. If a transistor dissipates 3 W of power under no-signal conditions, the maximum ac load power is 1.5 W when an ac signal drives the transistor. Conversely, if you are trying to build a class A amplifier that will deliver 30 W of ac load power, you will need a transistor that can dissipate 60 W of power under no-signal conditions.

EXAMPLE 10-8.
The data sheet of a 2N4143 says $P_{D(max)}$ is 310 mW for an *ambient* (surrounding) temperature of 25°C. For temperatures greater than 25°C, you are to *derate* $P_{D(max)}$ by 2.81 mW for each degree above 25°C. If the transistor is used in a class A amplifier operating up to 75°C, what is the maximum ac load power it can deliver?

SOLUTION.
When the ambient temperature increases, it forces the internal transistor temperature to increase; this reduces the amount of heat the transistor can safely dissipate. For this reason, data sheets normally include a derating factor similar to the one given.

When the ambient temperature rises from 25°C to 75°C, the change is 50°C. Therefore, we subtract

$$(50)2.81 \text{ mW} = 140 \text{ mW}$$

from the 25°C rating of 310 mW to get

$$P_{D(\text{max})} = 310 \text{ mW} - 140 \text{ mW}$$
$$= 170 \text{ mW at } 75°C$$

At best, therefore, we can bias the transistor so that it dissipates 170 mW under no-signal conditions.

When the ac signal drives the transistor, we can get a maximum ac load power of

$$P_{\text{load}} = 0.5(170 \text{ mW}) = 85 \text{ mW}$$

This assumes a Q point in the center of the ac load line.

10-6. LARGE-SIGNAL GAIN AND IMPEDANCE

We need to modify the approach of Chap. 9 (small-signal amplifiers) when we analyze power amplifiers. The first thing to realize is that $r'_e = 25 \text{ mV}/I_E$ is not useful with a power amplifier because the current and voltage swings are too large; we have to use *large-signal ac emitter resistance*, designated R'_e.

How to get R'_e

Because R'_e is a large-signal characteristic of a power transistor, we need information from the data sheet to get its value; no simple formula like $25 \text{ mV}/I_E$ exists for the value of R'_e. Instead, we have to use the *transconductance curve* shown on the data sheet of a power transistor.[2]

The transconductance curve is a graph of I_C versus V_{BE} similar to Fig. 10-11a. Some data sheets have I_C along the vertical axis and V_{BE} along the horizontal axis, but this convention is not standardized, and you often see V_{BE} along the vertical axis and I_C along the horizontal (Fig. 10-11b). Either way, it makes no difference, because we are interested in the changes in V_{BE} and I_C.

We define R'_e as the ratio of a *large change* in V_{BE} to a *large change* in I_E. In symbols,

$$R'_e = \frac{\Delta V_{BE}}{\Delta I_E} \qquad \text{for large changes} \qquad (10\text{-}13a)$$

Since V_{BE} and I_E are involved, R'_e is similar to r'_e, the difference being that R'_e is for large-signal operation while r'_e is for small-signal operation. As usual, we take advantage of the approximate equality between I_C and I_E. Because of this equality, we can rewrite Eq. (10-13a) as

[2] A good reference with transconductance curves and other transistor characteristics is *The Semiconductor Data Book*, Motorola Semiconductor Products, Inc., 1970.

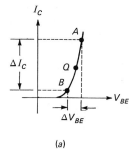

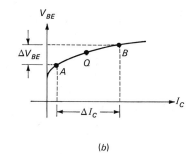

Figure 10-11. Bipolar transconductance curves.

(a)

(b)

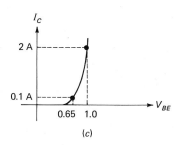

(c)

$$R'_e \cong \frac{\Delta V_{BE}}{\Delta I_C} \qquad\qquad (10\text{-}13b)\;***$$

Equation (10-13a) gives us the physical meaning of R'_e; it is a large change in base-emitter voltage divided by a large change in emitter current. Because of this, R'_e is like an *average* ac emitter resistance that we can use for large-signal operation. Equation (10-13b), on the other hand, gives us a way to calculate the approximate value of R'_e using information on the data sheet of a power transistor.

As an example, Fig. 10-11c shows part of the transconductance curve of a 2N3789. Suppose we want the value of R'_e between an I_C of 0.1 A and 2 A. Then, we read a change in V_{BE} of 0.35 V. And,

$$R'_e = \frac{\Delta V_{BE}}{\Delta I_C} = \frac{0.35}{2 - 0.1} = 0.184 \ \Omega$$

Large-signal formulas

The derivations of large-signal formulas are similar to those given in Chap. 9 for small-signal amplifiers. The difference is that R'_e is used instead of r'_e, and β_{dc} instead of β. In other words, when we want a large-signal formula for gain or input impedance, all we do is replace r'_e by R'_e and β by β_{dc} in the ideal small-signal formula.

As an example, the base-driven prototype has a voltage gain from base to collector of

$$\frac{v_c}{v_b} = \frac{r_C}{r_E + r_e'}$$

for an ideal small-signal amplifier, and

$$\frac{v_c}{v_b} = \frac{r_C}{r_E + R_e'}$$

for the same amplifier used with large signals. Also, the base-driven amplifier has an input impedance of

$$z_{\text{in(base)}} = \beta(r_E + r_e')$$

in the small-signal case, and

$$z_{\text{in(base)}} = \beta_{\text{dc}}(r_E + R_e')$$

for the large-signal case. Finally, since power gain is the product of current gain and voltage gain, we may write

$$\frac{p_c}{p_b} = \beta \frac{r_C}{r_E + r_e'}$$

for the small-signal case, and

$$\frac{p_c}{p_b} = \beta_{\text{dc}} \frac{r_C}{r_E + R_e'} \tag{10-14}$$

for large signals.

This will be our approach. We will derive formulas for the *ideal small-signal* case. When we need large-signal formulas, we will replace r_e' by R_e', and β by β_{dc}. The resulting large-signal formulas are more accurate than ideal formulas for two reasons: First, the value of R_e' is based on information from the data sheet. Second, R_e' includes the effects of base-spreading resistance r_b' because V_{BE} is the voltage from the base lead to the emitter lead.

Points A and B

In calculating the value of R_e' from the transconductance curve, you need to use end points A and B in Fig. 10-12b. These you get from the ac load line of the amplifier you are analyzing. Presumably, the amplifier is a power amplifier where everything is optimized for maximum power output. Because of this, the usual case is a power amplifier whose Q point is centered and whose signal swing uses most of the ac load line.

In Fig. 10-12a, we have shown the A and B points just short of saturation and cutoff. The reason is that voltage gain drops off rapidly as the instantaneous operating point approaches saturation. Also, the transconductance curve on a data sheet often does not show I_C's all the way down to cutoff. For these reasons, we use A and B points that are 10 percent in from saturation and cutoff. That is, we will base our large-signal calculations on a maximum collector current of $1.9I_{CQ}$ and a minimum collector current

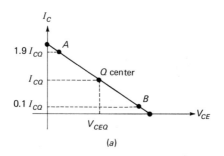

*Figure 10-12. Arbitrary end points
for reading transconductance
curves and h_{FE} curves.*

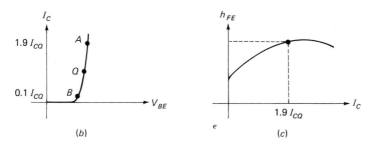

of $0.1I_{CQ}$. If a power amplifier has an I_{CQ} of 1 A, we use a maximum of 1.9 A and a minimum of 0.1 A when we calculate R_e' in Fig. 10-12b.

The value of β_{dc} or h_{FE} to use in the large-signal formulas is the value you read at $1.9I_{CQ}$. Specifically, Fig. 10-12c is similar to the graph given on a typical data sheet. After you locate $1.9I_{CQ}$, the corresponding h_{FE} is the value of β_{dc} to use in all large-signal formulas.

A crude estimate of R_e' and β_{dc}

There will be times when you do not have a data sheet. In this case, you can always use the following method. The peak-to-peak ac voltage across the emitter diode of most power transistors is in tenths of a volt. As a rough estimate, assume the peak-to-peak ac voltage is 0.5 V. Then,

$$R_e' = \frac{\Delta V_{BE}}{\Delta I_E} \cong \frac{0.5}{1.9I_{CQ} - 0.1I_{CQ}} \cong \frac{0.5}{2I_{CQ}}$$

or

$$R_e' \cong \frac{0.25}{I_{CQ}} = \frac{250 \text{ mV}}{I_{CQ}} \qquad (10\text{-}15)\,{}^{***}$$

Because the actual ΔV_{BE} may not equal 0.5 V peak-to-peak, the values of R_e' you get with Eq. (10-15) are only rough estimates that may be off more than a factor of 2; the true value of R_e' may be half or twice that of Eq. (10-15). When a rough approximation like this suits the analysis, Eq. (10-15) is useful. (Note the similarity to 25 mV/I_E.)

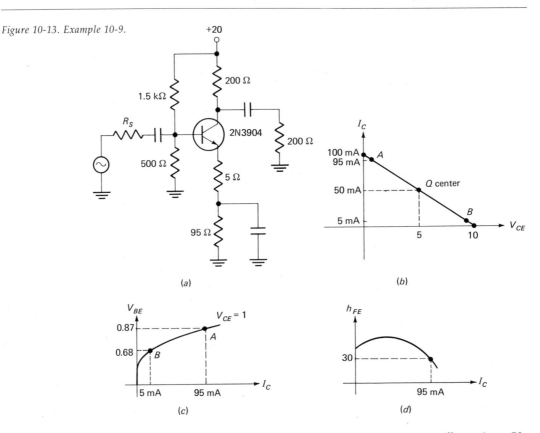

Figure 10-13. Example 10-9.

(a)

(b)

(c)

(d)

When we need an estimate for the β_{dc} of a power transistor, we will use $\beta_{dc} = 50$. Unquestionably, this is very crude because of the large variations in β_{dc}. But without a data sheet, we can hardly expect accuracy.

EXAMPLE 10-9.
Calculate the large-signal voltage and power gain from base to collector in Fig. 10-13a. Also, work out the value of $z_{in(base)}$.

SOLUTION.
By ideal methods of dc analysis, we can calculate $I_{CQ} \cong 50$ mA and $V_{CEQ} \cong 5$ V. Furthermore, with Eq. (10-9), we can prove the Q point is approximately centered on the load line. For this reason, we can draw the load line shown in Fig. 10-13b.

Ten percent of I_{CQ} is 5 mA. Therefore, points A and B correspond to collector currents of 95 mA and 5 mA. When you look at the data sheet of a 2N3904, you will find a transconductance curve like Fig. 10-13c. The V_{BE} values corresponding to I_C's of 5 mA and 95 mA are 0.68 V and 0.87 V. So,

$$R'_e = \frac{0.87 - 0.68}{0.095 - 0.005} \cong 2.1 \ \Omega$$

The data sheet of a 2N3904 also shows h_{FE} has a value of 30 when I_C is 95 mA (Fig. 10-13d). This is the value of β_{dc} to use in all large-signal formulas.

Now, we are ready to make the calculations. The large-signal voltage gain equals

$$\frac{v_c}{v_b} \cong \frac{r_C}{r_E + R'_e} = \frac{100}{5 + 2.1} \cong 14$$

The large-signal power gain is

$$\frac{p_c}{p_b} \cong \beta_{dc} \frac{v_c}{v_b} \cong 30(14) = 420$$

And the large-signal input impedance equals

$$z_{in(base)} \cong \beta_{dc}(r_E + R'_e) \cong 30(5 + 2.1) = 213 \ \Omega$$

EXAMPLE 10-10.
Calculate the power gain and input impedance of the base in Fig. 10-14a.

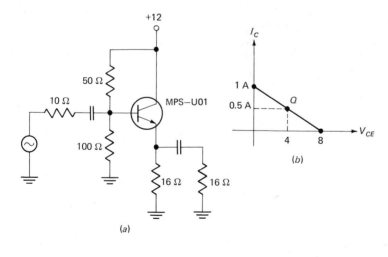

Figure 10-14. Example 10-10.

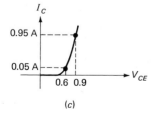

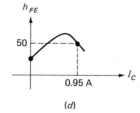

SOLUTION.

By methods used several times before, you can calculate $I_{CQ} \cong 0.5$ A and $V_{CEQ} \cong 4$ V. Also, you will find the Q point is centered on the ac load line as shown in Fig. 10-14*b*.

When you examine the data sheet for an MPS-U01, you see a transconductance graph like Fig. 10-14*c* and a dc current-gain graph like Fig. 10-14*d*. With Fig. 10-14*c*, we get

$$R_e' = \frac{0.9 - 0.6}{0.95 - 0.05} \cong 0.33 \ \Omega$$

and with Fig. 10-14*d* we get $\beta_{dc} = 50$.

The large-signal power gain of the emitter follower is

$$\frac{p_c}{p_b} \cong \beta_{dc} \frac{v_c}{v_b} \cong \beta_{dc} \frac{r_E}{r_E + R_e'}$$

$$= 50 \frac{8}{8 + 0.33} = 48$$

The second factor in the foregoing product is the voltage gain, which you can see is almost unity, typical of an emitter follower.

The input impedance of the base equals

$$z_{in(base)} \cong \beta_{dc}(r_E + R_e') = 50(8 + 0.33)$$
$$= 416 \ \Omega$$

EXAMPLE 10-11.

Repeat Examples 10-9 and 10-10 but assume no data sheets are available.

SOLUTION.

Example 10-9. After finding $I_{CQ} = 50$ mA, we calculate a rough value of R_e' by using Eq. (10-15) as follows.

$$R_e' \cong \frac{0.25}{I_{CQ}} = \frac{0.25}{0.05} = 5 \ \Omega$$

Then, the voltage gain is

$$\frac{v_c}{v_b} \cong \frac{r_C}{r_E + R_e'} = \frac{100}{5 + 5} = 10$$

Using a rough value of 50 for β_{dc}, we calculate the power gain as

$$\frac{p_c}{p_b} \cong \beta_{dc} \frac{v_c}{v_b} = 50(10) = 500$$

And the input impedance is

$$z_{in(base)} \cong \beta_{dc}(r_E + R_e') = 50(10) = 500 \ \Omega$$

Example 10-10. After finding $I_{CQ} = 0.5$ A, we get

$$R'_e \cong \frac{0.25}{0.5} = 0.5 \ \Omega$$

The power gain is

$$\frac{p_c}{p_b} \cong 50 \frac{8}{8 + 0.5} \cong 47$$

and the input impedance is

$$z_{in(base)} \cong 50(8 + 0.5) = 425 \ \Omega$$

By comparing these results with those found in Examples 10-9 and 10-10, you can see close agreement for some quantities but large error for others. Whether or not you can use rough answers depends on your particular application. For preliminary analysis or troubleshooting, rough answers like the ones found in this example may be all right. Also, such answers may be acceptable when *negative feedback* is used (Chap. 18).

Problems

10-1.   Plot the Q point for the amplifier of Fig. 10-15a.

10-2.   Locate the I_{CQ} and V_{CEQ} for Fig. 10-15b.

10-3.   The transistor of Fig. 10-15c has $\beta_{dc} = 100$. What are the values of I_{CQ} and V_{CEQ}?

10-4.   Plot the Q point for the amplifier shown in Fig. 10-15d.

10-5.   What values do I_{CQ} and V_{CEQ} have in Fig. 10-15e?

10-6.   Referring to Fig. 10-5b, work out the saturation current and cutoff voltage for the amplifier of Fig. 10-15a.

10-7.   Draw the ac load line for the circuit of Fig. 10-15b.

10-8.   In Fig. 10-15c, what is the value of cutoff voltage? Of saturation current? (Use a β_{dc} of 100.)

10-9.   In Fig. 10-15d, what is the minimum acceptable BV_{CEO} for the transistor? The minimum acceptable $I_{C(max)}$ rating?

10-10. Draw the ac load line for the emitter follower shown in Fig. 10-15e.

10-11. The transformer of Fig. 10-15f reflects an impedance of $n^2 R_L$ into the collector circuit. Draw the ac load line for the amplifier.

10-12. An amplifier has an I_{CQ} of 2 A. If $r_C + r_E = 10 \ \Omega$, what is the voltage change when the instantaneous operating point swings from the Q point to the cutoff point?

10-13. Use Eq. (10-9) to check whether or not the Q point of Fig. 10-15a is centered.

10-14. Is the Q point of Fig. 10-15b centered?

10-15. For some β_{dc} between 100 and 300, the amplifier of Fig. 10-15c will have a cen-

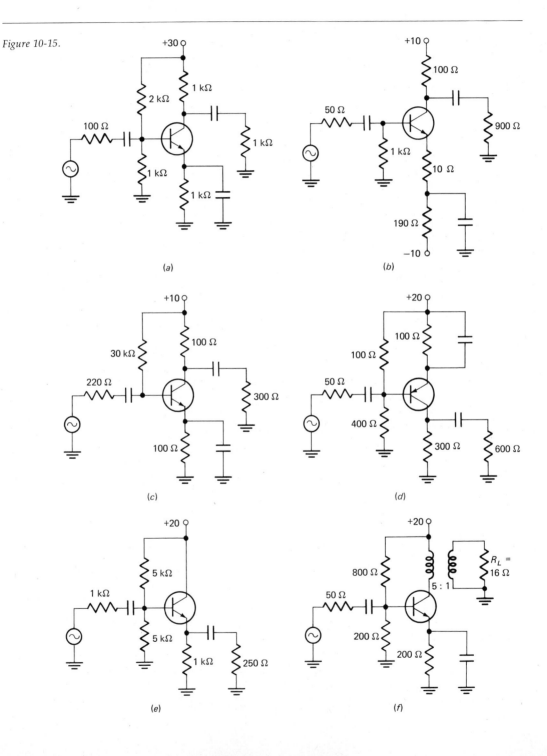

Figure 10-15.

tered Q point. Find this value of β_{dc} by using a trial-and-error approach (neglect 0.7 V).

10-16. If we change the 600-Ω resistor in Fig. 10-15d to another value, we can get a centered Q point. What value do we need?

10-17. By changing the turns ratio in Fig. 10-15f, we can get a centered Q point. What turns ratio do we need?

For all remaining problems the Q point is centered and the signal is large enough to use the entire ac load line.

10-18. How much ac load power is there in Fig. 10-16a?

10-19. What is value of ac load power in the emitter follower of Fig. 10-16b? And the value of P_{DQ}?

10-20. Calculate the ac load power for the swamped amplifier of Fig. 10-16c. What is the value of P_{DQ}?

10-21. Work out the ac load power for the phase splitter of Fig. 10-16d. What is lowest acceptable $P_{D(max)}$ rating?

10-22. The ideal transformer of Fig. 10-16e steps up the load impedance by n^2. What is the ac load power going into the primary winding? How much ac load power does the 3-Ω resistor receive?

10-23. A transistor has a $P_{D(max)}$ rating of 87 W at 25°C. If used in a class A amplifier, what is the maximum ac load power?

10-24. A 2N3719 has a $P_{D(max)}$ rating of 1 W at 25°C, and a derating factor of 5.72 mW/°C rise. What is the $P_{D(max)}$ rating at 75°C? At 150°C?

10-25. What $P_{D(max)}$ rating does the transistor of Fig. 10-16a require?

10-26. What is lowest acceptable $P_{D(max)}$ rating for the transistor in Fig. 10-16c?

10-27. The voltage across a 16-Ω loudspeaker is 8 V. How much power does this load receive? If the amplifier driving the load is a class A amplifier, what is the minimum acceptable $P_{D(max)}$ rating for the transistor?

10-28. Calculate the value of R'_e given the curve of Fig. 10-17a and the indicated points.

10-29. What is the value of R'_e between the pair of points given in Fig. 10-17b?

10-30. If $1.9I_{CQ} = 10$ A, what is β_{dc} in Fig. 10-17c?

10-31. Over the indicated range shown in Fig. 10-17d, what is the value of R'_e?

Repeat: for all remaining problems the Q point is centered and the signal is large enough to use the entire ac load line.

10-32. The R'_e of the transistor in Fig. 10-16a is 1 Ω. Work out the large-signal voltage and power gain from base to collector using a β_{dc} of 50.

10-33. No data sheet is available for the transistor of Fig. 10-16a. Work out the approximate large-signal voltage and power gain from base to collector. What is the large-signal input impedance looking into the base?

10-34. $R'_e = 0.2$ Ω and $\beta_{dc} = 100$ for the transistor of Fig. 10-16d. What is the large-signal voltage gain from base to collector? From base to emitter? And what is the large-signal input impedance looking into the base?

10-35. The transistor of Fig. 10-16c has an $R'_e = 1$ Ω and a $\beta_{dc} = 75$. Calculate the large-

Figure 10-16.

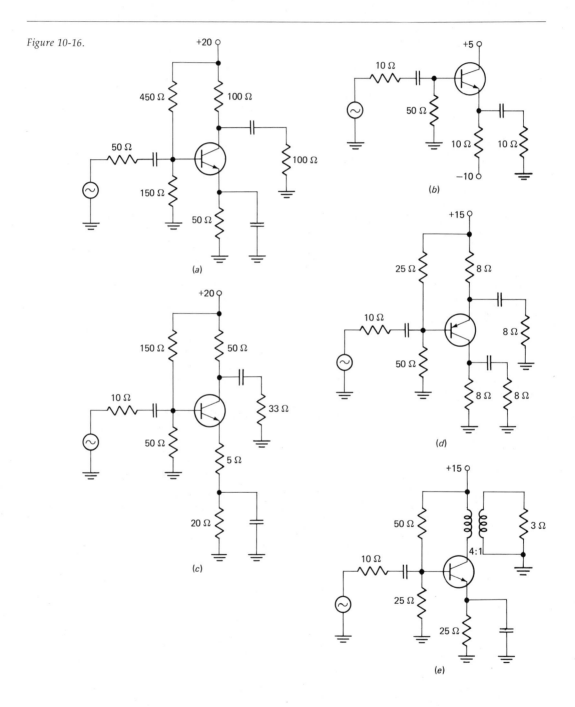

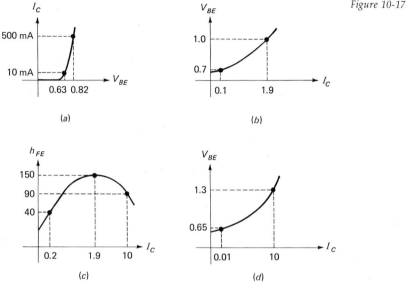

Figure 10-17.

signal voltage and power gain from base to collector, and the input impedance looking into the base. Also, what is the input impedance when the base-biasing resistors are included?

10-36. The amplifier of Fig. 10-16b has the transconductance curve given by Fig. 10-17b and the h_{FE} curve of Fig. 10-17c. Calculate the large-signal voltage and power gain from base to emitter. What is the large-signal input impedance of the base $z_{in(base)}$?

10-37. The data sheet is not available for the transistor of Fig. 10-16e. Work out a rough value for large-signal voltage gain from base to collector. What is the approximate power gain from base to collector? The input impedance looking into the base?

11. Class B Push-Pull Amplifiers

Class A is the common way to run a transistor in linear circuits because it leads to the simplest and most stable biasing circuits. But it requires a $P_{D(\max)}$ rating of twice the load power; it also has a quiescent or no-signal current drain of 50 percent $I_{C(\text{sat})}$ when the Q point is centered. In the earlier stages of a system, the $P_{D(\max)}$ rating and no-signal current drain are usually small enough to accept. But near the end of many systems, the $P_{D(\max)}$ rating and the no-signal current drain become so large we can no longer use class A amplifiers.

The *class B push-pull amplifier* is a two-transistor circuit with these outstanding advantages: the $P_{D(\max)}$ rating drops to one-fifth of the load power and the no-signal current drain to around 1 percent of $I_{C(\text{sat})}$. The first advantage is important when large amounts of load power are needed as in communication transmitters, etc. The second advantage is desirable in battery-powered systems like transistor radios.

For load power up to approximately 10 W, you can usually find a satisfactory integrated-circuit class B amplifier. Above 10 W or so, the discrete class B circuit may be used.

11-1. THE BASIC IDEA OF PUSH-PULL ACTION

Before we discuss the idea of push-pull action, we will define other classes of operation and will show the ac load line for class B circuits.

Figure 11-1. Collector-current waveforms. (a) *Class A.* (b) *Class B.* (c) *Class C.*

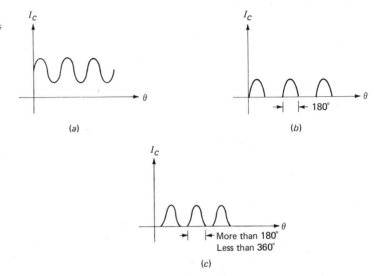

(a)

(b)

(c)

Class A, B, and AB

We already know the transistor of a class A amplifier remains FR-biased throughout the cycle; this means the collector current of a class A amplifier flows for 360° as shown in Fig. 11-1*a*.

In a class B circuit, the transistor stays FR-biased only for *half the cycle.* During the other half cycle, the transistor is RR-biased, or cut off. This means collector current flows for 180° in each transistor of a class B circuit (see Fig. 11-1*b*).

Class AB is between class A and class B. The transistor of a class AB circuit is FR-biased for more than half a cycle but less than the whole cycle. In other words, collector current flows for more than 180° but less than 360° (Fig. 11-1*c*).

The ac load line for class B

Figure 11-2 shows the ac load line for one transistor in a class B circuit. Neglecting I_{CBO}, the Q point is at cutoff and has coordinates of

$$I_{CQ} = 0$$

and
$$V_{CEQ} = V_{CE\text{(cutoff)}}$$

When the ac signal comes in, the instantaneous operating point swings from Q to saturation as shown. This produces half cycles of current and voltage. By letting one transistor handle the positive half cycle and another transistor the negative half cycle, we can get a final output signal that is a complete sine wave.

Especially important, we need a formula for the saturation current in a class B

Figure 11-4. Examples 11-1 and 11-2.

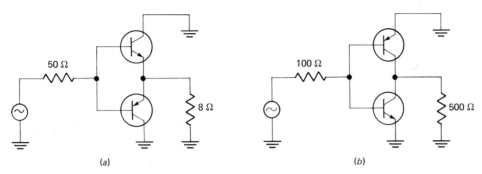

(a)   (b)

EXAMPLE 11-1.

Figure 11-4a shows the ac equivalent circuit for a class B push-pull emitter follower. In the dc equivalent circuit (not shown) each transistor has a $V_{CEQ} = 10$ v. Calculate the saturation current.

SOLUTION.

Equation (11-1) is the key. In Fig. 11-4a, $r_C = 0$ and $r_E = 8$ Ω. With Eq. (11-1),

$$I_{C(sat)} = \frac{V_{CEQ}}{r_C + r_E} = \frac{10}{8} = 1.25 \text{ A}$$

EXAMPLE 11-2.

Figure 11-4b shows the ac equivalent circuit for a class B push-pull CE amplifier. In the dc equivalent circuit (not shown) each transistor has a $V_{CEQ} = 50$ V. What is the saturation current?

SOLUTION.

The lower transistor amplifies the positive half cycle of source voltage; the upper one takes care of the negative half cycle. As you can see, $r_C = 500$ Ω and $r_E = 0$. With Eq. (11-1),

$$I_{C(sat)} = \frac{V_{CEQ}}{r_C + r_E} = \frac{50}{500} = 0.1 \text{ A}$$

11-2. NONLINEAR DISTORTION

One of the critical things about a class B amplifier is the Q point. In the ideal class B circuit, the Q point is at cutoff. But in a practical class B amplifier, the Q point is slightly above cutoff. This section tells you why.

Crossover distortion

Figure 11-5a shows a class B push-pull amplifier. Suppose no bias at all is applied to the emitter diodes. Then, the incoming ac signal has to rise to about 0.7 V to overcome the barrier potential. Because of this, essentially no current flows through Q_1 when the signal is less than 0.7 V. The action on the other half cycle is complementary; the other transistor does not turn on until the negative half cycle of source voltage is more negative than approximately -0.7 V. For this reason, if no bias at all is applied to the emitter diodes, the output of a class B push-pull amplifier looks like Fig. 11-5b.

The signal of Fig. 11-5b is distorted; it no longer is a sine wave because of the clipping action between half cycles. Since this clipping occurs between the time one transistor shuts off and the other comes on, we call it *crossover distortion*.

To eliminate crossover distortion, we need to apply a slight forward bias to each emitter diode. This means locating the Q point slightly above cutoff as shown in Fig. 11-5c. As a guide, an I_{CQ} from 1 to 5 percent of $I_{C(\text{sat})}$ is enough to eliminate crossover distortion, the exact value usually determined by experiment with the particular circuit. We will refer to this slight forward bias on each emitter diode as *trickle bias*.

Strictly speaking, we have class AB operation because each transistor is FR-biased for slightly more than half a cycle. But this is hair splitting, and most people still call the circuit a class B push-pull amplifier. We will do the same.

Figure 11-5. Biasing a push-pull circuit. (a) Ac equivalent of emitter follower. (b) Crossover distortion. (c) Ac load line with trickle bias. (c) Emitter-diode curve with trickle bias.

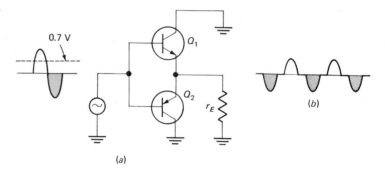

(a)

(b)

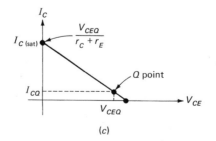

(c)

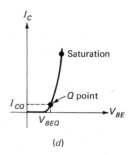

(d)

The remaining distortion

Figure 11-5d shows a typical transconductance curve. With trickle bias, the Q point is slightly above the knee as shown. The part of the curve from Q up to saturation is not quite a straight line. Because of this, changes in V_{BE} do not produce exactly proportional changes in I_C. In other words, a perfect half sine of V_{BE} swinging from Q upward does not result in a perfect half sine wave of I_C. Since I_C flows through the load resistance to produce the output voltage, the output waveform is slightly different from the input waveform. We call this kind of distortion *nonlinear distortion.*

With trickle bias, the Q point is slightly above the knee, so that the almost-linear portion of the curve from Q to saturation is used. Because of this, the nonlinear distortion may be so low you cannot see it on an oscilloscope. Ideally, we neglect the nonlinear distortion of a class B push-pull amplifier when it has trickle bias. Later, when we talk about harmonics, we will take a closer look at the amount of nonlinear distortion that remains in a class B amplifier.

11-3.   SETTING UP THE Q POINT

The most difficult thing about a class B amplifier is setting up a stable Q point. The problem is much more difficult than in a class A circuit, the reason being that V_{BE} becomes a crucial quantity in a class B circuit. We can no longer treat it as 0.7 V. In a class B circuit, we need its exact value.

One-supply emitter follower

The *class B push-pull emitter follower* is one of the most important class B circuits. For this reason, we will concentrate on it and extend the results to other class B circuits.

Figure 11-6a shows the *dc equivalent circuit* for a class B push-pull emitter follower. The two transistors need to be *fully complementary,* meaning similar transconductance curves, maximum ratings, etc. For instance, the 2N3904 and 2N3906 are complementary, the first being an *npn* transistor and the second a *pnp*; both have almost the same maximum ratings, transconductance curves, and so on. Complementary pairs like these are commercially available for almost any application.

In Fig. 11-6a, the collector and emitter currents flow down through the *npn* and *pnp* transistors. Because of the series connection, the I_{CQ}'s of the transistors are equal. Note also the V_{CEQ} of each transistor is half the supply voltage V_{CC} because the transistors have the same characteristics.

To set up a trickle bias, we need to select a value of R_1 and R_2 that just turns on each emitter diode. The exact value of quiescent base-emitter voltage V_{BEQ} becomes important because even a tenth-volt difference can produce a large difference in collector current. For this reason, we either have to look at the transconductance curve, or we have to find the correct V_{BEQ} by experiment.

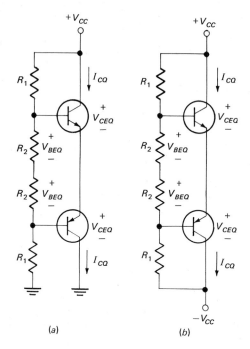

Figure 11-6. Dc equivalent circuits of push-pull emitter follower. (a) One power supply. (b) Two power supplies.

(a)

(b)

If you have a data sheet, look at the transconductance curve and read the V_{BE} value in the vicinity of 1 to 5 percent $I_{C(\text{sat})}$. The value of V_{BE} in this range is what you have to set up with the voltage divider of Fig. 11-6a. Approximately,

$$V_{BEQ} \cong \frac{R_2}{R_1 + R_2} \frac{V_{CC}}{2} \qquad \text{(one supply)} \qquad (11\text{-}2a)$$

Equation (11-2a) is an approximation because the quiescent base current lowers the voltage across R_2 slightly. For the approximation to be close,

$$R_2 \ll \frac{0.7}{I_{BQ}} \qquad (11\text{-}2b)$$

For instance, if $I_{CQ} = 10$ mA and $\beta_{dc} = 50$, then $I_{BQ} = 0.2$ mA. And, $0.7/I_{BQ} = 3.5$ kΩ. Therefore, R_2 must be much smaller than 3.5 kΩ for Eq. (11-2a) to be valid.

If you do not have a data sheet for the transistor, you can build the circuit of Fig. 11-6a and find the correct V_{BEQ} by experiment. But, beware! You can easily burn out the transistors by exceeding their maximum collector-current rating. Therefore, if you try to find V_{BEQ} experimentally, slowly raise V_{BEQ} from 0.6 V to higher values. You can do this by adjusting resistance values or by increasing V_{CC}. When I_{CQ} is between 1 and 5 percent of $I_{C(\text{sat})}$, you have the correct trickle bias.

The lower I_{CQ} is, the better from the standpoint of no-signal current drain. On the other hand, you have to get rid of the crossover distortion. Therefore, the final setting of the Q point in the 1 to 5 percent $I_{C(sat)}$ range will depend on the particular transistor type, the circuit design, and the amount of allowable distortion.

Two-supply emitter follower

Figure 11-6b shows how to use two supplies to set up trickle bias in a class B push-pull emitter follower. Again, the collector and emitter currents flow down through the *npn* and *pnp* transistors. When two equal and opposite supplies are used like this, the dc voltage from emitter to ground is ideally zero. Also, the dc voltage from the middle of the voltage divider to ground is ideally zero.

In Fig. 11-6b, the dc voltage across the upper R_1 and R_2 equals V_{CC}. As a result,

$$V_{BEQ} \cong \frac{R_2}{R_1 + R_2} \, V_{CC} \qquad \text{(two supplies)} \tag{11-3}$$

Again, the quiescent base current lowers the voltage across R_2. For Eq. (11-3) to be a good approximation, R_2 must be much smaller than $0.7/I_{BQ}$.

EXAMPLE 11-3.
Set up trickle bias for the circuit of Fig. 11-7a, given an $R_2 = 240 \ \Omega$, $I_{C(sat)} = 100$ mA, and the transconductance curve of Fig. 11-7b.

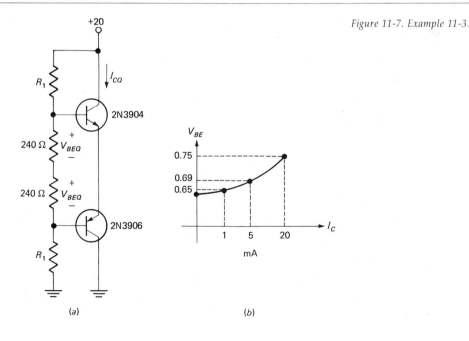

Figure 11-7. Example 11-3.

(a)   (b)

SOLUTION.

With an $I_{C(sat)}$ of 100 mA, we need an I_{CQ} from 1 to 5 mA to eliminate crossover distortion. On the transconductance curve of Fig. 11-7b, the corresponding V_{BE} values are 0.65 and 0.69 V. We will use the average value, 0.67 V. With Eq. (11-2a),

$$0.67 = \frac{240}{R_1 + 240} \, 10$$

Solving for R_1 gives

$$R_1 = 3.34 \text{ k}\Omega$$

The nearest standard size of 3.3 kΩ is close enough.

The narrow range of V_{BE} in Fig. 11-7b shows you how rapidly I_C changes. If resistance values are not accurate or if power-supply voltage changes, V_{BE} can easily increase to 0.75 V; this produces an I_{CQ} of 20 mA as shown in Fig. 11-7b. This gives you an idea of how critical and sensitive the biasing of a class B amplifier is. With silicon, the barrier potential decreases approximately 2.5 mV per degree rise; the required V_{BEQ} changes approximately the same amount. This means changes in temperature upset the trickle bias in a class B circuit. A later section tells you how to compensate for temperature change.

11-4. THE COMPLETE CLASS B EMITTER FOLLOWER

Figure 11-8a shows a complete circuit for the class B push-pull emitter follower. Resistors R_1 and R_2 set up the trickle bias described in the preceding section. On the positive half cycle of ac source voltage, the voltage increases on the left end of the input capacitor; this forces the voltage on the right end of the capacitor to increase. This increasing voltage is fed through each R_2 to the bases, turning on the upper transistor and shutting off the lower one. The emitter voltage of the upper transistor now follows the base voltage. The rising voltage on the left end of the output capacitor is coupled to the right end of the capacitor; as a result, the positive half cycle appears across the final load resistor R_L.

During the negative half cycle of source voltage, the voltage on the left end of the input capacitor decreases; this decreasing voltage is coupled through each R_2 into the bases. The *npn* transistor turns off, but the *pnp* comes on. As before, the emitter-follower action results in the output voltage following the input voltage. In this way, the negative half cycle of output voltage appears across R_L.

The ac analysis for any circuit like Fig. 11-8a is straightforward. Start by visualizing all capacitors shorted and V_{CC} reduced to zero. This results in the ac equivalent circuit of Fig. 11-8b. Because the push-pull action is complementary on each half cycle, we need analyze only half a cycle.

Figure 11-8c shows the circuit redrawn for the positive half cycle of source voltage.

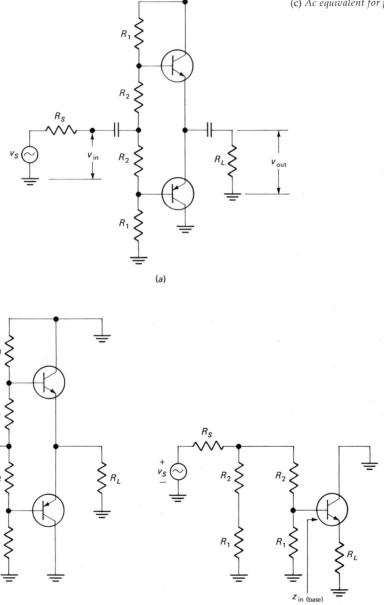

Figure 11-8 Class B push-pull emitter follower. (a) Complete circuit. (b) Ac equivalent circuit. (c) Ac equivalent for positive half cycle.

$+V_{CC}$

R_1

R_2

R_S

v_S

v_{in}

R_2

R_1

R_L

v_{out}

(a)

R_1

R_2

R_S

v_S

R_2

R_L

R_1

(b)

R_S

v_S

R_2

R_2

R_1

R_1

R_L

$z_{in\,(base)}$

(c)

The large-signal input impedance of the base is

$$z_{\text{in(base)}} \cong \beta_{\text{dc}}(r_E + R_e')$$

Since $r_E = R_L$ in Fig. 11-8c,

$$z_{\text{in(base)}} \cong \beta_{\text{dc}}(R_L + R_e') \qquad\qquad (11\text{-}4a)$$

In a typical circuit, R_L is much greater than R_e', so that

$$z_{\text{in(base)}} \cong \beta_{\text{dc}}R_L \qquad \text{for } R_L \gg R_e' \qquad\qquad (11\text{-}4b)$$

To get the large-signal voltage gain from base to emitter, we use

$$\frac{v_e}{v_b} = \frac{r_E}{r_E + R_e'}$$

In Fig. 11-8c, $r_E = R_L$ so that

$$\frac{v_e}{v_b} = \frac{R_L}{R_L + R_e'} \qquad\qquad (11\text{-}5a)$$

or

$$\frac{v_e}{v_b} \cong 1 \qquad \text{for } R_L \gg R_e' \qquad\qquad (11\text{-}5b)$$

The large-signal power gain from base to emitter is

$$\frac{p_e}{p_b} = \beta_{\text{dc}}\frac{v_e}{v_b} \qquad\qquad (11\text{-}6a)$$

or

$$\frac{p_e}{p_b} \cong \beta_{\text{dc}} \qquad \text{for } R_L \gg R_e' \qquad\qquad (11\text{-}6b)$$

When equal and opposite power supplies are available, the complete class B push-pull emitter follower looks like Fig. 11-9. As previously discussed, the dc voltages at points A and B are ideally zero with respect to ground. In a practical circuit, small differences between the upper and lower halves of the circuit may make these dc voltages slightly different from zero. Nevertheless, the voltages are close enough to zero that we can leave out the input and output coupling capacitors. This is a decided advantage because the amplifier can handle frequencies all the way down to zero.

EXAMPLE 11-4.
The push-pull emitter follower of Fig. 11-10a has $\beta_{\text{dc}} = 25$ and $R_e' = 0.1\ \Omega$. Calculate the voltage and power gain from base to emitter. What is the input impedance of the base?

SOLUTION.
The R_L in Fig. 11-10a is 8 Ω, which is much greater than the R_e' of 0.1 Ω. Therefore, the

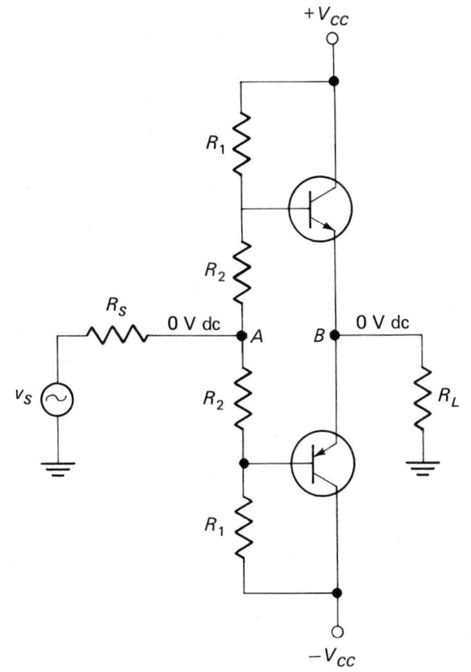

Figure 11-9. Two-supply push-pull emitter follower.

simplified emitter-follower formulas apply as follows:

$$\frac{v_e}{v_b} \cong 1$$

$$\frac{p_e}{p_b} \cong \beta_{dc} = 25$$

and

$$z_{in(base)} \cong \beta_{dc} R_L = 25(8) = 200 \ \Omega$$

EXAMPLE 11-5.

Figure 11-10b shows a two-supply class B emitter follower with Darlington pairs used instead of single transistors. The Darlington beta is 2500 and the R'_{e2} of the output transistor in each Darlington pair is 0.1 Ω. Calculate the approximate voltage gain, power gain, and input impedance of the base.

SOLUTION.

You often see Darlington pairs used in class B amplifiers. Among other things, this allows you to use larger R_1 and R_2 in each voltage divider. Since each Darlington pair

Figure 11-10. Examples 11-4 and
11-5.

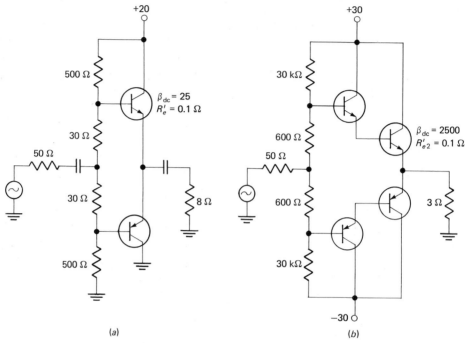

(a)　　　　　　　　　　　　(b)

acts approximately like a single transistor with a β_{dc} of 2500 and an R'_e of 0.2 Ω, we
proceed as follows.

In Fig. 11-10b, R_L is 3 Ω, which is much larger than an R'_e of 0.2 Ω. Therefore, we
can use the simplified emitter-follower formulas, that is,

$$\frac{v_e}{v_b} \cong 1$$

$$\frac{p_e}{p_b} \cong \beta_{dc} = 2500$$

and
$$z_{in(base)} \cong \beta_{dc} R_L = 2500(3) = 7500 \ \Omega$$

11-5. POWER RELATIONS

At the beginning of this chapter, we said a class B push-pull amplifier required a much
lower $P_{D(max)}$ rating. This section discusses the $P_{D(max)}$ rating and other power relations
in the class B circuit.

Load power

Figure 11-11 shows the ideal ac load line for a class B circuit. It is ideal because we have shown the Q point at cutoff and have neglected $V_{CE(\text{sat})}$. In a practical amplifier, the Q point is slightly above cutoff and the saturation point does not quite touch the vertical axis.

The main idea right now is this. Figure 11-11 shows the largest current and voltage waveforms we can get with one transistor of a class B circuit; the other transistor produces the dotted half cycle. For this reason, the ac load voltage has a peak value of V_{CEQ}, or an rms value of $0.707V_{CEQ}$. Also, the ac load current has a peak value of $I_{C(\text{sat})}$, equivalent to an rms value of $0.707I_{C(\text{sat})}$. The average load power for the maximum swing shown in Fig. 11-11 is

$$
\begin{aligned}
P_{\text{load}} &= V_{\text{rms}}I_{\text{rms}} \\
&= 0.707V_{CEQ} \times 0.707I_{C(\text{sat})}
\end{aligned}
$$

or

$$
P_{\text{load}} = 0.5V_{CEQ}I_{C(\text{sat})} \qquad \text{(maximum)} \tag{11-7}
$$

When analyzing circuits, you can figure out the value of V_{CEQ} by looking at the diagram; the value of $I_{C(\text{sat})}$ is easy to calculate with Eq. (11-1). And with Eq. (11-7), you can work out the largest average load power. As an example, the V_{CEQ} of Fig. 11-10a is half the supply voltage; so

$$
V_{CEQ} = \frac{V_{CC}}{2} = \frac{20}{2} = 10 \text{ V}
$$

Figure 11-10a has $r_C = 0$ and $r_E = 8 \ \Omega$; therefore, with Eq. (11-1),

$$
I_{C(\text{sat})} = \frac{10}{8} = 1.25 \text{ A}
$$

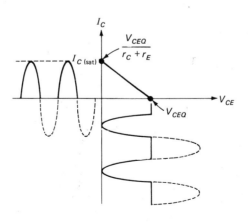

Figure 11-11. Deriving power relations for class B push-pull.

Finally, with Eq. (11-7), we get

$$P_{\text{load}} = 0.5(10)1.25 = 6.25 \text{ W}$$

Power dissipation

During the half cycle when a transistor is off, it dissipates no power. But when it conducts, it does dissipate power. The average power over one complete cycle is P_D. Finding this average power dissipation requires averaging the *instantaneous* power dissipation throughout the cycle. Specifically, on the ac load line of Fig. 11-11, the instantaneous power dissipation p equals

$$p = V_{CE} I_C$$

Each different operating point along the load line has a different amount of power. To get the *average* power dissipation over the whole cycle requires taking the average of all instantaneous power dissipations.

The Appendix uses calculus to find the average power dissipated by the transistor *over one complete cycle*. Here are the results:

$$P_D = 0.068 V_{CEQ} I_{C(\text{sat})} \quad \text{(100 percent load line)} \tag{11-8}$$

This equation gives the average power dissipated by one transistor in a class B circuit when the signal is large enough to use *the entire ac load line*. For instance, suppose the V_{CEQ} and $I_{C(\text{sat})}$ of Fig. 11-11 are 10 V and 1.25 A. Then, if the signal swings over 100 percent of the load line, each transistor dissipates an average power of

$$P_D = 0.068(10)1.25 = 0.85 \text{ W}$$

The average power dissipation given by Eq. (11-8) is valid only when the entire load line is used. If the signal is not large enough to use the entire load line, the average power dissipation changes. The Appendix proves the worst case occurs when 63 percent of the load line is used. The average power dissipation for this worst case is

$$P_D \cong 0.1 V_{CEQ} I_{C(\text{sat})} \quad \text{(worst case)} \tag{11-9}***$$

For instance, with the same V_{CEQ} and $I_{C(\text{sat})}$ used a moment ago, the worst-case dissipation is

$$P_D = 0.1(10)1.25 = 1.25 \text{ W}$$

In designing a class B push-pull amplifier, we have to expect the worst case sooner or later. In other words, we cannot count on always having a signal large enough to use the entire ac load line. Because of this, the $P_{D(\text{max})}$ rating of each transistor in a class B circuit must equal or be greater than the worst-case dissipation, that is,

$$P_{D(\text{max})} \geqslant 0.1 V_{CEQ} I_{C(\text{sat})} \tag{11-10}$$

The maximum load power in a class B amplifier is $0.5V_{CEQ}I_{C(sat)}$, which is five times the worst-case dissipation. Therefore, another useful relation is

$$P_{D(max)} \geq \frac{P_{load(max)}}{5} \qquad (11\text{-}11)^{***}$$

EXAMPLE 11-6.
To deliver a maximum load power of 100 W, what $P_{D(max)}$ rating does each transistor in a class B push-pull amplifier need?

SOLUTION.
With Eq. (11-11),

$$P_{D(max)} \geq \frac{100}{5} = 20 \text{ W}$$

[Note: the transistor of a class A amplifier would need a $P_{D(max)}$ rating of at least 200 W to deliver 100 W of load power.]

EXAMPLE 11-7.
How much load power can the transistors of Fig. 11-12 produce? What $P_{D(max)}$ rating do they need?

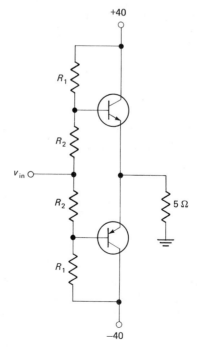

+40

Figure 11-12. Example 11-7.

R_1

R_2

v_{in}

R_2

$5\,\Omega$

R_1

−40

SOLUTION.

In this two-supply emitter follower, $V_{CEQ} = 40$ V. The saturation current is

$$I_{C(\text{sat})} = \frac{40}{5} = 8 \text{ A}$$

With Eq. (11-7), the maximum load power is

$$P_{\text{load}} = 0.5(40)8 = 160 \text{ W}$$

And with Eq. (11-9), the worst-case power dissipation is

$$P_D = 0.1(40)8 = 32 \text{ W}$$

Each transistor needs a $P_{D(\text{max})}$ of at least 32 W.[1]

11-6. TEMPERATURE EFFECTS

Class B amplifiers have much worse temperature effects than class A. The main problem is the change in V_{BE} with temperature. Here is what happens with a simple voltage divider. At room temperature, all is well because the correct V_{BEQ} sets up an I_{CQ} from 1 to 5 percent of $I_{C(\text{sat})}$. But when the temperature increases, the barrier potential decreases approximately 2.5 mV per degree. As a result, I_{CQ} can become objectionably large. This offsets one of the main advantages of class B operation, namely, the small no-signal current drain. (2.5 mV per degree is a good approximation for silicon.)

Thermistors

When the expected temperature range is large, you have to compensate for the changes in V_{BE}. One way is with *thermistors,* resistances that decrease with increasing temperature. In other words, a thermistor has a *negative temperature coefficient.* Thermistors are commercially available with room-temperature resistances from less than 4 Ω to more than 10 kΩ.

Here is the idea of thermistor compensation. In Fig. 11-13a, the circled R_2's are thermistors. We select values for R_1 and R_2 to set up a trickle bias at room temperature. Then, as the temperature increases, the barrier potential decreases about 2.5 mV per degree; the resistance of the thermistors also decreases, and this reduces the V_{BEQ} applied to each emitter diode. If the correct thermistor is used, it compensates for the changes in barrier potential.

Often, the thermistor resistance changes too fast with increasing temperature, and the emitter diode gets less than enough voltage. One way to improve the compensation is to add some resistance in series with the thermistors as shown in Fig. 11-13b. The added resistance will decrease the rate of change of R_2 with temperature. In this way, it

[1] To increase $P_{D(\text{max})}$, the transistor is often mounted in thermal contact with the chassis, which acts like a heat sink.

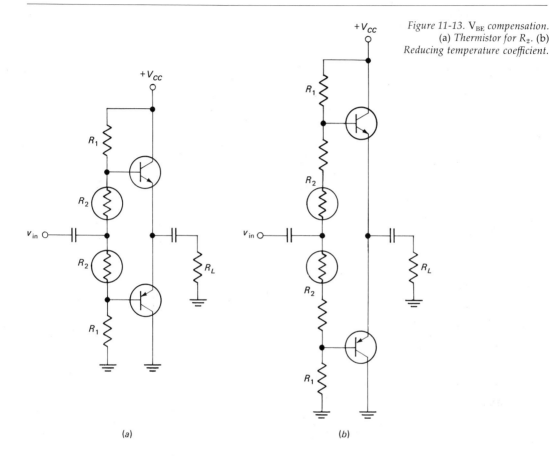

Figure 11-13. V_{BE} compensation. (a) Thermistor for R_2. (b) Reducing temperature coefficient.

(a)

(b)

is possible to compensate (at least partially) for the changes in barrier potential over a wide temperature range.

Diode compensation

Better yet, we can compensate by using the very device causing the trouble: a diode. Figure 11-14 shows the idea behind *diode compensation*. We select rectifier diodes whose curves are similar to the curves of the emitter diodes. This allows us to set up the correct V_{BEQ} at room temperature. When the temperature increases, the barrier potential of the external diodes decreases; automatically, the voltage fed to the emitter diodes decreases. In this way, the emitter diodes receive approximately the correct V_{BEQ} over a wide temperature range.

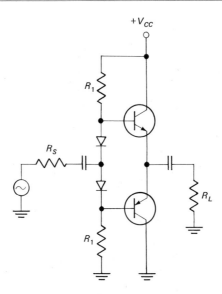

Figure 11-14. Diode compensation for V$_{BE}$ *changes.*

In discrete circuits, the success of diode compensation boils down to finding a compensating diode whose curve is a close match to the emitter-diode curve. This is not an easy thing to do, and you usually have to settle for partial compensation.

With integrated circuits, the same chip of semiconductor material is used for the compensating diode and the emitter diode. Because of this, the diode curves are almost identical; diode compensation is the main method used in integrated circuits.

Changes in diode voltage versus current

But, whether the circuit is discrete or integrated, another problem exists. Figure 11-15a shows the curve for an emitter diode; Fig. 11-15b is for a compensating diode. The horizontal distance between the room-temperature curve and the elevated-temperature curve depends on how much current is flowing. Near the knee, the change in V_{BE} is approximately the same as the change in barrier potential, around 2.5 mV per degree (Fig. 11-15a). But further up the curve, the change in V_{BE} may be as low as 1 mV per degree (Fig. 11-15b). With a circuit like Fig. 11-14, *the compensating diodes need more quiescent current* than the emitter diodes; otherwise, the ac signal is not coupled properly. For this reason, even though the compensating and emitter diodes have identical curves, the best you can do in mass production is a partial compensation for V_{BE} changes.

EXAMPLE 11-8.

The β_{dc} of each Darlington pair in Fig. 11-16 is 2500. Calculate $z_{in(base)}$ and the *quiescent diode current* I_{DQ}.

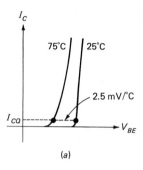

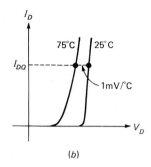

Figure 11-15. Diode temperature coefficient. (a) Emitter diode. (b) Compensating diode.

SOLUTION.

$$z_{\text{in(base)}} \cong \beta_{\text{dc}} R_L = 2500(10) = 25 \text{ k}\Omega$$

and applying Ohm's law to the diode branch,

$$I_{DQ} = \frac{40 - 4V_{BEQ}}{4000} \cong \frac{40}{4000} = 10 \text{ mA}$$

Note that R_1 is 2 kΩ, which is less than one-tenth $z_{\text{in(base)}}$. We have not discussed the size of R_1 yet, but to avoid clipping the input signal, the current through the com-

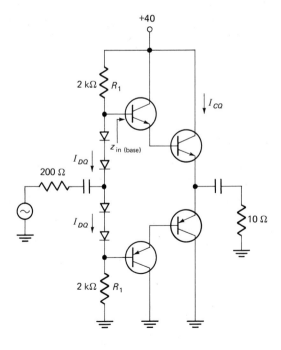

Figure 11-16. Darlington push-pull emitter follower.

pensating diodes must flow throughout the cycle. For this to happen, I_{DQ} must be large enough to allow for the increased base current and the change in current through R_1 at the signal peak. By an advanced derivation, the design rule that guarantees enough I_{DQ} is $R_1 \ll z_{in(base)}$. Therefore, whenever you use diode compensation, you should try to keep R_1 much less than $z_{in(base)}$; this prevents the compensating diodes from turning off before reaching the peaks of the input signal.

EXAMPLE 11-9.
Calculate the no-signal current drain, maximum load power, and $P_{D(max)}$ ratings needed in Fig. 11-16. Use a trickle bias of I_{CQ} equals 2 percent $I_{C(sat)}$.

SOLUTION.
First, we draw the ac load line. To do this, note V_{CEQ} equals 20 V. Since $r_C + r_E = 10\ \Omega$, $I_{C(sat)} = 2$ A. Figure 11-17 shows the ac load line.
The *no-signal current drain* $I_{d(no\ sig)}$ from the supply is

$$I_{d(no\ sig)} \cong I_{DQ} + I_{CQ}$$

In the preceding example, we found $I_{DQ} = 10$ mA. I_{CQ} is 2 percent of $I_{C(sat)}$, which equals 40 mA. Therefore,

$$I_{d(no\ sig)} = 10\ \text{mA} + 40\ \text{mA} = 50\ \text{mA}$$

With Eq. (11-7), the maximum load power equals

$$P_{load} = 0.5(20)2 = 20\ \text{W}$$

The worst-case transistor dissipation is one-fifth of P_{load}; therefore, each output transistor needs a rating of at least

$$P_{D(max)} = \frac{20\ \text{W}}{5} = 4\ \text{W}$$

The first transistor in each Darlington pair can have a lower $P_{D(max)}$ rating than 4 W because the current is smaller by a factor of β_2. Assuming a β_{dc} of 50, the first transistor in each Darlington pair needs a $P_{D(max)}$ rating of at least

$$P_{D(max)} = \frac{4\ \text{W}}{50} = 80\ \text{mW}$$

Figure 11-17. Load line.

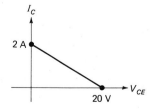

11-7. COMPLEMENTARY PAIRS

When you use Darlington pairs in a class B amplifier, you need four compensating diodes instead of two (see Fig. 11-18a). This usually makes it harder to compensate over a large temperature range. For this reason, pairs of transistors are sometimes connected as shown in Fig. 11-18b.

Now we are using complementary transistors. The upper pair acts like an *npn* transistor. The lower pair acts like a *pnp* transistor. Because of this, you often will see these *complementary pairs* used instead of Darlington pairs. The action of these complementary pairs is similar to the action of Darlington pairs. In Fig. 11-18c,

$$z_{\text{in(base)}} \cong \beta_{\text{dc}} R_L$$

where β_{dc} is the product of $\beta_1 \beta_2$. Also, the quiescent base current for each input transistor is

$$I_{BQ} = \frac{I_{CQ}}{\beta_{\text{dc}}}$$

One advantage of complementary pairs over Darlington pairs is simpler diode compensation. In Fig. 11-18c, we need only two compensating diodes instead of four. Besides reducing the diode count, this usually results in better compensation over a large temperature range.

A second advantage is that we have to compensate for V_{BE} changes only in the *first* transistor. Under changing signal conditions, the power dissipation in a transistor changes; this changes the internal temperature of the transistor. Because the first transistor dissipates much less power than the second, the V_{BE} changes caused by a change in signal level are much smaller in the first transistor. Because of this, compensation is better with complementary pairs.

11-8. OTHER CLASS B AMPLIFIERS

We have concentrated on the class B push-pull emitter follower because it is one of the most important class B circuits. Most of the ideas and formulas discussed earlier apply to other class B circuits.

Complementary CE push-pull amplifier

Figure 11-19a shows two CE amplifiers connected in a push-pull circuit. Resistors R_1 and R_2 set up trickle bias as before. The positive half cycle of source voltage turns on the lower transistor. During this time, the lower transistor acts like a CE amplifier. Because of this, the voltage gain, power gain, and input impedance are found by the usual methods of analysis. For Fig. 11-19a, here are the formulas of interest:

Figure 11-18. (a) Darlington push-pull circuit. (b) Complementary pairs. (c) Push-pull emitter follower using complementary pairs.

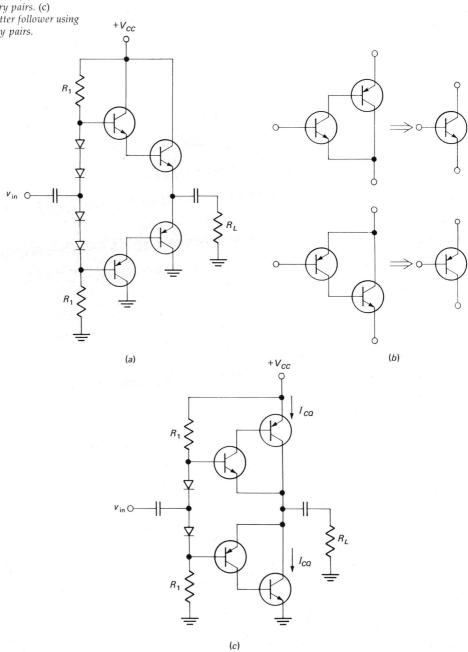

(a)

(b)

(c)

Figure 11-19. (a) Push-pull CE amplifier. (b) Push-pull swamped amplifier.

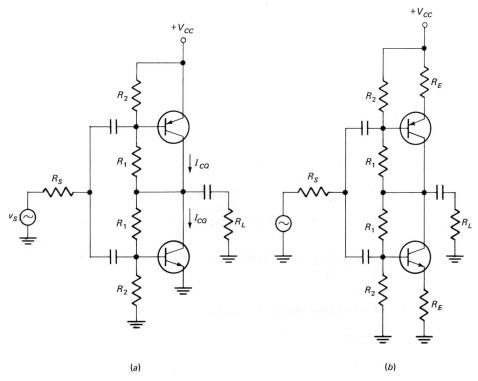

(a)   (b)

$$\frac{v_c}{v_b} \cong \frac{R_L}{R'_e}$$

$$\frac{p_c}{p_b} = \beta_{dc} \frac{v_c}{v_b}$$

and $\qquad z_{in(base)} \cong \beta_{dc} R'_e$

The advantage of a push-pull CE amplifier is voltage gain as well as power gain; the disadvantage is more nonlinear distortion than with an emitter follower. The nonlinear distortion is often bad enough that you have to swamp the emitter diodes as shown in Fig. 11-19b. This reduces the distortion because the nonlinearity of the emitter diodes is the cause of the nonlinear distortion; by swamping the emitter diodes, you reduce the nonlinear effect the emitter diodes have on voltage gain. With heavy swamping,

$$\frac{v_c}{v_b} \cong \frac{R_L}{R_E}$$

$$\frac{p_c}{p_b} \cong \beta_{\text{dc}} \frac{R_L}{R_E}$$

and
$$z_{\text{in(base)}} \cong \beta_{\text{dc}} R_E$$

Transformer-coupled push-pull amplifiers

The increasing availability of complementary transistors has done away with the use of transformers in many applications. But there are still applications where a circuit like Fig. 11-20a is used. As far as dc operation is concerned, you can see that voltage divider R_1 and R_2 forward-biases each emitter diode. As usual, you select values to set up a trickle bias. Also note that both transistors are npn in Fig. 11-20a.

With the dots shown, the positive half cycle of source voltage turns on the upper transistor, producing a negative-going voltage on the collector. Because of the dots on the output transformer, the load voltage is negative-going.

The negative half cycle of source voltage results in complementary action; the lower transistor turns on, producing a positive-going load voltage.

Because of the center-tapped windings, here are the reflected impedances. The base sees a reflected source impedance of

$$r_B = \frac{R_S}{4n^2}$$

The collector sees a reflected ac load resistance of

$$r_C = \left(\frac{n}{2}\right)^2 R_L$$

With these reflected impedances, we can draw the ac equivalent circuit of Fig. 11-20b. R_2 is in series with $R_S/4n^2$. Because of this, the equivalent Thevenin circuit is a voltage source with an impedance of $R_2 + R_S/4n^2$.

Figure 11-20c shows the simplified ac equivalent circuit for the positive half cycle of source voltage; the other half cycle is complementary. Figure 11-20c is the base-driven prototype we have been analyzing for several chapters; we can easily find voltage gain, power gain, input impedance, etc., using the methods of earlier chapters.

Each transistor in Fig. 11-20a has a V_{CEQ} approximately equal to the supply voltage (there is a small voltage loss in each half of the output primary). Because the reflected ac load resistance is

$$r_C = \left(\frac{n}{2}\right)^2 R_L$$

we can draw the ac load line as shown in Fig. 11-20d. With this load line, we can calculate load power, dissipation power, etc.

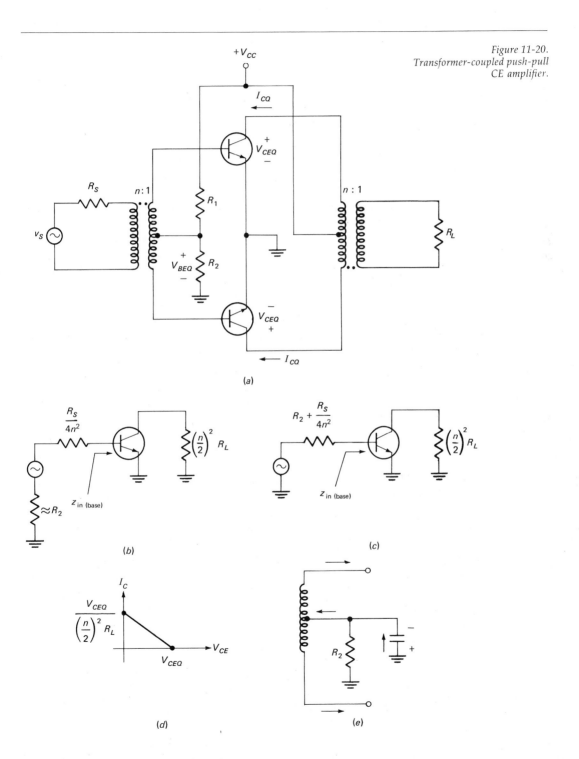

Figure 11-20.
Transformer-coupled push-pull
CE amplifier.

Unwanted clamping

Do you recall the unwanted clamping discussed in Chap. 5? The idea was that a coupling capacitor driving an unbalanced diode load would charge until it prevented the signal from reaching the load. In other words, whenever a capacitor works into an unbalanced load (more charging current than discharging current), the capacitor voltage can build up and produce unwanted clamping.

The class B push-pull amplifier is susceptible to unwanted clamping because each half of the circuit acts like an unbalanced load. Unless we are careful about how we use coupling and bypass capacitors, the dc voltages on these capacitors can change when a signal is present; this will disturb the circuit operation.

For example, resistor R_2 in Fig. 11-20a is part of the biasing network. Since it is in series with the center tap, ac current flows through R_2; this means we are losing some ac signal in R_2. Why not bypass R_2 as shown in Fig. 11-20e? This will put the center tap at ac ground and prevent loss of signal voltage and power. It won't work! You will get unwanted clamping. On each half cycle, the ac current flows *into* the center tap. Therefore, the bypass capacitor will charge negatively on both half cycles. Because of this, the negative voltage builds up until the input signal is clamped off. In other words, the negative voltage across the capacitor eventually prevents the transistors from turning on. This is why we do not bypass R_2 in Fig. 11-20a.

Whenever you add a capacitor to a class B circuit, remember you may get unwanted clamping. To avoid unwanted clamping, the charging and discharging current should be equal. If this condition is satisfied, no unwanted clamping occurs. If not, you may get unwanted clamping action; it depends on the exact circuit.

EXAMPLE 11-10.
In Fig. 11-21a, what is the voltage gain from base to collector? The input impedance of the base? ($R'_e = 1\ \Omega$.)

SOLUTION.
The swamping resistors in each emitter provide heavy swamping. Therefore,

$$\frac{v_c}{v_b} = \frac{R_L}{R_E} = \frac{500}{50} = 10$$

and the input impedance is

$$z_{\text{in(base)}} = \beta_{\text{dc}} R_E = 100(50) = 5\ \text{k}\Omega$$

EXAMPLE 11-11.
What is maximum load power in Fig. 11-21a? What $P_{D(\text{max})}$ rating do the transistors need?

SOLUTION.
The V_{CEQ} of each transistor is 10 V. Therefore,

Figure 11-21. Examples 11-10 and 11-11.

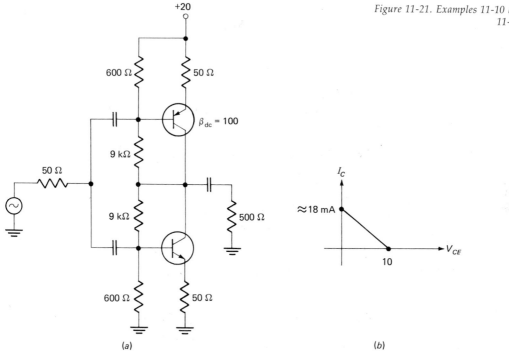

(a)

(b)

$$I_{C(\text{sat})} = \frac{V_{CEQ}}{r_C + r_E} = \frac{10}{500 + 50} \cong 18 \text{ mA}$$

Figure 11-21b shows the ac load line. If the signal uses the entire ac load line, the maximum load power equals

$$P_{\text{load}} = 0.5 V_{CEQ} I_{C(\text{sat})} = 0.5(10)0.018$$
$$= 90 \text{ mW}$$

The transistors need a $P_{D(\text{max})}$ rating of at least one-fifth of this, or

$$P_{D(\text{max})} = 18 \text{ mW}$$

If you had to temperature-compensate a push-pull CE amplifier like Fig. 11-21a, you would use a diode in the place of each 600-Ω resistor. You would also have to reduce the 9-kΩ resistors to a value much less than $\beta_{\text{dc}} r_E$, as discussed in Example 11-8.

EXAMPLE 11-12.
Calculate the voltage gain from base to collector and the maximum load power in Fig. 11-22a. ($R'_{e2} = 0.1 \ \Omega$.)

Fig. 11-22. Example 11-12.

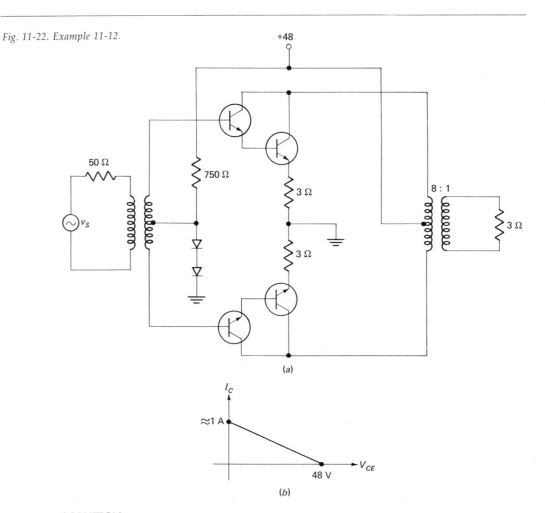

(a)

(b)

SOLUTION.

The reflected ac load resistance seen by each collector is

$$r_C = \left(\frac{n}{2}\right)^2 R_L = (4)^2 3 = 48 \ \Omega$$

The voltage gain from base to collector is

$$\frac{v_c}{v_b} = \frac{r_C}{r_E + R'_e} = \frac{48}{3 + 0.2} = 15$$

Figure 11-22b shows the ac load line found by

$$V_{CEQ} = 48 \ V$$

and
$$I_{C(\text{sat})} = \frac{V_{CEQ}}{r_C + r_E} = \frac{48}{48 + 3} \cong 1 \text{ A}$$

The maximum load power equals

$$P_{\text{load}} = 0.5 V_{CEQ} I_{C(\text{sat})} \cong 0.5(48)1 = 24 \text{ W}$$

Problems

11-1. Figure 11-23a is the ac equivalent of a class B push-pull amplifier. If each transistor has a $V_{CEQ} = 28$ V, what does $I_{C(\text{sat})}$ equal?

11-2. In Fig. 11-23a, $I_{C(\text{sat})}$ equals 2.25 A. What is the V_{CEQ} across each transistor?

11-3. What value does $I_{C(\text{sat})}$ have in Fig. 11-23b when

1. $V_{CEQ} = 10$ V
2. $V_{CEQ} = 25$ V
3. $V_{CEQ} = 80$ V

11-4. In Fig. 11-23c, what does $I_{C(\text{sat})}$ equal if V_{CEQ} is 24 V?

11-5. The transformer of Fig. 11-23d steps up the impedance by n^2. If each transistor has a V_{CEQ} of 25 V, what is the value of $I_{C(\text{sat})}$?

11-6. The 8:1 turns ratio of Fig. 11-23e is for the total primary winding. What does $I_{C(\text{sat})}$ equal if $V_{CEQ} = 50$ V?

11-7. A class B push-pull amplifier has $V_{CEQ} = 16$ V. Figure 11-23a is its ac equivalent circuit. By experiment, you find it takes 2 percent of $I_{C(\text{sat})}$ to eliminate crossover distortion. How much is this standby current?

11-8. The V_{CEQ} in Fig. 11-23b is 30 V. If it takes 4 percent of $I_{C(\text{sat})}$ to eliminate crossover distortion, how much current does the amplifier drain from the supply when no ac signal is present?

11-9. In the dc equivalent circuit of Fig. 11-24a, what is the approximate voltage across the upper 680-Ω resistor?

11-10. In Fig. 11-24a, $I_{CQ} = 5$ mA and $\beta_{\text{dc}} = 50$. What is the value I_{BQ}? What value does $0.7/I_{BQ}$ have? Is condition (11-2b) satisfied? What is the total current drain from the supply?

11-11. We want to use a V_{CC} of 30 V instead of 20 V in Fig. 11-24a. To set up a V_{BEQ} of 0.68 V, we can change the 9.3-kΩ resistors. What new value should we use in the place of these resistors?

11-12. How much voltage is there across the upper 220-Ω resistor of Fig. 11-24b? How much current does the voltage divider draw from the supply (neglect I_{BQ})?

11-13. Each Darlington pair of Fig. 11-24b has a β_{dc} of 3000. $I_{C(\text{sat})}$ is 20 A, and we need 1 percent of this to eliminate crossover distortion. What is the value of I_{BQ}? The value of $0.7/I_{BQ}$? Is condition (11-2b) satisfied? What is the total current drain from the supply?

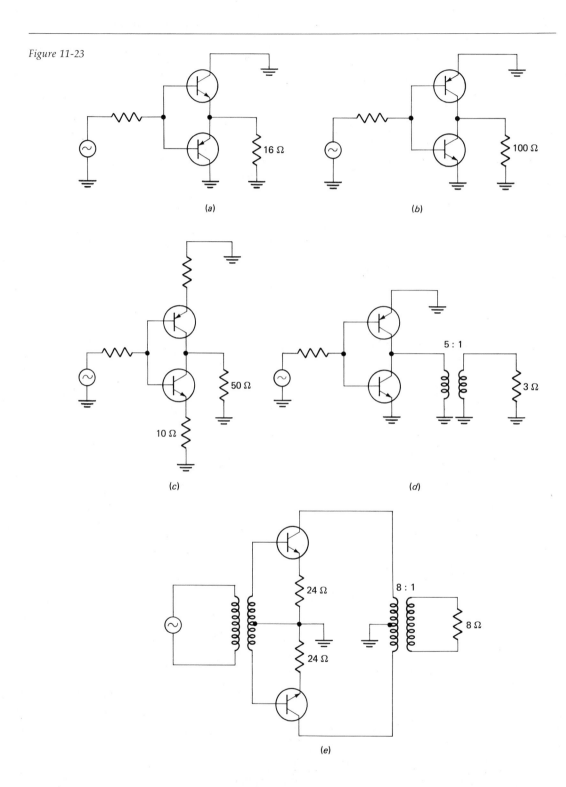

Figure 11-23

(a)

(b)

(c)

(d)

(e)

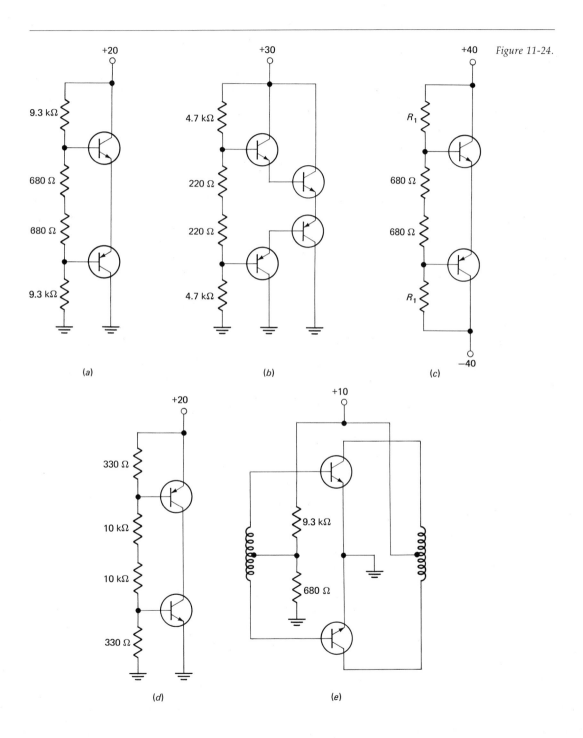

Figure 11-24.

(a)

(b)

(c)

(d)

(e)

11-14. In the two-supply circuit of Fig. 11-24c, we need a V_{BEQ} of 0.685 V to set up the correct trickle bias. What value of R_1 do we have to use? If you use a 10 percent tolerance resistor for R_1 and R_2, what is the maximum possible V_{BEQ} you can get in mass production?

11-15. In the CE circuit of Fig. 11-24d, what is the voltage across the upper 330-Ω resistor? Across the 10-kΩ resistor? If the bias is normal, are the transistors silicon or germanium?

11-16. Figure 11-24e shows part of a transformer-coupled class B push-pull amplifier. The 680-Ω resistor provides a forward bias for each npn transistor. How much voltage is there across this 680-Ω resistor? If each transistor has an I_{CQ} of 1.5 mA, how much no-signal current drain is there on the power supply?

11-17. What is the large-signal input impedance of the base of the conducting transistor in Fig. 11-25a? The power gain from base to emitter? The voltage gain from base to emitter?

11-18. In Fig. 11-25b, what is the input impedance of the base? The power gain from base to emitter? How much dc voltage is there across the 20-Ω load resistor?

11-19. The Darlington pairs of Fig. 11-25c have $\beta_{dc} = 5000$ and $R'_e = 0.1\ \Omega$. What is the approximate input impedance of the base? The power gain from base to emitter?

11-20. Some of the ac source voltage in Fig. 11-25a is dropped across resistances between the source and the base. If the source has a peak voltage of 10 V, what is the peak voltage at the base of the conducting transistor?

11-21. To deliver a maximum load power of 50 W, what $P_{D(\text{max})}$ rating do the transistors of a class B push-pull amplifier need? What does P_D equal for 100 percent swing? 63 percent swing?

11-22. The ac signal of Fig. 11-25a uses the entire ac load line. What is the maximum load power? The power dissipation of each transistor? If the signal is reduced, what is the worst-case power dissipation in each transistor?

11-23. Calculate the maximum load power in Fig. 11-25b. What is the minimum $P_{D(\text{max})}$ rating of each transistor?

11-24. The data sheets of a particular set of complementary transistors say the $P_{D(\text{max})}$ rating is 12 W for a case temperature of 25°C (the case is the body or housing of the transistor). The derating factor is 68 mW per degree (derating factor was discussed in Example 10-8). Will these transistors be able to dissipate the worst-case P_D in Fig. 11-25b when the case temperature is 25°C? If so, what is the highest case temperature at which these transistors will still be able to dissipate the worst-case P_D?

11-25. What is the maximum load power in Fig. 11-25c? The second transistor in each Darlington pair has to handle the worst-case power dissipation. What $P_{D(\text{max})}$ rating does it need? The first transistor in each Darlington pair has a signal current that is smaller than load current by a β of 50. What is the worst-case power dissipation for this first transistor?

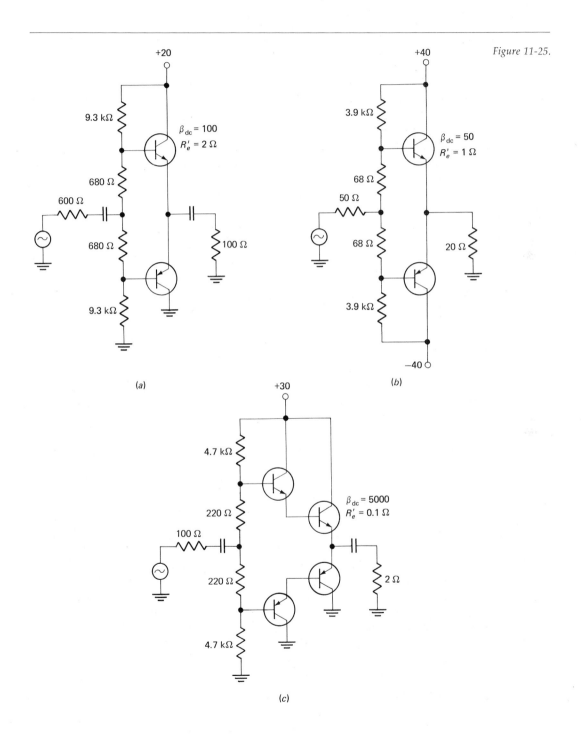

Figure 11-25.

(a)

(b)

(c)

11-26.   The ac source of Fig. 11-25c supplies only a small amount of power which you can neglect in this problem. The V_{CC} supply, on the other hand, must supply power for the voltage divider, for each transistor dissipation, and for the final load resistance. If the load power is maximum, how much dc current drain is there from the supply?

11-27.   A thermistor has a resistance of 1000 Ω at 25°C. If it has a negative-temperature coefficient of -1 percent per degree Celsius, what is its resistance at 35°C?

11-28.   Figure 11-26a shows a voltage divider with a thermistor resistance of 680 Ω at 25°C. If the temperature rises to 26°C, and if the thermistor changes -1 percent, how much does the voltage across the thermistor change?

11-29.   In Fig. 11-26b, an ordinary 430-Ω resistor is in series with a thermistor whose resistance is 250 Ω at 25°C. If the thermistor changes -1 percent for each degree rise, what is the decrease in V_{BEQ} for each degree rise?

11-30.   Figure 11-26c shows a diode supplying V_{BEQ}. Figure 11-26d is the diode curve. How much voltage is there across the diode? If we change the 1-kΩ resistor to a 2-kΩ resistor, how much voltage is there across the diode?

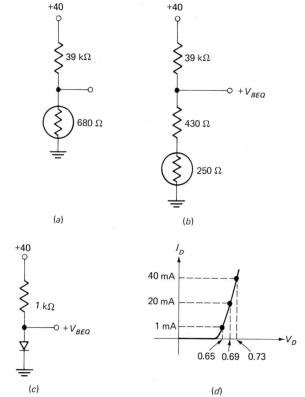

Figure 11-26.

Figure 11-27.

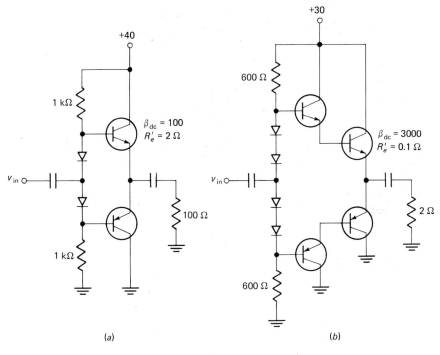

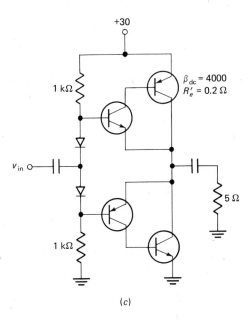

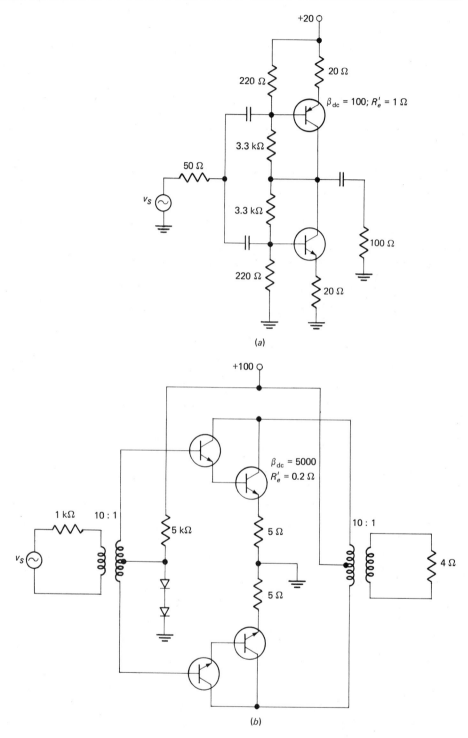

Figure 11-28.

(a)

(b)

11-31. The supply voltage of Fig. 11-26c is changed from 40 V to 25 V. To set up a V_{BEQ} of 0.69 V, what new value do we need for the series resistor? (Fig. 11-26d is the curve of the diode.)

11-32. How much current is there in the compensating diodes of Fig. 11-27a? What is the value of $z_{in(base)}$? How much smaller is R_1 than $z_{in(base)}$?

11-33. What is the approximate current in the compensating diodes of Fig. 11-27b? The value of $z_{in(base)}$?

11-34. In Fig. 11-27a and b, I_{CQ} equals 1 percent of $I_{C(sat)}$.

1. What is the total no-signal current drain in Figs. 11-27a?
2. In Fig. 11-27b?

11-35. What does $z_{in(base)}$ equal in Fig. 11-27c? If we use 1 percent of $I_{C(sat)}$ to eliminate crossover distortion, how much I_{CQ} is there? What is the total current drain with no signal?

11-36. In Fig. 11-28a, what is the voltage gain from base to collector? The input impedance of the base?

11-37. What is the maximum load power in Fig. 11-28a? What $P_{D(max)}$ rating do the transistors need?

11-38. If we use an I_{CQ} of 5 percent $I_{(sat)}$ in Fig. 11-28a, what is the total no-signal current drain on the V_{CC} supply?

11-39. In Fig. 11-28b, what is the reflected impedance seen by each collector? What is the reflected source impedance seen by each base? What is the voltage and power gain from base to collector?

11-40. What is the voltage gain from source to final output in Fig. 11-28b?

11-41. In Fig. 11-28b, what is the maximum power delivered to $r_C + r_E$? How much power does the 4-Ω resistor receive?

12. Class C Power Amplifiers

A *class C amplifier* can deliver more load power than a class B amplifier. To amplify a sine wave, however, it has to be *tuned* to the sine-wave frequency. Because of this, the tuned class C amplifier is a *narrowband* circuit; it can amplify only the resonant frequency and those frequencies near it.

To avoid large inductors and capacitors in the resonant circuit, class C amplifiers have to operate at *radio frequencies* (above 20 kHz). Therefore, even though it is the most efficient of all classes, class C is useful only when we want to amplify a narrow band of radio frequencies (RF).

Class C power amplifiers normally use *RF power transistors;* these have characteristics optimized for RF signals. Typical $P_{D(\max)}$ ratings for RF power transistors are from 1 W to over 75 W.

12-1. BASIC CLASS C ACTION

Class C means collector current flows for less than 180°. In a practical class C circuit, the current flows for much less than 180°, and looks like the narrow pulses of Fig. 12-1*a*. As will be discussed later, when narrow *current* pulses like these drive a high-Q resonant circuit, the *voltage* across the circuit is almost a perfect sine wave.

Clamping

Before discussing tuned circuits, we need the basic idea of a class C *untuned* circuit (Fig. 12-1*c*). Without an ac input signal, no collector current flows because the emitter

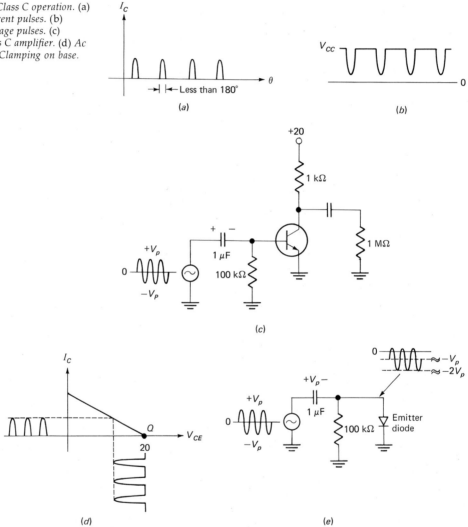

Figure 12-1. Class C operation. (a) Collector current pulses. (b) Collector voltage pulses. (c) Untuned class C amplifier. (d) Ac load line. (e) Clamping on base.

diode is unbiased (not even trickle bias).[1] Therefore, the Q point is at cutoff as shown in Fig. 12-1*d*.

What happens when the ac signal comes in? Section 5-11 was about *clamping action*. The idea in a clamper is to charge a coupling capacitor to approximately the peak voltage of the input signal. This is precisely what we do in Fig. 12-1*c*. The input coupling capacitor, the 100-kΩ base resistor, and the emitter diode form a negative

[1] We are neglecting the small effects of I_{CBO}.

clamper as shown in Fig. 12-1e. The first few positive cycles turn on the emitter diode and charge the capacitor with the plus-minus polarity shown. On the negative half cycles, the only discharge path is through the 100-kΩ resistor. As long as the period T of the input signal is much smaller than the RC discharging time constant, the capacitor loses only a small amount of its charge. Therefore, after a few cycles, the capacitor voltage builds up to approximately the peak voltage of the input signal. This produces the familiar negatively clamped waveform shown in Fig. 12-1e.

To replace the small amount of capacitor charge lost during each cycle, the base voltage swings above zero and turns on the emitter diode briefly at each positive peak. The *conduction angle* is much less than 180°. This is why the collector current waveform is a train of narrow pulses as shown in Fig. 12-1a. When these current pulses flow through the ac collector resistance, they produce the negative-going pulses of Fig. 12-1b.

In terms of the ac load line, here is what happens. At each positive peak of the base voltage, the emitter diode turns on briefly. This forces the instantaneous operating point to move from cutoff toward saturation. In the process, we get the narrow current and voltage pulses shown in Fig. 12-1d.

Emitter breakdown

Now is a good time to mention BV_{EBO}, the breakdown voltage from emitter to base with an open collector (equivalent to zero collector current). Since the clamped waveform of Fig. 12-1e reaches a negative peak of approximately $-2V_p$, this is the maximum reverse voltage across the emitter diode. Unless the emitter diode can withstand this voltage, it will break down. Data sheets for RF power transistors list BV_{EBO} or its equivalent $V_{BE(max)}$ to indicate the breakdown voltage of the emitter diode. This rating has to be greater than $2V_p$ (approximately); otherwise, the emitter diode breaks down.

As an example, the data sheet of a 2N3950 gives a $V_{BE(max)}$ rating of 4 V. If we use this transistor in Fig. 12-1c, the emitter diode does not break down as long as the peak-to-peak input voltage is less than approximately 4 V. But when the peak-to-peak voltage is greater than 4 V, the emitter diode will break down.

$V_{BE(max)}$ ratings are typically less than 5 V for RF power transistors. Often, you may have an input signal whose peak-to-peak value is greater than 5 V. In a case like this, add a diode in series with the base (Fig. 12-2a) or in series with the emitter (Fig. 12-2b). On the positive half cycle, both diodes conduct and the capacitor charges as before. On the negative half cycle, the rectifier diode with its larger breakdown voltage prevents the emitter diode from breaking down. (Almost any rectifier diode has a breakdown voltage greater than 25 V.)

EXAMPLE 12-1.
In Fig. 12-2a, at what the frequency does the discharge time constant equal 10 times the period?

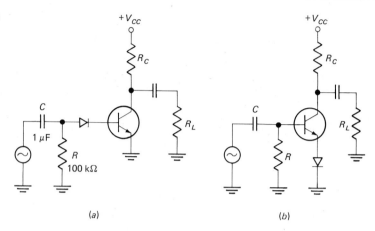

Figure 12-2. Avoiding breakdown of emitter diode.

(a)   (b)

SOLUTION.

The discharge time constant equals

$$RC = 10^5 (10^{-6}) = 0.1 \text{ s}$$

One-tenth of this is 0.01 s, and the frequency is

$$f = \frac{1}{0.01} = 100 \text{ Hz}$$

12-2. OTHER CLASS C RELATIONS

Figure 12-3a shows an untuned class C amplifier including an emitter resistor. If we Thevenize the base circuit, the Thevenin impedance is the parallel of C and R_B. Therefore, Fig. 12-3b is an alternative form of untuned class C amplifier; you can use the RC network of Fig. 12-3a or b, whichever is more convenient in a particular application.

Saturation current and average base voltage

As already mentioned, the Q point is at cutoff as shown in Fig. 12-3c.[2] Again, the general load line of Fig. 10-5b applies, so that the saturation current equals

$$I_{C(\text{sat})} \cong I_{CQ} + \frac{V_{CEQ}}{r_C + r_E}$$

Since $I_{CQ} = 0$ and $V_{CEQ} = V_{CC}$,

$$I_{C(\text{sat})} \cong \frac{V_{CC}}{r_C + r_E} \qquad \text{(untuned class C)} \qquad (12\text{-}1)$$

[2] This is approximate because of the small effects of I_{CBO}.

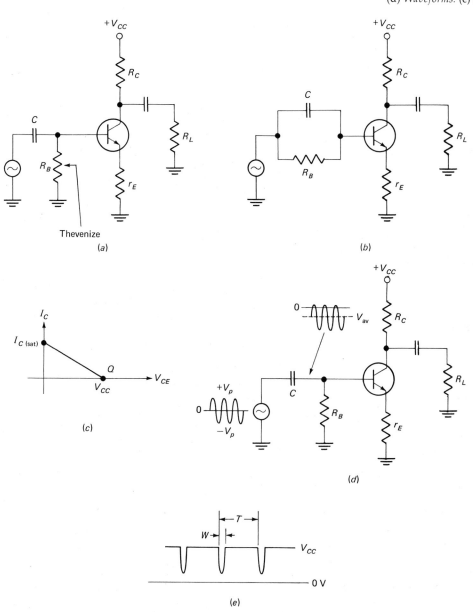

Figure 12-3. Untuned class C amplifier. (a) *Circuit.* (b) *Equivalent circuit.* (c) *Load line.* (d) *Waveforms.* (e) *Duty cycle.*

This is a close approximation; for a more accurate value, you have to subtract $V_{CE(\text{sat})}$ from V_{CC} in the numerator.

When a signal is present, we get the voltage waveforms of Fig. 12-3d. Ideally, the capacitor voltage is

$$V_C \cong V_p$$

But to a second approximation, we must allow for the drop across the emitter diode; therefore, a more accurate value for capacitor voltage is

$$V_C \cong V_p - 0.7 \tag{12-2}$$

The base voltage is a negatively clamped waveform as shown. Ideally, the average voltage of this waveform is

$$V_{\text{av}} \cong -V_p$$

To a second approximation, the waveform actually goes above the zero level by about 0.7 V; this means the average base voltage is 0.7 V greater than $-V_p$. In symbols,

$$V_{\text{av}} \cong -V_p + 0.7 \tag{12-3}$$

Base-voltage check

When working with class C circuits, one of the troubleshooting checks is the average base voltage given by Eq. (12-3). As an example, if the input signal has a peak of 4 V,

$$V_{\text{av}} = -4 + 0.7 = -3.3 \text{ V}$$

With a high-impedance dc voltmeter, you can measure the dc voltage from base to ground. If you get a reading around -3.3 V, you know the clamper is working. You must use a high-impedance dc voltmeter when measuring this base voltage to avoid changing the RC time constant.

This voltmeter check is useful when you do not have an oscilloscope. But with an oscilloscope, you can look at the base-voltage waveform and can see whether or not the clamper is working properly.

Duty cycle

The brief turn-on of the emitter diode at each positive peak produces narrow pulses of collector current. When this current flows through r_C, we get the negative-going voltage pulses shown in Fig. 12-3e. With pulses like these, we often refer to the *duty cycle,* defined as

$$\text{Duty cycle} = \frac{W}{T} \times 100 \text{ percent} \tag{12-4}^{***}$$

where W is the width of each pulse and T the period (see Fig. 12-3e). For instance, suppose an oscilloscope shows a pulsed waveform with each pulse being 0.2 μs wide and the period being 1.6 μs. Then, the duty cycle of the waveform is

$$\text{Duty cycle} = \frac{0.2}{1.6} \times 100 \text{ percent} = 12.5 \text{ percent}$$

The duty cycle affects transistor power dissipation. Each pulse of Fig. 12-3e appears when the transistor is conducting. Between pulses, the transistor is off. The smaller the duty cycle, the less the average power dissipation of the transistor.

There are many applications in pulse and digital work where you may want narrow voltage pulses like Fig. 12-3e. But most RF applications require a sine-wave output. The next section talks about this.

EXAMPLE 12-2.
Draw the ac load line for the untuned class C amplifier of Fig. 12-4a.

SOLUTION.
V_{CEQ} equals 30 V because under no-signal conditions the collector current is zero and the collector voltage equals V_{CC}. Ac resistance r_C is the parallel of 100 Ω and 20 kΩ, which is 100 Ω to a close approximation. The ac emitter resistance r_E is adjustable from 0 to 5 Ω. Even when $r_E = 5$ Ω, $r_C + r_E$ is approximately equal to 100 Ω. So,

$$I_{C(\text{sat})} \cong \frac{V_{CEQ}}{r_C + r_E} \cong \frac{30}{100} = 0.3 \text{ A}$$

Incidentally, by adjusting r_E, you can change the size of the collector current pulses.

Figure 12-4b shows the ac load line. If the operating point swings over the entire load line, the current pulses will have a peak value of 0.3 A and the voltage pulses a peak of 30 V (negative-going).

EXAMPLE 12-3.
The frequency of the input signal in Fig. 12-4a is 250 kHz. If the pulses of current look like Fig. 12-4c, what is the duty cycle of the current waveform?

SOLUTION.
First, get the period of the input signal.

$$T = \frac{1}{f} = \frac{1}{250(10^3)} = 4 \ \mu s$$

Since the input signal produces the current waveform of Fig. 12-4c, the period of the current waveform equals 4 μs. The pulse width is given as 0.1 μs. With Eq. (12-4), the duty cycle is

$$\text{Duty cycle} = \frac{0.1}{4} \times 100 \text{ percent} = 2.5 \text{ percent}$$

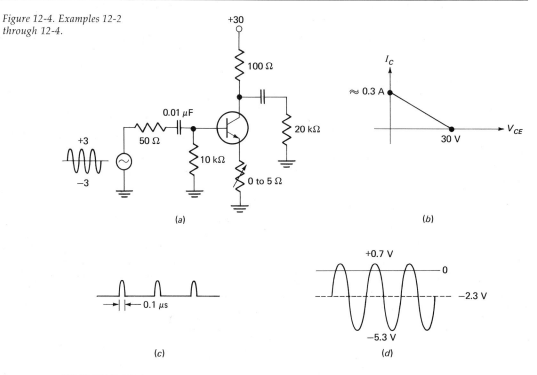

Figure 12-4. Examples 12-2 through 12-4.

(a)

(b)

(c)

(d)

EXAMPLE 12-4.

If you use an oscilloscope to look at voltage waveforms, what voltage waveform will you see from base to ground in Fig. 12-4a?

SOLUTION.

The input circuit is a negative clamper. Ideally, you see a negatively clamped sine wave swinging from the 0 V to −6 V. To a second approximation, however, we must include the 0.7 V or thereabouts needed to turn on the emitter diode. The peak-to-peak voltage of the base waveform is still 6 V, the same as the input. But the base voltage must rise to approximately 0.7 V at each positive peak. In other words, the 6-V peak-to-peak sine wave has a positive peak of approximately 0.7 V as shown in Fig. 12-4d. The average value of −2.3 V and the negative peak of −5.3 V are also shown.

12-3. THE TUNED CLASS C AMPLIFIER

How do we get sine waves from narrow current pulses? By using a resonant circuit in the collector. Figure 12-5a shows one of several possible tuned circuits. As before, the base circuit acts like a negative clamper, and the emitter diode turns on briefly at each positive peak. This results in narrow collector current pulses as shown in Fig. 12-5b.

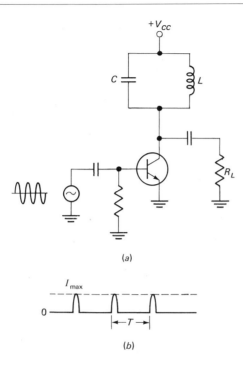

$+V_{CC}$

C L

R_L

(a)

I_{max}

0

$\leftarrow T \rightarrow$

(b)

The *fundamental frequency* of a pulse train like Fig. 12-5b is defined as

$$f_1 = \frac{1}{T}$$

(12-5)***

where f_1 is called the fundamental or *first harmonic*. The topic of *harmonics* is important enough to merit in-depth study; for this reason, we postpone detailed discussion of harmonics until later. For now, all we need to know is that a pulse waveform like Fig. 12-5b is made up of *many sine-wave frequencies*. The fundamental or first harmonic is the lowest frequency. As an example, if T equals 10 μs in Fig. 12-5b, the fundamental or first harmonic equals 100 kHz. This is the lowest frequency in the pulsed waveform.

When narrow current pulses like Fig. 12-5b drive a resonant circuit, we can produce an almost perfect sine wave of voltage. To get a sine wave with the fundamental frequency, here is what we do. First, the resonant frequency has to equal the fundamental frequency of the pulsed waveform. Second, the tuned circuit must have a high Q (greater than 10) to get an almost perfect sine wave of voltage. As an example, if the pulse waveform of Fig. 12-5b has a period of 10 μs, the fundamental frequency is 100 kHz. By tuning the LC tank of Fig. 12-5a to 100 kHz, we can get an almost perfect sine wave of voltage across the resonant tank when the Q is greater than 10.

A satisfactory explanation of how a resonant circuit produces a sine wave of voltage from pulses of current is given later when we discuss harmonics. The best we can do at the moment is describe the circuit action in terms of capacitor and inductor interaction. The next section does this.

EXAMPLE 12-5.

In the tuned class C amplifier of Fig. 12-6a, the input frequency is 1 MHz. The resonant tank is tuned to the fundamental frequency of the collector current pulses.

1. What is the value of L?
2. If the inductor has a series resistance of 10 Ω at 1 MHz, what is the Q of the resonant circuit?

SOLUTION.

As usual, the input circuit is a clamper; the clamped waveform on the base turns on the emitter diode briefly at each positive peak. The result is narrow collector current pulses. Since the input signal produces these pulses, the pulses must have the same fundamental frequency as the input signal, 1 MHz.

1. The resonant tank must be tuned to 1 MHz. Therefore, we use

$$f \cong \frac{1}{2\pi \sqrt{LC}}$$

which is the approximate formula for resonant frequency whenever Q is greater than 10. Substituting the known values gives

$$10^6 \cong \frac{1}{2\pi \sqrt{L(500)10^{-12}}}$$

Solving for L gives

$$L = 50.7 \ \mu H$$

2. Figure 12-6b is part of the ac equivalent circuit. We have shorted the output coupling capacitor and have reduced V_{CC} to zero. This results in a resonant tank with narrow current pulses driving it. The inductive reactance of the coil is

$$X_L = 2\pi f L = 2\pi (10^6) 50.7 (10^{-6}) = 318 \ \Omega$$

The Q of the coil is

$$Q_{\text{coil}} = \frac{X_L}{R_{\text{series}}} = \frac{318}{10} = 31.8$$

As discussed in basic circuit-theory books, the equivalent parallel resistance of the coil is

$$R_{\text{parallel}} \cong Q_{\text{coil}} X_L = 31.8(318) \cong 10 \ k\Omega$$

Figure 12-6. Example 12-6.

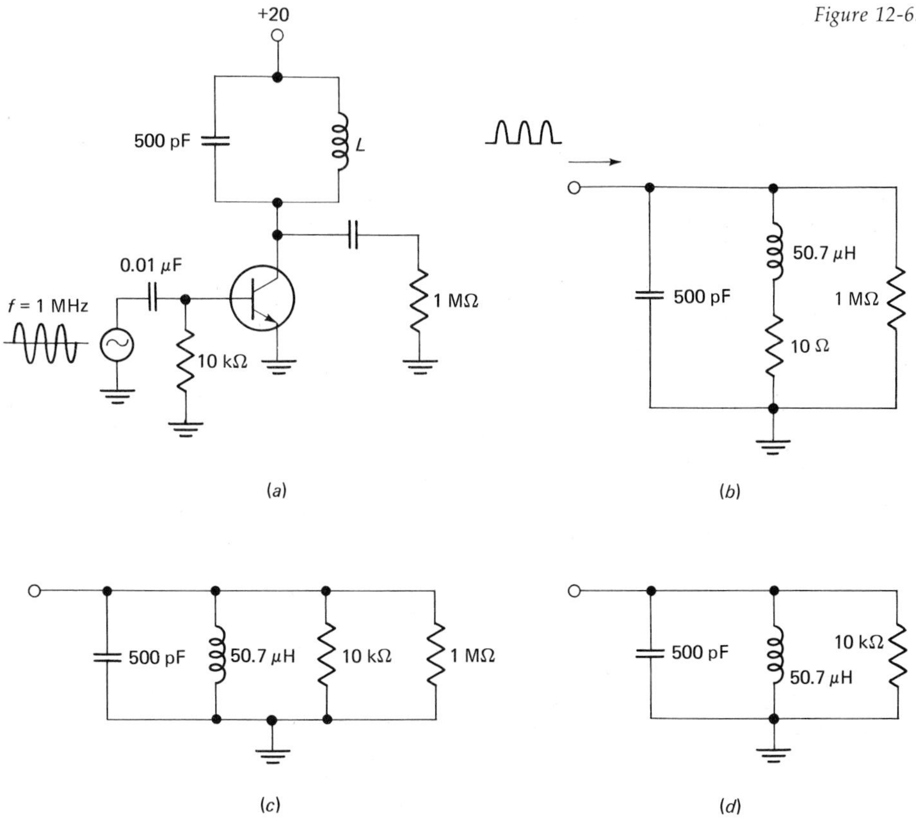

(This is a close approximation when Q is greater than 10.)

We now draw the equivalent circuit of Fig. 12-6c. All we have done is to represent the 10 Ω of series resistance in the coil by its parallel equivalent of 10 kΩ. In other words, because Q_{coil} is greater than 10, the circuit of Fig. 12-6c behaves almost the same as the circuit of Fig. 12-6b. Now it is plain what effect the 1-MΩ load resistor has; almost none. The 1 MΩ is so large compared to the 10 kΩ we can neglect it and use the ac equivalent circuit of Fig. 12-6d.

Figure 12-6d is the parallel resonant circuit covered in any basic textbook. Since the 1-MΩ load resistor is large enough to neglect, the Q of the resonant circuit equals the Q of the coil;

$$Q_{circuit} = Q_{coil} = 31.8$$

With a high Q like this, we can expect an almost perfect sine wave of voltage to appear across the tank circuit.

12-4.   TUNED CLASS C ACTION

This section describes how a tuned resonant circuit produces a sine wave of voltage when driven by current pulses. The proof comes later; all you are supposed to get from this discussion is the interaction between the capacitor and the inductor of a resonant circuit in a class C amplifier.

Figure 12-7a shows a current pulse driving a resonant tank. If the pulse is narrow, the inductor looks like a high impedance and the capacitor like a low impedance. For this reason, most of the current charges the capacitor. We will assume the capacitor charges to 10 V during the pulse.

After the pulse ends, the circuit looks like Fig. 12-7b. The charged capacitor will discharge through the coil and the load resistor. As the current in the coil builds up, the capacitor voltage decreases until it reaches 0 V (Fig. 12-7c). The magnetic field around the inductor now collapses and induces a minus-plus voltage across the coil as shown in Fig. 12-6d. This forces current to flow into the capacitor and charges it with the opposite polarity shown.

After the magnetic field has completely collapsed, the inductor current drops to zero; at this instant, the capacitor has fully charged in the opposite direction. The maximum voltage on the capacitor will be less than the 10 V we started with because of the

Figure 12-7. Tuned-circuit action. (a) Initial capacitor charging. (b) Capacitor discharges. (c) Inductor current maximum. (d) Collapsing inductor field charges capacitor. (e) Capacitor discharges. (f) Inductor current maximum. (g) At end of cycle.

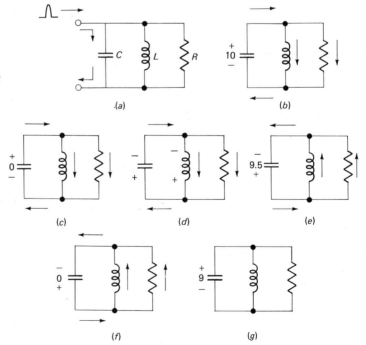

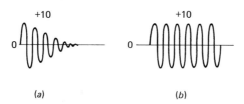

Figure 12-8. (a) *Single input pulse.* (b) *Train of input pulses.*

energy absorbed by the load resistor. For the sake of explanation, we will assume the capacitor voltage reaches only 9.5 V.

The capacitor cannot hold this charge; it must start to discharge through the inductor and load resistor as shown in Fig. 12-7e. The action is now complementary. That is, the magnetic field builds up with the opposite polarity. The field reaches its maximum value when the capacitor has completely discharged as shown in Fig. 12-7f. Again, the magnetic field collapses, inducing a voltage across the coil which charges the capacitor to a positive voltage. Because of energy losses in the load resistor, the maximum positive voltage on the capacitor will be less than the initial 10 V; arbitrarily, we have shown it as 9 V in Fig. 12-7g.

If only a single current pulse drives a tank circuit, the capacitor will charge and discharge repeatedly, the maximum voltage decreasing slightly each cycle, until the waveform dies out as shown in Fig. 12-8a. On the other hand, if a train of narrow current pulses drives the circuit, each pulse recharges the capacitor to the full voltage; in this way, we can get almost a perfect sine wave like Fig. 12-8b.

12-5. POWER RELATIONS

In a tuned class C amplifier like Fig. 12-9a, the quiescent or no-signal collector voltage is

$$V_{CEQ} = V_{CC}$$

When a signal is present, it forces the total collector voltage to swing above and below this voltage. The collector voltage can drop no lower than $V_{CE(sat)}$. This is why the maximum output voltage waveform of a tuned class C amplifier looks like Fig. 12-9b.

As discussed in Example 12-5, the series resistance of the inductor can be replaced by its equivalent parallel resistance as shown in Fig. 12-9c. This $R_{parallel}$ is in shunt with R_L; the product-over-sum of these two resistances is the equivalent ac load resistance r_L shown in Fig. 12-9d.

What is the maximum load power in a tuned class C amplifier? At most, the output voltage can have a peak-to-peak value of approximately $2V_{CC}$ as shown in Fig. 12-9b; the equivalent rms voltage is $0.707V_{CC}$. Therefore, the maximum load power equals

$$P_{load} = \frac{V_{rms}^2}{r_L} = \frac{(0.707V_{CC})^2}{r_L}$$

Figure 12-9. Deriving power relations. (a) Tuned circuit. (b) Maximum possible voltage waveform (c) Ac equivalent circuit for tank (d) Simplified tank circuit.

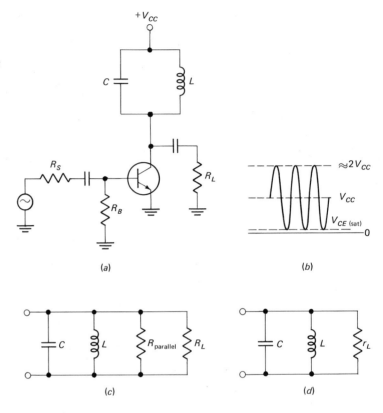

(a)

(b)

(c)

(d)

or
$$P_{\text{load}} = 0.5 \frac{V_{CC}^2}{r_L} \quad \text{(100 percent load line)} \quad (12\text{-}6)$$

The average power dissipation of the transistor depends on the duty cycle and the signal swing. The mathematics is too complicated to show here (it's in the Appendix), but it can be proved that a peak-to-peak output voltage of $2V_{CC}$ results in

$$P_D \cong 0.5 V_{CE(\text{sat})} \frac{V_{CC}}{r_L} \quad \text{(100 percent load line)} \quad (12\text{-}7a)$$

This is lowest possible dissipation for maximum output signal; you get this low dissipation only when the *entire ac load line is used,* and when the *duty cycle is less than 10 percent.*

When the signal does not use the entire load line, the transistor power dissipation may be much higher than the value given by Eq. (12-7a). The worst case occurs when only half the load line is used, that is, when the ac output voltage in Fig. 12-9a has a peak value of $V_{CC}/2$. In this case,

$$P_D = 0.125 \frac{V_{CC}^2}{r_L} \quad \text{(50 percent load line)} \quad \text{(12-7b)}$$

Again, r_L is the total ac load resistance across the tank circuit; it includes the coil losses.[3]

To take advantage of the high efficiency of a class C amplifier, you have to use most of the load line. In other words, in tuned class C circuits you have to drive the base hard enough to swing the output signal over most of the load line. In this way, you may approach the best-case dissipation given by Eq. (12-7a).

The ratio of the best-case P_D to the maximum load power is useful. With Eqs. (12-6) and (12-7a), we get a ratio of

$$\frac{P_D}{P_{\text{load}}} = \frac{V_{CE(\text{sat})}}{V_{CC}}$$

or

$$P_D = \frac{V_{CE(\text{sat})}}{V_{CC}} P_{\text{load}} \quad \text{(100 percent load line)} \quad \text{(12-8)}***$$

With this equation, we can compare class C with class A and B. As you recall,

$$P_D = 2P_{\text{load}} \quad \text{for class A}$$

and

$$P_D = 0.2P_{\text{load}} \quad \text{for class B push-pull}$$

To do better than class B push-pull, the ratio $V_{CE(\text{sat})}/V_{CC}$ must be less than 0.2; this is easy to accomplish. Since $V_{CE(\text{sat})}$ is usually less than 1 V, all we need is a V_{CC} more than 5 V; then we get less P_D than with a class B push-pull amplifier. Therefore, for a given $P_{D(\text{max})}$ rating, the tuned class C amplifier can deliver the most load power.[4]

EXAMPLE 12-6.
Calculate the maximum load power in Fig. 12-10. What $P_{D(\text{max})}$ does the transistor need for maximum load power if $V_{CE(\text{sat})}$ is 0.75 V?

SOLUTION.
The peak-to-peak collector voltage in Fig. 12-10 is 100 V maximum, equivalent to a peak of 50 V, or an rms value of approximately 35 V. The reflected load is 200 Ω. With Eq. (12-6),

$$P_{\text{load}} = 0.5 \frac{V_{CC}^2}{r_L} = 0.5 \frac{50^2}{200} = 6.25 \text{ W}$$

[3] This assumes a resonant tank exactly tuned to the fundamental frequency. If the tuning is off, the power dissipation can be much higher than given by Eq. (12-7a) or (12-7b).

[4] Equation (12-8) is the best you can do. Normally, you would have to include a safety factor to allow for mistuning and other effects. A good source of information is the data sheet for the RF transistor.

Figure 12-10. Example 12-6.

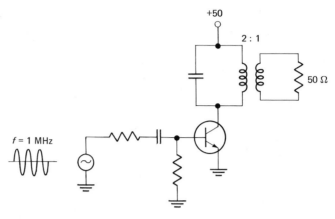

For this condition, the power dissipation equals

$$P_D = \frac{V_{CE(\text{sat})}}{V_{CC}} P_{\text{load}} = \frac{0.75}{50} 6.25 = 93.8 \text{ mW}$$

Therefore, the transistor needs a $P_{D(\text{max})}$ rating of at least 93.8 mW, provided the entire load line is used; more, if not.

12-6. FREQUENCY MULTIPLIERS

When the output signal of a tuned class C amplifier is maximum (a peak-to-peak value of approximately $2V_{CC}$), the class C amplifier is the most efficient of all classes; it delivers more load power for a given transistor dissipation than any other class. But remember, you get this efficiency only at a single frequency (or frequencies close to it).

One important application for the high efficiency of a class C amplifier is an *oscillator*. This circuit produces an output sine wave without an input signal. A later chapter discusses oscillators, and at that time you will again meet the class C amplifier.

Another application for tuned class C circuits is with *frequency multipliers*. The idea is to tune the resonant tank to a *harmonic* or multiple of the input frequency. Fig. 12-11a shows narrow current pulses driving a tuned circuit. The narrow current pulses have a fundamental frequency of 1 MHz. In an ordinary tuned class C amplifier, we tune the resonant circuit to the fundamental frequency. Then, each current pulse recharges the capacitor once per output cycle (Fig. 12-11b).

Suppose we change the resonant frequency of the tank to 2 MHz; this is the second multiple or second harmonic of 1 MHz. The capacitor and inductor will interact at a 2-MHz rate and produce an output voltage of 2 MHz as shown in Fig. 12-11c. Now, the 1-MHz current pulses recharge the capacitor on every other output cycle.

If we tune the tank to 3 MHz, the third harmonic of 1 MHz, we get a 3-MHz output signal (Fig. 12-11d). In this case, the 1-MHz current pulses recharge the capacitor every

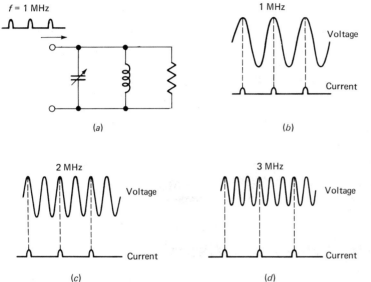

Figure 12-11. Frequency multipliers. (a) Current pulses drive resonant tank. (b) Tuned to fundamental. (c) Tuned to second harmonic. (d) Tuned to third harmonic.

third output cycle. As long as the Q of the resonant circuit is high, the output voltage looks almost like a perfect sine wave.

Figure 12-12 shows one way to build a frequency multiplier. The input signal has a frequency of f; this signal is negatively clamped at the base. The resulting collector current is a train of narrow current pulses with a fundamental frequency f. By tuning the resonant tank to the nth harmonic, we get an output voltage whose frequency is nf.

In Fig. 12-12, the current pulses recharge the capacitor every nth output cycle. For this reason, the output power decreases when we tune to higher harmonics; the higher

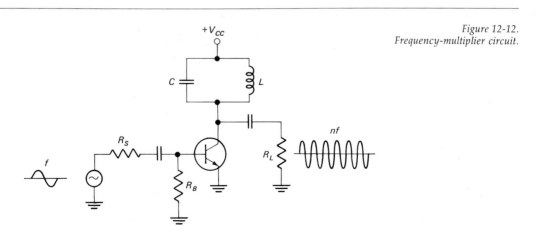

Figure 12-12. Frequency-multiplier circuit.

Figure 12-13

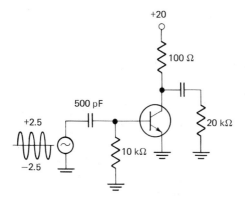

n is, the lower the output power. Because of this decreasing efficiency at higher harmonics, a tuned class C frequency multiplier like Fig. 12-12 is ordinarily used only for lower harmonics like the second or third. For n greater than 3, we usually turn to more efficient semiconductor devices.[5]

Problems

12-1. In Fig. 12-13, the period of the input signal is much smaller than the discharging time constant of the base circuit. What is the peak-to-peak base voltage? Ideally, what is the maximum negative base voltage? To a second approximation, what is the largest negative base voltage? What BV_{EBO} does the transistor need?

12-2. For good clamping action in Fig. 12-13, we want the discharging time constant to be at least ten times the period of the input signal. What is the lowest input frequency satisfying this condition?

12-3. In Fig. 12-13, what is the value of $I_{C(sat)}$? And V_{CEQ}?

12-4. If you use a high-impedance voltmeter to measure the dc voltage from base to ground in Fig. 12-13, what voltage do you read? (Give the ideal and second-approximation answers.)

12-5. If the output voltage of Fig. 12-13 has a period of 1 μs and a pulse width of 0.005 μs, what is the duty cycle of the output waveform?

12-6. Figure 12-14a shows another form of the base clamping circuit. What is the discharging time constant? For this time constant to be at least ten times the input period, what is the lowest frequency we can use?

12-7. Sometimes the RC circuit of a clamper is moved to the emitter circuit as shown in Fig. 12-14b. After the first few input cycles, the capacitor voltage reaches approximately V_p. Between positive input peaks, the capacitor discharges through the 1-kΩ resistor. If we want the RC time constant of the emitter circuit

[5] The *step-recovery* diode is outstanding for generating any harmonic above the third.

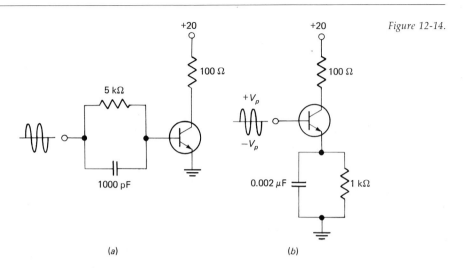

+20

100 Ω

5 kΩ

1000 pF

(a)

+20

100 Ω

$+V_p$

$-V_p$

0.002 μF

1 kΩ

(b)

Figure 12-14.

to be at least ten times the input period, what is lowest permitted input frequency?

12-8. The current waveform of Fig. 12-15a has a period of 2 μs. What is the fundamental frequency of this current waveform?

12-9. Figure 12-15b shows a current waveform made up of sawtooth pulses. The duty cycle and fundamental frequency are still defined by Eqs. (12-4) and (12-5). Calculate the duty cycle and fundamental frequency.

12-10. Figure 12-16a shows a tuned class C amplifier. What is the resonant frequency of the tank circuit? For the discharging time constant of the base clamper to be ten times the period of the input signal, what value should R have?

12-11. The inductor of Fig. 12-16a has a series resistance of 2 Ω at the resonant frequency of the tank. What is the Q of the coil? What value does the equivalent parallel resistance of the coil have?

12-12. In Fig. 12-16a, the coil has a Q of 100 at the resonant frequency of the tank. What is the series resistance of this coil? The equivalent parallel resistance of the coil? If R_L equals 10 kΩ, what is the total parallel resistance across the tank circuit as resonance? The Q of the resonant circuit equals the total parallel resistance across the tank divided by X_L. What value does this Q have?

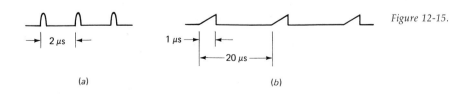

2 μs

1 μs

20 μs

(a)

(b)

Figure 12-15.

Figure 12-16.

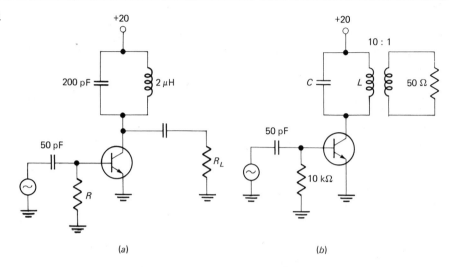

(a)   (b)

12-13. Figure 12-16b shows transformer coupling to the final load resistor. What is the reflected load resistance seen by the collector? If the coil has an X_L of 100 Ω and a Q_{coil} of 50, what is the Q of the entire tank circuit?

12-14. In Fig. 12-16a, what is the value of V_{CEQ}? What is the largest possible peak-to-peak voltage across the tank?

12-15. What is the maximum load power in Fig. 12-16a when r_L equals 1 kΩ?

12-16. If the r_L in Fig. 12-16b equals 2.5 kΩ, what is the maximum load power? How much of this total load power does the 50-Ω load resistor receive?

12-17. The transistor of Fig. 12-16a has a $V_{CE(sat)}$ of 0.8 V. For a low duty cycle and 100 percent load-line swing, what is the power dissipation in the transistor if the load power equals 100 mW?

12-18. An RF power transistor has a $P_{D(max)}$ rating of 2 W at the highest temperature it will operate at. If the transistor has a $V_{CE(sat)}$ of 1 V, what is the maximum possible load power with a V_{CC} supply of 30 V?

12-19. A tuned class C amplifier has a V_{CC} of 40 V, an r_L of 50 Ω, and a $V_{CE(sat)}$ of 0.6 V. How much load power can this amplifier deliver?

12-20. A transistor type has a $V_{CE(sat)}$ of 0.5 V and a $P_{D(max)}$ rating of 1 W. Given a V_{CC} supply of 20 V, what is the maximum load power you can get with a class A amplifier? Class B push-pull? Tuned class C?

12-21. What is the fundamental frequency of Fig. 12-15a? The second harmonic? The third harmonic?

12-22. What is the tenth harmonic of Fig. 12-15b? The fiftieth?

12-23. If we wanted to use the circuit of Fig. 12-16a as a *frequency doubler* ($n = 2$), what input frequency would we use?

12-24. For the circuit of Fig. 12-16a to act as a frequency tripler ($n = 3$), what input frequency do we need?

12-25. If Fig. 12-16a is used in an X4 frequency multiplier, what value should R have to make the discharging time constant of the base clamper ten times the period of the clamped signal?

12-26. In a tuned class C amplifier, the size of the current pulses driving the tuned circuit depends on the duty cycle. With an advanced derivation, the current pulses have an amplitude of

$$I_{max} = \frac{\pi}{4k} \frac{V_{CEQ}}{r_C + r_E}$$

where k equals W/T, the duty cycle divided by 100 percent. Calculate I_{max} for a duty cycle of 5 percent a V_{CEQ} of 10 V, and an $r_C + r_E$ of 100 Ω.

13. Field-Effect Transistors

The bipolar transistor is the backbone of electronics. But in some applications the *field-effect transistor* (FET) is preferred. This chapter describes FETs and prepares us for the FET circuits of the next chapter.

13-1. THE JFET

The *junction* FET, or JFET, is a unipolar transistor; it needs only majority carriers to work. It is much easier to understand than the bipolar transistor.

JFET regions

Figure 13-1a shows part of a JFET. The lower end is called the *source* and the upper end the *drain;* the piece of semiconductor between the source and drain is known as the *channel.* Since n material is used in Fig. 13-1a, the majority carriers are conduction-band electrons. Depending on voltage V_{DD} and the resistance of the channel, we get a certain amount of current.

By embedding two p regions in the sides of the channel, we get the *n-channel* JFET of Fig. 13-1b. Each of these p regions is called a *gate*. When the manufacturer connects a separate external lead to each gate, we call the device a *dual-gate* JFET. The main use of a dual-gate JFET is with a *mixer,* a special circuit to be discussed in Chap. 21.

Throughout this chapter we concentrate on the *single-gate* JFET, a device whose gates are internally connected. A single-gate JFET has only one external gate lead, as shown in Fig. 13-1c. When using this symbol, we must remember the two p regions have the same potential because they are internally connected.

Figure 13-1. (a) Part of JFET. (b) Dual-gate JFET. (c) Single-gate JFET.

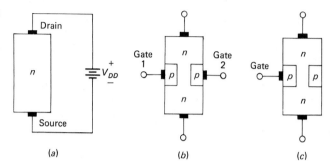

Biasing the JFET

Figure 13-2a shows the normal polarities for biasing an n-channel JFET. The idea is to apply a negative voltage between the gate and the source; this *reverse-biases* the gate. Since the gate is reverse-biased, only a negligibly small current flows in the gate lead. To a first approximation, gate current is zero.

The name *field effect* is related to the depletion layers around each pn junction. Figure 13-2b shows these depletion layers. The current from source to drain must flow through the narrow channel between depletion layers. The size of these depletion layers determines the width of the conducting channel.

In Fig. 13-2a the more negative we make the gate, the narrower the conducting channel becomes. In this way, the voltage on the gate can determine the amount of current between the source and the drain. The key difference between a JFET and a bipolar transistor is this: the gate is reverse-biased whereas the base is forward-biased. This crucial difference means the JFET acts like a *voltage-controlled* device; ideally, input voltage alone controls the output current. This is different from a bipolar where we need input voltage and current to control output current.

We can summarize the first major difference between the JFET and the bipolar in terms of resistance. The input resistance of a JFET ideally approaches infinity. To a sec-

Figure 13-2. (a) Normal biasing of JFET. (b) Depletion layers.

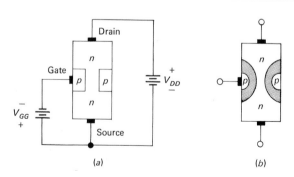

ond approximation, it is well into the megohms, depending on the particular JFET type. Therefore, in applications where high input resistance is needed, the JFET is preferred.[1]

The price we pay for larger input resistance is smaller control over output current. In other words, a JFET is less sensitive to changes in input voltage than a bipolar transistor. In almost any JFET an input change of 0.1 V produces less than a 10-mA change in output current. But in a bipolar transistor a 0.1-V change easily produces more than a 10-mA change in output current. As will be discussed later, this means a JFET has less voltage gain than a bipolar.

Schematic symbol

Figure 13-3a shows the schematic symbol for a JFET. As a memory aid, think of the thin vertical line (Fig. 13-3b) as the channel; the source and drain connect to this line. Also, an arrow is on the gate; *this arrow points to the n material.* In this way, you can remember that Fig. 13-3a represents an *n*-channel JFET.

In many JFETs the source and drain are interchangeable; you can use either end as the source and the other end as the drain. For this reason, the JFET symbol of Fig. 13-3a is symmetrical; the gate points to the center of the channel. When you use the symmetrical JFET symbol, you have to label the terminals as done in Fig. 13-3a or use abbreviations as shown in Fig. 13-3c. Alternatively, you can use the nonsymmetrical symbol of Fig. 13-3d; here, the gate is drawn closer to the source.

[1] Input impedance is the same as input resistance at low frequencies where reactance is negligible. Chapter 16 discusses high-frequency effects.

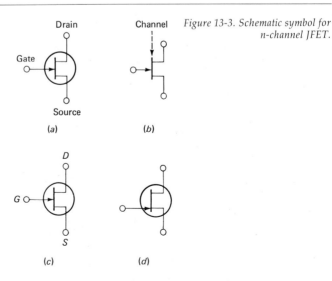

Figure 13-3. Schematic symbol for n-channel JFET.

Figure 13-4. Schematic symbol for p-channel JFET.

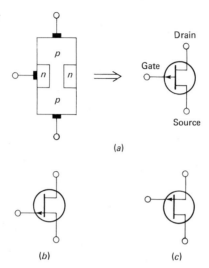

(a)

(b)   (c)

Figure 13-4*a* shows the *p*-channel JFET and its schematic symbol. In all FET symbols the arrow points to the *n* material; since the arrow points away from the channel, we recognize the symbol of Fig. 13-4*a* as a *p*-channel JFET. The *p*-channel JFET is complementary to the *n*-channel JFET. For this reason, we can concentrate on the *n*-channel JFET and extend the results to *p*-channel JFETs by the theory of complements discussed earlier (Sec. 7-8).

Figure 13-4*b* shows the alternative symbol for a *p*-channel JFET. The gate arrow is closer to the source. When using *n*-channel and *p*-channel JFETs in the same circuit diagram, we can draw the *p*-channel JFETs upside down as shown in Fig. 13-4*c*.

13-2. JFET DRAIN CURVES

With a circuit like Fig. 13-5*a* we can get *drain curves* for the JFET. Conventional current flows down the channel from drain to source.

Pinchoff voltage

As already discussed, the gate is reverse-biased for normal operation. Included in this is the special case of zero gate voltage. In other words, we can reduce V_{GS} to zero as shown in Fig. 13-5*b*; this special case is called the *shorted-gate* condition.

Figure 13-5*c* is the graph of drain current versus drain voltage for the shorted-gate condition. Notice the similarity to a collector curve. The drain current rises rapidly at first, but then levels off. In the region between V_P and $V_{DS(\max)}$ the drain current is almost constant. When the drain voltage is too large, the JFET breaks down as shown. As

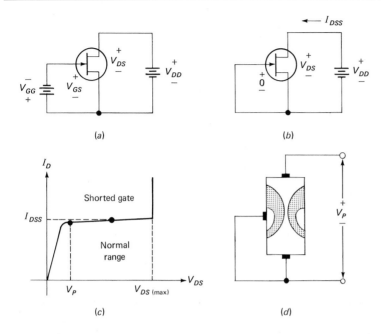

with a bipolar transistor, the normal region of operation is along the almost horizontal part of the curve; in this region the JFET acts like a current source. We can symbolize this normal operating range by

$$V_P < V_{DS} < V_{DS(\text{max})} \qquad (13\text{-}1)^{***}$$

The *pinch-off voltage* V_P is roughly the drain voltage above which drain current becomes essentially constant. For drain voltages less than V_P the channel is wide and the drain current is proportional to drain voltage. When the drain voltage equals V_P, the channel has become narrow and the depletion layers *almost touch*, as shown in Fig. 13-5d. The depletion layers are thicker at the top because the voltage across the *pn* junctions increases as we approach the drain end. The small passage between the depletion layers tends to limit or pinch off the current. With further increases in drain voltage the tendency to increase drain current is offset by narrowing of the channel; in the normal range, an increase in drain voltage results in only a slight increase in drain current, as shown in Fig. 13-5c.

Shorted-gate drain current

In Fig. 13-5c the drain current in the normal range is labeled I_{DSS}; the subscripts stand for *D*rain to *S*ource with *S*horted gate. Since the drain current is almost constant in the normal region of operation, I_{DSS} represents the approximate drain current for any drain voltage between V_P and $V_{DS(\text{max})}$. Furthermore, since I_{DSS} is measured under shorted-gate conditions, I_{DSS} is the maximum drain current we can get for normal operation of a JFET.

Any data sheet for a JFET gives the value of I_{DSS}. As an example, the data sheet of an MPF102 (Motorola) lists a minimum I_{DSS} of 2 mA and a maximum I_{DSS} of 20 mA. This means the shorted-gate drain current of some MPF102s may be as low as 2 mA or as high as 20 mA. Stated another way, if we use a curve tracer to look at the drain curves, the highest drain curve of an MPF102 can be anywhere from 2 to 20 mA.

Gate-source cutoff voltage

Drain curves look very much like collector curves. For instance, Fig. 13-6 shows drain curves for a typical JFET. The highest curve is for $V_{GS} = 0$, the shorted-gate condition. The pinch-off voltage is approximately 4 V, and the breakdown voltage is 30 V. So, the normal operating region for this particular JFET is

$$4\ \text{V} < V_{DS} < 30\ \text{V}$$

As we see, I_{DSS} is 10 mA for V_{DS} of 15 V. Ideally, the drain current is 10 mA for any V_{DS} between 4 and 30 V.

A negative gate voltage leads to other drain curves. As we see, a V_{GS} of -1 V drops the drain current to about 5.62 mA. A V_{GS} of -2 V reduces drain current to around 2.5 mA, and so on. The bottom curve is especially important; a V_{GS} of -4 V reduces the drain current to approximately zero. We call this voltage the *gate-source cutoff voltage* and symbolize it by $V_{GS(\text{off})}$.

JFETs have a wide range of $V_{GS(\text{off})}$ values. Data sheets specify $V_{GS(\text{off})}$ at an arbitrarily small value of drain current. For instance, the data sheet of an MPF102 lists a maximum $V_{GS(\text{off})}$ of -8 V for a drain current of 2 nA. Since this value of drain current is so small compared to useful values of drain current, we can think of $V_{GS(\text{off})}$ as the gate-source voltage that cuts off drain current.

At $V_{GS} = V_{GS(\text{off})}$, the depletion layers touch; this explains why the drain current is approximately zero. As we saw earlier, V_P is the value of drain voltage that pinches off the current for the shorted-gate condition. Because of this,

$$V_P = |V_{GS(\text{off})}| \qquad \text{(13-2)}^{***}$$

Data sheets do not list V_P, but they do give $V_{GS(\text{off})}$, which is equivalent. For example, if a data sheet gives a $V_{GS(\text{off})}$ of -4 V, we immediately know V_P equals 4 V.

Equation (13-1) pins down the normal range of V_{DS}. Now we can tie down the normal range of V_{GS}. Since the shorted-gate condition gives us the highest drain curve and $V_{GS(\text{off})}$ produces the lowest drain curve, the normal range of V_{GS} is

$$V_{GS(\text{off})} < V_{GS} < 0 \qquad \text{(13-3}a\text{)}$$

When V_{GS} is in this range, I_D must be in the interval

$$0 < I_D < I_{DSS} \tag{13-3b}$$

As an example, in Fig. 13-6 the normal range of drain voltage is between 4 and 30 V, the normal range of gate voltage is between −4 and 0 V, and the normal range of drain current is between 0 and 10 mA.

EXAMPLE 13-1.
At 25°C, a 2N5486 has a gate current of 1 nA when $V_{GS} = -20$ V. If the ambient temperature is raised to 100°C, the gate current is 0.2 μA when $V_{GS} = -20$ V. Calculate the dc resistance from the gate to the source at each temperature.

SOLUTION.
At 25°C,

$$R_{GS} = \frac{V_{GS}}{I_G} = \frac{20}{10^{-9}} = 20{,}000 \text{ M}\Omega$$

At 100°C,

$$R_{GS} = \frac{20}{0.2(10^{-6})} = 100 \text{ M}\Omega$$

Therefore, even at a high temperature the input resistance of a JFET is very high.

13-3. THE TRANSCONDUCTANCE CURVE

As we saw earlier, the transconductance curve relates the output current to the input voltage. With a bipolar transistor the transconductance curve is the graph of I_C versus V_{BE}; with a JFET it is graph of I_D versus V_{GS}. For instance, by reading the values of I_D

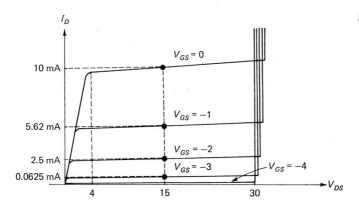

Figure 13-6. Typical set of drain curves.

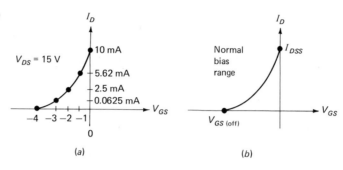

Figure 13-7. Transconductance curves.

(a)

(b)

and V_{GS} in Fig. 13-6, we can plot the transconductance curve shown in Fig. 13-7a. In general, the transconductance curve of any JFET will look like Fig. 13-7b.

Curve is parabolic

The transconductance curve in Fig. 13-7b is part of a *parabola*. It can be shown that the equation of the transconductance curve is

$$I_D = I_{DSS}\left[1 - \frac{V_{GS}}{V_{GS(off)}}\right]^2 \qquad (13\text{-}4)^{***}$$

This equation is accurate to within a few percent for any JFET.[2]

As an example, suppose a JFET has an I_{DSS} of 4 mA and a $V_{GS(off)}$ of −2 V. By substitution into Eq. (13-4), we get

$$I_D = 0.004\left(1 + \frac{V_{GS}}{2}\right)^2$$

With this equation we can calculate the drain current for any gate voltage in the normal range.

Most data sheets do not give drain curves or transconductance curves. Instead, you get the values of I_{DSS} and $V_{GS(off)}$. By substituting these values into Eq. (13-4), you can then calculate the approximate drain current for any condition you are interested in.[3]

Square law

Square law is another name for parabolic. This is why JFETs are often called square-law devices. For reasons to be discussed in Chap. 21, the square-law property

[2] The derivation is too advanced for this book. If interested, you can find it in Sevin, L. J.: *Field-effect Transistors*, McGraw-Hill Book Company, New York, 1965, pp. 1–23.

[3] For accurate calculations, do not use the $V_{GS(off)}$ given on the data sheet; instead, calculate $V_{GS(off)}$ with Eq. (14-18).

gives the JFET another major advantage over the bipolar, at least in special circuits called *mixers*.

EXAMPLE 13-2.
A JFET has $I_{DSS} = 16$ mA and $V_{GS(off)} = -8$ V. Calculate the value of I_D for each of these: $V_{GS} = -2$ V, -4 V, and -6 V.

SOLUTION.
Substitute the given I_{DSS} and $V_{GS(off)}$ into Eq. (13-4) to get

$$I_D = 0.016 \left(1 + \frac{V_{GS}}{8}\right)^2$$

With this equation, calculate the drain currents as follows:

When $V_{GS} = -2$ V,

$$I_D = 0.016(1 - 0.25)^2 = 9 \text{ mA}$$

When $V_{GS} = -4$ V

$$I_D = 0.016(1 - 0.5)^2 = 4 \text{ mA}$$

When $V_{GS} = -6$ V

$$I_D = 0.016(1 - 0.75)^2 = 1 \text{ ma}$$

EXAMPLE 13-3.
Plot I_D/I_{DSS} versus $V_{GS}/V_{GS(off)}$.

SOLUTION.
We can rewrite Eq. (13-4) as

$$\frac{I_D}{I_{DSS}} = \left[1 - \frac{V_{GS}}{V_{GS(off)}}\right]^2$$

By substituting ¼, ½, ¾, and 1 for $V_{GS}/V_{GS(off)}$, we can calculate values of ¹⁄₁₆, ¼, ⁹⁄₁₆, and 1 for I_D/I_{DSS}. Figure 13-8*a* summarizes these results; it applies to any JFET, and we often will refer to it.

EXAMPLE 13-4.
Figure 13-8*b* shows a JFET biasing circuit (to be analyzed in the next chapter). If I_D is approximately half of I_{DSS}, what is the approximate value of V_{GS}?

SOLUTION.
In Fig. 13-8*a*, a current ratio of ⁹⁄₁₆ is the nearest value we can find for the condition I_D equal to half of I_{DSS}. The corresponding voltage ratio is

Figure 13-8. Examples 13-2 through 13-4.

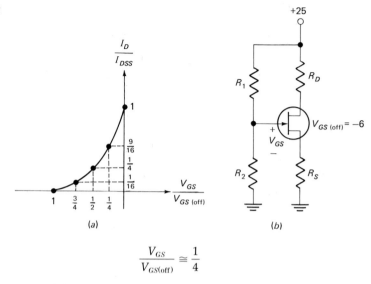

(a)

(b)

$$\frac{V_{GS}}{V_{GS(\text{off})}} \cong \frac{1}{4}$$

The JFET in Fig. 13-8b has a $V_{GS(\text{off})}$ of -6 V; therefore, to set up an I_D of half I_{DSS}, we need a V_{GS} of about -1.5 V.

This is a useful result. We often bias JFETs at about half the I_{DSS} value; this means we have a V_{GS} of about one-fourth the $V_{GS(\text{off})}$ value.

13-4. DEPLETION-ENHANCEMENT MOSFETS

The *metal-oxide semiconductor* FET or MOSFET has a source, gate, and drain. Gate voltage controls the drain current; the main difference between a JFET and a MOSFET is that we can apply positive gate voltages and still have essentially zero gate current.

MOSFET regions

Let's look at the parts of a MOSFET. To begin with, there's an *n* region with source and drain as shown in Fig. 13-9a. As before, a positive voltage applied to the drain-source terminals forces conduction-band electrons to flow from source to drain. Unlike the JFET, the MOSFET has only a single *p* region as shown in Fig. 13-9b. We call this region the *substrate*. This *p* region constricts the channel between source and drain so that only a small passage remains at the left side of Fig. 13-9b. Electrons flowing from source to drain must pass through this narrow channel.

A thin layer of metal oxide (usually silicon dioxide) is deposited over the left side of the channel as shown in Fig. 13-9c. This metal oxide is an *insulator*. Finally, a metallic gate is deposited on the insulator (Fig. 13-9d). Because the gate is insulated from the channel, a MOSFET is also known as an *insulated-gate* FET (IGFET).

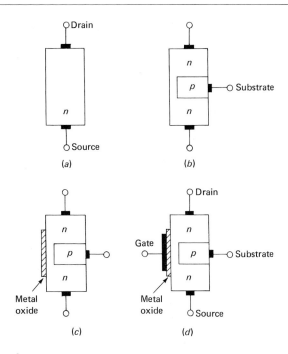

Figure 13-9. Parts of MOSFET.
(a) The n channel. (b) Adding the
substrate. (c) Adding the metal
oxide. (d) Adding the gate.

Depletion mode

How does the MOSFET of Fig. 13-10*a* work? In some applications a voltage is applied to the substrate to control channel current. But throughout this book we always connect the substrate to the source. As usual, the V_{DD} supply forces conduction-band electrons to flow from source to drain. These electrons flow through the narrow channel to the left of the *p* substrate.

As before, the gate voltage can control the resistance of the *n* channel. But since the gate is insulated from the channel, we can apply either a positive or a negative voltage to the gate. In Fig. 13-10*a* we have shown a negative gate voltage. The simplest way to visualize conduction in the channel is this. Think of the gate as one plate of a capacitor; the metal oxide acts like a dielectric and the *n* channel like the other plate. From basic theory, we know charges on a capacitor plate induce opposite charges on the other plate. Therefore, a negative gate voltage means many conduction-band electrons are distributed along the gate as shown in Fig. 13-10*b*; these negative charges repel conduction-band electrons in the *n* channel, leaving a layer of positive ions in part of the channel (Fig. 13-10*b*). In other words, we have depleted the *n* channel of some of its conduction-band electrons.

The more negative the gate voltage, the greater the depletion of conduction-band electrons in the *n* channel. With enough negative gate voltage we can cut off the current between source and drain. Therefore, with negative gate voltage the action of a

Figure 13-10. (a) *Depletion-mode operation.* (b) *Depleting the channel.* (c) *Enhancement-mode operation.* (d) *Enhancing channel conductivity.*

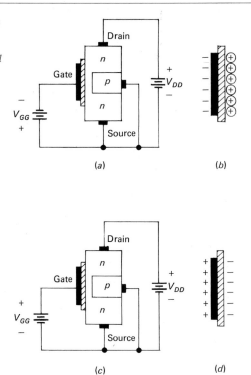

MOSFET is similar to the JFET. Because the action with a negative gate depends on depleting the channel of conduction-band electrons, we call negative-gate operation the *depletion mode*.

Enhancement mode

Since the gate of a MOSFET is insulated from the channel, we can apply a positive voltage to the gate as shown in Fig. 13-10c. Again, the gate acts like a capacitor plate. But this time, the positive charges on the gate induce negative charges in the *n* channel (see Fig. 13-10d). These negative charges are conduction-band electrons drawn into the channel. Because these conduction-band electrons are added to the conduction-band electrons already in the channel, the total number of conduction-band electrons in the channel has increased. In other words, a positive gate voltage increases or *enhances* the conductivity of the channel. The more positive the gate voltage, the greater the conduction from source to drain.

Operation of the MOSFET with a positive gate voltage depends on enhancing channel conductivity. For this reason, positive-gate operation (Fig. 13-10c) is called the *enhancement mode*.

Because of the insulating layer, negligible gate current flows in either mode of operation. In fact, the input resistance of a MOSFET is incredibly high, typically from 10,000 MΩ to over 10,000,000 MΩ.

The device of Fig. 13-10c is an *n*-channel MOSFET; the complementary device is the *p*-channel MOSFET. We will concentrate on the *n*-channel MOSFET and extend the results to the *p*-channel MOSFET by complementing currents and voltages.

MOSFET curves

Figure 13-11a shows typical drain curves for an *n*-channel MOSFET. $V_{GS(\text{off})}$ represents the negative gate voltage that cuts off drain current. For V_{GS} less than zero we get depletion-mode operation. On the other hand, for V_{GS} greater than zero we have enhancement-mode operation.

Figure 13-11b is the transconductance curve of a MOSFET. I_{DSS} represents the drain-source current with a shorted gate. But now, the curve extends to the right of the origin as shown. The curve is still parabolic, and we use the square-law equation described earlier. That is,

$$I_D = I_{DSS}\left[1 - \frac{V_{GS}}{V_{GS(\text{off})}}\right]^2 \tag{13-5}$$

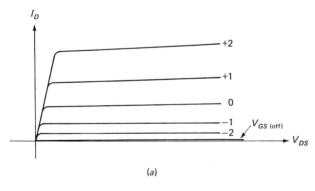

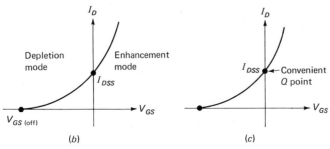

Figure 13-11. (a) *Drain curves.* (b) *Transconductance curve.* (c) *Convenient Q point.*

This is identical to the square-law equation of a JFET; the value of V_{GS}, however, can be positive or negative.

MOSFETs with a transconductance curve like Fig. 13-11c are easier to bias than JFETs. The reason is this. If we want, we can use the Q point shown in Fig. 13-11c. With this Q point, $V_{GS} = 0$ and $I_D = I_{DSS}$. Setting up a V_{GS} of zero is easy; it requires no dc voltage on the gate. The resulting circuit is simple and is discussed in the next chapter.

Because the MOSFET of Fig. 13-10 can operate in either the depletion mode or the enhancement mode, we call it a *depletion-enhancement* MOSFET. Since this kind of MOSFET conducts when $V_{GS} = 0$, it also is known as a *normally on* MOSFET.

Schematic symbol

Figure 13-12a shows the schematic symbol for a *normally on* MOSFET. The gate appears like a capacitor plate. Just to the right of the gate is the thin vertical line representing the channel; the drain lead comes out the top of the channel and the source lead connects to the bottom. The arrow is on the substrate and points to the n material; therefore, we have an n-channel MOSFET.

When an external lead is connected to the substrate, we have a four-terminal device. As already indicated, some applications drive the substrate with a voltage to have added control over channel current. Often, the manufacturer internally connects the substrate to the source; this results in a three-terminal device whose schematic symbol is shown in Fig. 13-12b.

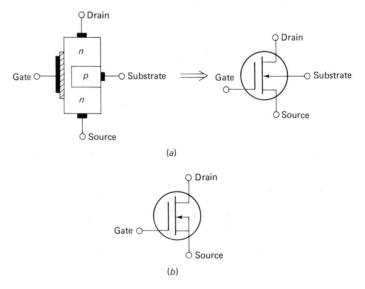

Figure 13-12. Schematic symbol for n-channel normally on MOSFET.

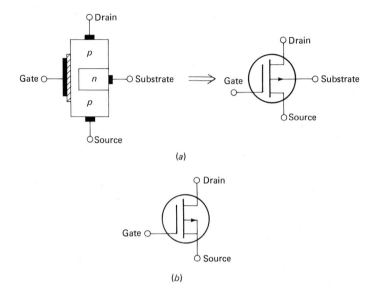

Figure 13-13. Schematic symbol for p-channel normally on MOSFET.

Figure 13-13*a* shows a *p*-channel MOSFET and its schematic symbol. As before, the way to remember the polarity of the MOSFET is by remembering the arrow points to the *n* material; since the arrow points away from the channel, we have a *p*-channel MOSFET. Figure 13-13*b* is the schematic symbol when the substrate is connected to the source.

13-5. ENHANCEMENT MOSFETS

There is one other kind of MOSFET, the *enhancement-only* type. As its name implies, it operates in the enhancement mode only. This kind of MOSFET is important in digital circuits.

Creating the inversion layer

Figure 13-14*a* shows the different parts of an enhancement-only MOSFET. Notice the substrate extends all the way to the metal oxide; structurally, there no longer is an *n* channel between the source and drain.

How does it work? Figure 13-14*b* shows normal biasing polarities. When $V_{GS} = 0$, the V_{DD} supply tries to force conduction-band electrons to flow from source to drain, but the *p* substrate has only a few thermally produced conduction-band electrons. Aside from these minority carriers and some surface leakage, the current between source and drain is zero. For this reason, an enhancement-only MOSFET is also called a *normally off* MOSFET.

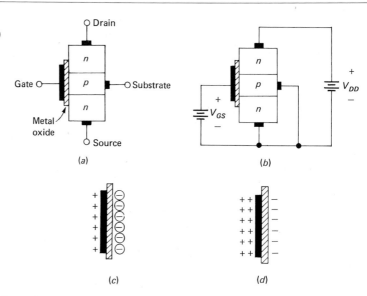

Figure 13-14. (a) Enhancement-only MOSFET. (b) Normal biasing. (c) Creating negative ions. (d) Creating the n-type inversion layer.

To get significant current, we have to apply enough positive voltage to the gate. What happens is rather complicated, but here is an easy way to understand the action. The gate again acts like one plate of a capacitor, the metal oxide like a dielectric, and the p substrate like the other plate. For lower gate voltages the positive charges in Fig. 13-14c induce negative charges in the p substrate. These charges are negative ions, produced by valence electrons filling holes in the p substrate. If we further increase the gate voltage, the additional positive charges on the gate can put conduction-band electrons into orbit around the negative ions (see Fig. 13-14d). In other words, when the gate is positive enough, it can create a thin layer of conduction-band electrons that stretch all the way from the source to the drain.

The created layer of conduction-band electrons is next to the metal oxide. This layer no longer acts like a p-type semiconductor. Instead, it appears like an n-type semiconductor. For this reason, the layer of p material touching the metal oxide is called an *n-type inversion layer*.

The threshold voltage

The minimum gate-source voltage that creates the n-type inversion layer is called the *threshold voltage* $V_{GS(th)}$. When V_{GS} is less than $V_{GS(th)}$, zero current flows from source to drain. But when V_{GS} is greater than $V_{GS(th)}$, an n-type inversion layer connects the source to the drain and we get significant current.

Threshold voltages depend on the particular type of MOSFET; $V_{GS(th)}$ can vary from less than a volt to more than 5 V. The 3N169 is an example of an enhancement-only MOSFET; it has a maximum threshold voltage of 1.5 V.

Enhancement-only curves

Figure 13-15a shows a set of curves for an enhancement-only MOSFET. The lowest curve is the $V_{GS(th)}$ curve. For gate voltages greater than the threshold value we get the higher curves.

Figure 13-15b is the transconductance curve. The curve is parabolic or square-law; the vertex of the parabola is at $V_{GS(th)}$. Because of this, the equation for the parabola is different from before; it now equals

$$I_D = K[V_{GS} - V_{GS(th)}]^2 \tag{13-6}$$

where K is a constant of proportionality that depends on the particular MOSFET.

Data sheets usually give the coordinates for one point on the transconductance curve as shown in Fig. 13-15b; after you substitute $I_{D(on)}$, $V_{GS(on)}$, and $V_{GS(th)}$ into Eq. (13-6), you can solve for the value of K. For instance, if an enhancement-only MOSFET has $I_{D(on)} = 8$ mA, $V_{GS(on)} = 5$ V, and $V_{GS(th)} = 3$ V, its transconductance curve looks like Fig. 13-15c. When we substitute these values into Eq. (13-6), we get

$$0.008 = K(5 - 3)^2 = 4K$$

or $$K = 0.002$$

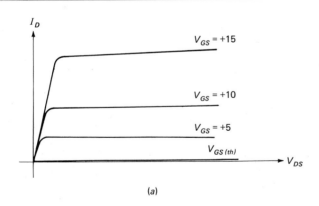

(a)

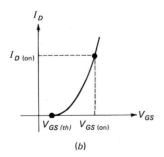

(b)

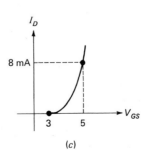

(c)

Figure 13-15. (a) Drain curves. (b) Transconductance curve. (c) Example.

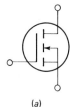 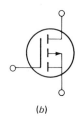

Figure 13-16. (a) *n-channel normally off MOSFET.* (b) *p-channel normally off MOSFET.*

(a)   (b)

Therefore, the transconductance equation of Fig. 13-15c is

$$I_D = 0.002(V_{GS} - 3)^2$$

Schematic symbol

When $V_{GS} = 0$, the enhancement-only MOSFET is off because no conducting channel exists between source and drain. The schematic symbol of Fig. 13-16a has a broken channel line to indicate this normally off condition. As we know, a gate voltage greater than the threshold voltage creates an n-type inversion layer that connects the source to the drain. The arrow points to this inversion layer, which acts like an n channel when the device is conducting. For this reason, we have an n-channel enhancement-only MOSFET.

Figure 13-16b shows the schematic symbol for the complementary MOSFET, the p-channel enhancement-only MOSFET.

13-6. SUMMARY

There are three fundamental types of FETs: the JFET, the depletion-enhancement MOSFET, and the enhancement-only MOSFET. The JFET operates only in the depletion mode (Fig. 13-17a), the depletion-enhancement MOSFET in either mode (Fig. 13-17b), and the enhancement-only MOSFET only in the enhancement mode (Fig. 13-17c).

All are square-law devices, meaning their transconductance curve is parabolic. The transconductance equations of Fig. 13-17a and b are the same because each parabola has its vertex at $V_{GS(off)}$. The equation for Fig. 13-17c, however, differs because the vertex is at $V_{GS(th)}$.

Two of the main advantages of FETs are high input impedance and the square-law property. Other advantages of FETs as well as their disadvantages are discussed in the next chapter.

EXAMPLE 13-5.

A 3N169 is an n-channel enhancement-only MOSFET. It has a gate leakage current of 100 pA for a gate voltage of 35 V and an ambient temperature of 125°C. Calculate the dc resistance from gate to source.

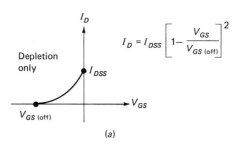

$$I_D = I_{DSS}\left[1 - \frac{V_{GS}}{V_{GS\,(off)}}\right]^2$$

Depletion only

I_{DSS}

$V_{GS\,(off)}$

(a)

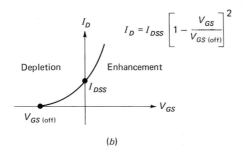

$$I_D = I_{DSS}\left[1 - \frac{V_{GS}}{V_{GS\,(off)}}\right]^2$$

Depletion   Enhancement

I_{DSS}

$V_{GS\,(off)}$

(b)

Figure 13-17. Transconductance curves. (a) JFET. (b) Normally on MOSFET. (c) Normally off MOSFET.

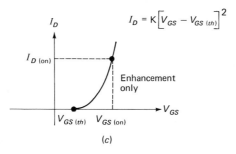

$$I_D = K\left[V_{GS} - V_{GS\,(th)}\right]^2$$

$I_{D\,(on)}$

Enhancement only

$V_{GS\,(th)}$ $V_{GS\,(on)}$

(c)

SOLUTION.
The dc resistance is

$$R_{GS} = \frac{V_{GS}}{I_G} = \frac{35}{100(10^{-12})} = 350{,}000 \text{ M}\Omega$$

Problems

13-1. At room temperature the 2N4220 (an *n*-channel JFET) has a reverse gate current of 0.1 nA for a reverse gate voltage of 15 V. Calculate the dc resistance from gate to source.

13-2. When the ambient temperature is 150°C, a JFET has a reverse gate current of 100 nA for a reverse gate voltage of 15 V. What is the dc resistance from gate to source at this elevated temperature?

13-3. A JFET data sheet indicates $V_{GS(off)}$ has a minimum value of -2 V and a maximum value of -4.5 V. What is the minimum pinch-off voltage we can expect with this type of JFET? If the $V_{DS(max)}$ of this JFET is 25 V, in what range should V_{DS} be for the JFET to act almost like a current source (use the larger pinch-off voltage)?

13-4. If a JFET has the drain curves of Fig. 13-18a, what value does I_{DSS} have? To ensure operation on the almost flat part of the top curve, what range should V_{DS} be kept in?

13-5. A JFET has an I_{DSS} of 9 mA and a $V_{GS(off)}$ of -3 V. What is the transconductance equation for this JFET? How much drain current is there when $V_{GS} = -1.5$ V?

13-6. Write the transconductance equation for the JFET whose curve is shown by Fig. 13-18b. How much drain current is there when $V_{GS} = -4$ V? When $V_{GS} = -2$ V?

13-7. Calculate the value of drain current for a $V_{GS} = -4.5$ V in Fig. 13-18b.

13-8. If a JFET has a transconductance curve like Fig. 13-18c, what value does I_{DSS} have? $V_{GS(off)}$? Pinch-off voltage V_P?

13-9. Write the equation for the JFET transconductance curve of Fig. 13-18c.

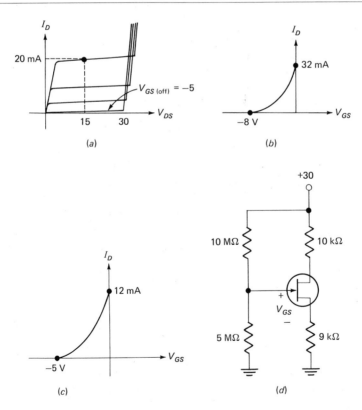

Figure 13-18.

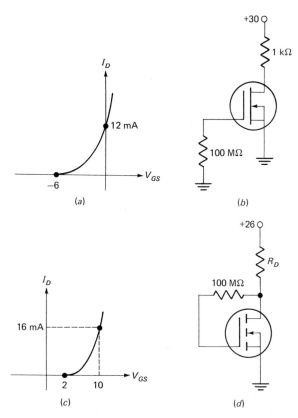

Figure 13-19.

13-10. If a JFET has a square-law curve like Fig. 13-18c, how much drain current is there when $V_{GS} = -1$ V?

13-11. A JFET has a drain current of 5 mA. If $I_{DSS} = 10$ mA and $V_{GS(off)} = -6$ V, what is the *approximate* value of V_{GS}? Of V_P?

13-12. The JFET of Fig. 13-18d has an $I_{DSS} = 2$ mA and a $V_{GS(off)} = -2$ V. The I_D in the circuit is 1 mA. What is the approximate value of V_{GS}? Since 1 mA is flowing down through the drain and source resistors, how much voltage is there across each of these resistors?

13-13. An *n*-channel depletion-enhancement MOSFET has an I_{DSS} of 8 mA and a $V_{GS(off)}$ of -4 V. How much drain current is there when $V_{GS} = -1$ V? And when $V_{GS} = 1$ V?

13-14. A MOSFET has the transconductance curve of Fig. 13-19a. What is the value of V_P? Of I_{DSS}? Of I_D when V_{GS} is -1 V?

13-15. The normally on MOSFET of Fig. 13-19b has the transconductance curve of Fig. 13-19a. $V_{GS} = 0$. How much drain current is there? What is the value of V_{DS}? The FET is operating above the knee of the $V_{GS} = 0$ drain curve. Why?

13-16. An enhancement-only MOSFET has the transconductance curve of Fig. 13-19c. What value does the constant K in Eq. (13-6) have?

13-17. If a normally off MOSFET has the curve of Fig. 13-19c, how much drain current is there when $V_{GS} = 3$ V? $V_{GS} = 4$ V?

13-18. The MOSFET shown in Fig. 13-19d has the transconductance curve of Fig. 13-19c. Negligible gate current flows. If $V_{DS} = 10$ V, what is the value of V_{GS}? How much drain current is there? What value of R_D satisfies the given values?

14. FET Circuit Analysis

This chapter is about small-signal FET amplifiers, where the signal swing is less than 10 percent of the ac load line. After discussing biasing circuits, we go into ac analysis. The formulas we derive for dc and ac action should be easy to understand and remember; they are similar to formulas for the bipolar transistor.

14-1. GATE BIAS

Gate bias is like base bias. Figure 14-1a shows the prototype for gate bias. V_{GG} and R_G may be discrete components or Thevenin quantities. Regardless, whenever we can reduce a FET biasing circuit to the form of Fig. 14-1a, we have it in gate-bias prototype form. Figure 14-1b is an equivalent form of gate bias.

As mentioned in the preceding chapter, normal biasing of a JFET requires

$$V_P < V_{DS} < V_{DS(\text{max})}$$

and
$$0 < I_D < I_{DSS}$$

When we analyze or design a JFET circuit, we must be sure these conditions are satisfied. In all the JFET circuits that follow, we assume these conditions are true; otherwise, the JFET is saturated or cut off, conditions of no interest in this chapter.

The derivation for drain current is analogous to the derivation for collector current. In Fig. 14-1a we start by summing voltages around the gate loop to get

$$-V_{GG} + I_G R_G + V_{GS} + I_D R_S = 0$$

or
$$I_D = \frac{V_{GG} - V_{GS} - I_G R_G}{R_S} \tag{14-1}$$

Figure 14-1. Gate-biased prototype.

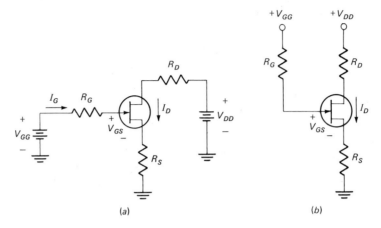

(a)   (b)

In a practical circuit the first two terms of the numerator swamp out the last term. That is, since gate current I_G is very small,

$$V_{GG} - V_{GS} \gg I_G R_G \qquad (14\text{-}2)$$

Therefore, the practical formula for drain current becomes

$$I_D = \frac{V_{GG} - V_{GS}}{R_S} \qquad \text{(third)} \qquad (14\text{-}3)$$

This is our third approximation. Gate-source voltage V_{GS} is a negative quantity when a JFET is normally biased; therefore, the numerator is actually larger than V_{GG}. V_{GS} is analogous to the V_{BE} of a transistor. Like V_{BE}, it changes with temperature and from one JFET to another. Therefore, a designer tries to swamp out V_{GS} by using a large value of V_{GG}. In this way, drain current becomes

$$I_D \cong \frac{V_{GG}}{R_S} \qquad \text{for } V_{GG} \gg |V_{GS}| \qquad (14\text{-}4)^{***}$$

This will be our ideal approximation. For instance, if $V_{GS} = -2$ V, V_{GG} must be much greater than 2 V to swamp out V_{GS}. In Eq. (14-4), V_{GG} and R_S are independent of the JFET. This means I_D is independent of the JFET characteristics.

As with bipolars, Thevenizing the gate circuit often reduces the actual circuit to gate-biased prototype form. Because of this, we can use Eqs. (14-1) through (14-4) to analyze many JFET biasing circuits.

EXAMPLE   14-1.
Work out the drain current in Fig. 14-2 as follows:

1. Get an ideal answer.
2. Let $V_{GS} = -2$ V and use the third approximation.

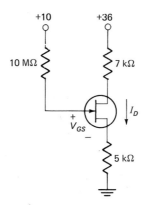

Figure 14-2. Examples 14-1 and 14-2.

SOLUTION.

1. With Eq. (14-4) the ideal drain current is

$$I_D \cong \frac{V_{GG}}{R_S} = \frac{10}{5000} = 2 \text{ mA}$$

2. Substituting $V_{GS} = -2$ V into Eq. (14-3) gives

$$I_D = \frac{V_{GG} - V_{GS}}{R_S} = \frac{10 - (-2)}{5000} = 2.4 \text{ mA}$$

EXAMPLE 14-2.

Calculate the ideal value of V_{DS} of Fig. 14-2.

SOLUTION.

The way to get V_{DS} is to find the drain-to-ground voltage V_D and subtract the source-to-ground voltage V_S.

$$V_D = V_{DD} - I_D R_D = 36 - 0.002(7000) = 22 \text{ V}$$

and
$$V_S = I_D R_S = 0.002(5000) = 10 \text{ V}$$

Therefore,

$$V_{DS} = V_D - V_S = 22 - 10 = 12 \text{ V}$$

14-2.   PRACTICAL BIASING CIRCUITS

In a system you may have only one positive supply and one negative supply. There-fore, a separate V_{GG} supply as shown in Fig. 14-2 is impractical. Usually, the V_{GG} voltage represents the Thevenin voltage driving the gate.

Voltage-divider bias

Figure 14-3*a* shows one of the better ways to bias a JFET. The idea is analogous to voltage-divider bias of a bipolar transistor. After Thevenizing the gate circuit, we have the prototype form of Fig. 14-3*b*. As mentioned, we would like to swamp out V_{GS}; this means we need a $V_{TH} \gg V_{GS}$.

How large is V_{GS} in typical JFET circuits? A study shows

$$-8 \text{ V} < V_{GS(\text{off})} < 0$$

for more than 90 percent of all JFETs. Many JFETs are biased in the vicinity of

$$I_D \cong \frac{I_{DSS}}{2}$$

In the preceding chapter we proved an I_D of half I_{DSS} meant a V_{GS} of about one-fourth $V_{GS(\text{off})}$. Therefore, as an approximate guide

$$-2 \text{ V} < V_{GS} < 0$$

To have a second approximation, we will use $V_{GS} = -2$ V. After substitution into Eq. (14-3),

$$I_D \cong \frac{V_{GG} + 2}{R_S} \qquad \text{(second)} \qquad (14\text{-}5)$$

In a voltage-divider bias circuit like Fig. 14-3, V_{TH} typically has to be much greater than 2 V to do an adequate swamping job on V_{GS}. Because of this, swamping is not as easy with JFETs as with bipolars; in the bipolar circuit we only have to swamp out 0.7 V.

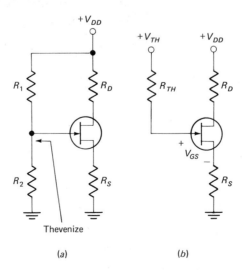

Figure 14-3. Voltage-divider bias (a) Actual circuit. (b) Thevenized form.

(a)

(b)

Therefore, voltage-divider biasing of JFETs gives good swamping only when large supply voltages are available. When only low-voltage supplies are on hand, we turn to *current-source bias* (Sec. 14-4).

Self-bias

Figure 14-4a shows *self-bias;* it is not as good a swamping circuit as voltage-divider bias. In fact, since $V_{GG} = 0$, there is no swamping. Here is how it works. The conventional current flowing down through R_S produces a source-to-ground voltage of

$$V_S = I_D R_S$$

Since negligible gate current flows, the gate terminal is at dc ground, that is,

$$V_G = 0$$

Summing voltages around the gate loop gives

$$V_{GS} + V_S = 0$$

or

$$V_{GS} = -V_S = -I_D R_S$$

The biasing voltage V_{GS} is set up entirely by the IR drop across R_S; this is why the circuit is called *self-bias.*

As an example, if $I_D = 1$ mA and $R_S = 1$ kΩ,

$$V_S = 0.001(1000) = 1 \text{ V}$$

This means the source terminal is $+1$ V with respect to ground. The gate terminal is 0 V with respect to ground because negligible gate current flows through R_G. Therefore, the source is $+1$ V with respect to the gate; equivalently, $V_{GS} = -1$ V.

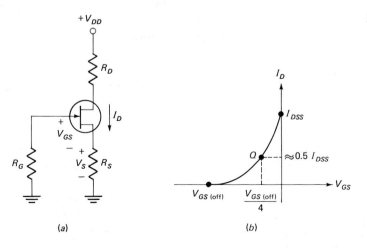

Figure 14-4. Self-bias. (a) Circuit. (b) Typical Q point.

(a)

(b)

In Fig. 14-4a, V_{GG} is zero. When we substitute this into the prototype formula (14-3), we get

$$I_D = -\frac{V_{GS}}{R_S} \tag{14-6}$$

Designers often bias a circuit like Fig. 14-4a to get an I_D of half I_{DSS}. This implies a V_{GS} of about one-fourth $V_{GS(off)}$ as shown in Fig. 14-4b. Therefore,

$$I_D \cong -\frac{V_{GS(off)}}{4R_S} \qquad \text{for } I_D \cong 0.5 I_{DSS} \tag{14-7}^{***}$$

If you are designing a self-biased circuit, you would use design-center values, either arithmetic or geometric averages (Sec. 7-7).

Simplicity is one of the few advantages of self-bias. It is not considered a good biasing circuit because V_{GS} has not been swamped. Unless there is a special reason, you avoid self-bias and use voltage-divider bias or one of the forms to be discussed.

EXAMPLE 14-3.
Calculate the drain current in Fig. 14-5a.

SOLUTION.
With Eq. (14-4),

$$I_D \cong \frac{V_{GG}}{R_S} = \frac{10}{10,000} = 1 \text{ mA}$$

Or, with Eq. (14-5),

$$I_D \cong \frac{V_{GG} + 2}{R_S} = \frac{10 + 2}{10,000} = 1.2 \text{ mA}$$

EXAMPLE 14-4.
The JFET of Fig. 14-5b has a $V_{GS(off)}$ of -5 V and the drain current is approximately half of I_{DSS}. Calculate the value of drain current.

SOLUTION.
With Eq. (14-7),

$$I_D = \frac{5}{4(2500)} = 0.5 \text{ mA}$$

14-3. SOURCE BIAS

Figure 14-6a shows *source bias* (analogous to emitter bias). The idea is to swamp out V_{GS}. Since most of V_{SS} appears across R_S, we get an I_D almost independent of JFET characteristics. Figure 14-6b is a simpler way to show the circuit.

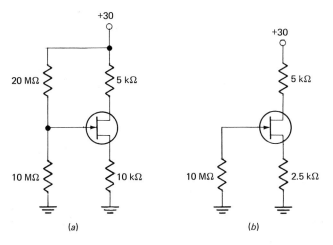

Figure 14-5. Examples 14-3 and 14-4.

Summing voltages around the gate loop gives

$$I_G R_G + V_{GS} + I_D R_S - V_{SS} = 0$$

or

$$I_D = \frac{V_{SS} - V_{GS} - I_G R_G}{R_S} \tag{14-8}$$

In any practical circuit

$$V_{SS} - V_{GS} \gg I_G R_G \tag{14-9}$$

because gate current I_G is very small. Therefore, the practical formula for drain current is

$$I_D = \frac{V_{SS} - V_{GS}}{R_S} \quad \text{(third)} \tag{14-10}$$

When no data sheets are available, we can use $V_{GS} = -2$ V as an estimate. This means

$$I_D \cong \frac{V_{SS} + 2}{R_S} \quad \text{(second)} \tag{14-11}$$

When V_{SS} is much greater than V_{GS},

$$I_D \cong \frac{V_{SS}}{R_S} \quad \text{(ideal)} \tag{14-12}***$$

The results are analogous to emitter bias. Therefore, in an ideal analysis of a source-biased circuit we visualize all of the V_{SS} voltage across R_S; this means a drain current equal to the ratio of V_{SS} to R_S. Since these quantities are independent of the JFET characteristics, we get an I_D that stays almost constant.

Figure 14-6. Source bias.

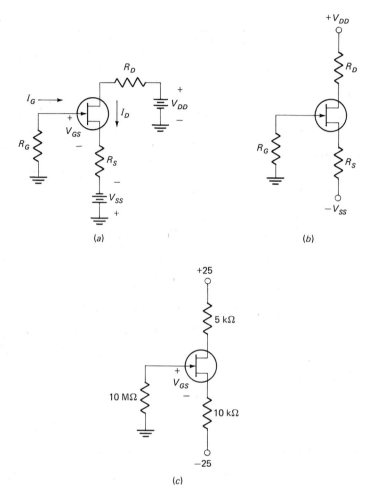

(a)

(b)

(c)

The swamping of V_{GS} can be very effective when large negative sources are available. For example, the drain current of Fig. 14-6c is ideally

$$I_D \cong \frac{25}{10,000} = 2.5 \text{ mA}$$

To a second approximation,

$$I_D \cong \frac{25 + 2}{10,000} = 2.7 \text{ mA}$$

14-4. CURRENT-SOURCE BIAS

When V_{GS} is small, it is easy to swamp out. But when large, it may not be possible to swamp out with available supplies. In other words, if you have only 10-V supplies or less to work with, you will have a hard time swamping out larger values of V_{GS}. In a case like this, *current-source bias* solves the problem.

Two supplies

Figure 14-7a shows how it's done when two supplies are available. Instead of driving the source terminal through a resistance R_S, we drive it with a bipolar transistor. As long as the bipolar is FR-biased, it acts like a current source and forces the JFET to have an I_D equal to I_C.

In Fig. 14-7a the bipolar transistor has an emitter current of

$$I_E \cong \frac{V_{EE} - 0.7}{R_E} \quad \text{(second)}$$

or

$$I_E \cong \frac{V_{EE}}{R_E} \quad \text{(ideal)}$$

The collector diode acts like a current source of approximately this value; therefore, it forces the drain current of the JFET to approximately equal I_E. The V_{GS} of the JFET automatically takes on whatever value is necessary to make $I_D = I_C$.

In a circuit like Fig. 14-7a you must make sure

$$I_C < I_{DSS} \tag{14-13a}$$

This guarantees V_{GS} has a negative value. In Fig. 14-7a you can sum voltages from gate to source to emitter to base and get

$$V_{GS} + V_{CE} - 0.7 = 0$$

or

$$V_{CE} = 0.7 - V_{GS} \tag{14-13b}$$

Since V_{GS} has a negative value, V_{CE} is greater than 0.7 V; this is enough to prevent most bipolars from being in the saturation region. In other words, for normal operation of Fig. 14-7a you also must satisfy

$$V_{CE(\text{sat})} < 0.7 - V_{GS} \tag{14-13c}$$

Here is an example. The bipolar of Fig. 14-7b has an emitter current of approximately

$$I_E \cong \frac{10}{10,000} = 1 \text{ mA}$$

Figure 14-7. Current-source bias with two power supplies.

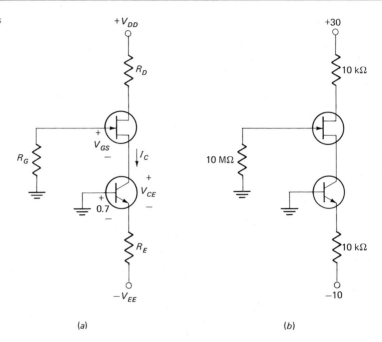

(a)

(b)

This forces the drain current to equal approximately 1 mA. V_{GS} is a slave variable in this action; it has to take on whatever value is necessary to set up 1 mA. As long as the JFET has an I_{DSS} greater than 1 mA, V_{GS} will be negative. If $V_{GS} = -2$ V, then with Eq. (14-13b),

$$V_{CE} = 0.7 - (-2) = 2.7 \text{ V}$$

well above the $V_{CE(\text{sat})}$ of any bipolar transistor. Even when V_{GS} is close to zero, V_{CE} is more than 0.7 V, which is enough to keep most bipolars from saturating; this is essential; otherwise, the bipolar does not act like a current source.

Current-source bias like Fig. 14-7a is swamping at its best. V_{GS} and its variations are almost completely out of the picture. The only significant variable is the V_{BE} of the bipolar. We know it varies slightly from one transistor to the next, and with temperature change. But these changes in V_{BE} are only a tenth volt or so. Therefore, with a circuit like Fig. 14-7a we can have an almost solid value of I_D even with a low value of V_{EE}.

The extra expense for the bipolar in Fig. 14-7a is usually negligible compared to the cost of a JFET. Most FETs cost more than bipolars. Besides, the bipolar of Fig. 14-7a is not part of the ac circuit because we would bypass the collector to ground. All the bipolar has to do is set up a dc collector current. Because of this, almost any small-signal inexpensive bipolar transistor will do an adequate job.

One supply

When you do not have a negative supply voltage, you can still use current-source bias as shown in Fig. 14-8a. In this circuit the voltage divider (R_1 and R_2) sets up voltage-divider bias on the bipolar transistor. That is, almost all the voltage across R_2 appears across the R_E resistor. This fixes the emitter current to a value essentially independent of the JFET characteristics. Again, the collector diode acts like a current source, forcing the drain current to equal the collector current.

Especially note, in Fig. 14-8a you do not ground the bottom of R_G. You must connect it to the base of the bipolar; this is necessary to reverse-bias the collector diode.

Figure 14-8b shows a concrete example. The Thevenin base voltage is 10 V. Most of this appears across R_E and sets up

$$I_E \cong \frac{10}{10,000} = 1 \text{ mA}$$

The collector diode now forces approximately 1 mA of current to flow through the JFET. When this flows through the drain resistor (8 kΩ), it produces an 8-V drop and makes the drain voltage 22 V with respect to ground. Since the base is 10 V to ground, the gate also must be 10 V to ground (no gate current). Assuming a V_{GS} of -2 V, the collector is 12 V with respect to ground; this is more than enough to reverse-bias the collector diode.

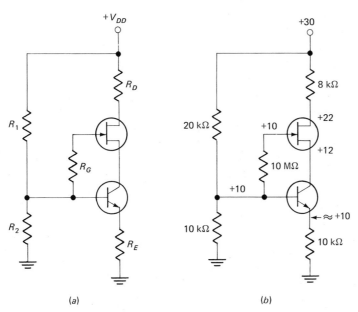

(a)

(b)

Figure 14-8. Current-source bias with one power supply.

To a second approximation, the emitter current in Fig. 14-18b is

$$I_E \cong \frac{10 - 0.7}{10,000} = 0.93 \text{ mA}$$

As a result, the drain voltage is slightly higher than 22 V. But the point is clear. Let the temperature change, or plug any JFET you like into the circuit; as long as I_C is less than I_{DSS}, the bipolar acts like a current source and you get a solid I_D value.

14-5. p-CHANNEL JFET BIASING

For normal bias of a p-channel JFET, all voltages and currents are opposite those of an n-channel JFET. Again, we take advantage of the complementary nature of n-channel and p-channel JFETs. There is no need to derive separate formulas for the p-channel JFET. If we understand n-channel JFET biasing, all we do to get the corresponding p-channel JFET circuit is complement voltages and currents.

Figure 14-9 shows biasing circuits that are complementary to the n-channel JFET circuits discussed in preceding sections. Here, we recognize voltage-divider bias, self-bias, source bias, two-supply current-source bias, and one-supply current-source bias.

As with bipolars, many people draw p-channel JFETs upside down; this unifies and simplifies schematic diagrams. Figure 14-10 shows how the p-channel JFET biasing circuits look when drawn upside down. In each case we can identify the drain end by the R_D resistor connected to it. Whenever there is any doubt about which end is the drain, you can label each terminal or you can use the alternative schematic symbol discussed earlier (Fig. 13-4c).

In Fig. 14-10 we again recognize voltage-divider bias, self-bias, source bias, two-supply current-source bias, and one-supply current-source bias. (If you are having any trouble understanding these upside-down circuits, review Secs 7-9 and 7-10.)

EXAMPLE 14-5.
Calculate the drain current in Fig. 14-11a. And the drain-to-ground voltage.

SOLUTION.
We are using the alternative symbol for the JFET; the gate is closer to the source than in the symmetrical symbol. We have voltage-divider bias of an upside-down p-channel JFET. Ideally, all of the 10 V across the 10-MΩ resistor appears across the 2-kΩ resistor. This sets up a drain current of

$$I_D \cong \frac{10}{2000} = 5 \text{ mA} \qquad \text{(ideal)}$$

When this current flows through the 1-kΩ drain resistor of Fig. 14-11a, it produces a drop of

$$V_D = I_D R_D = 0.005(1000) = 5 \text{ V}$$

This is the drain-to-ground voltage.

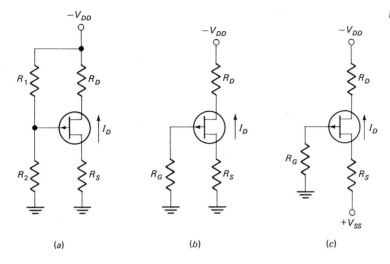

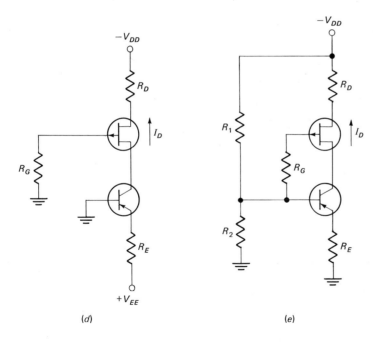

Figure 14-9. Biasing p-channel JFETs.

*Figure 14-10. Upside-down
p-channel JFET circuits.*

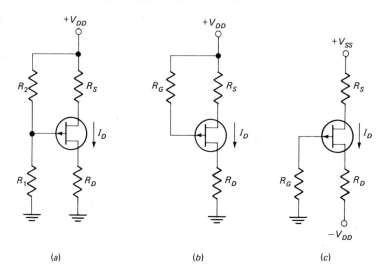

(a)   (b)   (c)

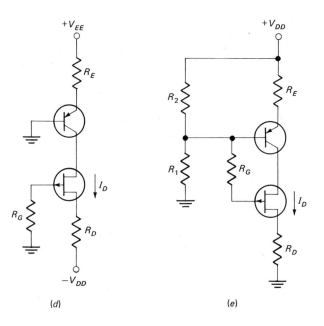

(d)   (e)

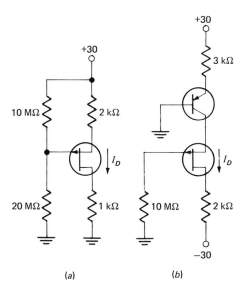

Figure 14-11. Examples 14-5 and 14-6.

(a)   (b)

To a second approximation, we can assume $V_{GS} = -2$ V. Then,

$$I_D \cong \frac{10 + 2}{2000} = 6 \text{ mA}$$

and

$$V_D \cong 0.006(1000) = 6 \text{ V}$$

EXAMPLE 14-6.
Calculate the ideal values of I_D and V_D in Fig. 14-11b.

SOLUTION.
The bipolar has an ideal emitter current of

$$I_E \cong \frac{30}{3000} = 10 \text{ mA}$$

Therefore, I_D approximately equals 10 mA. When this flows down through the 2-kΩ drain resistor, we get a drain-to-ground voltage of

$$V_D = I_D R_D - 30 = 0.01(2000) - 30 = -10 \text{ V}$$

14-6. BIASING MOSFETS

To begin with, the gate-biased and source-biased prototypes (Figs. 14-1 and 14-6) apply to MOSFETs as well as JFETs. Therefore, the same prototype formulas apply:

$$I_D = \frac{V_{GG} - V_{GS}}{R_S} \quad \text{(gate bias)}$$

and
$$I_D = \frac{V_{SS} - V_{GS}}{R_S} \quad \text{(source bias)}$$

Practical biasing circuits

With depletion-enhancement (DE) MOSFETs, V_{GS} can be negative or positive. But with enhancement-only (E) MOSFETs, V_{GS} has to be greater than $V_{GS(th)}$, a positive quantity. With these restrictions in mind, you can examine each of the biasing circuits discussed earlier and decide which ones MOSFETs work in. Table 14-1 summarizes the biasing circuits you can use with different kinds of FETs.

TABLE 14-1.   FET BIASING CIRCUITS

	VOLTAGE-DIVIDER	SELF-BIAS	SOURCE-BIAS	CURRENT-SOURCE
JFET	Yes	Yes	Yes	Yes
DE MOSFET	Yes	D only	Yes	D only
E MOSFET	Yes	No	Yes	No

Here is what Table 14-1 is all about. Any kind of FET will work in a voltage-divider bias circuit. But in a self-biased circuit, V_{GS} must be negative; therefore, DE MOSFETs will work only in the depletion mode; E MOSFETs cannot work in a self-bias circuit. Any kind of MOSFET works in a source-biased circuit because V_{GS} can be positive or negative. When it comes to current-source bias, the FET must operate in the depletion mode; this is the only way to ensure reverse bias on the collector diode, at least in the basic circuits of Figs. 14-7a and 14-8a.

Zero bias of a DE MOSFET

Since a DE MOSFET can operate in either the depletion or enhancement mode, we can set its Q point at $V_{GS} = 0$ as shown in Fig. 14-12a. Then, an input ac signal to the gate can produce variations above and below the Q point as shown. Being able to use $V_{GS} = 0$ is an advantage when it comes to biasing. For one thing, if a circuit is designed for a center value of $V_{GS} = 0$, voltage-divider bias with smaller supply voltages may be adequate to swamp out variations in V_{GS}.

In addition, we can use the unique biasing circuit of Fig. 14-12b with a DE MOSFET; this simple circuit has no applied gate or source voltage; therefore, $V_{GS} = 0$ and $I_D = I_{DSS}$. It follows that

$$V_{DS} = V_{DD} - I_{DSS}R_D$$

As long as V_{DS} is greater than V_P, operation is on the almost flat part of the $V_{GS} = 0$ drain curve. The circuit of Fig. 14-12b is unique with DE MOSFETs; you cannot get normal bias if you try using a bipolar, a JFET, or an E MOSFET in this circuit.

Aside from its simplicity, the circuit has this advantage: we can *direct-couple* the ac

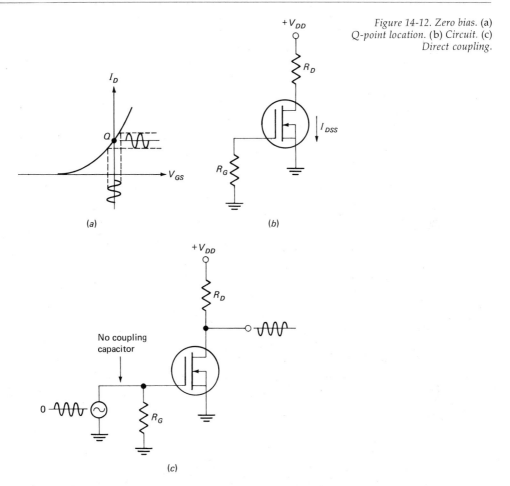

Figure 14-12. Zero bias. (a) Q-point location. (b) Circuit. (c) Direct coupling.

signal into the gate as shown in Fig. 14-12c. This allows operation down to zero frequency. In some applications, this can be important.

Drain-feedback bias of an E MOSFET

Figure 14-13a shows *drain-feedback bias,* a type of bias you can use with E MOSFETs. With negligible gate current, no voltage appears across R_G; therefore, $V_{GS} = V_{DS}$. Typically, V_{DS} is kept above 10 V to ensure operation well above the pinch-off voltage. Like collector-feedback bias, the circuit of Fig. 14-13a tends to compensate for changes in FET characteristics. Specifically, if I_D tries to increase for some reason, V_{DS} decreases; this reduces V_{GS}, which partially offsets the original increase in I_D.

Figure 14-13b shows the Q point on the transconductance curve. V_{GS} equals V_{DS}, and the corresponding I_D equals $I_{D(on)}$, a value of drain current well above the threshold

Figure 14-13. Drain-feedback bias.

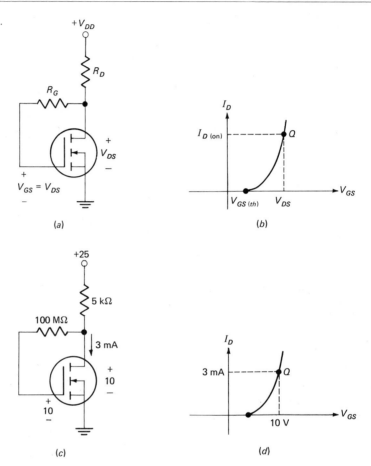

(a)

(b)

(c)

(d)

point. To assist you, data sheets for E MOSFETs usually give a value of $I_{D(on)}$ for $V_{GS} = V_{DS}$. This helps in setting up the Q point. In design, all you do is select a value of R_D that sets up the specified V_{DS}.

For instance, suppose the data sheet of an E MOSFET specifies $I_{D(on)} = 3$ mA when $V_{GS} = V_{DS} = 10$ V. If we have a 25-V supply to work with, we can select an R_D of 5 kΩ as shown in Fig. 14-13c. When 3 mA of drain current flows, V_{DS} equals 10 V and so too does V_{GS}. Therefore, the E MOSFET is operating at its specified *on* point (Fig. 14-13d).

In a circuit like Fig. 14-13c different MOSFETs or a varying temperature may cause I_D to differ from 3 mA. But the changes are partially offset by the feedback from drain to gate. If I_D tries to increase above 3 mA, V_{DS} drops below 10 V; this lowers V_{GS}, which reduces the attempted change in I_D. The overall effect is a smaller increase in I_D than would take place without the feedback.

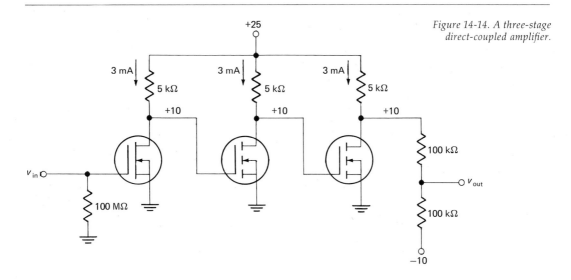

Figure 14-14. A three-stage direct-coupled amplifier.

Dc amplifier

A *dc amplifier* is one that can operate all the way down to zero frequency without a loss of gain. One way to build a dc amplifier, or dc amp, is to leave out all coupling and bypass capacitors.

Figure 14-14 shows a dc amp using MOSFETs. The input stage is a DE MOSFET with zero bias; this allows us to direct-couple into the gate. The second and third stages use E MOSFETs; each gate gets its V_{GS} from the drain of the preceding stage. The design of Fig. 14-14 uses MOSFETs with drain currents of 3 mA. For this reason, each drain runs at 10 V with respect to ground. We tap the final output voltage from between the 100-kΩ resistors. Since the lower resistor is returned to −10 V, the quiescent output voltage is 0 V. When an ac signal drives the amplifier, regardless of how low its frequency, we get an amplified output voltage.

There are other ways of designing dc amplifiers. The beauty of Fig. 14-14 is its simplicity.

Complementary MOS

Given an *n*-channel MOSFET biasing circuit, we can get the *p*-channel MOSFET circuit by reversing all voltages and currents as we have done for bipolars and JFETs. *Complementary MOS* (CMOS) is the combined use of *n*-channel and *p*-channel MOSFETs; many digital integrated circuits use complementary MOS.

EXAMPLE 14-7.
The data sheet of a 2N3796 (a DE MOSFET) has these maximum ratings:

1. $V_{DS(\text{max})} = 25$ V
2. $V_{GS(\text{max})} = \pm 30$ V
3. $I_{D(\text{max})} = 20$ mA
4. $P_{D(\text{max})} = 300$ mW at $T_A = 25°$C; derate 1.7 mW per degree.

What do these ratings mean?

SOLUTION.
Ratings like these are similar to bipolar ratings, but a few comments may help.

1. $V_{DS(\text{max})}$ is the maximum voltage you can apply between the drain and source without breakdown. You must keep V_{DS} less than $V_{DS(\text{max})}$ for normal operation.
2. $V_{GS(\text{max})}$ is the maximum voltage you can apply between the gate and source without destroying the thin layer of metal oxide. Breakdown of the thin insulating layer is always destructive; you can throw the MOSFET away if it happens. The polarity is immaterial; a gate voltage greater than $+30$ V or more negative than -30 V destroys the metal-oxide layer.

    Aside from directly applying a V_{GS} greater than 30 V, you can destroy the insulating layer in more subtle ways. If you remove or insert a MOSFET into a circuit while the power is on, transient voltages may be large enough to ruin the MOSFET. Besides transient voltage, static voltage may destroy a MOSFET. Pick it up often and you may deposit enough static electrical charge on the gate to exceed the $V_{GS(\text{max})}$ rating. This is the reason MOSFETs are shipped with a wire ring around the leads. You remove the ring after the MOSFET is connected in the circuit.
3. $I_{D(\text{max})}$ is the maximum continuous drain current you can have without risking damage; in this case, not more than 20 mA of continuous drain current should flow.
4. The $P_{D(\text{max})}$ rating of 300 mW means the MOSFET can dissipate up to 300 mW when the ambient temperature is 25°C. For higher ambient temperatures, you subtract 1.7 mW for each degree above 25°C.

14-7. AC EQUIVALENT CIRCUITS

The complete ac equivalent circuit of a JFET or MOSFET includes lead inductances, internal capacitances, and so on. We neglect the lead inductances and a few other minor effects and start with the approximate equivalent circuit of Fig. 14-15a. C_{gd} is the capacitance between the gate and the drain, C_{gs} the capacitance from gate to source, and r_{gs} the ac resistance from gate to source. On the output side, C_{ds} and r_{ds} are the ac capacitance and resistance from drain to source.

Low-frequency model

At lower frequencies the X_C of each capacitor becomes high enough to neglect, and the equivalent circuit simplifies to Fig. 14-15b. Resistance r_{gs} is high in JFETs; in

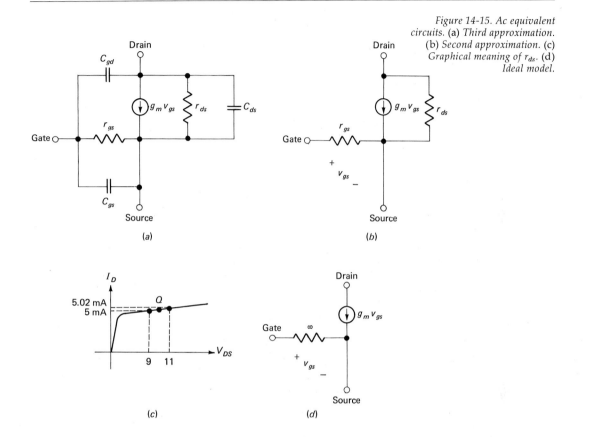

MOSFETs it becomes incredibly large. Therefore, in our ideal model we will treat r_{gs} as infinite.

Resistance r_{ds} is the ac resistance from drain to source and equals

$$r_{ds} = \frac{\Delta V_{DS}}{\Delta I_D} \qquad \text{for constant } V_{GS} \qquad (14\text{-}14)$$

Above the knee of any drain curve, the change in I_D is small for a given change in V_{DS}; therefore, r_{ds} has large values, typically from 10 kΩ to 1 MΩ. As an example, if we have the drain curve of Fig. 14-15c, the value of r_{ds} in the given interval is

$$r_{ds} = \frac{11 - 9 \text{ V}}{5.02 - 5 \text{ mA}} = 100 \text{ k}\Omega$$

Because r_{ds} is usually high, we neglect it in our ideal model.

Figure 14-15d shows the ideal ac equivalent circuit for a FET (J or MOS). The ac input voltage appears across an infinite resistance; the drain acts like a perfect current source with a value of

$$i_d = g_m v_{gs} \qquad \text{(ideal)} \qquad (14\text{-}15)$$

(Notice the use of lowercase letters and subscripts for ac quantities.) In this equation, g_m is a constant of proportionality.

Transconductance

The quantity g_m is called the *transconductance*. By rearranging Eq. (14-15),

$$g_m = \frac{i_d}{v_{gs}}$$

Since ac current and voltage are equal to changes in total current and voltage, we can rewrite this equation as

$$g_m = \frac{\Delta I_D}{\Delta V_{GS}} \qquad \text{for constant } V_{DS} \qquad (14\text{-}16)$$

Figure 14-16 brings out the meaning of g_m in terms of changes in I_D and V_{GS}. When we have a transconductance curve, we can select two nearby points like A and B. The ratio of the change in I_D to the change in V_{GS} gives us the g_m value between these two points. If we select another pair of points further up the curve at C and D, we get more of a change in I_D for a given change in V_{GS}; therefore, g_m has a larger value further up the curve.

On a data sheet for JFETs and DE MOSFETs you are usually given the value of g_m for $V_{GS} = 0$, that is, the value of g_m between points like C and D in Fig. 14-16. We will designate this value of g_m as g_{m0} to indicate it is measured at $V_{GS} = 0$. By deriving the slope of the transconductance curve at other points, we can prove any g_m equals

$$g_m = g_{m0} \left[1 - \frac{V_{GS}}{V_{GS(\text{off})}} \right] \qquad (14\text{-}17)^{***}$$

Figure 14-16. Graphical meaning of transconductance.

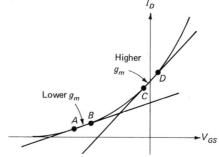

This equation lets us calculate any g_m value in terms of the given g_{m0} value.[1] Sometimes, the data sheet of a JFET or DE MOSFET gives you a curve showing g_m versus I_D; in this case, you don't need Eq. (14-17) and can read the value of g_m from the curve.

The data sheets for most E MOSFETs include a curve showing the value of g_m for different I_D. Incidentally, g_m is often designated y_{fs} (it stands for the forward transfer admittance). For example, the data sheet of a 2N4222 lists a typical y_{fs} of 4500 μmhos; this is identical to g_m. Transconductance has the units of mhos because Eq. (14-16) is the ratio of current to voltage.

A more accurate value of $V_{GS(\text{off})}$

In a transconductance curve like Fig. 14-17a, drain current changes rapidly when $V_{GS} = 0$, but slowly when V_{GS} approaches $V_{GS(\text{off})}$. Since there is some leakage current in the channel, it is difficult to get an accurate measurement of $V_{GS(\text{off})}$. For this reason, don't use the $V_{GS(\text{off})}$ given on a data sheet when you are calculating I_D with the transconductance Eq. (13-5). Instead, calculate an accurate value of $V_{GS(\text{off})}$ with

$$V_{GS(\text{off})} = -\frac{2I_{DSS}}{g_{m0}} \qquad (14\text{-}18)^{***}$$

This equation is derived by working out the formula for the tangent line through I_{DSS} (done in the Appendix). Since I_{DSS} and g_{m0} can be accurately measured, we can calculate accurate values for $V_{GS(\text{off})}$ with data-sheet values for I_{DSS} and g_{m0}.

Alternatively, if you have a transconductance graph, you can get $V_{GS(\text{off})}$ in another way. One of the basic properties of a parabola is that its vertex is twice as far from the origin as the tangent-line intercept shown in Fig. 14-17a. By drawing a tangent line to the curve at I_{DSS}, you can read the value of the intercept; doubling this value gives $V_{GS(\text{off})}$. For instance, in Fig. 14-17b the tangent line cuts through the horizontal axis at −2; therefore, $V_{GS(\text{off})} = -4$ V.

---

[1] If interested in derivation, see the Appendix.

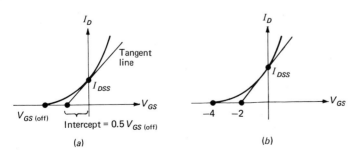

Figure 14-17. Graphical method to get accurate $V_{GS(\text{off})}$.

EXAMPLE 14-8.

A FET has an I_{DSS} of 10 mA and a g_{m0} of 4000 μmhos. Calculate the $V_{GS(off)}$ and the g_m for an I_D of approximately half I_{DSS}.

SOLUTION.
With Eq. (14-18),

$$V_{GS(off)} = -\frac{2(0.01)}{0.004} = -5 \text{ V}$$

When I_D is half I_{DSS}, V_{GS} is approximately one-fourth of $V_{GS(off)}$. In this case, $V_{GS} \cong -1.25$ V. Substituting this into Eq. (14-17) along with the given g_{m0} results in

$$g_m = 0.004 \left(1 - \frac{1.25}{5}\right) = 0.003 \text{ mho} = 3000 \text{ } \mu\text{mhos}$$

14-8. AC ANALYSIS

Ac analysis of FET amplifiers is straightforward. In the usual way, you start by shorting all coupling and bypass capacitors; then you reduce the dc-supply voltages to zero. You simplify this ac equivalent as much as possible by applying Thevenin's theorem. Often, the circuit will be in one of the ac prototype forms shown in Fig. 14-18. This basic prototype form is called *gate-driven* and is analogous to the bipolar base-driven prototype.

Prototype formulas

When we replace any of FETs in Fig. 14-18 by the ideal model, we get the equivalent circuit of Fig. 14-19. Voltage v_g is the ac voltage from gate to ground, v_{gs} the ac voltage from gate to source, and v_d the ac voltage from drain to ground. In Fig. 14-19 the positive half cycle of input voltage forces drain current to flow up through r_D; this produces the negative half cycle of output voltage; in other words, we get phase inversion. The ac output voltage is

$$v_d = i_d r_D \tag{14-19a}$$

and the ac voltage from source to ground is

$$v_s = i_d r_S \tag{14-19b}$$

Since the polarity of v_s is plus-minus, this voltage is in phase with the input signal.

If we sum ac voltages in the gate loop of Fig. 14-19, we get

$$-v_g + v_{gs} + i_d r_S = 0$$

or

$$v_g = v_{gs} + i_d r_S \tag{14-19c}$$

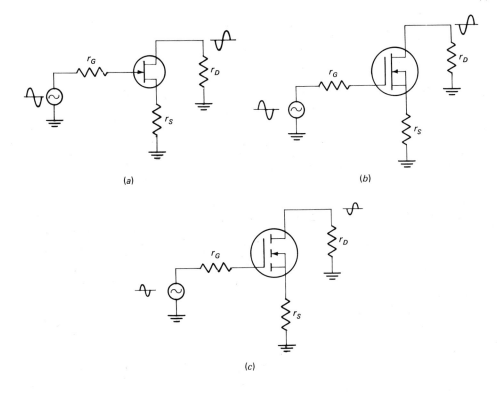

Figure 14-18. Ac gate-driven prototypes. (a) JFET. (b) DE MOSFET. (c) E MOSFET.

(a)

(b)

(c)

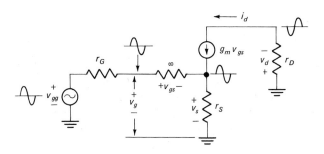

Figure 14-19. Gate-driven amplifier.

With Eq. (14-15), $v_{gs} \cong i_d/g_m$; therefore, Eq. (14-19c) becomes

$$v_g \cong \frac{i_d}{g_m} + i_d r_S$$

or

$$v_g \cong i_d \left(r_S + \frac{1}{g_m} \right) \qquad (14\text{-}20)$$

Voltage gain

Now, we can express the voltage gain in terms of circuit quantities. With Eqs. (14-19a) and (14-20) the voltage gain from gate to drain is

$$\frac{v_d}{v_g} \cong \frac{i_d r_D}{i_d (r_S + 1/g_m)}$$

or

$$\frac{v_d}{v_g} \cong \frac{r_D}{r_S + 1/g_m} \qquad \text{(ideal)} \qquad (14\text{-}21)^{***}$$

For the special case of $r_S = 0$, this becomes

$$\frac{v_d}{v_g} \cong g_m r_D \qquad \text{(ideal, } r_S = 0) \qquad (14\text{-}21a)$$

Or, if r_S is large enough to swamp out $1/g_m$ in Eq. (14-21),

$$\frac{v_d}{v_g} \cong \frac{r_D}{r_S} \qquad \text{(ideal, } r_S \gg 1/g_m) \qquad (14\text{-}21b)$$

Source follower

In the prototypes of Fig. 14-18, r_D may equal zero and the output signal may be taken from the source terminal. In this case, we have a *source follower,* analogous to the emitter follower. With Eqs. (14-19b and 14-20) the voltage gain from gate to source is

$$\frac{v_s}{v_g} \cong \frac{i_d r_S}{i_d (r_S + 1/g_m)}$$

or

$$\frac{v_s}{v_g} \cong \frac{r_S}{r_S + 1/g_m} \qquad \text{(ideal)} \qquad (14\text{-}22a)^{***}$$

If r_S is much greater than $1/g_m$,

$$\frac{v_s}{v_g} \cong 1 \qquad \text{(ideal, } r_S \gg 1/g_m) \qquad (14\text{-}22b)$$

The source follower acts like an emitter follower; its voltage gain is less than unity. Furthermore, a source follower has a very high input impedance and is often used at the front end of measuring instruments like voltmeters.

The basic bipolar amplifier circuits are the common-emitter amplifier, the swamped amplifier, and the emitter follower. The corresponding FET circuits are the common-source (CS) amplifier, the swamped amplifier, and the source follower. With the equations just derived, we can calculate the voltage gain for any of these FET amplifiers.

Ac load line

Since the gate-driven prototypes of Fig. 14-18 have the same form as a base-driven prototype, the ac load line of a FET amplifier will be the same except for the change in subscripts as follows:

(emitter)	$E \longrightarrow S$	(source)
(base)	$B \longrightarrow G$	(gate)
(collector)	$C \longrightarrow D$	(drain)

Therefore, the general load line derived earlier (Fig. 10-5b) still applies if we change the subscripts as shown in Fig. 14-20.

I_{DQ} and V_{DSQ} are the coordinates of the quiescent point. The upper end of the ac load line has a current of

$$I_{D(\text{sat})} = I_{DQ} + \frac{V_{DSQ}}{r_D + r_S}$$

and the lower end of the load line has a voltage of

$$V_{DS(\text{cutoff})} = V_{DSQ} + I_{DQ}(r_D + r_S)$$

In the bipolar amplifier the instantaneous operating point can swing to within a few-tenths volt of the upper end of the ac load line. In the FET amplifier, however, the signal swing cannot get past the upper limit shown in Fig. 14-20; typically, this upper limit has a voltage coordinate of several volts. In other words, when a FET amplifier

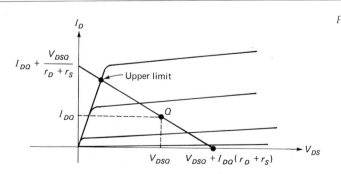

Figure 14-20. FET ac load line.

saturates, it usually has several volts across it rather than a few-tenths volt like a bipolar.

EXAMPLE 14-9.

Figure 14-21a shows a common-source amplifier. What is the voltage gain from gate to drain? The input impedance?

SOLUTION.

Visualize the ac equivalent circuit. Then, r_D equals 5 kΩ, r_S is zero, and g_m equals 5000 μmhos. With Eq. (14-21a),

$$\frac{v_d}{v_g} \cong g_m r_D = 0.005(5000) = 25$$

The input impedance neglecting capacitive effects approximately equals the value of the gate-return resistor R_G because $z_{in(gate)}$ approaches infinity. So,

$$z_{in} \cong R_G = 10 \text{ M}\Omega$$

EXAMPLE 14-10.

What is the voltage gain from gate to drain in Fig. 14-21b?

SOLUTION.

Visualize the ac equivalent circuit and you can see

$$r_D = 10{,}000 \parallel 40{,}000 = 8 \text{ k}\Omega$$

$$r_S = 400 \ \Omega$$

and $$g_m = 5000 \ \mu\text{mhos}$$

With prototype formula (14-21)

$$\frac{v_d}{v_s} \cong \frac{r_D}{r_S + 1/g_m} = \frac{8000}{400 + 1/0.005}$$

$$= \frac{8000}{400 + 200} = 13.3$$

Note that r_S is twice as large as $1/g_m$; the swamping is only partially effective in this case.

EXAMPLE 14-11.

The E MOSFET of Fig. 14-21c uses drain-feedback bias. Calculate the voltage gain from gate to source for this source-follower circuit.

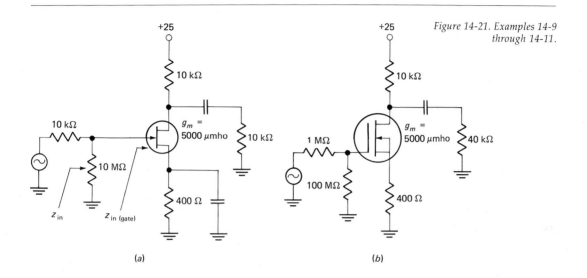

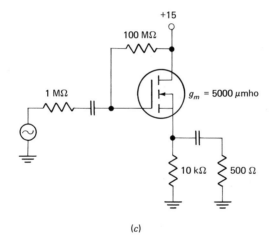

Figure 14-21. Examples 14-9 through 14-11.

SOLUTION.

The ac resistance seen by the source terminal of the MOSFET is

$$r_S = 10{,}000 \parallel 500 \cong 500\ \Omega$$

With Eq. (14-22a),

$$\frac{v_s}{v_g} \cong \frac{500}{500 + 1/0.005} = 0.714$$

14-9. FETS VERSUS BIPOLARS

Which is better: a FET or a bipolar? This is impossible to answer without knowing the particular application, the amount of money you have to work with, and other factors. All we do in this section is indicate some obvious and not-so-obvious differences between the two.

Input impedance

There is no question about this. The FET is an outstanding high-impedance device. To get input impedances greater than a megohm, you normally turn to the FET. The best you can do with a bipolar is a marginal design. For this reason, FETs are the natural choice at the front end of voltage-measuring instruments where you often need high input impedances like 10 MΩ or more.

As a guide, the input resistance looking into the gate of a JFET is typically from 100 MΩ to over 10,000 MΩ. With a MOSFET the input resistance is from about 10,000 MΩ to over 10,000,000 MΩ.

Voltage gain

When it comes to voltage gain, the FET is inferior to the bipolar. If you examine the equations for g_m and r_e', you will conclude the g_m of a bipolar transistor equals

$$g_{m(\text{bipolar})} = \frac{1}{r_e'}$$

With this, you can compare bipolars and FETs. For instance, when $I_E = 1$ mA, $r_e' = 25$ Ω (ideal). Therefore, the g_m of a bipolar transistor is

$$g_m = \frac{1}{25} = 0.04 \text{ mho} = 40,000 \ \mu\text{mhos}$$

At higher dc emitter currents the g_m increases. For I_E equals 5 mA, r_e' is ideally 5 Ω, and

$$g_m = \frac{1}{5} = 0.2 \text{ mho} = 200,000 \ \mu\text{mhos}$$

Given the same ac load resistance in a CE amplifier and a CS amplifier, the voltage gain of the CE amplifier is

$$A_{\text{bipolar}} = g_{m(\text{bipolar})} r_L$$

and the voltage gain of the CS amplifier is

$$A_{\text{FET}} = g_{m(\text{FET})} r_L$$

By taking the ratio of these gains,

$$\frac{A_{\text{bipolar}}}{A_{\text{FET}}} = \frac{g_{m(\text{bipolar})}}{g_{m(\text{FET})}}$$

Since the g_m of the bipolar is much larger than the g_m of a typical FET, we get much more voltage gain with a bipolar.

As an example, almost any FET has a g_m under 10,000 μmhos. Arbitrarily, we can use 5000 μmhos as an average. When I_E is 1 mA, a bipolar has a g_m of 40,000 μmhos and will produce eight times as much voltage gain as the FET. When $I_E = 5$ mA, the bipolar g_m is 200,000 μmhos, and the bipolar produces 40 times as much gain as the FET.

So, there's no question on this point. When you need lots of voltage gain, bipolars are way in front.

Square-law advantage

When used as a small-signal amplifier, the output voltage from a FET is linearly related to the input voltage because only a small part of the square-law curve is used. But with larger signals, more of the curve is used and nonlinear distortion appears. In an amplifier this nonlinear distortion is unwanted. But in another circuit called a *mixer* (Chap. 21) this square-law nonlinear distortion has a tremendous advantage; therefore, the FET is a better choice than the bipolar for mixer applications. Almost any communications receiver uses a mixer; you will find the FET mixer in many of the newer designs.

Cost

There is a wide range in the cost of bipolars and FETs. But on the average, bipolars cost less than FETs. What often happens in the design of discrete circuits is this. The more expensive FET is used near the front end of a system to give high input impedance, serve as a mixer, or perform some function it does especially well. But elsewhere in the system you find the bipolar used for voltage gain, power gain, and other functions it is better suited for. In general, unless the FET has a clear-cut advantage, use the bipolar transistor.

Low-supply voltages

In linear circuits we want the bipolar or FET to operate on the flat part of the collector or drain curves. Typically, this requires a V_{CE} greater than a volt. But with the FET this requires a V_{DS} greater than the pinch-off voltage, which can be several volts. Because of this, when you have low-supply voltages to work with (like a 6-V battery in a transistor radio), the bipolar transistor makes the design and operation of amplifier circuits much easier.

Large-scale integration (LSI)

Integrated circuits use mostly bipolars and MOSFETs. There are fewer steps in making an MOS IC than a bipolar IC. Also, a manufacturer can pack more MOSFETs on a chip than bipolars. Therefore, when it comes to large-scale integration (more than a hundred components), the MOSFET is widely used. The fewer manufacturing steps and greater density of integrated components almost always result in MOS/LSI packages that cost less than bipolar/LSI packages.

Summary

There are other considerations, and to make a sensible choice you have to take everything into account. We have listed a few advantages and disadvantages of FETs and bipolars. As far as usage is concerned, you can expect the following. With discrete circuits you will see a heavy use of bipolars, a moderate use of JFETs, and a limited use of MOSFETs. On the other hand, with integrated circuits the bipolars and MOSFETs are heavily used; JFETs, much less.

Problems

14-1. In Fig. 14-22a, $V_{GS} = -1$ V. Calculate

1. The ideal drain current and drain-to-ground voltage
2. The third-approximation value of drain current and drain-to-ground voltage

14-2. If $V_{GS} = -2.5$ V in Fig. 14-22a, what is the value of V_{DS}?

14-3. Calculate the ideal and second-approximation values of drain current in Fig. 14-22b. Also, the values of V_{DS}.

14-4. Ideally, what is the power dissipation in the FET of Fig. 14-22b?

14-5. If the JFET of Fig. 14-22b has a $V_{GS(off)}$ of -5 V, what is its transconductance equation? What is value of drain current?

14-6. The JFET of Fig. 14-22c operates with an I_D of about half I_{DSS}. Calculate the approximate value of I_D. What voltage is there from drain to ground? From source to ground?

14-7. In Fig. 14-23a, V_{GS} is negligibly small compared to the V_{SS} supply. Calculate the value of drain current. How much drain-to-ground voltage is there?

14-8. If $V_{GS} = -5$ V in Fig. 14-23a, what does I_D equal? V_{DS}?

14-9. Neglect V_{BE} in Fig. 14-23b. How much drain current is there? If $V_{GS} = -2$ V, what does V_{DS} equal?

14-10. In Fig. 14-23c calculate the approximate value of I_E. What does the drain-to-ground voltage equal? If V_{GS} is -3 V, what does V_{DS} equal?

14-11. Calculate the following ideal voltages with respect to ground: gate voltage, source voltage, and drain voltage in Fig. 14-24a.

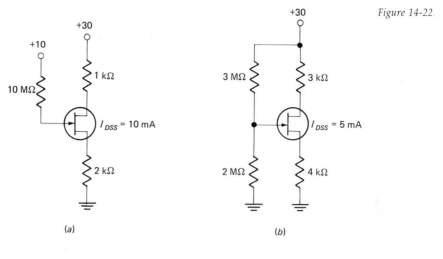

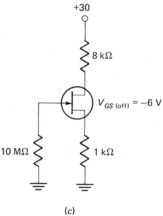

Figure 14-22

14-12. What are the ideal values of I_D and V_{DS} in Fig. 14-24b?

14-13. The *n*-channel MOSFET of Fig. 14-25a is a 2N3797 with these specifications: $I_{DSS(min)} = 2$ mA, $I_{DSS(typ)} = 2.9$ mA, and $I_{DSS(max)} = 6$ mA. Calculate the typical value of V_{DS}. What is lowest value of V_{DS}?

14-14. The 3N170 of Fig. 14-25b has an $I_{D(on)}$ of 10 mA when $V_{GS} = V_{DS} = 10$ V. Select a value of R_D that sets up this value of drain current.

14-15. Figure 14-25c shows part of a digital circuit called a *flip-flop*. $V_{GS(th)} = 2$ V. When $V_{GS} = 10$ V, $I_D = 1$ mA. Suppose no drain current flows through T_1.

1. What is the voltage from the drain of T_1 to ground?

Figure 14-23.

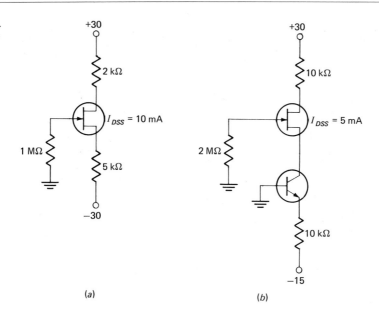

(a)

(b)

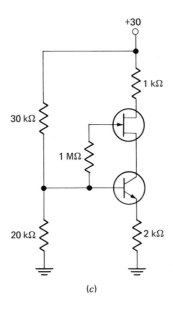

(c)

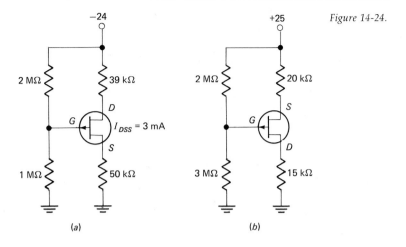

(a)

(b)

Figure 14-24.

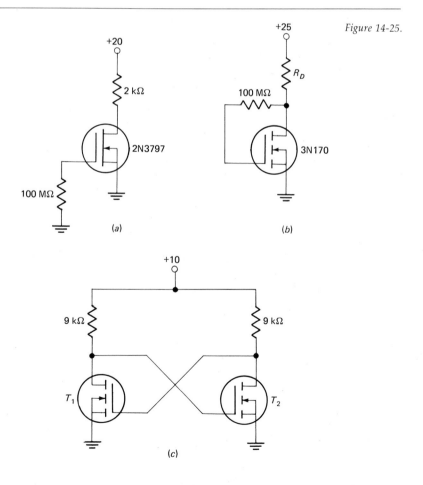

(a)

(b)

+10

9 kΩ

9 kΩ

T₁

T₂

(c)

Figure 14-25.

Figure 14-26.

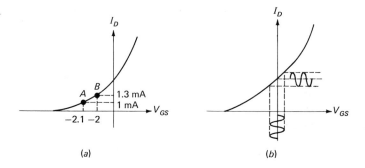

(a)   (b)

2. How much voltage is applied to the gate of T_2?
3. How much voltage is there on the drain of T_2?
4. Why is T_1 off?

14-16. Calculate the value of g_m between A and B on the curve of Fig. 14-26a.

14-17. In Fig. 14-26b suppose the sine waves have these peak-to-peak values: 0.6 mA and 0.25 V. Calculate the g_m between the indicated points.

14-18. A FET has a g_{m0} of 5000 μmhos and $V_{GS(off)}$ of -4 V. Calculate the g_m at -1 V and at -3 V.

14-19. A 2N3822 has a maximum g_{m0} of 6500 μmhos and a $V_{GS(off)}$ of -6 V. If it is biased at $I_D = 0.5I_{DSS}$, what is the value of g_m?

14-20. A FET has an I_{DSS} of 10 mA and a g_{m0} of 10,000 μmhos. Calculate its $V_{GS(off)}$.

14-21. Suppose a 2NXYZ (fictitious type) has a maximum g_{m0} of 5000 μmhos and a minimum g_{m0} of 2000 μmhos. If $V_{GS}/V_{GS(off)}$ equals 0.25, what is the range in g_m?

14-22. The JFET of Fig. 14-27a has a g_m of 1500 μmhos. What is the voltage gain from gate to drain? The impedance of the gate is much higher than R_G; what is the input impedance of the stage?

14-23. The FET of Fig. 14-27b has a g_m of 1000 μmhos. How much voltage gain is there from gate to drain?

14-24. In Fig. 14-27c the g_{m0} of the FET is 3000 μmhos and the I_{DSS} is 2 mA. What is the approximate value of g_m for the given bias conditions? What is the voltage gain from gate to drain?

14-25. If g_m equals 2000 μmhos in the source follower of Fig. 14-27d, what is the voltage gain from gate to source?

14-26. The JFET of Fig. 14-27d has a g_{m0} of 6000 μmhos and an I_{DSS} of 16 mA. What is the value of g_m at the Q point? The voltage gain from gate to source? The input impedance (neglect gate current)?

14-27. The 1-MΩ source impedance of Fig. 14-28a is unusually high, but this is no problem because the input impedance of the MOSFET amplifier is 100 MΩ at the input frequency. The input frequency is also unusually low, only 0.01 Hz;

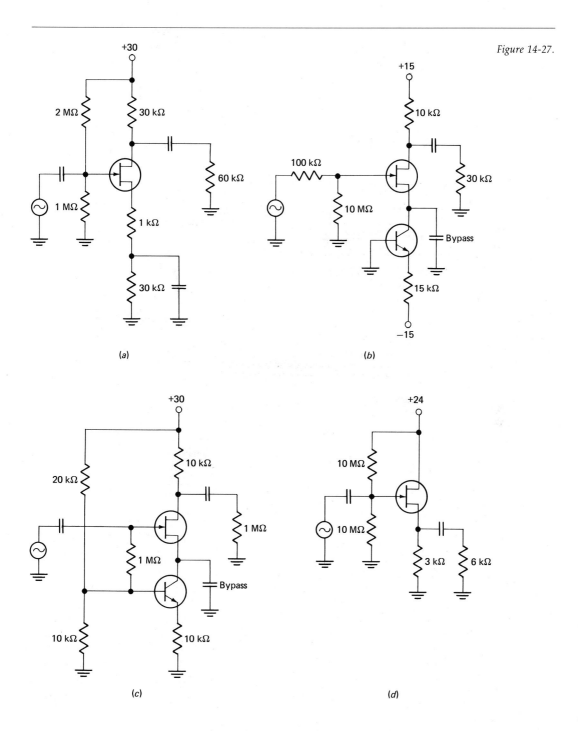

Figure 14-27.

Figure 14-28.

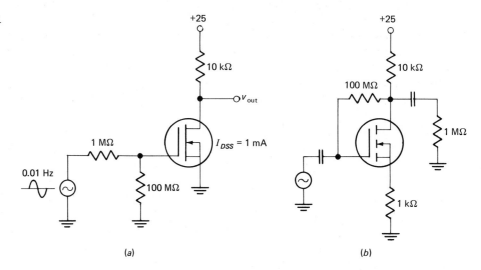

(a)   (b)

however, this causes no problem because there are no coupling and bypass capacitors. If the MOSFET has a g_m of 2000 μmhos, what is the voltage gain from gate to drain?

14-28. The input signal of Fig. 14-28a has a peak value of 0.1 V. If the MOSFET has a g_m of 2000 μmhos, what is the value of drain voltage at the positive peak of input voltage?

14-29. In Fig. 14-28b the MOSFET has a g_m of 3000 μmhos. Calculate the voltage gain from gate to drain. If the g_m changes to 5000 μmhos, what will the gain equal?

Figure 14-29

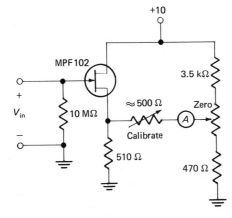

14-30. In Fig. 14-14 each MOSFET has a g_m of 2000 μmhos. What is the voltage gain from the input to the output of this three-stage amplifier?

14-31. Figure 14-29 shows a simple FET dc voltmeter. With an MPF102 or similar JFET the voltmeter reads input dc voltages up to at least 2.5 V. The zero adjust is set as in any voltmeter; the calibrate adjust is set periodically to give full-scale deflection when $V_{in} = 2.5$ V. A calibrate adjustment like this takes care of variations from one FET to another and FET aging effects.

1. If the drain current through the 510-Ω resistor equals 4 mA, how much dc voltage is there from source to ground?
2. If no current flows through the ammeter, what voltage does the wiper tap off the zero adjust?
3. If an input voltage of 2.5 V produces 1-mA full-scale deflection, how much deflection does 1.25 V produce (assume linearity)?
4. What is the input impedance of the voltmeter if the gate current is negligible?

15. Decibels and Miller's Theorem

Thevenin's theorem is marvelous; it enables us to reduce complicated circuits to simple prototypes we can analyze. This chapter introduces two more powerful tools: *decibels* and *Miller's theorem*. Decibels are based on the logarithms learned in trigonometry; these logarithms simplify gain calculations. Miller's theorem applies to circuits like collector-feedback bias discussed earlier. When applicable, Miller's theorem can reduce a complicated circuit into two simple circuits.

15-1. DECIBEL POWER GAIN

Suppose the input power to an amplifier is p_1 and the output power is p_2. The power gain G equals

$$G = \frac{p_2}{p_1}$$

So, if p_2 is 100 W and p_1 is 1 W, we have a power gain of 100.

Nothing prevents us from taking the logarithm of G. For instance, if $G = 100$,

$$\log G = \log 100 = 2$$

(Throughout this book we always talk about logarithms to the base 10.) If $G = 1000$, then

$$\log G = \log 1000 = 3$$

Or, if $G = 20,000$,

$$\log 20,000 = \log (2 \times 10^4) = \log 2 + \log 10^4$$
$$= 0.30103 + 4 = 4.30103$$

To keep G distinct from its logarithm, we will use G' for log G. In symbols,

$$G' = \log G \qquad\qquad\qquad (15\text{-}1)^{***}$$

Since G stands for the power gain, we need another name for G'; the name for G' is the *bel power gain* (in honor of Bell). So, if a circuit has a power gain of 100, it has a bel power gain of

$$G' = \log G = \log 100 = 2$$

Since G is the ratio of two powers, the unit of watts cancels out, so that G is dimensionless. G' also is dimensionless, but to make sure it is never confused with G, we will attach the label *bel* (abbreviated B) to all answers for G'. Therefore, if $G = 100$,

$$G' = \log G = \log 100 = 2 \text{ B}$$

When an answer is in bels like this, we automatically know it represents the bel power gain and not the ordinary power gain.

Heavily used values

Bel power gain is nothing more than the logarithm of the ordinary power gain. You can always look up accurate logarithms in a table of logarithms or on a slide rule. But most of the calculations in this book can be done by remembering the following approximate logarithms:

$\log 1 = 0$
$\log 2 = 0.3$
$\log 4 = 0.6$
$\log 8 = 0.9$
$\log 10 = 1$
$\log 10^2 = 2$
$\log 10^n = n$

The tens are the easiest to remember since the logarithm of 10^n equals n. The logarithms of 2, 4, and 8 are also easy when you notice the arithmetic progression 0.3, 0.6, and 0.9. (The latter are close approximations for the exact logarithms: 0.30103, 0.60206, and 0.90309.)

Also important to remember are these logarithmic relations:

$$\log xy = \log x + \log y$$

$$\log \frac{x}{y} = \log x - \log y$$

and

$$\log x^n = n \log x$$

All these are familiar from earlier courses. These relations along with the numerical values listed are heavily used not only in this book but throughout the electronics industry.

Prefixes

0.001 volt can be written as 1 millivolt, or as 1000 microvolts, or in many other ways. So too can we express bel power gain in equivalent ways. For instance, if $G = 100$, then

$$G' = \log 100 = 2 \text{ bels (B)} = 20 \text{ decibels (dB)} = 200 \text{ centibels (cB)}$$
$$= 2000 \text{ millibels (mB)}$$

and so on. The only prefix used in practice is *deci* (one-tenth, or 10^{-1}). We abbreviate decibel as dB.

EXAMPLE 15-1.
Calculate the bel power gain for each of the following: $G = 2$, 4, and 8.

SOLUTION.
When $G = 2$,

$$G' = \log 2 = 0.3 \text{ B} = 3 \text{ dB}$$

When $G = 4$,

$$G' = \log 4 = 0.6 \text{ B} = 6 \text{ dB}$$

When $G = 8$,

$$G' = \log 8 = 0.9 \text{ B} = 9 \text{ dB}$$

Notice the simple progression here. When the power gain doubles, the bel power gain increases 3 dB. This is always true because if you start with G_1 and double it, you have $G = 2G_1$ and

$$G' = \log 2G_1 = \log 2 + \log G_1 = 3 \text{ dB} + \log G_1$$

The factor of 2 in front of G_1 adds 0.3 B or 3 dB to the answer.
Therefore, if we cause the power gain to double, we can say either

1. We have doubled the power gain.
2. We have increased the bel power gain by 3 dB.

Both are equivalent statements.

EXAMPLE 15-2.
Work out the power gain for each of these: $G' = 3$ dB, 40 dB, and 43 dB.

SOLUTION.
When $G' = 3$ dB or 0.3 B,

$$G = \text{antilog } 0.3 = 2$$

In other words, we work in the opposite direction, from the logarithm back to the original number.
When $G' = 40$ dB or 4 B,

$$G = \text{antilog } 4 = 10^4 = 10{,}000$$

When $G' = 43$ dB or 4.3 B,

$$G = \text{antilog } 4.3 = 2 \times 10^4 = 20{,}000$$

Alternatively, we could have converted 43 dB by realizing it is 3 dB more than 40 dB. As we saw in the preceding example, an increase of 3 dB in bel power gain means the ordinary power gain has doubled. Since a bel power gain of 40 dB is equivalent to a power gain of 10,000, a bel power gain of 43 dB is equivalent to a power gain of 20,000.

15-2. POWER GAIN OF CASCADED STAGES

Figure 15-1*a* shows blocks representing two stages in an amplifier. The input power to the first stage is p_1. The output power p_2 of the first stage goes into the second stage. The final output power is p_3.

Suppose we want to calculate the overall power gain, the ratio

$$G = \frac{p_3}{p_1}$$

There are two ways we can go about this: a direct method working with ordinary power and an indirect method using bel power gain.

Direct method

The power gain of the first stage is

$$G_1 = \frac{p_2}{p_1}$$

and the power gain of the second stage is

$$G_2 = \frac{p_3}{p_2}$$

The total power gain is

$$G = \frac{p_3}{p_1}$$

which is equal to

$$G = \frac{p_3}{p_1} \frac{p_2}{p_2} = \frac{p_2}{p_1} \frac{p_3}{p_2}$$

or
$$G = G_1 G_2 \qquad\qquad\qquad (15\text{-}2)^{***}$$

This proves the total power gain of cascaded stages equals the product of each stage gain. No matter how many stages there are, we can find the total power gain by multiplying the individual stage gains.

As an example, Fig. 15-1b shows a first-stage power gain of 100 and a second-stage power gain of 200. The total power gain is

$$G = 100 \times 200 = 20{,}000$$

Indirect method

We know we can get the answer to a multiplication problem by the indirect method of adding the logarithms of factors and taking the antilog of the sum. Sometimes, this indirect method saves time and we use it.

Since bel power gain is nothing more than the logarithm of ordinary power gain,

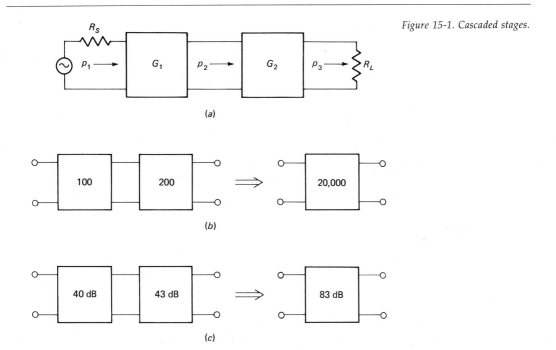

Figure 15-1. Cascaded stages.

we can calculate the total power gain by adding the bel power gain of each stage and taking the antilog of the sum. Specifically, starting with G,

$$G = G_1 G_2$$

Taking the logarithm of both sides gives

$$\log G = \log G_1 G_2 = \log G_1 + \log G_2$$

With Eq. (15-1) we rewrite this as

$$G' = G_1' + G_2' \qquad\qquad (15\text{-}3)^{***}$$

where G' is the total bel power gain, G_1' is the bel power gain of the first stage, and G_2' is the bel power gain of the second stage. This equation says the total bel power gain of two cascaded stages equals the sum of the bel power gains of each stage. It follows that no matter how many stages there are, the total bel power gain equals the sum of the individual bel power gains. Once we have found the total bel power gain, we can convert back to ordinary power gain if desired.

As an example, Fig. 15-1c shows the same two-stage amplifier as Fig. 15-1b, except the gains are expressed in decibels. The total bel power gain is

$$G' = 40 \text{ dB} + 43 \text{ dB} = 83 \text{ dB} = 8.3 \text{ B}$$

We can leave the answer like this, or convert it back to ordinary power gain as follows:

$$G = \text{antilog } G' = \text{antilog } 8.3$$
$$= 2 \times 10^8 = 200{,}000{,}000$$

An answer in decibels has the advantage of being compact and easy to write. In the foregoing example it is easier to write 83 dB instead of 200,000,000 or 2×10^8.

Simplified measurements

Another advantage of decibels is they simplify measurements. For example, many *microwave* instruments measure power (microwave refers to frequencies from 1000 MHz to 100,000 MHz). The meter on such instruments often resembles Fig. 15-2a. The upper scale is marked in milliwatts. Suppose we measure the input power to a stage like Fig. 15-2b; the needle (solid line) will indicate 0.25 mW. When we measure the output power in Fig. 15-2b, the needle moves to full scale (dashed line), indicating an output power of 1 mW. Therefore, the power gain equals 4.

The lower scale is called the *dBm scale*. As shown in Fig. 15-2a, 0 dBm is equivalent to 1 mW, -3 dBm to 0.5 mW, -6 dBm to 0.25 mW, and so on. These values come about as follows. 1 mW is convenient to use as a *power reference* because power levels in microwave systems lie on both sides of this value. Suppose we let **P** stand for the *ratio of power to 1 mW*. In symbols,

$$\mathbf{P} = \frac{p}{p_{\text{ref}}} = \frac{p}{1 \text{ mW}}$$

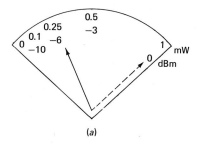

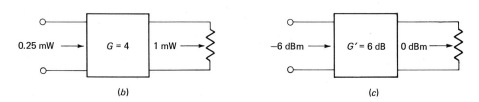

Figure 15-2. Meaning of dBm.

Taking the logarithm of this power ratio gives

$$\mathbf{P}' = \log \frac{p}{1 \text{ mW}} \tag{15-4}$$

By working out the values of $\mathbf{P}'$ for different values of p, we can calibrate the meter face in terms of dB with respect to 1 mW.

As an example, if $p = 0.5$ mW,

$$\mathbf{P}' = \log \frac{0.5 \text{ mW}}{1 \text{ mW}} = \log 0.5 = \log \frac{1}{2}$$

$$= \log 1 - \log 2 = -0.3 \text{ B} = -3 \text{ dB}$$

To make sure we remember the milliwatt reference, we add the letter m to get

$$\mathbf{P}' = -3 \text{ dBm}$$

In this way, we can mark the lower scale in Fig. 15-2a with the correct values of dBm.

If we use the dBm scale to measure the input and output power in Fig. 15-2b, we will read −6 dBm for the input power and 0 dBm for the output power (Fig. 15-2c). Since the needle moves from −6 dBm to 0 dBm in Fig. 15-2a, the amplifier has a bel power gain of 6 dB.

Simplified system design

Data sheets often specify devices in terms of their bel power gains. There is a good reason for this. When we cascade devices as shown in Fig. 15-3a, we can add the bel

Figure 15-3. Additive property of bel power gain.

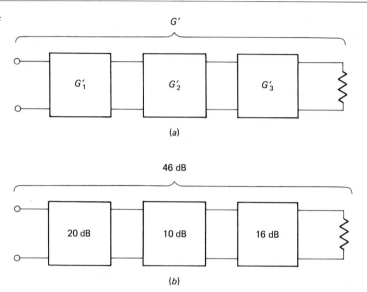

(a)

(b)

power gains to get the total bel power gain. For instance, in Fig. 15-3*b* the total bel power gain equals 46 dB, the sum of the individual bel power gains. This is usually easier and faster than multiplying ordinary power gains. To anyone designing a complicated system with many blocks, using decibels is a decided advantage because it reduces the labor of calculation.

EXAMPLE 15-3.

A bipolar amplifier has a power gain of 0.5 from source to base, and a power gain of 40,000 from base to collector. Calculate the individual bel power gains and the total bel power gain.

SOLUTION.

The bel power gain from source to base is

$$G_1' = \log 0.5 = \log \frac{1}{2} = \log 1 - \log 2$$

$$= -0.3 \text{ B} = -3 \text{ dB}$$

The bel power gain from base to collector is

$$G_2' = \log 40{,}000 = \log (4 \times 10^4) = 4.6 \text{ B} = 46 \text{ dB}$$

By adding the individual bel power gains, we get a total bel power gain of

$$G' = -3 \text{ dB} + 46 \text{ dB} = 43 \text{ dB}$$

15-3. BEL VOLTAGE GAIN

In Fig. 15-4a the voltage across the amplifier input terminals equals v_1 and the voltage across the output terminals is v_2. Therefore, we have a voltage gain from input to output of

$$A = \frac{v_2}{v_1}$$

The input power to the amplifier is

$$p_1 = \frac{v_1^2}{R_1}$$

and the output power is

$$p_2 = \frac{v_2^2}{R_2}$$

So, the power gain from input to output equals

$$G = \frac{p_2}{p_1} = \frac{v_2^2/R_2}{v_1^2/R_1} = \left(\frac{v_2}{v_1}\right)^2 \frac{R_1}{R_2}$$

Since the ratio v_2/v_1 is the voltage gain A, we can rewrite the power gain as

$$G = A^2 \frac{R_1}{R_2} \qquad\qquad (15\text{-}5a)$$

Impedance-matched case

In many systems (microwave, telephone, etc.) we get the special case of $R_1 = R_2$. For this condition,

$$G = A^2 \qquad \text{for } R_1 = R_2 \qquad\qquad (15\text{-}5b)$$

For instance, if $A = 100$, $G = 10,000$. Figure 15-4b shows a system with these gains. Each block has an input impedance of R, and each block sees a load impedance of R. We call systems like this *impedance-matched systems;* each stage has a power gain equal to the square of its voltage gain.

Taking the logarithm of both sides of Eq. (15-5b) gives

$$\log G = \log A^2 = 2 \log A$$

or
$$G' = 2 \log A \qquad \text{(impedance match only)} \qquad\qquad (15\text{-}6)$$

With this equation we can calculate the bel power gain without having to calculate the ordinary power gain. For example, if an impedance-matched amplifier has a voltage gain of 100, it has a bel power gain of

$$G' = 2 \log A = 2 \log 100 = 4 \text{ B} = 40 \text{ dB}$$

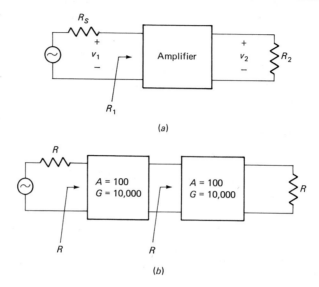

Figure 15-4. (a) Relation of power and voltage gains. (b) Power and voltage gain in impedance-matched system.

Nonmatched case

Through the years since bel power gain was first defined, voltage gain has become more useful than power gain, at least in many branches of electronics. The reason is simple enough. It is easier to measure voltage than power. With an amplifier like Fig. 15-4a we can easily measure the input and output voltages.

Because of this, a second kind of bel gain has come into use: the *bel voltage gain*. The formula for bel voltage gain is

$$A' = 2 \log A \qquad (15\text{-}7)^{***}$$

where A' is the bel voltage gain and A is the ordinary voltage gain. So, if $A = 100$, the bel voltage gain equals

$$A' = 2 \log A = 2 \log 100 = 4 \text{ B} = 40 \text{ dB}$$

Or, if $A = 20{,}000$,

$$A' = 2 \log A = 2 \log 20{,}000 = 2(4.3)$$
$$= 8.6 \text{ B} = 86 \text{ dB}$$

The reason for including a coefficient of 2 in Eq. (15-7) is to ensure bel power gain and bel voltage gain are equal in impedance-matched systems. In other words, with Eq. (15-7) we can rewrite Eq. (15-6) as

$$G' = A' \qquad \text{(impedance match only)} \qquad (15\text{-}8)^{***}$$

where $G' = \log G$ and $A' = 2 \log A$. Whenever the impedances are matched, we can use

G' or A' interchangeably because they are equal. For instance, an impedance-matched amplifier with a voltage gain of 100 has a power gain of 10,000; equivalently, this amplifier has a bel voltage gain of 40 dB and a bel power gain of 40 dB.

When the impedances are not matched, the bel power gain and bel voltage gain are no longer equal. Therefore, we have to calculate each by its own formula. In other words, if we want bel power gain, we have to use

$$G' = \log G$$

and for bel voltage gain, we must use

$$A' = 2 \log A$$

Both types of bel gain are widely used. Bel power gain predominates in communications, microwaves, and other systems where power is important. Bel voltage gain seems to be far in front, however, in areas of electronics where it is convenient to measure voltage rather than power.

Since bel voltage gain is based on logarithms, the additive property of cascaded stages holds. That is, the total bel voltage gain of cascaded stages equals the sum of the individual bel voltage gains. In symbols,

$$A' = A'_1 + A'_2 + \cdots \tag{15-9}$$

Also, many voltmeters have a decibel scale. When you measure the input and output voltages of an amplifier, the difference in the dB readings is the bel voltage gain of the amplifier.

As an example, the two-stage amplifier of Fig. 15-5a has a first-stage voltage gain of 100 and a second-stage voltage gain of 200. The total voltage gain is

$$A = A_1 A_2 = 100 \times 200 = 20{,}000$$

The first stage has a bel voltage gain of

$$A'_1 = 2 \log 100 = 4 \text{ B} = 40 \text{ dB}$$

and the second stage,

$$A'_2 = 2 \log 200 = 2(2.3) = 4.6 \text{ B} = 46 \text{ dB}$$

The total bel voltage gain is

$$A' = 40 \text{ dB} + 46 \text{ dB} = 86 \text{ dB}$$

If we use the decibel scale of a voltmeter when measuring input and output voltages, the needle will show an increase of 40 dB for the first stage, 46 dB for the second stage, and 86 dB for the entire amplifier (see Fig. 15-5b).

EXAMPLE 15-4.
The JFET in Fig. 15-6a has a g_m of 4000 μmhos. Calculate the bel voltage gain from gate to drain.

Figure 15-5. Additive property of bel voltage gain.

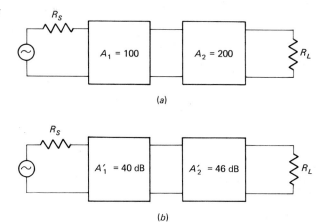

(a)

(b)

SOLUTION.

The drain sees an ac resistance of

$$r_D = 10,000 \parallel 10,000 = 5 \text{ k}\Omega$$

The voltage gain from gate to drain is

$$A = \frac{v_d}{v_g} = g_m r_D = 0.004(5000) = 20$$

The bel voltage gain equals

$$A' = 2 \log A = 2 \log 20 = 2.6 \text{ B} = 26 \text{ dB}$$

EXAMPLE 15-5.

Figure 15-6b shows a two-stage amplifier. What is the total voltage gain in decibels?

SOLUTION.

We found the voltage gain of the second stage in the preceding example: 26 dB. The collector of the first stage sees approximately 10 kΩ of ac load resistance. Since the I_E of the first stage is approximately 1 mA, r'_e is about 25 Ω. The voltage gain of the first stage therefore equals

$$A_1 = \frac{r_C}{r'_e} = \frac{10,000}{25} = 400$$

And, the bel voltage gain of the first stage is

$$A'_1 = 2 \log 400 = 5.2 \text{ B} = 52 \text{ dB}$$

The total bel voltage gain is the sum of individual bel voltage gains; in this case,

$$A' = A'_1 + A'_2 = 52 \text{ dB} + 26 \text{ dB} = 78 \text{ dB}$$

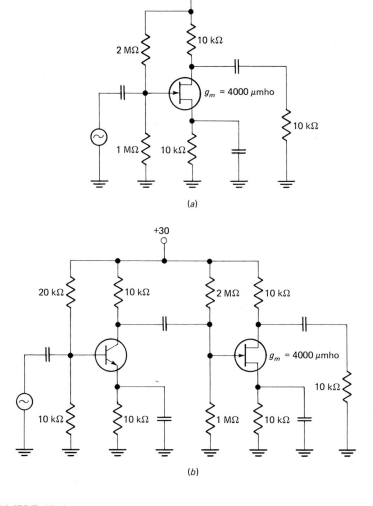

Figure 15-6. Examples 15-4 and 15-5.

(a)

(b)

EXAMPLE 15-6.
What is the total bel voltage gain of the three-stage amplifier shown in Fig. 15-7?

SOLUTION.
The three stages have identical biasing resistors. Approximately, 1 mA of dc emitter current flows in each transistor. Ideally, r_e' is 25 Ω in each stage. Therefore, given a β of 100 for each transistor,

$$z_{\text{in(base)}} = 100(25) = 2500 \ \Omega$$

in all stages.

Figure 15-7. Example 15-6.

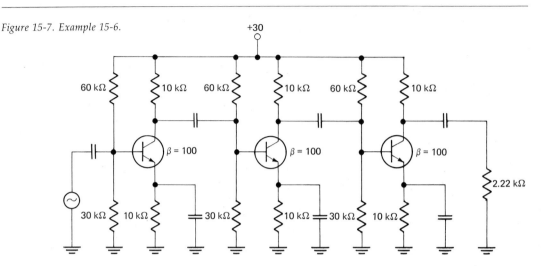

The first collector sees a 10-kΩ resistor shunted by the input impedance of the second stage, which is

$$z_{in} = 60{,}000 \parallel 30{,}000 \parallel 2500 = 2220 \ \Omega$$

Therefore, the collector of the first stage sees

$$r_C = 10{,}000 \parallel 2220 = 1820 \ \Omega$$

And, the voltage gain from base to collector for the first stage is

$$A_1 = \frac{r_C}{r'_e} = \frac{1820}{25} \cong 73$$

The first-stage bel voltage gain equals

$$A'_1 = 2 \log 73 = 2(1.86) \cong 3.7 \text{ B} = 37 \text{ dB}$$

The collectors of the second and third stages see exactly the same ac resistance as the collector of the first stage; therefore, each of these stages has the same bel voltage gain. The total bel voltage gain of the three-stage amplifier is the sum of the individual bel voltage gains:

$$A' = A'_1 + A'_2 + A'_3 \cong 3(37 \text{ dB}) = 111 \text{ dB}$$

15-4. MILLER'S THEOREM

Figure 15-8a shows an inverting amplifier. The dot convention is similar to that used with a transformer; when the upper input terminal goes positive, so too does the lower

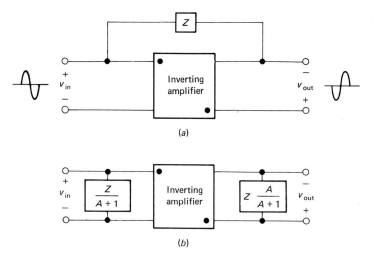

Figure 15-8. Miller theorem.

output terminal. This implies the output voltage is 180° out of phase with the input voltage.

In Fig. 15-8a an impedance Z is connected from the output to the input. Many circuits have an impedance like this, and it produces a feedback effect similar to collector-feedback bias discussed earlier. *Miller's theorem* helps us with circuits like these because it splits the impedance into two components as shown in Fig. 15-8b. In other words, the Miller theorem says Fig. 15-8b is equivalent to Fig. 15-8a.

To be precise, when an inverting amplifier has a voltage gain A, the Miller theorem says we can replace impedance Z by an input Miller impedance of

$$Z_{\text{in(Miller)}} = \frac{Z}{A + 1} \qquad (15\text{-}10a)$$

and an output Miller impedance of

$$Z_{\text{out(Miller)}} = Z\,\frac{A}{A + 1} \qquad (15\text{-}10b)$$

These Miller impedances appear across the input and output terminals as shown in Fig. 15-8b. This is useful because we can lump these Miller impedances into the input impedance and the load resistance of the amplifier. More about this later.

Z **purely resistive**

We will soon prove the Miller theorem, but before we do, let's make sure we understand it by examining two special cases. Figure 15-9a shows the first special case,

Figure 15-9. Miller theorem for the resistive case.

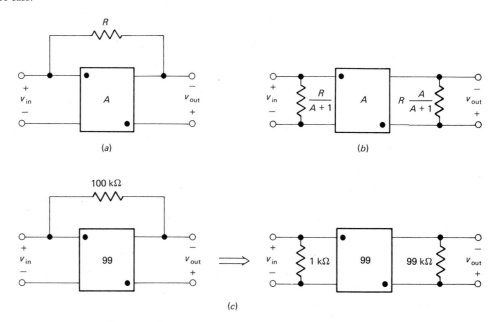

(a)

(b)

(c)

$Z = R$. The Miller theorem says we can split this R into two parts: an input part and an output part as shown in Fig. 15-9b. In symbols,

$$R_{\text{in(Miller)}} = \frac{R}{A + 1} \qquad\qquad (15\text{-}11a)^{***}$$

and

$$R_{\text{out(Miller)}} = R\,\frac{A}{A + 1} \qquad\qquad (15\text{-}11b)^{***}$$

Here is a concrete example. Fig. 15-9c shows an inverting amplifier with a voltage gain of 99 and a resistance of 100 kΩ from input to output. With the Miller theorem we can split this into

$$R_{\text{in(Miller)}} = \frac{100\ \text{k}\Omega}{99 + 1} = 1\ \text{k}\Omega$$

and

$$R_{\text{out(Miller)}} = (100\ \text{k}\Omega)\,\frac{99}{99 + 1} = 99\ \text{k}\Omega$$

The 1-kΩ component appears across the input, and the 99-kΩ component across the output as shown.

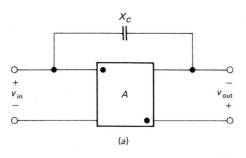

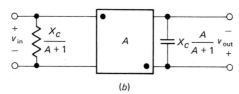

Figure 15-10. Miller theorem for the capacitive case.

Z **purely capacitive**

The only other special case we are interested in is the capacitive case, $Z = X_C$ as shown in Fig. 15-10*a*. With the Miller theorem the reactance splits into the two components of Fig. 15-10*b*.

Often, we prefer to work directly with capacitance instead of reactance. Note that

$$\frac{X_C}{A+1} = \frac{1}{2\pi fC(A+1)} = \frac{1}{2\pi f C_{in(Miller)}}$$

where

$$C_{in(Miller)} = C(A+1) \qquad (15\text{-}12a)^{***}$$

On the output side,

$$X_C \frac{A}{A+1} = \frac{1}{2\pi fC}\frac{A}{A+1} = \frac{1}{2\pi f C_{out(Miller)}}$$

where

$$C_{out(Miller)} = C\frac{A+1}{A} \qquad (15\text{-}12b)^{***}$$

Figure 15-11*a* and *b* summarize these results for the purely capacitive case. The Miller theorem says a capacitance C from input to output appears like a *much larger capacitance* across the input terminals. For instance, if $C = 10$ pF and $A = 99$, then

$$C_{in(Miller)} = (10 \text{ pF})(99+1) = 1000 \text{ pF}$$

and

$$C_{out(Miller)} = (10 \text{ pF})\frac{99+1}{99} = 10.1 \text{ pF}$$

Figure 15-11*c* illustrates these results.

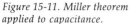

Figure 15-11. Miller theorem applied to capacitance.

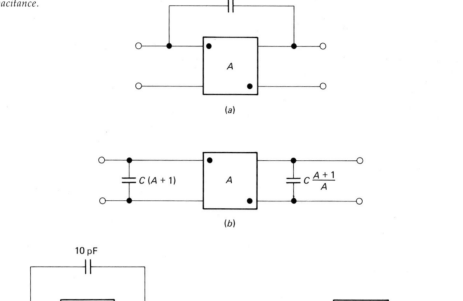

Useful tool

The Miller theorem is a powerful tool. It is like the Thevenin theorem because it eliminates awkward problems and deepens our insight into circuit action. The two most important cases of the Miller theorem are the resistive case (Fig. 15-9) and the capacitive case (Fig. 15-11). Whenever an R or C connects the input and output terminals of an amplifier, we can apply the Miller theorem to the ac equivalent circuit. From now on, *Millerize* a circuit means to split the R or C into its Miller components in the ac equivalent circuit.

EXAMPLE 15-7.

The voltage gain from base to collector equals 200 in Fig. 15-12a. Millerize the circuit.

SOLUTION.

This is the resistive case. The 1-MΩ resistor connects the input of the CE amplifier to the output. With Miller's theorem,

$$R_{in(Miller)} = \frac{R}{A+1} = \frac{10^6}{200+1} \cong 5 \text{ k}\Omega$$

and
$$R_{out(Miller)} = R\,\frac{A}{A+1} = 10^6\,\frac{200}{200+1} \cong 1 \text{ M}\Omega$$

Note that the output Miller component is approximately equal to the original resistance R whenever the voltage gain is large.

Figure 15-12b shows the ac equivalent circuit before Millerizing. Fig. 15-12c is the Miller equivalent circuit. Because the 1-MΩ Miller component is so much greater than

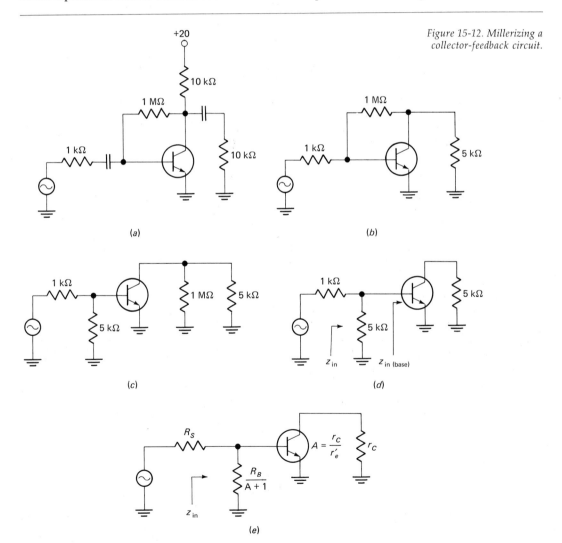

Figure 15-12. Millerizing a collector-feedback circuit.

the 5-kΩ load resistor, we can simplify the circuit to Fig. 15-12d. Especially note how the input Miller resistance appears in parallel with $z_{in(base)}$. For this reason, the z_{in} of the amplifier is the parallel of the 5-kΩ Miller component and $z_{in(base)}$.

As discussed in Chap. 7, the R_B biasing resistor of a collector-feedback biased circuit equals $\beta_{dc}R_C$ for midpoint bias. Because of this, the output Miller component is much greater than r_C in a practical circuit. For this reason, a Millerized collector-feedback ac circuit almost always appears as shown in Fig. 15-12e. The input Miller resistance equals

$$R_{in(Miller)} = \frac{R_B}{A + 1}$$

where $A = r_C/r'_e$, and the input impedance equals

$$z_{in} = R_{in(Miller)} \parallel z_{in(base)}$$

EXAMPLE 15-8.
Figure 15-13a shows a CE amplifier with an r'_e of 25 Ω. The coupling and bypass capacitors look like ac shorts, but not the 3-pF capacitor. Millerize and simplify the ac circuit.

SOLUTION.
Figure 15-13b shows the ac equivalent circuit. The voltage gain equals

$$A = \frac{r_C}{r'_e} = \frac{5000}{25} = 200$$

Therefore, the Miller components are

$$C_{in(Miller)} = C(A + 1) = (3 \text{ pF})(200 + 1) \cong 600 \text{ pF}$$

and
$$C_{out(Miller)} = C\frac{A + 1}{A} = 3 \text{ pF}\frac{200 + 1}{200} \cong 3 \text{ pF}$$

Figure 15-13c shows the Millerized circuit. As a final step in simplification, we can Thevenize the base circuit to the left of the input Miller capacitance. 1 kΩ in parallel with 20 kΩ is approximately 950 Ω; Fig. 15-13d shows the Millerized and Thevenized ac circuit for the original circuit of Fig. 15-13a.

15-5. PROOF OF MILLER'S THEOREM

Figure 15-14a shows an inverting amplifier with an impedance Z from input to output. In this section we prove Z can be split into the two Miller components. We will simplify the derivation by using a $z_{in(amp)}$ of infinity and leaving the output terminals unloaded as shown.

Because of the 180° phase inversion, the ac voltage across Z equals

$$V = V_1 - (-V_2) = V_1 + V_2$$

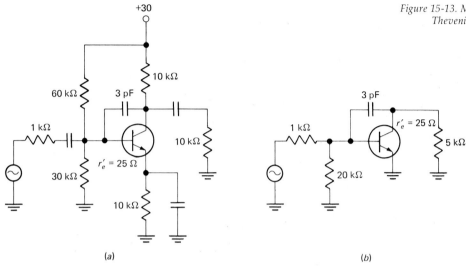

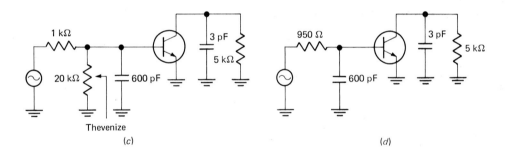

For instance, if $V_1 = 1$ V and $V_2 = 100$ V, V would equal 101 V. The ac current through Z is

$$I = \frac{V_1 + V_2}{Z} \qquad (15\text{-}13)$$

which we can rewrite as

$$I = \frac{V_1 + AV_1}{Z} = \frac{V_1(1 + A)}{Z}$$

We can rearrange this into

$$\frac{V_1}{I} = \frac{Z}{A + 1} \qquad (15\text{-}13a)$$

Figure 15-14. Proof of Miller's theorem.

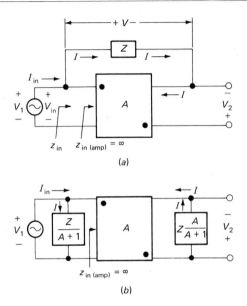

(a)

(b)

In Fig. 15-14a, $z_{\text{in(amp)}}$ is infinite; therefore, $I_{\text{in}} = I$ and $z_{\text{in}} = V_{\text{in}}/I_{\text{in}} = V_1/I$. This allows us to rewrite Eq. (15-13a) as

$$z_{\text{in}} = \frac{Z}{A+1}$$

This is the input impedance the source sees in Fig. 15-14a.

In the Millerized circuit of Fig. 15-14b the source sees exactly the same impedance; therefore, the input side of Fig. 15-14b has the same values of voltage and current as the original circuit of Fig. 15-14a.

To prove the second half of the Miller theorem, we proceed as follows. Since $A = V_2/V_1$, $V_1 = V_2/A$. This allows us to write Eq. (15-13) as

$$I = \frac{V_2/A + V_2}{Z} = \frac{V_2(1/A + 1)}{Z}$$

which can be rearranged into

$$\frac{V_2}{I} = \frac{A}{A+1} Z \tag{15-14}$$

In Fig. 15-14a the output voltage is negative and current flows back into the amplifier. On the alternate half cycle the output voltage goes positive and current flows out of the amplifier. On each half cycle the amplifier sees a load impedance of

$$z_{\text{load}} = \frac{V_2}{I}$$

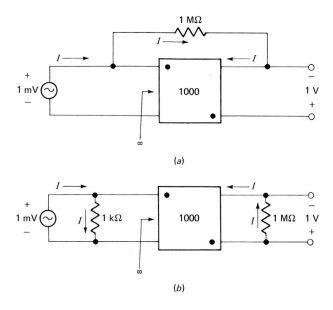

Figure 15-15. Example 15-9.

Since this ratio is the same as the left member of Eq. (15-14),

$$z_{\text{load}} = \frac{A}{A+1} Z$$

In the Millerized circuit of Fig. 15-14b the amplifier sees exactly the same load impedance; therefore, the output side of Fig. 15-14b has exactly the same values of voltage and current as the original circuit of Fig. 14a. In other words, the Millerized circuit is equivalent to the original circuit as far as the voltages and currents are concerned. Because of this, we can use either circuit in our calculations.

EXAMPLE 15-9.

A numerical proof of the Miller theorem may be easier to understand. Prove that Fig. 15-15b is the approximate Miller equivalent circuit for Fig. 15-15a.

SOLUTION.

To prove equivalence, all we have to do is prove I has the same value in each circuit. In Fig. 15-15a,

$$I = \frac{V_1 + V_2}{Z} = \frac{0.001 + 1}{10^6} = \frac{1.001}{10^6} \cong 1 \ \mu A$$

In Fig. 15-15a the amplified output voltage is much greater than the input voltage. For this reason, the voltage across the 1-MΩ resistor has approximately the same value as the output voltage. On the positive half cycle of source voltage, conventional current flows from the source through the 1-MΩ resistor and back into the output terminal of the amplifier.

Figure 15-15b shows the approximate Miller equivalent circuit. With 1 mV across the input, a current

$$I = \frac{0.001}{1000} = 1 \ \mu A$$

flows through the input Miller resistance. And, on the output side a current of

$$I = \frac{1}{10^6} = 1 \ \mu A$$

flows through the output Miller resistance. Since the currents in Fig. 15-15a and b are equal, we have proved the Miller theorem for this particular example.

15-6. ACCURATE TRANSISTOR CIRCUIT ANALYSIS

The ideal transistor is adequate for much of the analysis we have to do. But there are times when we need accurate answers. This section shows you how to get them.

Effect of α

One of the first approximations we made was

$$\alpha \cong 1$$

With modern transistors α is almost always greater than 0.95, and usually more than 0.98. Therefore, we expect it to have only a minor effect on voltage gain and other quantities. Now, we show precisely what that effect is.

Figure 15-16a is the base-driven prototype, and Fig. 15-16b is the same circuit with

Figure 15-16. Effect of α on base-driven amplifier.

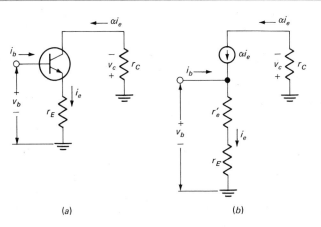

(a)   (b)

the ideal model. But this time we take the effect of α into account. The ac collector voltage equals

$$v_c = \alpha i_e r_C$$

In Fig. 15-16b the ac base voltage equals

$$v_b = i_e(r_E + r'_e)$$

The ratio of these two voltages is the voltage gain:

$$\frac{v_c}{v_b} = \frac{\alpha i_e r_C}{i_e(r_E + r'_e)}$$

or

$$\frac{v_c}{v_b} = \frac{\alpha r_C}{r_E + r'_e} \qquad (15\text{-}15a)$$

Now you can see the effect of α on voltage gain; if α is 0.95, the voltage gain is 95 percent of the ideal voltage gain; or, if α is 0.98, the voltage gain is 98 percent of the ideal voltage gain.

The input impedance of the base is slightly different from the ideal value found earlier. The ac base voltage

$$v_b = i_e(r_E + r'_e)$$

can be written

$$v_b = (i_c + i_b)(r_E + r'_e) = (\beta + 1)i_b(r_E + r'_e)$$

or

$$z_{\text{in(base)}} = \frac{v_b}{i_b} = (\beta + 1)(r_E + r'_e) \qquad (15\text{-}15b)$$

Equation (6-4b) relates dc alpha and beta; it also applies to ac alpha and beta. With Eq. (6-4b),

$$z_{\text{in(base)}} = \frac{\beta(r_E + r'_e)}{\alpha} \qquad (15\text{-}15c)$$

This brings out the effect of alpha on ideal input impedance. If α is 0.95, the input impedance of the base is 5 percent larger than the ideal impedance. When α is 0.98, $z_{\text{in(base)}}$ is only 2 percent larger than the ideal input impedance.

With Eq. (15-15a) we can improve the accuracy of the voltage-gain calculation; with Eq. (15-15b) or (15-15c), we can improve the accuracy of the input-impedance calculation.

Effect of base-spreading resistance

The next improvement in accuracy is the effect of r'_b. We first discussed r'_b in Sec. 6-4; as you recall, its value can get up to 1000 Ω, but usually it is from 50 to 150 Ω. This section derives the effect of r'_b on voltage gain and input impedance.

Figure 15-17a shows a base-driven prototype with r_b'. Alternatively, we can visualize r_b' in series with the base lead as shown in Fig. 15-17b. In either case, the ac collector voltage equals

$$v_c = \alpha i_e r_C$$

and the ac base voltage is

$$v_b = i_b r_b' + i_e (r_E + r_e')$$

which we can rewrite as

$$v_b = \frac{i_e}{\beta + 1} r_b' + i_e (r_E + r_e') \tag{15-15d}$$

By taking the ratio of v_c to v_b and simplifying, we get

$$\frac{v_c}{v_b} = \frac{\alpha r_C}{r_E + r_e' + r_b'/(\beta + 1)} \tag{15-16a}$$

This is similar to Eq. (15-15a) except $r_b'/(\beta + 1)$ is added to the denominator.

By straightforward algebra we can rearrange Eq. (15-15d) to get these expressions for input impedance:

$$z_{\text{in(base)}} = (\beta + 1) \left(r_E + r_e' + \frac{r_b'}{\beta + 1} \right) \tag{15-16b}$$

or

$$z_{\text{in(base)}} = \frac{\beta}{\alpha} \left(r_E + r_e' + \frac{r_b'}{\beta + 1} \right) \tag{15-16c}$$

These expressions are similar to Eqs. (15-15b) and (15-15c) except $r_b'/(\beta + 1)$ is added after r_e'.

You can get some feeling for the effect of r_b' as follows. When $r_b' = 100 \ \Omega$ (the middle of its typical range) and $\beta = 50$,

---

Figure 15-17. Effect of α and r_b' on base-driven amplifier.

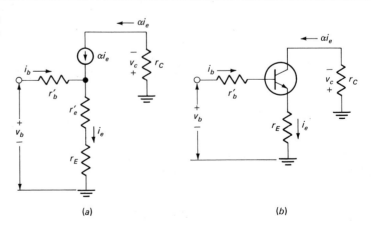

(a)   (b)

$$\frac{r_b'}{\beta + 1} = \frac{100}{50 + 1} \cong 2 \ \Omega$$

This small value is added to $r_E + r_e'$ in Eqs. (15-16a) and (15-16b). Therefore, the typical effect of r_b' is negligible in a small-signal amplifier. About the only case in which r_b' is important is in a power amplifier because both r_E and r_e' may approach zero. In a case like this, we fall back on *large-signal analysis* which includes r_b' (Sec. 10-6).

Effect of ac collector resistance

The reverse-biased collector diode has an ac resistance r_c'. This resistance is normally more than a megohm, and we can expect it to have a minor effect on voltage gain and input impedance.

Figure 15-18a shows a base-driven prototype with r_b' and r_c'. Alternatively, we can visualize the circuit as shown in Fig. 15-18b where r_c' connects the input and output of an amplifier. If A is the voltage gain from b' to the collector, we can Millerize the circuit as shown in Fig. 15-18c. The output Miller resistance is

$$r_c' \frac{A}{A + 1} \cong r_c' \qquad \text{for } A \gg 1$$

Therefore, to a close approximation the collector sees an ac resistance of

$$r_C'' = r_c' \parallel r_C \cong r_C \qquad (15\text{-}17a)$$

(Resistance r_c' is normally in the megohms; this is why the collector sees an ac resistance of r_C to a close approximation.)

The voltage gain from b' to collector in Fig. 15-18c equals

$$A = \frac{v_c}{v_b'} = \frac{\alpha r_C''}{r_E + r_e'} \qquad (15\text{-}17b)$$

and the impedance looking into the b' interface is

$$z' = \frac{r_c'}{A + 1} \,\bigg\|\, (\beta + 1)(r_E + r_e') \qquad (15\text{-}17c)$$

If we add r_b' to this, we get the input impedance looking into the b interface,

$$z_{\text{in(base)}} = r_b' + z' \qquad (15\text{-}17d)$$

We can use this accurate value of $z_{\text{in(base)}}$ to analyze source-loading effects (Secs. 9-5 and 9-6).

Besides $z_{\text{in(base)}}$ we also want the voltage gain from b to collector in Fig. 15-18c. To get this gain, note

$$\frac{v_c}{v_b} = \frac{v_b'}{v_b} \frac{v_c}{v_b'}$$

*Figure 15-18. Effect of α, r'_b, and
    r'_c on base-driven amplifier.*

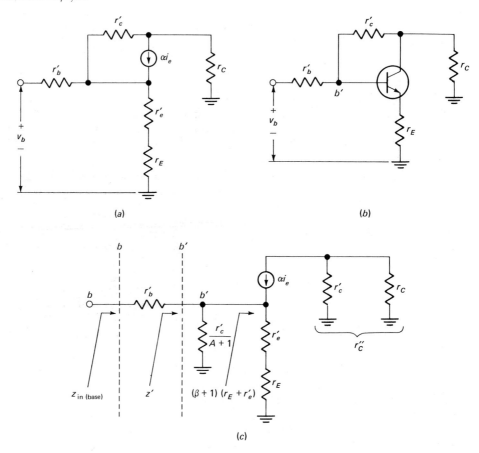

(a)   (b)

(c)

The first ratio on the right side is the voltage gain from b to b'. By the voltage-divider theorem,

$$\frac{v'_b}{v_b} = \frac{z'}{r'_b + z'}$$

Therefore, with Eq. (15-17b) we get

$$\frac{v_c}{v_b} = \frac{z'}{r'_b + z'} \frac{\alpha r''_C}{r_E + r'_e} \qquad (15\text{-}17e)$$

With Eqs. (15-17a) through (15-17e) you can accurately analyze base-driven circuits. But if you are after high accuracy, use the *h*-parameter method given in the next section. The main reason for taking you through the analysis of Fig. 15-18c is to indicate how r_b' and r_c' affect the operation of a base-driven transistor circuit. Some of the conclusions we can come to are these:

1. The output Miller resistance is usually so large we can neglect it even in a third approximation.
2. The input Miller resistance shunts the base of the ideal transistor. Therefore, it lowers the input impedance. Resistance r_b', on the other hand, is in series and tends to increase the input impedance. With Eq. (15-17d) you can calculate an accurate value of $z_{in(base)}$.
3. Both r_b' and input Miller resistance lower the voltage gain from the external base lead to the collector. Equation (15-17e) gives you the voltage gain.

EXAMPLE 15-10.
The voltage gain found with an ideal transistor is 200. If β equals 50, what is the voltage gain when we take α into account?

SOLUTION.
First, calculate α using

$$\alpha = \frac{\beta}{\beta + 1} = \frac{50}{50 + 1} = 0.98$$

Second, correct the voltage gain using Eq. (15-15a):

$$\frac{v_c}{v_b} = 0.98\,(200) = 196$$

EXAMPLE 15-11.
A bipolar-transistor stage has these values: $\alpha = 0.95$, $r_C = 5$ kΩ, $r_E = 0$, $r_e' = 25$ Ω, $r_b' = 100$ Ω, and $\beta = 50$. Calculate the ideal voltage gain, and the corrected voltage gain.

SOLUTION.
Ideally,

$$\frac{v_c}{v_b} = \frac{r_C}{r_E + r_e'} = \frac{5000}{25} = 200$$

We can include the effects of α and r_b' by using Eq. (15-16a):

$$\frac{v_c}{v_b} = \frac{0.95\,(5000)}{0 + 25 + 100/51} = 176$$

EXAMPLE 15-12.
Earlier, we said the $z_{in(base)}$ of a Darlington emitter follower may be limited to a couple of megohms. Show why this is true.

SOLUTION.

In Fig. 15-18c the voltage gain from base to collector is zero when r_C is zero. Therefore, in a Darlington emitter follower the input Miller resistance equals r_c'. The β of a Darlington pair is so high the impedance looking into the b' interface is

$$z' \cong r_c'$$

and the impedance looking into the b interface is

$$z_{\text{in(base)}} \cong r_b' + r_c' \cong r_c'$$

Resistance r_c' is the ac resistance of the reverse-biased collector diode; it has a typical value of a couple of megohms. This explains why the $z_{\text{in(base)}}$ of a Darlington emitter follower is typically limited to a couple megohms.

15-7. HYBRID PARAMETERS

Hybrid (h) parameters are easy to measure; this is the reason some data sheets specify low-frequency transistor characteristics in terms of four h parameters. This section tells you what h parameters are, how they are related to the r' parameters we have been using, and how to calculate exact answers with h parameters.

What they are

The four h parameters are transistor characteristics like input impedance and current gain measured under special conditions. For instance, Fig. 15-19a shows the ac equivalent circuit for measuring two of the four h parameters. Note the output is shorted since $r_C = 0$. The parameter h_{ie} is nothing more than the input impedance with a *shorted output.* In symbols,

$$h_{ie} = z_{\text{in(base)}} \qquad \text{for } r_C = 0 \tag{15-18a}$$

Another parameter designated h_{fe} is the current gain with a *shorted output:*

$$h_{fe} = \frac{i_c}{i_b} \qquad \text{for } r_C = 0 \tag{15-18b}$$

You can see this is identical to β. These are the only h parameters we can measure with Fig. 15-19a.

To get the other two h parameters, we use the ac equivalent circuit of Fig. 15-19b. Now we drive the collector with a voltage source v_c; this forces a current i_c to flow into the collector. The base is left *open*; that is, r_B equals infinity. The *reverse* voltage gain v_b/v_c with an open base is designated h_{re}. In symbols,

$$h_{re} = \frac{v_b}{v_c} \qquad \text{for } r_B = \infty \tag{15-18c}$$

And finally, if we take the ratio of the ac collector current to the ac collector voltage, we

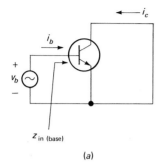
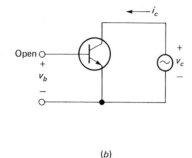

get the output admittance with an open base; this is designated h_{oe}:

$$h_{oe} = \frac{i_c}{v_c} \quad \text{for } r_B = \infty \tag{15-18d}$$

Relation of h parameters to r' parameters

Because the h parameters are easy to measure, data sheets for low-frequency small-signal transistors often give the four h parameters. When you want accurate values of the r' parameters, you can calculate them with these formulas:

$$\beta = h_{fe} \tag{15-19a}***$$

$$r'_c = \frac{\beta}{h_{oe}} \tag{15-19b}***$$

and
$$r'_b = h_{ie} - \beta r'_e \tag{15-19c}***$$

Resistance r'_e is found in the usual way: 25 mV/I_E.

h-parameter formulas

Derivation of h-parameter formulas is a special study of its own. Because we do not have the time to derive voltage gain and input impedance in terms of h parameters, we will state the formulas without proof.[1]

If you can reduce an amplifier to the base-driven prototype form of Fig. 15-20, and if you have the values of h_{ie}, h_{fe}, h_{re}, and h_{oe} at the quiescent point, you can calculate the exact voltage gain with

$$\frac{v_c}{v_b} = \frac{h_{fe} r_C}{h_{ie}[1 + (r_C + r_E)h_{oe}] + h_{fe} r_E - h_{fe} h_{re}(r_C + r_E)}$$

[1] Among many available references, see Cutler, P. C.: *Semiconductor Circuit Analysis*, McGraw-Hill Book Company, New York, 1964.

Figure 15-20. Base-driven prototype.

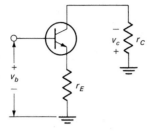

The input impedance of the base is

$$z_{\text{in(base)}} = h_{ie} + \frac{h_{fe}[r_E - h_{re}(r_C + r_E)]}{1 + h_{oe}(r_C + r_E)}$$

These equations are for any base-driven amplifier.

One special case is $r_E = 0$, the CE connection. For this case, the two formulas simplify to

$$\frac{v_c}{v_b} = \frac{h_{fe}}{h_{ie}(h_{oe} + 1/r_C) - h_{fe}h_{re}}$$

and

$$z_{\text{in(base)}} = h_{ie} - \frac{h_{fe}h_{re}}{h_{oe} + 1/r_C}$$

(Remember these are valid only for $r_E = 0$.)

On the other hand, when r_E is large enough to swamp out the transistor characteristics, both formulas reduce to

$$\frac{v_c}{v_b} \cong \frac{r_C}{r_E}$$

and

$$z_{\text{in(base)}} \cong h_{fe}r_E = \beta r_E$$

As you can see, these swamped-amplifier formulas are the same as those derived with the ideal-transistor approach.

EXAMPLE 15-13.

The data sheet for a small-signal transistor gives these parameters at $I_C = 10$ mA and $V_{CE} = 10$ V:

$$h_{ie} = 280 \ \Omega$$

$$h_{fe} = 75$$

$$h_{re} = 1.6 \times 10^{-4}$$

$$h_{oe} = 30 \times 10^{-6} \text{ mhos}$$

Work out the values of β, r_c', and r_b' using these h parameters.

SOLUTION.
With Eq. (15-19a),

$$\beta = h_{fe} = 75$$

With Eq. (15-19b),

$$r'_c = \frac{\beta}{h_{oe}} = \frac{75}{30(10^{-6})} = 2.5 \text{ M}\Omega$$

We use the standard formula to calculate r'_e:

$$r'_e = \frac{25 \text{ mV}}{10 \text{ mA}} = 2.5 \text{ } \Omega$$

Finally, we can get r'_b with Eq. (15-19c):

$$r'_b = h_{ie} - \beta r'_e = 280 - 75(2.5) = 92 \text{ } \Omega$$

Problems

15-1.   What is the bel power gain for each of these values of ordinary power gain: 1, 10, 100, 1000, 10,000, and 100,000?

15-2.   Calculate the bel power gain for $G = 2$, 20, 200, and 2000.

15-3.   Work out the bel power gain for $G = 16$, 16,000, and 1,600,000.

15-4.   Convert these bel power gains to ordinary power gains: 5 B, 30 dB, 66 dB, and 89 dB.

15-5.   Calculate an accurate value of bel power gain given an ordinary power gain of 3000.

15-6.   A three-stage amplifier has these power gains: $G_1 = 10$, $G_2 = 100$, and $G_3 = 1000$. What is the total power gain? What is the bel power gain of each stage? The bel power gain of the three-stage amplifier?

15-7.   A two-stage amplifier has these individual bel power gains: $G'_1 = 36$ dB and $G'_2 = -10$ dB. What is the total bel power gain? The total power gain?

15-8.   Work out the dBm value of each of these values of power:

1. $p = 1$ W
2. $p = 200$ mW
3. $p = 50$ μW

15-9.   The output power of a class B push-pull amplifier has a value of 49 dBm. Work out the value in watts.

15-10. In a two-stage amplifier the first stage has a power gain of $G_1 = 40$ and the second stage a power gain of $G_2 = 10,000$. What is the total power gain? And the total bel power gain?

Figure 15-21.

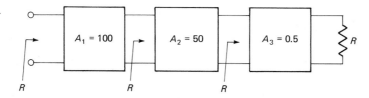

15-11. The power gain from a signal source to a base is 0.25; the power gain from base to collector is 5000. What is the total power gain in bels?

15-12. An impedance-matched amplifier has a voltage gain of 100. What is the voltage gain in decibels? The power gain in decibels?

15-13. Calculate the bel voltage gain for each of these values: $A = 2$, 20, 200, and 2000.

15-14. Work out the bel voltage gain for each of the following values: $A = 8$, 800, and 800,000.

15-15. An amplifier has a voltage gain of 5000. Express this in decibels.

15-16. A two-stage amplifier has an A_1 of 100 and an A_2 of 200. What is the total voltage gain? The total bel voltage gain?

15-17. In Fig. 15-21 convert each stage voltage gain to its bel voltage gain. What is the total bel voltage gain? The total ordinary voltage gain?

15-18. A FET amplifier has a g_m of 2000 μmhos and an r_C of 10 kΩ. No swamping resistor is used. What is the voltage gain from gate to drain? And the bel voltage gain?

15-19. The FET of Fig. 15-22 has a g_m of 4000 μmhos. The bipolar transistor has a β of 100 and an r'_e of 25 Ω.

Figure 15-22.

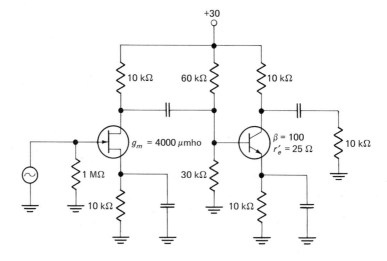

1. What is the bel voltage gain of the first stage from gate to drain?
2. What is the bel voltage gain of the second stage from base to collector?
3. Calculate the total bel voltage gain.

15-20. Suppose a 500-Ω swamping resistor is used in the second stage of Fig. 15-22. What is the total bel voltage gain in this case?

15-21. If a swamping resistor of 500 Ω is used in each stage of Fig. 15-7, what is the total bel voltage gain?

15-22. In Fig. 15-23a what is the value of the input Miller resistance? The output Miller resistance?

15-23. What value does the input Miller resistance of Fig. 15-23b have? The output Miller resistance?

15-24. Calculate the input and output Miller capacitances of Fig. 15-23c.

15-25. What are the values of input and output Miller capacitance in Fig. 15-23d?

15-26. In Fig. 15-24a the bipolar transistor has an r'_e of 5 Ω, and the voltage gain from base to collector equals 200. Calculate the input and output Miller resistances.

15-27. Work out the voltage gain for the FET amplifier of Fig. 15-24b. What are the values of input and output Miller capacitance?

15-28. What is the input Miller capacitance of Fig. 15-24c?

15-29. A CE amplifier has an r_C of 10 kΩ, an α of 0.98, an r'_e of 25 Ω, an r'_b of 200 Ω, and a β of 50. What is the voltage gain from base to collector? The input impedance? (Use all given data.)

15-30. Suppose you measure these values in Fig. 15-19a: $v_b = 5$ mV, $i_b = 2$ μA, and $i_c = 300$ μA. What are the values of h_{ie} and h_{fe}?

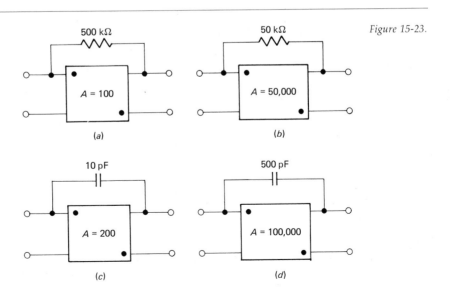

Figure 15-23.

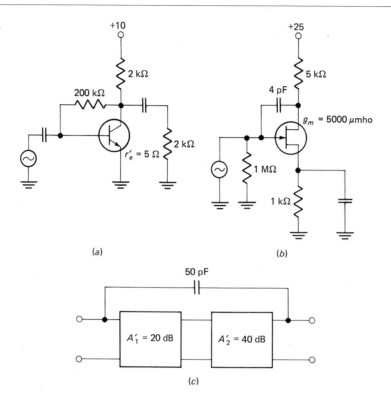

Figure 15-24.

(a)

(b)

(c)

15-31. In Fig. 15-19b $v_b = 1$ μV, $v_c = 10$ mV, and $i_c = 0.2$ μA. What are the values of h_{re} and h_{oe}?

15-32. A data sheet gives an h_{ie} of 2500 Ω for an I_C of 1 mA. If h_{fe} is 90 and r'_e is 25 Ω, what value does r'_b have?

15-33. A data sheet lists these h parameters at $I_C = 1$ mA: $h_{ie} = 2.7$ kΩ, $h_{fe} = 100$, $h_{re} = 2.5(10^{-4})$, and $h_{oe} = 25(10^{-6})$ mhos. Work out the values of β, r'_c, and r'_b.

15-34. A common-emitter amplifier has a transistor whose h parameters are those given in the preceding problem. If $r_c = 5$ kΩ,

1. Calculate the ideal voltage gain from base to collector. And the ideal input impedance of the base.
2. Calculate accurate values of voltage gain and input impedance using all four h parameters.

16. Frequency Effects

Now that we know about decibels and Miller's theorem, we can discuss frequency effects in an amplifier. This chapter explains the action of an amplifier outside the normal frequency range. We will find out what happens when coupling and bypass capacitors no longer look like short circuits. Furthermore, we will take into account the internal capacitances inside bipolars and FETs.

16-1. THE LAG NETWORK

The *lag network* of Fig. 16-1 is the key to analyzing high-frequency effects. Often, an actual circuit is more complicated than this, but with Thevenin's and Miller's theorems, we can reduce the circuit to the lag network shown. Note R is in series with C, which is across the output. As we will see, the output circuit of a transistor amplifier reduces to a lag network at higher frequencies. For this reason, we must discuss how a lag network affects the amplitude and phase angle of a signal.

Complex voltage gain

In Fig. 16-1 a sine wave drives the lag network. The frequency determines the capacitive reactance. By the well-known formula,

$$X_C = \frac{1}{2\pi f C}$$

Therefore, the higher the frequency, the smaller the X_C. A smaller X_C means a lower output voltage. So, we can conclude the following:

Figure 16-1. The lag network.

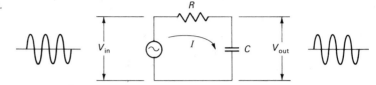

1. For very low frequencies, X_C approaches infinity or an open circuit; for this case, V_{out} equals V_{in}.
2. For very high frequencies, X_C goes to zero or a short circuit; in this case, V_{out} approaches zero.

We are interested in the *amplitude* and *phase angle* of the output voltage. To get formulas, we need to use complex numbers as done in basic circuit theory.[1] In terms of complex numbers the output voltage equals

$$\mathbf{V}_{out} = \mathbf{I}(-jX_C) = \frac{\mathbf{V}_{in}}{R - jX_C}(-jX_C)$$

Dividing both sides by V_{in} gives the complex voltage gain from input to output:

$$\frac{\mathbf{V}_{out}}{\mathbf{V}_{in}} = \frac{-jX_C}{R - jX_C} \tag{16-1}$$

Magnitude and angle

For our needs, it is better to split Eq. (16-1) into two equations: one for the magnitude of voltage gain, the other for phase angle. As it now stands, Eq. (16-1) is in rectangular form. To get the formulas we want, we must change it to polar form. When we do this,

$$\frac{\mathbf{V}_{out}}{\mathbf{V}_{in}} = \frac{X_C\,\underline{/-90°}}{\sqrt{R^2 + X_C^2}\,\underline{/-\arctan X_C/R}}$$

When an equation is in polar form like this, we can separate the magnitude and the angle to get two formulas:

$$\frac{V_{out}}{V_{in}} = \frac{X_C}{\sqrt{R^2 + X_C^2}} \tag{16-2}$$

and

$$\phi = -90° + \arctan\frac{X_C}{R} \tag{16-3}$$

These two equations are important. Equation (16-2) is the magnitude of the voltage

---

[1] A boldface letter like $\mathbf{V}$ stands for a complex quantity.

gain. In Fig. 16-1 if an ac voltmeter indicates a V_{in} of 2 V rms and a V_{out} of 0.5 V rms, the voltage gain equals 0.25. This is a pure number or magnitude; it tells nothing about the phase angle between input and output voltage. Unless otherwise indicated, when we say voltage gain, we mean *magnitude of voltage gain*.

Equation (16-3) is different. It pins down the phase angle of output voltage to input voltage. Whenever we know the values of X_C and R, we can calculate the phase angle.

Critical reactance

When the frequency is zero, X_C is infinite; with Eq. (16-3),

$$\phi = -90° + \arctan \infty = 0° \qquad (f = 0)$$

At the other extreme, when the frequency approaches infinity, X_C goes to zero. With Eq. (16-3),

$$\phi = -90° + \arctan 0 = -90° \qquad (f = \infty)$$

This tells the permissible range of phase angle. In symbols,

$$\phi = 0° \text{ to } -90° \qquad \text{(lag network)}$$

The midpoint in this range is $-45°$.

The *critical reactance* is the value of X_C that makes the phase angle equal $-45°$. When we examine Eq. (16-3), we can see that ϕ will equal $-45°$ when

$$\arctan \frac{X_C}{R} = 45°$$

In turn, arctan X_C/R equals $45°$ only when

$$X_C = R \qquad \text{(critical reactance)} \qquad\qquad (16\text{-}4a)^{***}$$

Figure 16-2a shows the special case of X_C equal to R. Since X_C equals R, phase angle ϕ equals $-45°$; this is why the output waveform of Fig. 16-2a starts $45°$ after the input waveform. If we change the frequency, we will change the value of X_C, and the phase angle will no longer equal $-45°$.

At the critical reactance the voltage gain of a lag network equals

$$\frac{V_{out}}{V_{in}} = \frac{X_C}{\sqrt{R^2 + X_C^2}} = \frac{R}{\sqrt{R^2 + R^2}} = \frac{1}{\sqrt{2}}$$

or

$$\frac{V_{out}}{V_{in}} = 0.707 \qquad \text{for } X_C = R \qquad\qquad (16\text{-}4b)^{***}$$

This says the voltage gain equals 0.707 when $X_C = R$. Figure 16-2b is an example of the critical-reactance condition. Put 1 V into a lag network when $X_C = R$ and you get out 0.707 V.

Figure 16-2. Critical reactance. (a) Phase angle is −45°. (b) Voltage gain is 0.707.

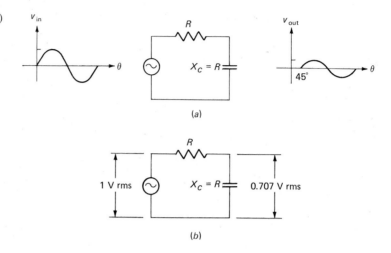

(a)

(b)

Critical frequency

You get the critical reactance at a single frequency. The formula for this frequency is easy to find as follows. When the reactance is critical,

$$X_C = R$$

or

$$\frac{1}{2\pi fC} = R$$

Solving for f gives

$$f = \frac{1}{2\pi RC}$$

We will call this value of frequency the *critical frequency* because it is the frequency that makes X_C equal to R.

EXAMPLE 16-1.
Calculate the critical frequency for the lag network of Fig. 16-3a. Show the output waveform for an input frequency equal to the critical frequency.

SOLUTION.
First, Thevenize the circuit to get it into prototype form. The Thevenin voltage equals half the source voltage; therefore, $V_{TH} = 10$ V peak. The Thevenin resistance is the parallel of two 10-kΩ resistances, or 5 kΩ.

Figure 16-3b shows the Thevenized circuit; we recognize this as the prototype of a lag network. Therefore, the critical frequency of this lag network equals

$$f = \frac{1}{2\pi RC} = \frac{1}{2\pi(5)10^3(80)10^{-12}} = 3.98(10^5)$$

or
$$f \cong 400 \text{ kHz}$$

At this frequency the output waveform is a sine wave with a peak of 7.07 V and a phase angle of $-45°$ (see Fig. 16-3b).

In the original network of Fig. 16-3a the output waveform is the same as shown in Fig. 16-3b. If we reduce the input frequency far below the critical frequency, the 80-pF capacitor appears like an open circuit. In this case the output waveform is a sine wave with a peak of 10 V and a phase angle of 0°.

On the other hand, raising the input frequency far above the critical frequency produces the same result in Fig. 16-3a and b: the peak value of the output waveform approaches zero and the phase angle approaches $-90°$.

16-2. PHASE ANGLE VERSUS FREQUENCY

To keep the critical frequency distinct from others, we will add the subscript c. With this notation the formula for critical frequency becomes

$$f_c = \frac{1}{2\pi RC} \qquad\qquad (16\text{-}5)^{***}$$

Nothing prevents us from dividing both sides of this equation by f to get

$$\frac{f_c}{f} = \frac{1}{2\pi RCf} = \frac{X_C}{R} \qquad\qquad (16\text{-}6)$$

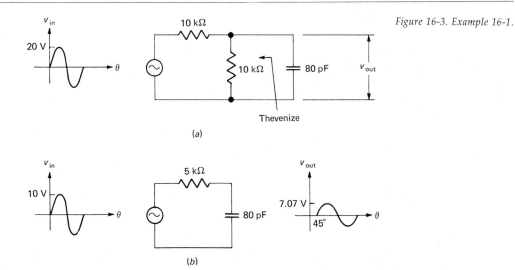

Figure 16-3. Example 16-1.

This useful ratio allows us to rewrite Eq. (16-3) as

$$\phi = -90° + \arctan \frac{f_c}{f} \tag{16-7}$$

Graph of angle

Equation (16-7) is easier to work with because the phase angle is related to input frequency f and critical frequency f_c. As an example, we will calculate a few values of phase angle for different frequencies.

When $f = 0.1f_c$, $f_c/f = 10$ and Eq. (16-7) gives

$\phi = -90° + \arctan 10 \cong -6°$

When $f = f_c$, $f_c/f = 1$ and

$\phi = -90° + \arctan 1 = -45°$

When $f = 10f_c$, $f_c/f = 0.1$ and

$\phi = -90° + \arctan 0.1 \cong -84°$

To these values we can add those found earlier:

When $f = 0$, $\phi = 0$

When $f = \infty$, $\phi = -90°$

These are enough values to graph *phase angle versus frequency* (Fig. 16-4a). At very low frequencies the angle is zero. It stays near zero until we get to $0.1f_c$; at this frequency the phase angle equals $-6°$. As the frequency increases, the lag angle increases. When the input frequency equals the critical frequency, the phase angle equals $-45°$. For an input frequency of ten times the critical frequency, the phase angle is $-84°$. Further increases in frequency produce little change because the limiting value is $-90°$.

Bode plot

A graph like Fig. 16-4a is called a *Bode plot* of phase angle. Knowing the phase angle is $-6°$ at $0.1f_c$ and $-84°$ at $10f_c$ is of little value except to indicate how close the phase angle is to its limiting values. Much more useful is the *ideal Bode plot* of Fig. 16-4b. This is the one to remember because it emphasizes these ideas:

1. When $f = 0.1f_c$, the phase angle is approximately zero.
2. When $f = f_c$, the phase angle is $-45°$.
3. When $f = 10f_c$, the phase angle is approximately $-90°$.

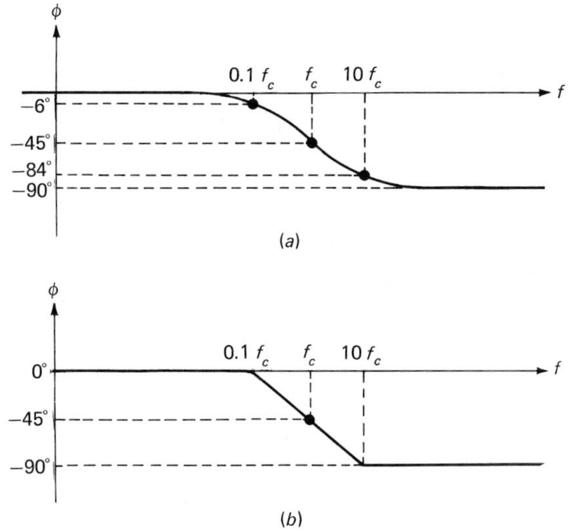

Figure 16-4. Bode plot of phase angle. (a) Exact. (b) Ideal.

The word *decade* is used to describe changes in frequency. Decade means a *factor of ten.* So, if we say the frequency has changed by a decade, we mean it has changed by a factor of ten. As an example, if we increase the frequency from 25 Hz to 250 Hz, we have increased it by a decade.

Another way to summarize the Bode plot of phase angle is this: at the critical frequency the phase angle equals $-45°$. A decade below the critical frequency the phase angle is approximately $0°$; a decade above the critical frequency the phase angle is approximately $-90°$.

16-3. BEL VOLTAGE GAIN VERSUS FREQUENCY

Logarithms make the analysis of lag networks easy when it comes to voltage gain. Before we take the logarithm of voltage gain, it will help if we rearrange Eq. (16-2) as follows:

$$A = \frac{V_{out}}{V_{in}} = \frac{X_C}{\sqrt{R^2 + X_C^2}} = \frac{1}{\sqrt{1 + R^2/X_C^2}}$$

With Eq. (16-6), we can write this as

$$A = \frac{1}{\sqrt{1 + f^2/f_c^2}} \tag{16-8}$$

Bel voltage gain

Now we are ready to get the bel voltage gain of a lag network. Referring to Eq. (16-8), the bel voltage gain is

$$A' = 2 \log A = 2 \log \frac{1}{\sqrt{1 + f^2/f_c^2}}$$

$$= -2 \log \sqrt{1 + f^2/f_c^2} = -\log (1 + f^2/f_c^2)$$

Each step is based on properties of logarithms; therefore, the bel voltage gain of a lag network equals

$$A' = -\log \left(1 + \frac{f^2}{f_c^2}\right) \qquad (16\text{-}9)$$

As an example, we will calculate some bel voltage gains:

When $f = 0.1f_c$, Eq. (16-9) gives

$$A' = -\log (1 + 0.01) = -\log 1.01$$
$$= -0.00432 \text{ B} = -0.0432 \text{ dB} \cong 0 \text{ dB}$$

When $f = f_c$,

$$A' = -\log (1 + 1) = -\log 2$$
$$= -0.30103 \text{ B} \cong -3 \text{ dB}$$

When $f = 10f_c$,

$$A' = -\log (1 + 100) = -\log 101$$
$$= -2.00432 \text{ B} \cong -20 \text{ dB}$$

The answers are rounded off to the nearest decibel; this is usual in preliminary analysis.

The bel voltage gains just calculated are interesting. Here is what we have found:

1. When the input frequency to a lag network is a decade below the critical frequency, the bel voltage gain is approximately zero.
2. When the input frequency equals the critical frequency, the bel voltage gain equals −3 dB (equivalent to an ordinary voltage gain of 0.707).
3. When the input frequency is a decade above the critical frequency, the bel voltage gain is approximately −20 dB.

As an example, the lag network of Fig. 16-5a has a critical frequency of 1.59 MHz. At this frequency an input sine wave with a peak of 1 V produces an output sine wave with a peak of 0.707 V; this is equivalent to the output voltage being down 3 dB from the input voltage. And at this critical frequency the output sine wave lags the input by 45° (Fig. 16-5b). If we decrease the frequency of the input signal from 1.59 MHz to 159 kHz, we have reduced frequency by a decade. In this case, the bel voltage gain is approximately 0 dB, meaning the output voltage almost equals the input voltage; the corresponding phase angle is approximately zero (Fig. 16-5c). On the other hand, if we

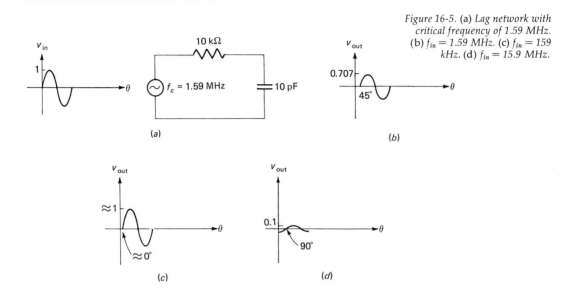

Figure 16-5. (a) *Lag network with critical frequency of 1.59 MHz.* (b) $f_{in} = 1.59$ *MHz.* (c) $f_{in} = 159$ *kHz.* (d) $f_{in} = 15.9$ *MHz.*

increase the input frequency to 15.9 MHz (one decade above the critical frequency), the bel voltage gain is −20 dB, so that the output voltage is one-tenth of the input voltage; furthermore, the output voltage is approximately 90° behind the input voltage (Fig. 16-5d).

Bode plot of bel voltage gain

In Eq. (16-9) when input frequency f is more than 10 times the critical frequency, the frequency term swamps out the 1. Because of this,

$$A' \cong -\log \frac{f^2}{f_c^2} \qquad \text{for } f \gg f_c$$

or

$$A' \cong -2 \log \frac{f}{f_c} \qquad \text{for } f \gg f_c \qquad\qquad (16\text{-}10)***$$

This last equation produces an incredible simplicity when we notice the following pattern:

When $f = 10f_c$,

$$A' = -2 \log 10 = -2 \text{ B} = -20 \text{ dB}$$

When $f = 100f_c$,

$$A' = -2 \log 100 = -4 \text{ B} = -40 \text{ dB}$$

When $f = 1000f_c$,

$$A' = -2 \log 1000 = -6 \text{ B} = -60 \text{ dB}$$

and so on. In other words, the bel voltage gain of a lag network drops 20 dB for each decade increase in frequency.

As an example, the lag network of Fig. 16-5a has these bel voltage gains:

f	A'
15.9 MHz	−20 dB
159 MHz	−40 dB
1590 MHz	−60 dB

and so on.

Whenever we can reduce a circuit to a lag network, we can calculate the critical frequency. Then, we immediately know the bel voltage gain is down 20 dB when the input frequency is one decade above the critical frequency; down 40 dB, two decades above the critical frequency; down 60 dB, three decades above the critical frequency, and so forth.[2]

Figure 16-6a shows the bel voltage gain of a lag network; this is called a *Bode plot of*

[2] If frequency *doubles,* voltage gain drops *6 dB.* See Prob. 16-16.

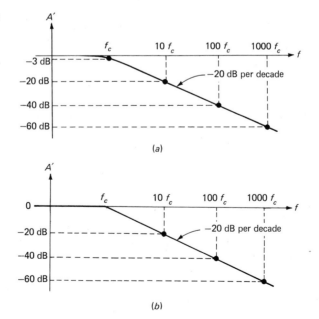

Figure 16-6. Bode plot of voltage gain. (a) *Exact.* (b) *Ideal.*

voltage gain. To see the graph over several decades of frequency, we have compressed the horizontal scale by showing equal distances between each decade in frequency. *Semilogarithmic* (or semilog) graph paper does the same thing; that is, the horizontal spacing is uniform between each decade. The advantage of marking off the horizontal scale in decades rather than units is this: well above the critical frequency the bel voltage gain drops off 20 dB for each decade increase in frequency; therefore, the graph of voltage gain will be a straight line with a slope of -20 dB per decade.

Figure 16-6b shows the *ideal Bode plot* of voltage gain. To a first approximation, we neglect the -3 dB at the critical-frequency point, and draw a straight line with a slope of -20 dB per decade. For all preliminary analysis, we use the ideal Bode plots for voltage gain and phase angle.

Break frequency

Another name for the critical frequency is *break frequency,* so called because the voltage gain in Fig. 16-6b breaks at f_c. That is, beyond the break frequency the voltage gain drops off at 20 dB per decade. *Cutoff frequency* and *corner frequency* have also been used for the critical frequency.

EXAMPLE 16-2.
Figure 16-7a shows a DE MOSFET circuit. The 50 pF of capacitance is part of the next stage. Calculate the critical frequency and describe the Bode plots.

SOLUTION.
The drain circuit can be reduced to a lag network as follows. Thevenize the circuit looking back into the drain. To do this, break the connection to the capacitor and calculate the Thevenin voltage and resistance. The voltage gain with no capacitor is

$$g_m r_D = 0.003(10,000) = 30$$

With a peak input of 1 mV, the Thevenin voltage at the drain is 30 mV peak. The drain of the FET acts ideally like a current source; therefore, the Thevenin resistance equals 10 kΩ.

Figure 16-7b shows the Thevenized drain circuit connected to the 50 pF of the next stage. We recognize this as a lag network. The critical frequency of this lag network is

$$f_c = \frac{1}{2\pi RC} = \frac{1}{2\pi(10^4)50(10^{-12})} = 318 \text{ kHz}$$

Because of amplifier phase inversion, the output signal at low frequencies is 180° out of phase with the input signal. In other words, when f_{in} is at least a decade below 318 kHz, the output voltage has peak of 30 mV and lags the input by 180° (see Fig. 16-7c). When the input signal has a frequency of 318 kHz, the output voltage is down 3 dB, which is equivalent to an ordinary voltage gain of 0.707; therefore, the output voltage has a peak of

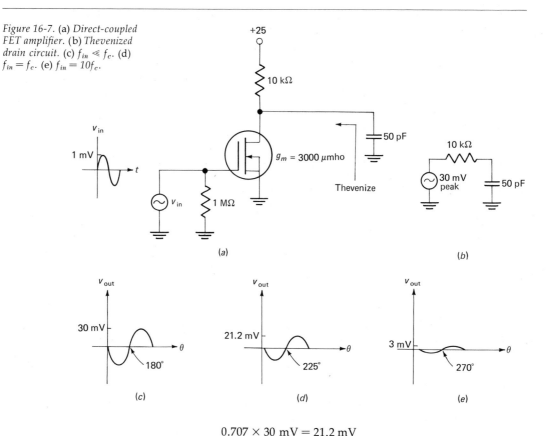

Figure 16-7. (a) Direct-coupled FET amplifier. (b) Thevenized drain circuit. (c) $f_{in} \ll f_c$. (d) $f_{in} = f_c$. (e) $f_{in} = 10f_c$.

$$0.707 \times 30 \text{ mV} = 21.2 \text{ mV}$$

and lags the input by 225° (the lag network adds an additional 45° to the existing 180° as shown in Fig. 16-7d). When f_{in} is increased to 3.18 MHz (one decade above the critical frequency), the output voltage has a peak of 3 mV and lags the input by approximately 270° (Fig. 16-7e). Beyond this point, each decade increase in frequency causes the output voltage to drop 20 dB, but the output signal still lags the input by 270°.

At low frequencies the ordinary voltage gain of the amplifier is 30; this corresponds to a bel voltage gain of approximately 30 dB. Therefore, we can describe the ideal Bode plot of voltage gain as follows. The bel voltage gain is 30 dB from 0 to 318 kHz; then it decreases 20 dB for each decade increase in frequency.

The ideal Bode plot of phase angle is this. The phase angle is −180° from 0 to 31.8 kHz (one decade below the critical frequency); the phase angle equals −225° at 318 kHz; and, it equals −270° beyond 3.18 MHz (a decade above f_c).

EXAMPLE 16-3.

Figure 16-8a shows the collector circuit of a transistor amplifier. The ac collector resistance is 10 kΩ. A capacitance of 50 pF is in shunt with r_C.

1. Prove that a current source driving a parallel RC circuit is equivalent to a lag network.
2. Calculate the critical frequency of the lag network.

SOLUTION.

1. To prove the collector circuit is equivalent to a lag network, we Thevenize the circuit to the left of point c. This gives a Thevenin voltage of $i_c r_C$ and a Thevenin resistance of r_C.

    Figure 16-8b shows the Thevenized collector circuit. Point c is still the collector node, and therefore the ac voltage from point c to ground is the same as the ac collector voltage in the original circuit. From now on, whenever we see a current source driving a shunt R and C (Fig. 16-8a), we will know it is equivalent to a lag network with the same values of R and C (Fig. 16-8b).

2. To get the critical frequency we use Eq. (16-5):

$$f_c = \frac{1}{2\pi RC} = \frac{1}{2\pi(10^4)50(10^{-12})} = 318 \text{ kHz}$$

This is the critical frequency of the collector lag network.

Each stage in a multistage transistor amplifier has a collector lag network. The one with the *lowest* critical frequency is most important because it determines the frequency where amplifier voltage gain starts to drop off.

Incidentally, when you connect an oscilloscope from collector to ground, you will be shunting from 10 to 50 pF across r_C. If the 50 pF in Fig. 16-8a represents the input capacitance of an oscilloscope, this capacitance is responsible for the loss of output voltage at input frequencies above 318 kHz.

The 50 pF of capacitance may be part of the next stage. If this is the case, the input capacitance of the next stage will be causing the voltage to drop at frequencies above 318 kHz.

EXAMPLE 16-4.

There are actually several lag networks in a transistor amplifier, and we will analyze them soon enough. Suppose the lag network of Fig. 16-8a is the one with the lowest critical frequency. If a CE amplifier has an r_e' of 25 Ω and its ac collector circuit looks like Fig. 16-8a, what does the ideal Bode plot of bel voltage gain look like?

SOLUTION.

The voltage gain of the CE stage is

$$A = \frac{r_C}{r_e'} = \frac{10,000}{25} = 400$$

The corresponding bel voltage gain is

$$A' = 2 \log A = 2 \log 400 = 5.2 \text{ B} = 52 \text{ dB}$$

Figure 16-8. Examples 16-3 and 16-4.

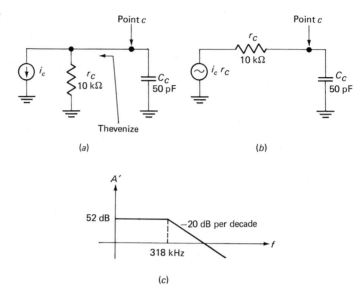

(a)

(b)

(c)

In the preceding example we calculated a critical frequency of 318 kHz. Therefore, Fig. 16-8c is the Bode plot of bel voltage gain. The bel voltage gain is 52 dB up to the break frequency; then, the bel voltage gain drops 20 dB per decade.

16-4. DC-AMPLIFIER RESPONSE

At very low frequencies the reactances of the coupling and bypass capacitors are no longer negligible. For this reason, coupling and bypass capacitors determine the lower end of the normal frequency range. On the other hand, the lag networks in an amplifier determine the upper end of the normal frequency range. In the normal range of frequencies the amplifier has its maximum voltage gain. To keep this voltage gain distinct from voltage gain outside the normal range, we will designate it as A_{mid}. For instance, in a CE amplifier A_{mid} equals r_C/r'_e (ideally). Or, in a swamped amplifier A_{mid} is r_C/r_E.

One dominant lag network

As mentioned earlier, we can design amplifiers with no coupling or bypass capacitors. An amplifier like this is called a *dc amplifier* (dc amp). Figure 16-9a shows the ideal Bode plot of bel voltage gain for a dc amp. As you can see, the dc amp has a bel voltage gain of A'_{mid} up to the break frequency f_2. Beyond this, the bel voltage gain drops 20 dB per decade.

The Bode plot of Fig. 16-9a assumes one dominant lag network. This is the reason

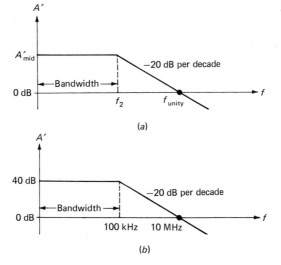

Figure 16-9. Dc-amplifier response.

the slope is -20 dB per decade. If another lag network has a critical frequency near the first, the Bode plot will break again at the critical frequency of the second lag network.

In the usual dc amplifier, however, the bel voltage gain drops 20 dB per decade until the graph crosses the horizontal axis at f_{unity}. The subscript *unity* means the ordinary voltage gain equals unity at this frequency. For instance, Fig. 16-9*b* shows the Bode plot for a dc amp whose low-frequency bel voltage gain is 40 dB (equivalent to an ordinary voltage gain of 100). The bel voltage gain breaks at 100 kHz (it's actually down 3 dB at this frequency). Ideally, the bel voltage gain drops 20 dB per decade until it crosses the horizontal axis at an f_{unity} of 10 MHz. Beyond this frequency, other lag networks will reach their critical frequencies, but we are not interested in them. (Chapter 19 will explain why the typical dc amplifier has one dominant lag network up to f_{unity}.)

Bandwidth

The *bandwidth* of a dc amplifier is the set of frequencies from zero to break frequency f_2. Any input signal with a frequency in this set receives the full voltage gain (ideally). For instance, if a dc amplifier has a Bode plot like Fig. 16-9*b*, any input signal whose frequency is between 0 and 100 kHz will receive an ideal bel voltage gain of 40 dB. Signals with frequencies outside the bandwidth of the amplifier receive less gain.

The *bandwidth value* B_v is not a set of frequencies; it is a single number. For a dc amplifier, the bandwidth value equals the highest frequency in the bandwidth, that is,

$$B_v = f_2 \quad \text{for dc amp} \tag{16-11}***$$

So, for the Bode plot of Fig. 16-9b we would say the bandwidth is the set of frequencies from 0 to 100 kHz, and the bandwidth value is

$$B_v = 100 \text{ kHz}$$

The bandwidth and bandwidth value are so closely related in a dc amplifier they are often used interchangeably. In other words, a practical person looking at Fig. 16-9b would say the dc amplifier has a bandwidth of 100 kHz. This is equivalent to saying the dc amplifier has a bandwidth value of 100 kHz, and therefore its bandwidth is the set of frequencies from 0 to 100 kHz.

16-5. RISETIME-BANDWIDTH RELATION

In testing a dc amplifier we can vary the frequency of the input signal until the bel voltage gain is down 3 dB from its normal value; the input frequency then equals the break frequency f_2 shown in Fig. 16-9a. But there is another way to find f_2. This section tells you about it.

Risetime

Given an RC circuit like Fig. 16-10a, basic circuit theory tells what happens after the switch is closed. If the capacitor is initially uncharged, the voltage will rise from 0 to V. The *risetime* T_R is the amount of time it takes the voltage to go from 0.1V (the 10 percent point) to 0.9V (the 90 percent point). If it takes 10 μs for the exponential waveform to go from the 10 percent point to the 90 percent point, the waveform has a risetime of 10 μs.

Instead of using a switch to apply the sudden step in voltage, we can use a square-wave generator. For instance, Fig. 16-10b shows a square-wave generator driving the same RC network as before. We have shown only the leading edge of one cycle. The risetime of the output waveform will be the same as before, given the same values of R and C.

Figure 16-10c shows how several cycles would look. As we see, the input voltage changes suddenly from one voltage level to another. The output voltage takes longer to make its transitions. It cannot suddenly step because the capacitor has to charge and discharge through the resistance. For this reason, it takes time for the output voltage to rise to the high voltage and to fall to the low voltage.

We are mainly interested in what happens at a single step. Once we have the output risetime produced by a single input step, we will have all the information we need for an important calculation.

Relation between T_R and RC

To derive an important result we first have to work out the relation between the risetime T_R and the time constant RC for the lag network of Fig. 16-10a. Basic courses

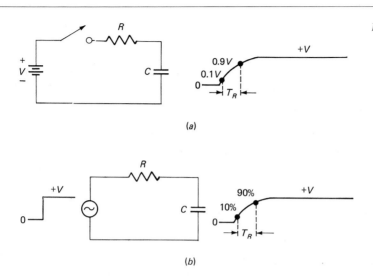

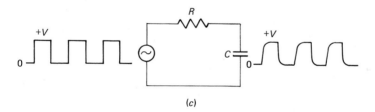

Figure 16-10. Risetime.

prove

$$v_{\text{out}} = V(1 - \epsilon^{-t/RC})$$

where V is the total step in voltage, t is the time after the switch is closed, and RC is the time constant of the lag network. At the 10 percent point, $v_{\text{out}} = 0.1V$. The value of t is unknown; let t_1 stand for this value of time and solve for t_1 as follows:

$$0.1V = V(1 - \epsilon^{-t_1/RC})$$

By rearranging, we get

$$\epsilon^{-t_1/RC} = 0.9$$

With a table of exponential values or with a slide rule,

$$\frac{t_1}{RC} \cong 0.1$$

or $\qquad\qquad t_1 \cong 0.1RC \qquad$ (time coordinate of 10 percent point)

Similarly, at the 90 percent point, $v_{out} = 0.9V$. If we let t_2 be the value of t, then

$$0.9V = V(1 - \epsilon^{-t_2/RC})$$

or

$$\epsilon^{-t_2/RC} = 0.1$$

With tables or slide rule, we find

$$t_2 \cong 2.3RC \qquad \text{(time coordinate of 90 percent point)}$$

The risetime T_R is the difference between t_2 and t_1; therefore, the formula for risetime is

$$T_R = t_2 - t_1 \cong 2.3RC - 0.1RC$$

or

$$T_R \cong 2.2RC \tag{16-12}***$$

As an example, if R equals 10 kΩ and C is 50 pF in Fig. 16-10a or b, then

$$RC = 10^4(50)10^{-12} = 0.5 \ \mu s$$

and the risetime of the output waveform equals

$$T_R = 2.2RC = 2.2(0.5 \ \mu s) = 1.1 \ \mu s$$

An important relation

Given a dc amplifier with one dominant lag network, we can find its break frequency with

$$f_2 = \frac{1}{2\pi RC}$$

According to Eq. (16-12), $T_R = 2.2RC$, or $RC = T_R/2.2$. When we substitute this into the formula for f_2, we get

$$f_2 = \frac{1}{2\pi T_R/2.2} = \frac{2.2}{2\pi T_R}$$

or

$$f_2 = \frac{0.35}{T_R} \qquad \text{(one lag network)} \tag{16-13}***$$

Why is this result important? *Because it relates sinusoidal and pulse operation of an amplifier.* This means we can test an amplifier either with an input sine wave or with an input pulse. For instance, suppose we have a dc amplifier and want its break frequency f_2. We can drive this amplifier with a step input as shown in Fig. 16-11a. If we measure a risetime of 1 μs, we can immediately predict a sinusoidal break frequency of

$$f_2 = \frac{0.35}{T_R} = \frac{0.35}{10^{-6}} = 350 \text{ kHz}$$

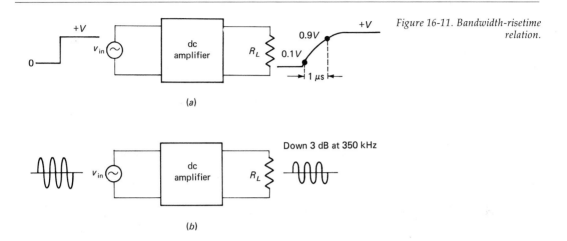

Down 3 dB at 350 kHz

(b)

Figure 16-11. Bandwidth-risetime relation.

We do not have to measure this break frequency. But if we did set up the same dc amplifier in the circuit of Fig. 16-11b, we would find the output voltage is down 3 dB when the input frequency is 350 kHz.

It works the other way too. You can measure f_2 directly with a setup like Fig. 16-11b. Once you have this sinusoidal break frequency, you can predict a risetime of

$$T_R = \frac{0.35}{f_2} = \frac{0.35}{350(10^3)} = 1\ \mu s$$

for a step input.

Almost always, whether you are after risetime or break frequency, it is faster and easier to use a step input and measure the risetime. Then, if desired, you can calculate the break frequency.

EXAMPLE 16-5.

Figure 16-12 shows a DE MOSFET stage. Because of the phase inversion, the leading edge of the output falls through V volts; the trailing edge rises through V volts. In a case like this, you can measure either the falltime or the risetime since they are equal.

1. What is the risetime of the output waveform?
2. If driven by a sine wave, at what frequency is the bel voltage gain down 3 dB from its normal value?

SOLUTION.

The drain ideally acts like a current source. As we saw in Example 16-3, a current source driving a parallel RC network is equivalent to a lag network. Since no other capacitances are given, the amplifier has only one lag network.

Figure 16-12. Example 16-5.

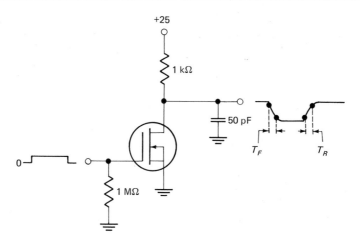

1.   The risetime of the output waveform is 2.2 times the RC time constant. So,

$$T_R = 2.2RC = 2.2(10^3)50(10^{-12}) = 0.11 \ \mu s$$

2.   You can calculate f_2 in either of two ways:

$$f_2 = \frac{0.35}{T_R} = \frac{0.35}{0.11(10^{-6})} = 3.18 \ \text{MHz}$$

or

$$f_2 = \frac{1}{2\pi RC} = \frac{1}{2\pi(10^3)50(10^{-12})} = 3.18 \ \text{MHz}$$

16-6.   HIGH-FREQUENCY FET ANALYSIS

Figure 16-13a is the ac equivalent circuit of common-source FET amplifier. Resistance r_D is the ac resistance seen by the drain. Resistance R_S is the ac resistance seen by the

Figure 16-13. Internal capacitances of a FET.

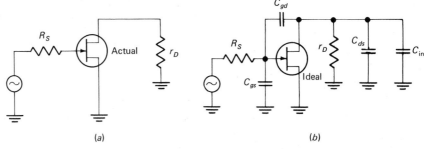

gate when looking back to the source; the biasing resistors are included in this, although they are usually large enough to neglect.

Ideally, the voltage gain from gate to drain is

$$A = \frac{v_d}{v_g} = g_m r_D \qquad (16\text{-}14)$$

This is the voltage gain in the midband of the amplifier. When the frequency of the input signal is high enough, however, the reactances in the amplifier are no longer negligible. In this section, we find the break frequencies for a FET amplifier that has an ac equivalent circuit like Fig. 16-13a.

Capacitances

There are capacitances in a FET amplifier that limit the high-frequency response. To begin with, the FET has internal capacitances between its three electrodes. C_{gs} is the internal capacitance between the gate and the source. C_{gd} is the capacitance between the gate and the drain and C_{ds} the capacitance between the drain and the source. Figure 16-13b shows these capacitances in the ac equivalent circuit.

If the output signal of the FET amplifier drives another amplifier stage, there is an additional capacitance C_{in} across the drain-ground terminals as shown in Fig. 16-13b. This input capacitance of the next stage includes *stray-wiring capacitance,* which is the capacitance between connecting wires and ground. As mentioned in Chap. 1, if a piece of AWG 22 wire is 1 ft long and is 1 in above a chassis, it has approximately 3.3 pF of capacitance with respect to ground. If the wire were only an inch long, it would have approximately 0.3 pF of capacitance to ground. Move this wire closer to the chassis and the capacitance increases; move it away from the chassis and the capacitance goes down. We can use 0.3 pF/in as a rough estimate for stray-wiring capacitance.

We also mentioned in Chap. 1 that a resistor has end-to-end capacitance. As a rough estimate, we can allow 1 pF for this capacitance. When necessary, you can measure this end-to-end capacitance with an ac bridge. Resistance r_D is the combined resistance of R_D (the drain biasing resistor) and other resistors in parallel with R_D in the ac equivalent circuit. The end-to-end capacitance of each of these discrete resistors can be added into C_{in}.

As we already know, all these capacitances produce negligible effects in the midband of the amplifier. But at high frequencies, ac current flows through these capacitances and changes the normal action of the FET amplifier. For instance, in Fig. 16-13b, C_{gs} is internal gate-to-source capacitance plus any stray-wiring capacitance from gate to ground. At a high enough frequency the X_C of this capacitance is very low. The ac current through this capacitance flows through R_S, producing a loss of ac voltage across R_S. Therefore, not as much ac voltage reaches the gate terminal. For this reason, the output voltage drops off.

Gate lag network

To find the break frequencies of the FET amplifier of Fig. 16-13b, we have to reduce the ac circuit until we reach prototype lag networks. The first step is to Millerize Fig. 16-13b. When we do this, we get an input Miller capacitance of

$$C_{\text{in(Miller)}} = C_{gd}(A + 1)$$

Since A is the voltage gain from the gate of the ideal FET to the collector, we can write

$$C_{\text{in(Miller)}} = C_{gd}(g_m r_D + 1)$$

Figure 16-14a shows this input Miller capacitance.

The output Miller capacitance is

$$C_{\text{out(Miller)}} = C_{gd}\frac{A}{A + 1} \cong C_{gd}$$

This is a reasonable approximation for A greater than 10; besides, the exact value is not important because it is only part of the total capacitance from drain to ground. Figure 16-14a shows the output Miller capacitance in shunt with C_{ds} and C_{in}.

Since the capacitances are in parallel, they can be added to get an equivalent capacitance for the gate circuit and for the drain circuit as shown in Fig. 16-14b. Looking directly into the gate, we see an almost infinite resistance; therefore, the gate circuit of Fig. 16-14b is in the form of a lag network. The critical frequency of this gate lag network is

$$f_c = \frac{1}{2\pi R_S C} \qquad \text{(gate lag network)} \qquad (16\text{-}15a)$$

where

$$C = C_{gs} + C_{gd}(g_m r_D + 1) \qquad (16\text{-}15b)$$

In the midband of the amplifier, maximum ac voltage reaches the gate (see Fig. 16-14b). But when the frequency of the input signal equals the critical frequency of the gate lag network, the ac gate voltage breaks and ideally drops 20 dB per decade. To a second approximation, the ac gate voltage is down 3 dB at the break frequency, down 20 dB a decade above the break frequency, and so on.

Drain lag network

Ideally, the drain acts like a current source. We already know a current source driving a parallel RC network is equivalent to a lag network (Example 16-3). Therefore the drain resistance and capacitance of Fig. 16-14b represent the R and C of a lag network. The critical frequency for this lag network is

$$f_c = \frac{1}{2\pi r_D C} \qquad \text{(drain lag network)} \qquad (16\text{-}16a)$$

where

$$C = C_{gd} + C_{ds} + C_{\text{in}} \qquad (16\text{-}16b)$$

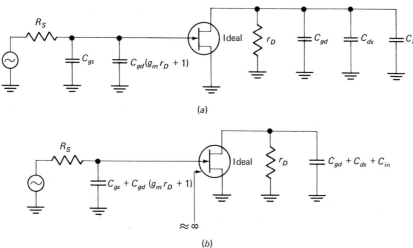

Figure 16-14. Gate and drain lag networks.

(a)

(b)

Either the gate lag network or drain lag network dominates in Fig. 16-14b; only by a coincidence would the two lag networks have the same break frequency. Therefore, we are interested in the lower of the break frequencies; this break frequency determines the upper end of the bandwidth. The voltage gain from signal source to final output in Fig. 16-14b equals $g_m r_D$ in the midband of the amplifier. This voltage gain breaks at the first critical frequency; it rolls off at 20 dB per decade until reaching the second break frequency; then, the voltage gain drops 40 dB per decade.

Capacitances on a data sheet

Unfortunately, data sheets do not list the capacitances of Fig. 16-15a. But they do usually give information for calculating these capacitances. The manufacturer typically will measure the FET capacitances under short-circuit conditions. For instance, C_{iss} is the input capacitance with an ac short across the output as shown in Fig. 16-15b. The manufacturer measures and lists this capacitance C_{iss} on his data sheet.

In Fig. 16-15b the shorted output means C_{gd} is in parallel with C_{gs} when viewed from the input terminals. In other words,

$$C_{iss} = C_{gs} + C_{gd} \qquad (16\text{-}17a)$$

C_{oss} is the capacitance looking back into the FET with an ac short across the input terminals (Fig. 16-15c). The short across the input means C_{gd} is in parallel with C_{ds}, so that

$$C_{oss} = C_{ds} + C_{gd} \qquad (16\text{-}17b)$$

Figure 16-15. Measuring FET capacitances.

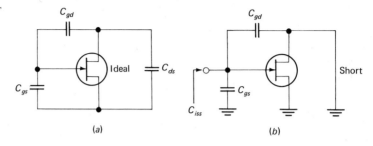

(a)   (b)

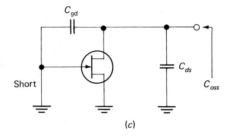

(c)

Finally, the manufacturer measures reverse capacitance C_{rss}, which equals

$$C_{rss} = C_{gd} \tag{16-17c}$$

By solving Eqs. (16-17a) through (16-17c) simultaneously, we can get these formulas:

$$C_{gd} = C_{rss} \tag{16-18a}$$

$$C_{gs} = C_{iss} - C_{rss} \tag{16-18b}$$

$$C_{ds} = C_{oss} - C_{rss} \tag{16-18c}$$

With these formulas we can calculate the capacitances needed to analyze the lag networks of a FET amplifier.

A final point. Often, a data sheet will list only C_{iss} and C_{rss} values. In a case like this, you can calculate C_{gd} and C_{gs}, but not C_{ds}. However, C_{ds} remains fairly constant for most FETs. A close approximation is

$$C_{ds} \cong 1 \text{ pF}$$

Whenever C_{oss} is not given, we can use this approximation for C_{ds}. (If necessary, you can measure C_{oss} as shown in Fig. 16-15c and calculate an accurate value for C_{ds}.)

EXAMPLE 16-6.

An MPF102 is an *n*-channel JFET. Its data sheet lists C_{iss} and C_{rss} as follows:

$$C_{iss} = 7 \text{ pF}$$

$$C_{rss} = 3 \text{ pF}$$

These are maximums or worst-case values. C_{oss} is not given. Calculate the C_{gd}, C_{gs}, and C_{ds}.

SOLUTION.
With Eqs. (16-18a) and (16-18b)

$$C_{gd} = C_{rss} = 3 \text{ pF}$$

$$C_{gs} = C_{iss} - C_{rss} = 7 \text{ pF} - 3 \text{ pF} = 4 \text{ pF}$$

Since C_{oss} is not given, we approximate C_{ds} as

$$C_{ds} \cong 1 \text{ pF}$$

EXAMPLE 16-7.
Calculate the upper break frequencies for the FET amplifier of Fig. 16-16a. Use a g_m of 4000 μmhos for the MPF102.

SOLUTION.
First, visualize the ac equivalent circuit at higher frequencies. This means all coupling and bypass capacitors appear as short circuits. The MPF102 has capacitances of 4 pF, 3 pF, and 1 pF between its electrodes as shown in Fig. 16-16b; these are the capacitances we calculated in the preceding example. Also, we will allow 4 pF for stray-wiring and end-to-end capacitances in the drain circuit.

The midband voltage gain is

$$A = g_m r_D = 0.004(5000) = 20$$

With this value we can Millerize the circuit to get Fig. 16-16c. Now, we can see the two lag networks. The gate lag network has these values of R and C:

$$R = 1 \text{ k}\Omega$$

$$C = 4 \text{ pF} + 63 \text{ pF} = 67 \text{ pF}$$

So, it has a critical frequency of

$$f_c = \frac{1}{2\pi RC} = \frac{1}{2\pi(10^3)67(10^{-12})} = 2.38 \text{ MHz} \cong 2.4 \text{ MHz}$$

The drain lag network has

$$R = 5 \text{ k}\Omega$$

$$C = 3 \text{ pF} + 1 \text{ pF} + 4 \text{ pF} = 8 \text{ pF}$$

and
$$f_c = \frac{1}{2\pi RC} = \frac{1}{2\pi(5)10^3(8)10^{-12}} = 3.98 \text{ MHz} \cong 4 \text{ MHz}$$

This break frequency is higher than the other. Therefore, the gate lag network is dominant; the upper end of the bandwidth is approximately 2.4 MHz.

EXAMPLE 16-8.
Calculate the bel voltage gain in Fig. 16-16a for the gate lag network, the ideal FET, and the drain lag network at 2.4 MHz. What is the total bel voltage gain at this frequency?

SOLUTION.
Since 2.4 MHz is the critical frequency of the gate lag network, its bel voltage gain is down 3 dB. In symbols,

$$A_1' = -3 \text{ dB}$$

The ordinary voltage gain of the ideal FET is

$$A_2 = g_m r_D = 0.004(5000) = 20$$

Figure 16-16. Examples 16-7 and 16-8.

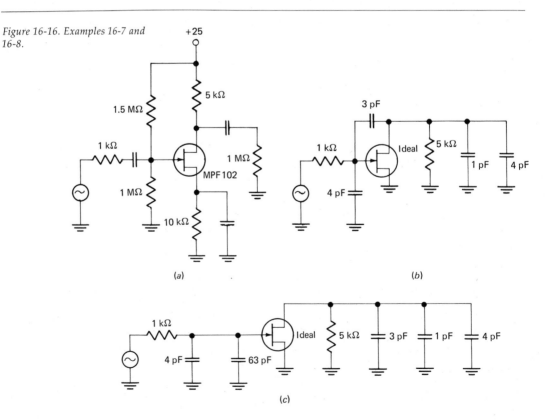

and the bel voltage gain is

$$A_2' = 2 \log 20 = 2.6 \text{ B} = 26 \text{ dB}$$

With Eq. (16-9) we can calculate the bel voltage gain of the drain lag network:

$$A_3' = -\log \left(1 + \frac{2.4^2}{4^2}\right) = -0.13 \text{ B} = -1.3 \text{ dB} \cong -1 \text{ dB}$$

Because the bel voltage gains of cascaded circuits are additive, we can get the total bel voltage gain of the amplifier:

$$A' = A_1' + A_2' + A_3' = -3 \text{ dB} + 26 \text{ dB} - 1 \text{ dB} = 22 \text{ dB}$$

Since the midband bel voltage gain is 26 dB, the bel voltage gain is down 4 dB at 2.4 MHz. Therefore, at a frequency slightly lower than 2.4 MHz the overall bel voltage gain will be down 3 dB. This frequency represents the exact upper end of the amplifier bandwidth. In all preliminary analysis, however, we approximate by using the lower critical frequency as the upper end of the bandwidth. So, with an amplifier like Fig. 16-16 the upper end of the bandwidth is ideally 2.4 MHz.

If the ideal answer is not good enough in your application, you can find the exact 3-dB frequency as follows. With Eq. (16-9) the bel voltage gain of the gate lag network is

$$A_1' = -\log \left(1 + \frac{f^2}{2.4^2}\right)$$

and the bel voltage gain of the drain lag network is

$$A_3' = -\log \left(1 + \frac{f^2}{4^2}\right)$$

where f is in MHz. By either trial and error or a simultaneous solution you can find the value of f that makes

$$A_1' + A_3' = -3 \text{ dB}$$

This value of f is the true 3-dB frequency of the amplifier. (If you work it out, you find f equals 1.9 MHz.)

16-7.   HIGH-FREQUENCY BIPOLAR ANALYSIS

Figure 16-17a shows the ac equivalent circuit for a CE amplifier. R_S represents the Thevenin resistance looking back from base to source; it includes biasing resistors in the base circuit. Resistance r_C is the ac resistance seen by the collector. The high-frequency analysis of a bipolar transistor amplifier is so similar to what we have just gone through for the FET that we can easily derive the critical frequencies for the input and output lag networks.

Figure 16-17. Internal
capacitances of a bipolar
transistor.

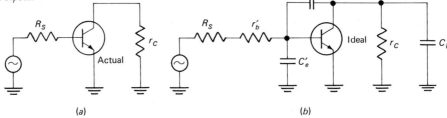

(a)　　　　　　　　　　　　(b)

Capacitances

When we first discussed the ac model of a transistor, we mentioned the internal capacitances across each depletion layer. C_e' is the emitter-diode capacitance; it is the capacitance of the forward-biased emitter diode. C_c' is the collector-diode capacitance; it is the capacitance of the reverse-biased collector diode (see Fig. 16-17b). Since the depletion layer of a forward-biased diode is much narrower than the depletion layer of a reverse-biased diode, C_e' is much larger than C_c'. For instance, in a 2N3904 with an I_E of 10 mA, C_e' is around 200 pF while C_c' is about 4 pF.

Figure 16-17b shows the ac equivalent circuit with these capacitances. The internal capacitance between the collector and the emitter (analogous to the C_{ds} in Fig. 16-13b) is small enough to neglect in a bipolar transistor. This is why we have only shown the C_{in} of the next stage. As before, we can include stray-wiring capacitance and resistor end-to-end capacitance in the value of C_{in}.

Resistance r_b' is too important in the high-frequency analysis to neglect. Here is the reason. At higher frequencies where significant ac currents flow through C_e' and C_c', additional current flows through r_b'. Because of this, there is an extra voltage drop across r_b'; this loss of signal voltage means the output voltage of the amplifier drops off.

Base lag network

To find the break frequencies for the bipolar amplifier of Fig. 16-17b, we have to reduce the circuit until we have two distinct lag networks. The first step is to get the input Miller capacitance. The voltage gain from the base of the ideal transistor to the collector is

$$A = \frac{r_C}{r_e'}$$

The input Miller capacitance therefore equals

$$C_{in(Miller)} = C_c'\left(\frac{r_C}{r_e'} + 1\right)$$

Figure 16-18a shows this input Miller capacitance.

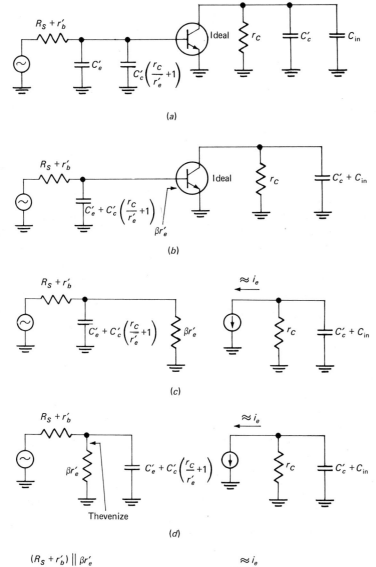

Figure 16-18. Reducing a bipolar amplifier to its base and collector lag networks.

The output Miller capacitance is approximately C_c' because the voltage gain A is ordinarily high in a CE amplifier. Figure 16-18a shows the output Miller capacitance in shunt with C_{in}.

The capacitances in Fig. 16-18a are in parallel; this allows us to add them to get an equivalent capacitance for the base circuit and for the collector circuit as shown in Fig. 16-18b. Unlike a FET, however, when we look into the base, we do not see an infinite resistance; instead, we see $\beta r_e'$. Therefore, we have to take an additional step to get to the lag networks. As proved in Sec. 9-5, we can split the ideal transistor into $\beta r_e'$ and a current source as shown in Fig. 16-18c.

In the base circuit of Fig. 16-18c the capacitance and $\beta r_e'$ are in parallel. Therefore, we can move $\beta r_e'$ to the left of the capacitance as shown in Fig. 16-18d. When we Thevenize Fig. 16-18d, we get a Thevenin resistance of

$$R_{TH} = (R_S + r_b') \| \beta r_e'$$

We don't have to bother with the Thevenin voltage because we are only after the R and C of the base lag network; this is all we need to calculate the critical frequency.

Figure 16-18e shows the final ac equivalent circuit. Aside from the complicated collection of symbols, the circuit is simple; it contains two lag networks. The base lag network has a critical frequency of

$$f_c = \frac{1}{2\pi RC} \tag{16-19a}$$

where
$$R = (R_S + r_b') \| \beta r_e' \tag{16-19b}$$

and
$$C = C_e' + C_c'\left(\frac{r_C}{r_e'} + 1\right) \tag{16-19c}$$

If the base lag network is dominant, we calculate the break frequency of the bipolar amplifier using Eqs. (16-19a) through (16-19c).

Collector lag network

Ideally, the collector acts like a current source; so, we recognize the collector circuit as a lag network. The critical frequency of this lag network is

$$f_c = \frac{1}{2\pi r_C(C_c' + C_{in})} \tag{16-20}$$

If the collector lag network has a lower critical frequency than the base lag network, it will determine the upper end of the amplifier bandwidth.

Capacitances on a data sheet

Again, it's a case of not getting exactly the capacitances C_e' and C_c' that we need. But enough information is usually given to allow us to calculate C_e' and C_c'. First, the collector-diode capacitance C'_c is the capacitance of the reverse-biased collector diode and depends on the dc voltage from collector to base. Data sheets usually list a value of capacitance at a specific V_{CB}, and we can use this as estimate for C'_c at other collector voltages. For instance, the 2N3904 has a C'_c of 4 pF at V_{CB} equals 5 V. If we operate at a different collector voltage, we can still use 4 pF as an estimate for C'_c.[3]

There is no standard designation for C'_c. Data sheets may list it by any of the following equivalent symbols: C_c, C_{cb}, C_{ob}, and C_{obo}. As an example, the data sheet of a 2N2330 gives a C_{ob} of 10 pF for $V_{CB} = 2$ V. The data sheet of an FT107B lists a C_{cb} of 4 pF at $V_{CB} = 5$ V.

The C'_e we are after is not usually listed on a data sheet because it is too difficult to measure directly. Instead, the manufacturer gives a value called the *current gain-bandwidth product* designated f_T. This particular quantity is the frequency where the current gain of a transistor drops to unity. Theoretically, one way to measure f_T is with a setup like Fig. 16-19. We increase the frequency of the input signal until the internal capacitances have made i_c decrease to the value of i_b. At this frequency (f_T), the current gain equals unity. The manufacturer actually measures f_T in a different way than this; but f_T still has the meaning we have given it.

By applying the definition of f_T to Fig. 16-18e, the Appendix proves this important relation:

$$C'_e + C'_c = \frac{1}{2\pi f_T r'_e} \tag{16-21a}$$

Almost always, C'_e is much greater than C'_c because it is the capacitance of a forward-biased diode. Therefore, a close approximation for C'_e is

$$C'_e \cong \frac{1}{2\pi f_T r'_e} \tag{16-21b}***$$

EXAMPLE 16-9.
The data sheet of a 2N3904 gives an f_T of 300 MHz at $I_E = 10$ mA. Work out the value of C'_e.

---

[3] If necessary, you can measure C'_c with an ac bridge.

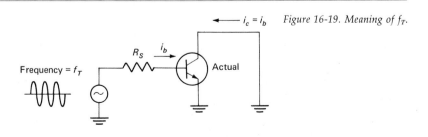

Frequency = f_T

R_S i_b

$i_c = i_b$

Actual

Figure 16-19. Meaning of f_T.

SOLUTION.

To use Eq. (16-21b) we need the value of r_e' under the test conditions given on the data sheet. Since I_E is 10 mA,

$$r_e' = \frac{25 \text{ mV}}{10 \text{ mA}} = 2.5 \ \Omega$$

Therefore, C_e' equals

$$C_e' \cong \frac{1}{2\pi f_T r_e'} = \frac{1}{2\pi (300) 10^6 (2.5)} = 212 \text{ pF}$$

This is the value of C_e' for an I_E of 10 mA.

EXAMPLE 16-10.

Suppose we use the 2N3904 of the preceding example in a CE amplifier with the following values: $R_S = 1 \text{ k}\Omega$, $r_b' = 100 \ \Omega$, $\beta r_e' = 250 \ \Omega$, $r_C = 1 \text{ k}\Omega$, $r_e' = 2.5 \ \Omega$, $C_c' = 4 \text{ pF}$, and $C_{\text{in}} = 5 \text{ pF}$. With the C_e' found in the preceding example, work out the break frequencies of the CE amplifier.

SOLUTION.

Refer to Fig. 16-18e to see how each given value fits into the lag networks. In the base lag network the equivalent R is

$$R = (R_S + r_b') \,||\, \beta r_e' = 1100 \,||\, 250 = 204 \ \Omega$$

The equivalent C is

$$C = C_e' + C_c' \left(\frac{r_C}{r_e'} + 1\right) = 212 \text{ pF} + 1604 \text{ pF} \cong 1820 \text{ pF}$$

So, the critical frequency of the base lag network is

$$f = \frac{1}{2\pi RC} = \frac{1}{2\pi (204) 1.82 (10^{-9})} \cong 430 \text{ kHz}$$

In the collector circuit of Fig. 16-18e the equivalent capacitance is

$$C = C_c' + C_{\text{in}} = 4 \text{ pF} + 5 \text{ pF} = 9 \text{ pF}$$

Therefore, the critical frequency of the collector lag network is

$$f_c = \frac{1}{2\pi RC} = \frac{1}{2\pi (10^3) 9 (10^{-12})} = 17.7 \text{ MHz}$$

No question about which lag network is dominant. The base lag network has a much lower critical frequency; therefore, it determines the upper end of the bandwidth. The voltage gain from source to output will break first at 430 kHz; it will roll off at a rate of 20 dB per decade until 17.7 MHz; then, the voltage gain breaks again and rolls off at 40 dB per decade.

Often, the base lag network dominates. But this is not necessarily always true. We may have a large capacitive load in the collector circuit, or the input capacitance of the next stage may be very large because of the Miller effect in the next stage. In a case like this, the collector lag network may dominate. It makes no difference to us; we know how to calculate both critical frequencies; the lower of the two is dominant.

16-8. THE LEAD NETWORK

Whenever you have a coupling capacitor, you have a *lead network*. Figure 16-20a shows the prototype form of a lead network. In the midband of an amplifier the coupling capacitor looks like an ac short, and the circuit acts like a voltage divider with a gain of

$$A_{mid} = \frac{V_{out}}{V_{in}} = \frac{R_L}{R_{TH} + R_L} \tag{16-22a}$$

At lower frequencies where X_C is no longer negligible, the current $\mathbf{I}$ equals

$$\mathbf{I} = \frac{\mathbf{V}_{in}}{R_{TH} + R_L - jX_C}$$

and the output voltage is

$$\mathbf{V}_{out} = \mathbf{I}R_L = \frac{\mathbf{V}_{in}R_L}{R_{TH} + R_L - jX_C}$$

Therefore, the complex voltage gain is

$$\mathbf{A} = \frac{\mathbf{V}_{out}}{\mathbf{V}_{in}} = \frac{R_L}{R_{TH} + R_L - jX_C} \tag{16-22b}$$

Magnitude and angle

Equation (16-22b) is the complex voltage gain. After we convert to polar form, we can split the complex voltage gain into its magnitude

$$A = \frac{R_L}{\sqrt{(R_{TH} + R_L)^2 + X_C^2}} \tag{16-23a}$$

and its phase angle

$$\phi = \arctan \frac{X_C}{R_{TH} + R_L} \tag{16-23b}$$

When X_C equals zero, ϕ equals zero (midband). On the other hand, when X_C approaches infinity, ϕ equals 90°. This gives us the permissible range of the phase angle:

$$0 \leq \phi \leq 90° \quad \text{(lead network)}$$

Figure 16-20. The lead network.
(a) Circuit. (B) Bode plot of phase
angle. (c) Bode plot of voltage
gain.

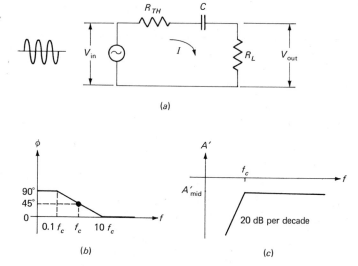

Now we see why it is called a lead network; at lower frequencies, the output voltage leads the input voltage.

Critical frequency

The critical reactance is the value of X_C that makes the phase angle equal 45°. When we examine Eq. (16-23b), it becomes apparent the condition for a 45° phase angle is

$$X_C = R_{TH} + R_L \qquad \text{(critical reactance)} \qquad (16\text{-}24a)\text{***}$$

Since $X_C = 1/2\pi fC$, we can solve for the critical frequency of a lead network to get

$$f_c = \frac{1}{2\pi(R_{TH} + R_L)C} \qquad (16\text{-}24b)$$

This is the frequency where the phase angle equals 45°. Also, it is the frequency where the voltage gain A is down 3 dB from its midband value.

Bode plots

By working out the phase angle for $0.1f_c$, f_c, and $10f_c$, we find phase angles of 6°, 45°, and 84°. Again, the 6° and 84° indicate how close the phase angle is to ideal values of 0° and 90°. For this reason, we can show the ideal Bode plot of phase angle as given in Fig. 16-20b. At the critical frequency the output voltage leads the input by 45°. A decade below the critical frequency the output voltage leads by approximately 90°, and a decade above by approximately 0°.

In the midband, Eq. (16-22a) gives a bel voltage gain of

$$A'_{\text{mid}} = 2 \log A_{\text{mid}}$$

where A_{mid} is the ordinary voltage gain of the voltage divider formed by R_{TH} and R_L (see Fig. 16-20a). For instance, if R_{TH} equals 10 kΩ and R_L is 30 kΩ,

$$A_{\text{mid}} = \frac{30{,}000}{10{,}000 + 30{,}000} = 0.75$$

and
$$A'_{\text{mid}} = 2 \log 0.75 = 2 \log (7.5 \times 10^{-1})$$
$$= 2(0.875 - 1) = -0.25 \text{ B} = -2.5 \text{ dB}$$

This is the midband bel voltage gain.

When the frequency of the input signal is below the midband, the bel voltage gain drops off. At the critical frequency it is down 3 dB from the midband bel voltage gain. A decade below the critical frequency, the bel voltage gain is down 20 dB from the midband bel voltage gain. For example, if the midband bel voltage gain is -2.5 dB, at the critical frequency the bel voltage gain is -5.5 dB; a decade below the critical frequency the bel voltage gain is -22.5 dB.

Figure 16-20c shows the ideal Bode plot of bel voltage gain. In the midband the bel voltage gain equals A'_{mid}. Below the critical frequency the bel voltage gain drops 20 dB per decade. Ideally, the bel voltage gain at the critical frequency equals A'_{mid} as shown. To a second approximation it is actually down 3 dB.

EXAMPLE 16-11.
In Fig. 16-21a calculate the critical frequencies of the coupling circuits. Assume the bypass capacitor looks like an ac short at these critical frequencies.

SOLUTION.
On the input side we work out the value of z_{in}.

$$z_{\text{in}} = 60{,}000 || 30{,}000 || 2500 \cong 2.2 \text{ k}\Omega$$

On the output side, looking back into the collector circuit, we see a Thevenin resistance of 10 kΩ (ideally). Therefore, we can show the equivalent ac circuit of Fig. 16-21b.

The base lead network has a critical frequency of

$$f_c = \frac{1}{2\pi(3.2)10^3(0.1)10^{-6}} = 498 \text{ Hz} \cong 500 \text{ Hz}$$

The collector lead network has a critical frequency of

$$f_c = \frac{1}{2\pi(40)10^3(0.1)10^{-6}} \cong 40 \text{ Hz}$$

The base lead network is the dominant one because it determines the lower end of the bandwidth.

Figure 16-21. Example 16-11.

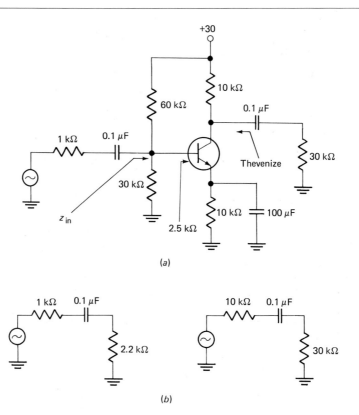

16-9. AC AMPLIFIER RESPONSE

The dc amplifier has no lower critical frequency; therefore, it works all the way down to zero frequency. The *ac amplifier,* on the other hand, has lead networks that place a lower limit on the bandwidth.

Bode plot of bel voltage gain

In a multistage ac amplifier there are many lead networks. To avoid certain problems discussed later (Chap. 19), one of the lead networks is often made dominant so that the bel voltage gain drops 20 dB per decade until the graph crosses the horizontal axis (see Fig. 16-22*a*).

In an ac amplifier the bandwidth is the set of frequencies between break frequencies f_1 and f_2 shown in Fig. 16-22*a*. Any input signal with a frequency in this set ideally receives the maximum amplifier voltage gain. The bandwidth value for an ac amplifier equals

$$B_v = f_2 - f_1 \qquad (16\text{-}25a)^{***}$$

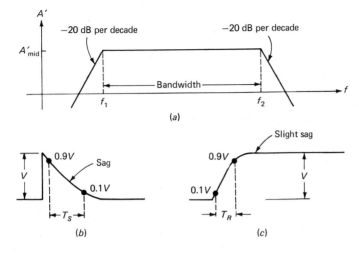

Figure 16-22. (a) *Ac-amplifier response*. (b) *Sagtime*. (c) *Risetime*.

In an ac amplifier with no resonant circuits the ac loads on the transistors (or FETs) appear resistive over many decades of frequency. For this reason, the upper break frequency f_2 is much larger than the lower break frequency f_1. Because of this, the bandwidth value of an untuned ac amplifier approximately equals

$$B_v \cong f_2 \quad \text{(untuned amp)} \tag{16-25b}$$

Measuring the break frequencies

If we drive an ac amplifier with a step voltage, we can measure the risetime as before and use

$$f_2 = \frac{0.35}{T_R} \quad \text{(one lag network)} \tag{16-26a}$$

to find the upper break frequency (see Fig. 16-22c).

By reducing the sweep speed of the oscilloscope, we can see the *sag* of Fig. 16-22b. This is caused by the dominant lead network, specifically by the charging of the dominant coupling capacitor. The *sagtime* T_S is the time between the 90 and 10 percent points shown in Fig. 16-22b. By a derivation almost identical to that given earlier for Eq. (16-26a),

$$f_1 = \frac{0.35}{T_S} \quad \text{(one lead network)} \tag{16-26b} ***$$

EXAMPLE 16-12.

Suppose you measure a sagtime of 7 ms when you drive an ac amplifier with a step voltage. What is the lower break frequency of the amplifier? (Assume one dominant lead network.)

SOLUTION.
With Eq. (16-26b),

$$f_1 = \frac{0.35}{0.007} = 50 \text{ Hz}$$

16-10. THE BYPASS CAPACITOR

At lower frequencies the coupling or bypass capacitor may dominate. In this section we assume the bypass capacitor causes the response to break before the coupling capacitor does.

Bipolar case

Figure 16-23*a* shows a base-driven prototype. When we Thevenize the circuit to the left of the bypass capacitor, we get the equivalent circuit of Fig. 16-23*b*. In the mid-

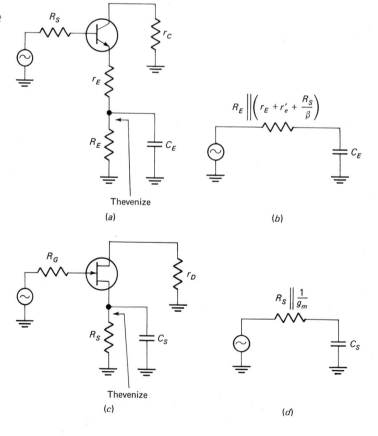

Figure 16-23. Bypass capacitor. (a) Bipolar circuit. (b) Equivalent bipolar circuit. (c) FET circuit. (D) Equivalent FET circuit.

band the capacitor looks like an ac short. But at lower frequencies the reactance is no longer negligible. Since R_E is usually negligibly large, the break frequency is

$$f_c \cong \frac{1}{2\pi(r_E + r'_e + R_S/\beta)C_E}$$ (16-27)

In this equation, R_S includes the shunting effects of biasing resistors seen when looking from the base to the source.

When analyzing an amplifier, you calculate the break frequencies produced by each bypass capacitor and coupling capacitor. The highest break frequency determines the lower end of the bandwidth.

An advantage of collector-feedback bias is that the emitter is at dc ground; therefore, it needs no bypass capacitor.

FET case

Figure 16-23c shows the gate-driven FET prototype. The Appendix proves the Thevenin resistance to the left of the bypass capacitor is R_S in parallel with $1/g_m$ (Fig. 16-23d). The break frequency of the bypass network is

$$f_c = \frac{1}{2\pi(R_S||1/g_m)C_S}$$ (16-28)

EXAMPLE 16-13.
Calculate the break frequency produced by the bypass capacitor of Fig. 16-21a. Use an r'_e of 25 Ω and a β of 100.

SOLUTION.
Resistance r_E is zero and R_S is approximately 1000 Ω. With Eq. (16-27),

$$f_c = \frac{1}{2\pi(25 + 1000/100)10^{-4}} \cong 45 \text{ Hz}$$

EXAMPLE 16-14.
A FET amplifier has an R_S of 1 kΩ, a g_m of 5000 μmhos, and a C_S of 10 μF. Calculate the break frequency.

SOLUTION.
The reciprocal of g_m is 1/0.005, or 200 Ω. With Eq. (16-28),

$$f_c = \frac{1}{2\pi(1000||200)10(10^{-6})} \cong 96 \text{ Hz}$$

Problems

16-1. The capacitance C in a lag network equals 100 pF. Calculate the critical frequency for each of these values of R: 100 Ω, 10 kΩ, and 1 MΩ.

16-2. Calculate the critical frequency for the lag network of Fig. 16-24a.

16-3. Work out the critical frequency for the circuit shown in Fig. 16-24b.

16-4. Figure 16-24c shows a voltage divider with 1-MΩ resistors. The 1-pF capacitance represents the end-to-end capacitance of the megohm resistor on the right. What critical frequency does the lag network have?

16-5. The FET of Fig. 16-25a has a g_m of 8000 μmhos. What is the voltage gain for low frequencies where X_C is negligible? At what frequency is the voltage gain down 3 dB? At what frequency is the phase shift of the lag network equal to $-45°$? What is the total phase shift from gate to drain at the critical frequency?

16-6. The g_m of the FET in Fig. 16-25b has a value of 5000 μmhos. If R_D equals 5 kΩ, what is the voltage gain for an input sine wave whose frequency is 10 kHz? What is the critical frequency?

16-7. The FET in Fig. 16-25b has a g_m of 5000 μmhos. Calculate the voltage gain and critical frequency for each of these:

1. $R_D = 1$ kΩ
2. $R_D = 2$ kΩ
3. $R_D = 4$ kΩ

Is the voltage gain directly or inversely proportional to the critical frequency?

16-8. Sketch the ideal Bode plot of bel voltage gain for the amplifier of Fig. 16-25a using a g_m of 10,000 μmhos. Also show the ideal Bode plot of phase angle.

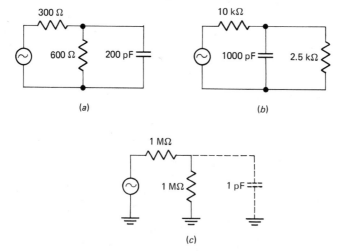

Figure 16-24.

(a)

(b)

(c)

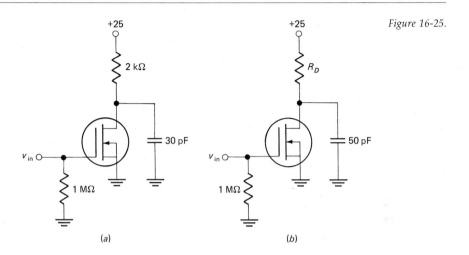

Figure 16-25.

16-9.   Figure 16-26a represents the collector circuit of a transistor. Calculate the break frequency of this lag network. What happens to the break frequency if we double the resistance?

16-10. The 40 pF in Fig. 16-26a is part of the circuit. If we connect an oscilloscope with an input capacitance of 50 pF across the 500-Ω resistor, what will the new critical frequency be? What is the critical frequency of the circuit when the oscilloscope is disconnected?

16-11. In Fig. 16-26b the only frequency effect we will consider is the collector lag network made up of r_C and the 10-pF capacitance. If r'_e is 25 Ω, what is the voltage gain v_c/v_b and the critical frequency for each of these:

1. $r_C = 1$ kΩ
2. $r_C = 2$ kΩ
3. $r_C = 10$ kΩ

Which of these r_C values produces the maximum voltage gain? Which is the one with the highest critical frequency?

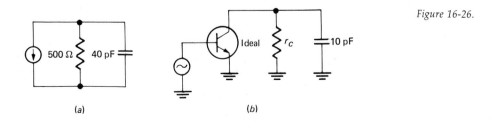

Figure 16-26.

Figure 16-27.

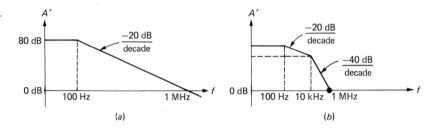

16-12. Sketch the ideal Bode plot of phase angle and bel voltage gain for the collector lag network of Fig. 16-26b with these values: $r'_e = 25 \ \Omega$ and $r_c = 5 \ k\Omega$.

16-13. For the Bode plot of Fig. 16-27a what is the midband bel voltage gain? What is the bandwidth value? What is the ideal bel voltage gain at a frequency of 10 kHz? What is the phase angle at 1 MHz?

16-14. Figure 16-27b shows an ideal Bode plot of bel voltage gain. At 10 kHz the gain breaks and then drops off 40 dB per decade. What is the ideal bel voltage gain at 10 kHz? At 100 Hz?

16-15. Given a Bode plot like Fig. 16-27b, what is the phase angle at 100 Hz? Phase angles are additive; what is the phase angle at 10 kHz? At 1 MHz?

16-16. An *octave* is a *factor of 2* in frequency, similar to a decade being a factor of 10. Saying the frequency increases an octave is equivalent to saying it doubles. Use Eq. (16-10) to prove the bel voltage gain drops 6 dB per octave when $f \gg f_c$. (Instead of 20 dB per decade, some people use the equivalent rate of 6 dB per octave.)

16-17. The amplifier of Fig. 16-28a has an A_{mid} of 100. If V_{in} equals 20 mV, what is the output voltage at the 100 percent point? The 90 percent point? What is the upper break frequency of the amplifier?

Figure 16-28.

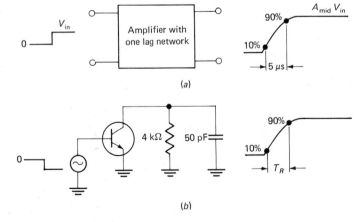

16-18. The negative input step voltage produces a positive-going output in the ac equivalent circuit of Fig. 16-28b. The only frequency effect to consider is the lag network in the collector. What risetime does the output waveform have?

16-19. A dc amplifier has a bel voltage gain of 60 dB and a break frequency of 10 kHz. If the bel voltage gain drops off 20 dB per decade until it equals 0 dB, what is the risetime of the output when a step input is used? If the dc amplifier is modified to get a bel voltage gain of 40 dB and a break frequency of 100 kHz, what will the new risetime be?

16-20. You have two data sheets for different dc amplifiers. The first shows a break frequency of 1 MHz. The second lists a risetime of 1 μs. Which amplifier has the greater bandwidth?

16-21. The data sheet for a 2N3797 lists these maximum values of capacitance: $C_{iss} = 8$ pF and $C_{rss} = 0.8$ pF. Calculate C_{gd} and C_{gs}.

16-22. The FET of Fig. 16-29 has these capacitances: C_{gs} is 6 pF, C_{gd} is 4 pF, and C_{ds} is 1 pF. Neglect all other capacitances. Calculate the critical frequencies of the gate and drain lag networks.

16-23. Same data as in the preceding problem, but add a stray wiring capacitance of 4 pF across the 6-kΩ resistor. Work out the critical frequencies.

16-24. Suppose we want the amplifier of Fig. 16-29 to have an f_2 break frequency of 20 kHz. One way to get this is to shunt a capacitor across the 6-kΩ resistor. Neglect all other capacitances and calculate the required C.

16-25. A 2N3300 has an f_T of 250 MHz for an I_E of 50 mA. Calculate the value of C'_e for this value of emitter current.

16-26. The bipolar transistor of Fig. 16-30 has these values: $r'_b = 50$ Ω, $\beta = 80$, $r'_e = 2.5$ Ω, $f_T = 200$ MHz, $C_{ob} = 3$ pF, and $C_{in} = 50$ pF. What is the break frequency of the base lag network? Of the collector lag network?

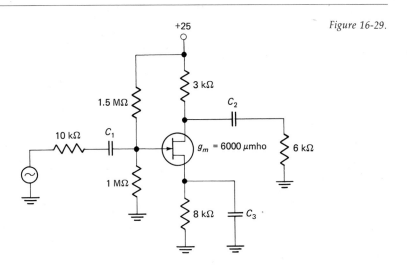

Figure 16-29.

Figure 16-30.

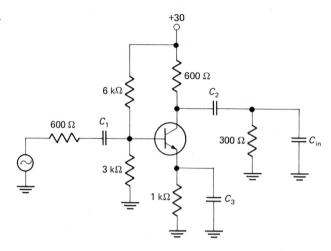

16-27. The beta-cutoff frequency f_β is the critical frequency of the base lag network in Fig. 16-18e for the special conditions of $r_C = 0$ and $R_S = \infty$. Substitute these values and determine the formula for f_β. After you have found f_β, use Eq. (16-21a) to get a relation between f_β and f_T.

16-28. The bipolar transistor of Fig. 16-30 has these values: $r_b' = 75\ \Omega$, $\beta = 20$, $r_e' = 1\ \Omega$, $f_T = 500$ MHz, $C_{ob} = 2$ pF, and $C_{in} = 100$ pF. Calculate the break frequencies for the base and collector lag networks.

16-29. In Fig. 16-29, $C_1 = 1\ \mu F$ and $C_2 = 2\ \mu F$. If the bypass capacitor looks like an ac short, what are the break frequencies of the input and output lead networks?

16-30. $C_1 = 1\ \mu F$ and $C_2 = 4\ \mu F$ in Fig. 16-30. The bypass capacitor acts like an ac short. Calculate the critical frequencies of the base and collector lead networks for a $\beta r_e'$ of 300 Ω.

16-31. An ac amplifier has a sagtime of 2 ms. If it has one dominant lead network, what is the break frequency of this network?

16-32. The coupling capacitors of Fig. 16-29 look like ac shorts. If C_3 equals 10 μF, what is the break frequency of the bypass network?

16-33. C_3 equals 200 μF in Fig. 16-30. If the coupling capacitors act like ac shorts, what break frequency does the bypass capacitor produce for a β of 50 and an r_e' of 2.5 Ω?

17. Integrated Circuits

As mentioned in Chap. 3, a manufacturer can produce an integrated circuit (IC) on a chip. ICs eliminate much of the drudgery in electronics work; you can find an IC for almost any routine application.

17-1. MAKING AN IC

Though we are mainly interested in using ICs, we should know something about how they are made.[1]

The p substrate

First, the manufacturer produces a p crystal several inches long and 1 to 2 in in diameter (Fig. 17-1a). He slices this into many thin *wafers* like Fig. 17-1b. One side of the wafer is lapped and polished to get rid of surface imperfections. This wafer is called the *p substrate*; it will be used as a chassis for the integrated components.

The epitaxial n layer

Next, the wafers are put in a furnace. A gas mixture of silicon atoms and pentavalent atoms passes over the wafers. This forms a thin layer of n-type semiconductor

---

[1] Two excellent books covering more details than we have time for are Deboo, G. J., and C. N. Burrous: *Integrated Circuits and Semiconductor Devices*, McGraw-Hill Book Company, New York, 1971, pp. 91–142. Stern, L.: *Fundamentals of Integrated Circuits*, Hayden Book Company, New York, 1968, pp. 56–115.

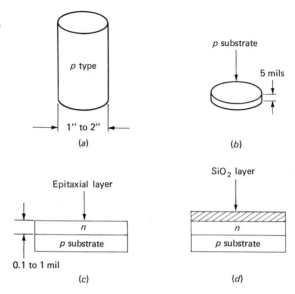

Figure 17-1. (a)p crystal. (b) Wafer. (c) Epitaxial layer. (d) Insulating layer.

on the heated surface of the substrate (see Fig. 17-1c). We call this thin layer an *epitaxial layer*. As shown in Fig. 17-1c, the epitaxial layer is about 0.1 to 1 mil thick.

The insulating layer

To prevent contamination of the epitaxial layer, pure oxygen is blown over the surface. The oxygen atoms combine with the silicon atoms to form the layer of silicon dioxide (SiO_2) shown in Fig. 17-1d. This glasslike layer of SiO_2 seals off the surface and prevents further chemical reactions; sealing off the surface like this is known as *passivation*.

Chips

Visualize the wafer subdivided into the areas shown in Fig. 17-2. Each of these areas will be a separate chip after the wafer is cut. But before the wafer is cut into chips, the manufacturer produces hundreds of circuits on the wafer, one on each area of Fig. 17-2. This simultaneous mass production is the reason for the low cost of integrated circuits.

Forming a transistor

Here is how an integrated transistor is formed. Part of the SiO_2 layer is etched off, exposing the epitaxial layer (see Fig. 17-3a). The wafer is then put into a furnace and

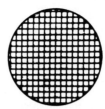

Figure 17-2. Cutting the wafer into chips.

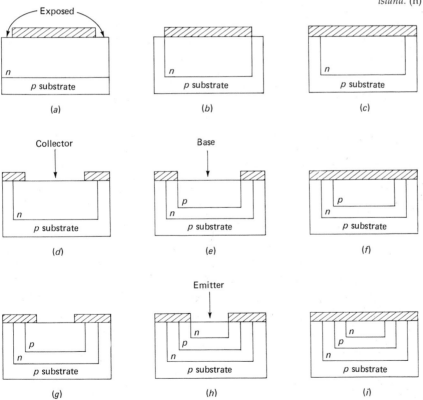

Figure 17-3. Steps in making an integrated transistor. (a) Exposed n layer. (b) Forming an n island. (c) Passivation. (d) Etching a window. (e) Forming the base. (f) Passivation. (g) Window over p island. (h) Forming the emitter. (i) Passivation.

trivalent atoms are diffused into the epitaxial layer. The concentration of trivalent atoms is enough to change the exposed epitaxial layer from n-type semiconductor to p-type. Therefore, we get an *island* of n material under the SiO_2 layer (Fig. 17-3b).

Oxygen is again blown over the wafer to form the complete SiO_2 layer shown in Fig. 17-3c. A hole is now etched in the center of the SiO_2 layer; this exposes the n-epitaxial layer (Fig. 17-3d). The hole in the SiO_2 layer is called a *window*. We are now looking down at what will be the collector of the transistor.

To get the base, we pass trivalent atoms through the window; these impurities diffuse into the epitaxial layer and form an island of p-type material (Fig. 17-3e). Then, the SiO_2 layer is reformed by passing oxygen over the wafer (Fig. 17-3f).

To form the emitter, we etch a window in the SiO_2 layer and expose the p island (Fig. 17-3g). By diffusing pentavalent atoms into the p island, we can form the small n island shown in Fig. 17-3h. We then passivate the structure by blowing oxygen over the wafer (Fig. 17-3i).

IC components

By etching windows in the SiO_2 layer, we can deposit metal to make electrical contact with the emitter, base, and collector. In this way, we have the integrated transistor of Fig. 17-4a.

To get a diode, we follow the same steps for a transistor up to the point where the p island has been formed and sealed off (Fig. 17-3f). Then, we etch windows to expose the p and n islands. By depositing metal through these windows, we make electrical contact with the cathode and anode of the integrated diode (see Fig. 17-4b).

By etching two windows above the p island of Fig. 17-3f, we can make metallic contact with the p island; this gives us an integrated resistor (Fig. 17-4c).

Transistors, diodes, and resistors are easy to integrate on a chip. For this reason, almost all integrated circuits use these components. Inductors and large capacitors are not practical to integrate on the surface of a chip.

Figure 17-4. Integrated components. (a)*Transistor.* (b) *Diode.* (c) *Resistor.*

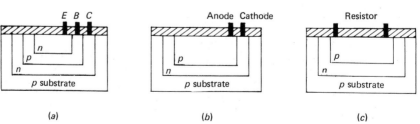

A simple example

To give you an idea of how a circuit is produced, look at the simple three-component circuit of Fig. 17-5a. To integrate this, we would simultaneously produce hundreds of circuits like this on a wafer. Each chip area would resemble Fig. 17-5b. The diode and resistor would be formed at the point mentioned earlier. At a later step, the emitter of the transistor would be formed. Then, we would etch windows and deposit metal to connect the diode, transistor, and resistor as shown in Fig. 17-5b.

Regardless of how complicated a circuit may be, it is mainly a process of etching windows, forming *p* and *n* islands, and connecting the integrated components.

The *p* substrate *isolates* the integrated components from each other. In Fig. 17-5b, depletion layers exist between the *p* substrate and the three *n* islands touching it. Because the depletion layers have essentially no current carriers, the integrated components are insulated from each other. This kind of insulation is known as *depletion-layer* or *diode isolation*.

Monolithic ICs

The integrated circuits we have described are called *monolithic* ICs. The word "monolithic" is from Greek and means "one stone." The word is apt because the integrated components are atomically part of one chip.

Monolithic ICs are by far the most common. But there are other kinds. *Thin-film* and *thick-film* ICs are larger than monolithic ICs but smaller than discrete circuits. With a thin- or thick-film IC, the passive components like resistors and capacitors are integrated simultaneously on a substrate. Then, *discrete* active components like tran-

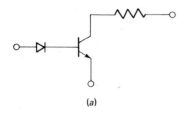

(a)

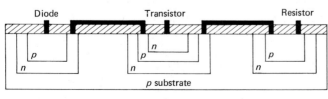

(b)

Figure 17-5. A simple integrated circuit. (a) Schematic diagram. (b) Integrated form.

sistors and diodes are connected to form a complete circuit. Therefore, commercially available thin- and thick-film circuits are combinations of integrated and discrete components.

Hybrid ICs combine two or more monolithic ICs in one package, or they may combine monolithic ICs with thin- or thick-film circuits.

SSI, MSI, and LSI

Figure 17-5*b* is an example of *small-scale integration* (SSI); only a few components have been integrated to form the complete circuit. As a guide, SSI refers to ICs with less than 12 integrated components.[2]

Medium-scale integration (MSI) refers to ICs that have from 12 to 100 integrated components per chip. *Large-scale integration* (LSI) refers to more than a hundred components. As mentioned earlier, it takes fewer steps to make an integrated MOSFET. Furthermore, a manufacturer can produce more MOSFETs on a chip than bipolar transistors. For this reason, MOS/LSI has become the largest segment of the LSI market.

17-2. THE DIFFERENTIAL AMPLIFIER

Transistors, diodes, and resistors are the only practical components in a monolithic IC. Capacitances have been integrated but usually are less than 50 pF. Therefore, IC designers cannot use coupling and bypass capacitors like a discrete-circuit designer. Instead, the stages of a monolithic IC have to be direct-coupled.

The *differential amplifier* (diff amp) is heavily used in linear ICs. A diff amp uses no coupling or bypass capacitors; all it requires are resistors and transistors, both easily integrated on a chip.

Emitter bias

Figure 17-6*a* shows the basic form of a differential amplifier (sometimes called a *difference amplifier*). There are two input signals and one output signal. Ideally, the circuit is symmetrical; each half is identical to the other half. With monolithic ICs we can approach this ideal symmetry because the integrated components are on the same chip and have almost identical characteristics.

How much dc emitter current is there in a diff amp? After shorting the ac sources as shown in Fig. 17-6*b*, we can split R_E into separate resistances of $2R_E$. This does not change the dc emitter current in either transistor because the parallel resistance still equals R_E. If the two halves are identical, no current can flow in the horizontal wire between emitters. Therefore, opening the wire cannot disturb the dc emitter currents (see Fig. 17-6*c*).

---

[2] The definitions of SSI, MSI, and LSI are often given in terms of *gates* rather than integrated components. Until you know what a digital gate is, you can use the number of integrated components.

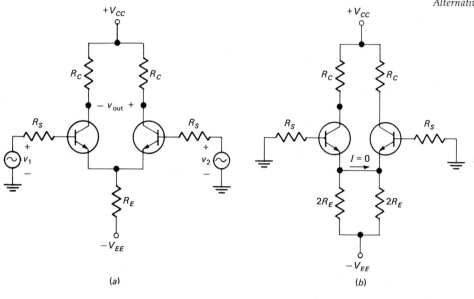

(a)

(b)

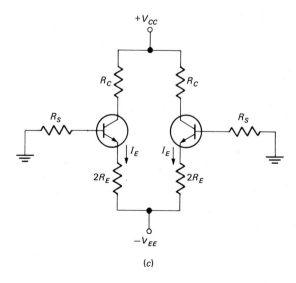

(c)

Now it is clear the dc equivalent circuit of a diff amp ideally is the same as two separate emitter-biased transistors. With Eq. (7-6),

$$I_E \cong \frac{V_{EE} - 0.7}{2R_E + R_S/\beta_{dc}} \qquad \text{(second)} \qquad\qquad (17\text{-}1)$$

This is the value of dc emitter current in each transistor. Often, the 0.7 V and the R_S/β_{dc} are negligible and we can use the ideal formula

$$I_E \cong \frac{V_{EE}}{2R_E} \qquad \text{(ideal)} \qquad\qquad (17\text{-}2)$$

An alternative viewpoint

Another way to visualize the dc action in Fig. 17-6a is this. Since each transistor is emitter-biased, the dc voltage from each emitter to ground is approximately zero. Therefore, almost all of the V_{EE} voltage is across R_E. Since both dc emitter currents flow down through R_E, it follows that

$$2I_E \cong \frac{V_{EE}}{R_E}$$

or

$$I_E \cong \frac{1}{2}\frac{V_{EE}}{R_E} \qquad \text{(ideal)} \qquad\qquad (17\text{-}2a)^{***}$$

For ideal analysis, therefore, visualize all of V_{EE} across R_E; the resulting current is shared equally by the transistors.

EXAMPLE 17-1.
Calculate the dc emitter current in each transistor of Fig. 17-7a.

SOLUTION.
The emitters are essentially at dc ground. Therefore, the total current in the common resistor is

$$2I_E \cong \frac{20}{10,000} = 2 \text{ mA}$$

The transistors share this current so that each has an emitter current of

$$I_E \cong 1 \text{ mA}$$

This is the ideal value of emitter current.

To a second approximation, the emitter-to-ground voltage is not zero. With Eq. (17-1),

$$I_E \cong \frac{20 - 0.7}{20,000 + 1000/100} = 0.965 \text{ mA}$$

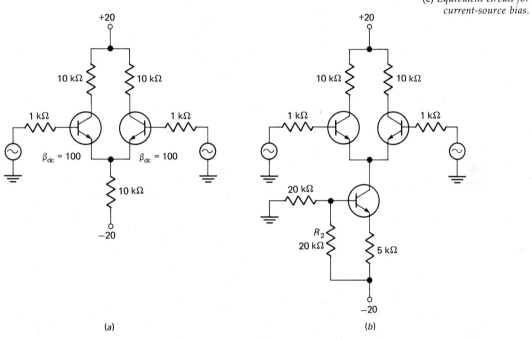

(a)

(b)

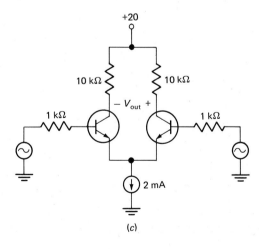

(c)

The larger we can make V_{EE} and R_E, the less important the 0.7 V and R_S/β_{dc}, and the more accurate the ideal formula.

EXAMPLE 17-2.
We can also use current-source bias to set up the dc emitter current as shown in Fig. 17-7b. Calculate the dc currents.

SOLUTION.
The bottom transistor uses voltage-divider bias. The voltage across R_2 is approximately 10 V. Most of this 10 V appears across the 5-kΩ resistor and produces a dc collector current of about

$$I_C \cong \frac{10}{5000} = 2 \text{ mA}$$

Because the two halves of the diff amp are identical, the current splits so that each gets 1 mA.

Since the lower transistor acts like a current source, we may draw the equivalent circuit of Fig. 17-7c; this is simpler and emphasizes that a constant current is being forced through the diff-amp transistors. Current-source bias is used a great deal in integrated circuits.

EXAMPLE 17-3.
Calculate the dc collector-to-ground voltages in Fig. 17-7c. Also, work out the dc voltage between collectors.

SOLUTION.
In Fig. 17-7c each transistor has a dc collector-to-ground voltage of

$$V_C = V_{CC} - I_C R_C \cong 20 - 0.001(10{,}000) = 10 \text{ V}$$

This is what you measure with a dc voltmeter between each collector and ground.

The dc output voltage in Fig. 17-7c equals the difference of two equal collector-to-ground voltages; therefore,

$$V_{\text{out}} = 0 \text{ V dc}$$

17-3.   AC ANALYSIS OF A DIFF AMP

Now that we have the main idea of dc action, let's find out what an ac signal does. In Fig. 17-8a, the output is taken between collectors. If the two halves of the diff amp are identical, the output voltage is zero when v_1 and v_2 are zero. If v_1 and v_2 change by exactly the same amount, the output voltage remains zero because of the symmetry. Only when there is a difference between v_1 and v_2 do we get an output voltage. When v_1 is more positive than v_2, more collector current flows through the transistor on the left; because of this, the output voltage has the polarity shown in Fig. 17-8a.

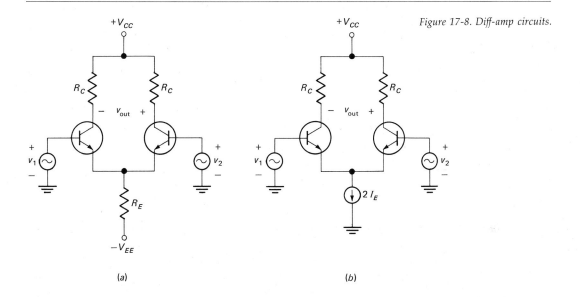

Figure 17-8. Diff-amp circuits.

(a)

(b)

A simple viewpoint of a diff amp is to visualize it as a bridge circuit. R_C and the left transistor are one side of the bridge; the other R_C and transistor are the right side. If the two input voltages are equal, the bridge is balanced and we get no output voltage. If the two input voltages are different, the bridge is unbalanced and we get an output voltage.

Output with only one input

For the moment, treat the source impedances as negligibly small (Fig. 17-8a). In a good diff-amp design, V_{EE} and R_E set up an almost constant current and we can visualize the diff amp by the equivalent circuit of Fig. 17-8b. If current-source bias is used, Fig. 17-8b becomes a highly accurate equivalent circuit.

To get the ac equivalent circuit, reduce all dc sources to zero; this means ac grounding the V_{CC} point and opening the $2I_E$ current source. Figure 17-9a shows the resulting circuit. The simplest way to analyze this ac circuit is with the superposition theorem. For small signals the operation is linear and we can calculate the effect of each source. The sum of the effects is the total effect.

For instance, by reducing v_2 to zero, we can calculate the output voltage produced by v_1 alone (Fig. 17-9b). Since the base of the second transistor is ac grounded, this transistor acts like an emitter-driven stage or CB connection. As derived in Sec. 9-10, the voltage gain of a CB stage is

$$\frac{v_c}{v_e} \cong \frac{r_C}{r'_e}$$

Figure 17-9. (a) Ac equivalent circuit of diff amp. (b) First step in superposition theorem. (c) Ideal ac equivalent circuit.

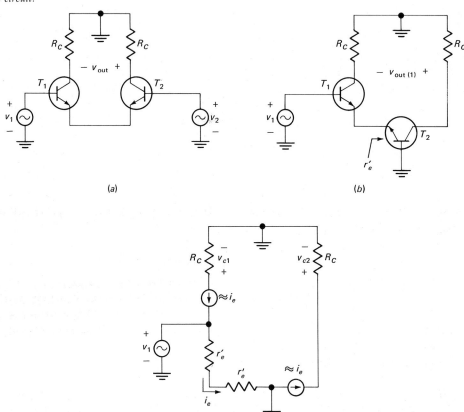

(a)

(b)

(c)

and the input impedance is

$$z_{in} \cong r'_e$$

To see how each r'_e affects the circuit action, we can replace Fig. 17-9*b* by 17-9*c*; the r'_e of one transistor is in series with the other. Furthermore, the ac input voltage is across both and sets up an ac emitter current of

$$i_e \cong \frac{v_1}{2r'_e}$$

In Fig. 17-9*c* the ac collector current of T_1 flows down, but the ac collector current of

T_2 flows up. The ac collector-to-ground voltage of T_2 therefore equals

$$v_{c2} \cong i_e R_C \cong \frac{v_1}{2r'_e} R_C$$

or

$$v_{c2} \cong \frac{R_C}{r'_e} \frac{v_1}{2}$$

This is a positive voltage with respect to ground during the positive half cycle of v_1.

The ac collector-to-ground voltage of T_1 is negative because of the complementary drop across R_C. Therefore, in Fig. 17-9c,

$$v_{c1} \cong -i_e R_C \cong -\frac{R_C}{r'_e} \frac{v_1}{2}$$

This is a negative voltage during the positive half cycle of v_1.

Individual outputs

The output voltage between collectors in Fig. 17-9b is the difference of v_{c2} and v_{c1}. This ac output voltage equals

$$v_{\text{out}(1)} = v_{c2} - v_{c1} \cong \frac{R_C}{r'_e} \frac{v_1}{2} - \left(-\frac{R_C}{r'_e} \frac{v_1}{2}\right)$$

or

$$v_{\text{out}(1)} \cong \frac{R_C}{r'_e} v_1 \tag{17-3a}$$

or

$$v_{\text{out}(1)} = A v_1$$

where A approximately equals R_C/r'_e. Voltage $v_{\text{out}(1)}$ is the ac voltage produced by v_1 alone.

In a similar way, we can reduce v_1 to zero and determine the output voltage produced by v_2 alone. When we do this, we get

$$v_{\text{out}(2)} \cong -\frac{R_C}{r'_e} v_2 \tag{17-3b}$$

or with A approximately equal to R_C/r'_e,

$$v_{\text{out}(2)} = -A v_2$$

This is the ac output voltage produced by v_2 alone. The minus sign means this component has the opposite polarity from that of Fig. 17-9a.

Superposed output

The total output voltage when *both* input signals *are present* simultaneously is the sum of the individual outputs:

$$v_{\text{out}} = v_{\text{out}(1)} + v_{\text{out}(2)}$$

or
$$v_{\text{out}} = Av_1 - Av_2 \quad \text{(two inputs)} \qquad (17\text{-}4)^{***}$$

where A approximately equals R_C/r_e'. Therefore, when we want the ac voltage out of a diff amp, we multiply each input voltage by A and take the difference.

Input impedance

Each signal acting alone sees an input impedance. For instance, in Fig. 17-9c the v_1 source sees $2r_e'$ in the emitter circuit. As usual, this emitter impedance is stepped up by the β factor, so that

$$z_{\text{in}} \cong 2\beta r_e' \quad \text{(ideal)} \qquad (17\text{-}5)^{***}$$

Similarly, when v_2 is acting alone, it sees an input impedance of $2\beta r_e'$.

When both signals are present simultaneously, the input impedance can be different from Eq. (17-5). We discuss this in the next chapter.

Inverting and noninverting inputs

In the diff amp of Fig. 17-10a the output voltage can drive another stage, possibly a diff amp. As we saw, a positive v_1 acting alone produces a positive output voltage. For this reason, the v_1 input is the *noninverting input*. On the other hand, a positive v_2 acting alone produces a negative output voltage. This is why the v_2 input is called the *inverting* input.

Differential input

Besides not having coupling and bypass capacitors, the diff amp has other advantages. If we like, we can ground one input and drive the other input. This special case is called *single-ended input* (similar to Fig. 17-9b).

Alternatively, we can drive the diff amp with a signal between the bases as shown in Fig. 17-10b. This is known as *double-ended* or *differential input*. When we have a differential input, v_{in} is the difference of v_1 and v_2. In symbols,

$$v_{\text{in}} = v_1 - v_2$$

With this viewpoint, Eq. (17-4) becomes

$$v_{\text{out}} = Av_1 - Av_2 = A(v_1 - v_2)$$

or
$$v_{\text{out}} = Av_{\text{in}} \quad \text{(Differential input)} \qquad (17\text{-}6)^{***}$$

where A equals R_C/r_e'.

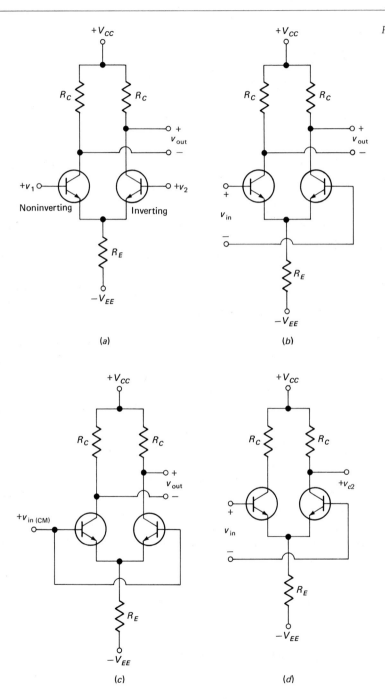

Figure 17-10. Ways to operate a diff amp. (a) Two inputs. (b) Differential input. (c) Common-mode input. (d) Single-ended output.

The point is this. If we have two separate signal sources, we think of the output voltage as $Av_1 - Av_2$. On the other hand, if we have only one signal source driving the diff amp in the differential mode, we think of the output as Av_{in}.

Common-mode input

Figure 17-10c illustrates the *common-mode input;* the same signal is applied to both inputs. If each half of the diff amp is identical, the ac output voltage will equal zero. About the only time we deliberately use a common-mode input signal is when we are testing the diff amp to see how well balanced the two halves are.

The *common-mode rejection ratio* is defined as

$$CM_{rej} = \frac{Av_{in(CM)}}{v_{out(CM)}} \tag{17-7a}$$

where the numerator is calculated and the denominator is measured. As an example, suppose $v_{in(CM)}$ is 1 V in Fig. 17-10c. Ideally, we should get nothing out, but there may be a small output signal because of nonsymmetry. Suppose A is 100 and $v_{out(CM)}$ is 0.01 V. Then, with Eq. (17-7a),

$$CM_{rej} = \frac{100(1\text{ V})}{0.01\text{ V}} = 10{,}000$$

On a data sheet, CM_{rej} is given in decibels. So, in this particular example,

$$CM'_{rej} = 2\log 10{,}000 = 8\text{ B} = 80\text{ dB}$$

The larger the value of CM_{rej}, the better the diff amp. Ideally, a common-mode input should produce zero output voltage; therefore, the ideal diff amp has a CM_{rej} of infinity.

We can rearrange Eq. (17-7a) to get

$$v_{out(CM)} = \frac{Av_{in(CM)}}{CM_{rej}} \tag{17-7b}$$

With this equation we can calculate how much common-mode output voltage occurs for common-mode input signals. Most forms of interference, static, induced voltages, etc., drive a diff amp in the common mode. A good diff amp has a large CM_{rej} value, which means it's virtually free of interference signals.

Single-ended output

Most linear ICs have one output pin that carries the signal with respect to ground. For this reason, a diff amp with a *single-ended output* (Fig. 17-10d) is often used in the later stages of a linear IC. Since only half the available output voltage is used, the voltage gain drops in half:

$$v_{c2} = \frac{A}{2} v_{\text{in}} \tag{17-8}$$

where A equals R_C/r'_e.

EXAMPLE 17-4.
Calculate the approximate output voltage for each diff amp in Fig. 17-11.

SOLUTION.
In all diff amps,

$$\frac{V_{EE}}{R_E} \cong \frac{10}{5000} = 2 \text{ mA}$$

Therefore, each transistor has a dc emitter current of about 1 mA; this means an r'_e of approximately 25 Ω. The ideal voltage gain is

$$A \cong \frac{R_C}{r'_e} = \frac{10{,}000}{25} = 400$$

In Fig. 17-11*a* we are using a single-ended noninverting input. Since v_2 is zero, Eq. (17-4) gives

$$v_{\text{out}} = Av_1 \cong 400(1 \text{ mV}) = 400 \text{ mV}$$

This ac output signal is in phase with the input signal.

We are applying a differential input signal in Fig. 17-11*b*. Therefore, using Eq. (17-6), we get

$$v_{\text{out}} = Av_{\text{in}} \cong 400(1 \text{ mV}) = 400 \text{ mV}$$

In Fig. 17-11*c* we still apply the input signal in the differential mode, but now we use a single-ended output. With Eq. (17-8),

$$v_{\text{out}} = \frac{A}{2} v_{\text{in}} \cong \frac{400}{2} 1 \text{ mV} = 200 \text{ mV}$$

EXAMPLE 17-5.
Suppose you connect the output of Fig. 17-11*a* to the input of Fig. 17-11*b*. Then, the double-ended output of the first diff amp is the differential input to the second diff amp. Not only will you get the desired input signal to the second diff amp, you also will get the power-supply ripple on each collector. This ripple goes into the second diff amp in the common mode.

1. If the differential input signal to the second diff amp is 1 mV, how much output voltage is there?
2. The ripple into each base of the second diff amp equals 1 mV. How much output ripple is there if CM_{rej} equals 10,000?

Figure 17-11. Examples 17-4 and 17-5.

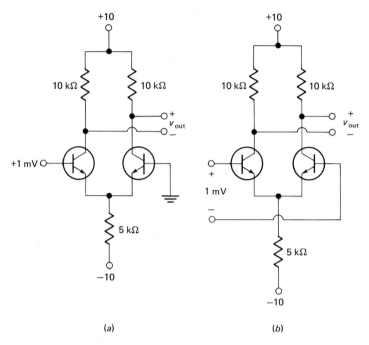

(a)   (b)

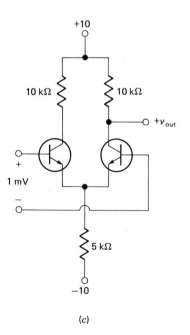

(c)

3. Calculate the signal-to-ripple ratio at the input to the second diff amp and at the output.

SOLUTION.

1. We already know the voltage gain is 400, found in the preceding example. Therefore, the output voltage of the second diff amp is

$$v_{out} = Av_{in} = 400(1 \text{ mV}) = 400 \text{ mV}$$

2. With Eq. (17-7b),

$$v_{out(CM)} = \frac{Av_{in(CM)}}{CM_{rej}} = \frac{400(1 \text{ mV})}{10,000} = 0.04 \text{ mV}$$

3. At the input to the second diff amp, the signal-to-ripple ratio is

$$\frac{v_s}{v_r} = \frac{1 \text{ mV}}{1 \text{ mV}} = 1 \qquad \text{(input)}$$

At the output of the second diff amp, the ratio is

$$\frac{v_s}{v_r} = \frac{400 \text{ mV}}{0.04 \text{ mV}} = 10,000 \qquad \text{(output)}$$

This final result shows how effectively the diff amp discriminates against ripple in the power supply. The input signal and ripple are equal going into the second diff amp, but the output signal is 10,000 times greater than the output ripple.

As mentioned, most forms of interference are applied with equal intensity to both sides of the diff amp, that is, in the common mode. For instance, externally produced noise from electric motors, neon signs, ignition systems, lighting, etc., induces common-mode voltages. The signal-to-noise ratio (common-mode noise) will therefore be improved by a factor of CM_{rej}.

17-4. CASCADED DIFF AMPS

Figure 17-12 will give you an idea of how diff amps can be cascaded. The first diff amp has noninverting and inverting inputs. We can drive this diff amp with two separate single-ended inputs or with one differential input. The second diff amp uses Darlington pairs to prevent excessive loading of the first diff-amp stage. The single-ended output of the second diff amp drives a Darlington emitter follower which produces the final output voltage.

Because no coupling or bypass capacitors are used, Fig. 17-12 is a dc amplifier. As a result, Fig. 17-12 can amplify signals with frequencies all the way down to zero. The midband voltage gain of the circuit is high, better than 10,000, because each diff amp has a voltage gain over 100.

Figure 17-12. Cascaded diff amps.

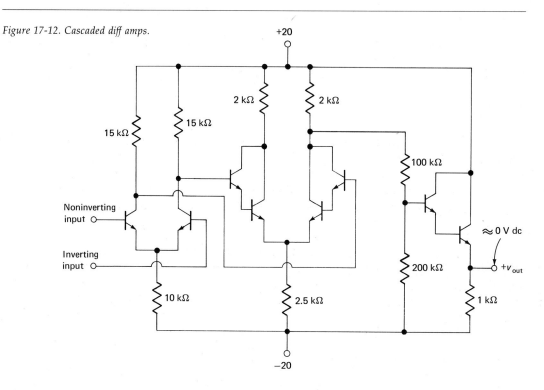

Schematic symbol

Given an amplifier like Fig. 17-12, we can save time by using a schematic symbol for the entire amplifier. Figure 17-13*a* shows a simple way to represent an amplifier with two inputs and one output. *A* is the unloaded voltage gain; this is the gain we get when no load resistor is used, or equivalently, the voltage gain when R_L is much greater than the Thevenin output impedance of the amplifier. The input and output voltages are with respect to the ground line.

Figure 17-13. Schematic symbols.

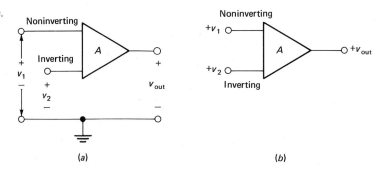

(a)   (b)

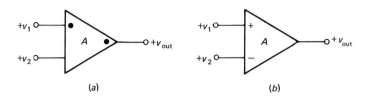

Figure 17-14. (a) *Dot convention.* (b) *The most common symbol.*

(a)   (b)

Most of the time we don't bother drawing the ground line; we simply draw the schematic symbol of an amplifier as shown in Fig. 17-13b. Whenever we see this symbol, we must remember the voltages are with respect to ground.

Figure 17-14a is a variation of the amplifier symbol. Instead of labeling the inputs as *noninverting* and *inverting,* we can use the dot convention; the dot on the upper input terminal and the dot on the output terminal mean these pins carry in-phase signal voltages.

Figure 17-14b shows the most widely used symbol. A positive input signal to the noninverting input produces a positive output voltage. For this reason, we mark the *noninverting input* with a *plus sign*. On the other hand, a positive input signal on the inverting input produces a negative output voltage; this is why the inverting input is marked with a negative sign.

Input impedance and Thevenin circuit

In Fig. 17-12 the input impedance equals $2\beta r'_e$ for a single-ended or differential input signal. By using Darlington pairs or FETs, we can step up this input impedance to high values. In any case, an amplifier has a z_{in} between its terminals as shown in Fig. 17-15.

Theoretically, we can Thevenize the output of any linear circuit. Therefore, we can

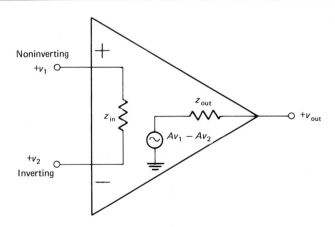

Figure 17-15. Input impedance and output Thevenin circuit.

visualize the output side of the amplifier by a Thevenin voltage source of

$$v_{TH} = Av_1 - Av_2$$

and by the Thevenin impedance of z_{out} shown in Fig. 17-15.

Figure 17-15 is important because it summarizes important values on data sheets, namely, z_{in}, z_{out}, and A. We need these values to analyze amplifier action under loaded conditions. For convenience, we don't normally show z_{in}, z_{out}, and A in schematic diagrams. Nevertheless, whenever we see a symbol like Fig. 17-14b, we should remember there is a z_{in} between the two input terminals and a Thevenin circuit looking back into the output terminal; the Thevenin output voltage is $Av_1 - Av_2$, and the Thevenin output impedance is z_{out}.

EXAMPLE 17-6.
Calculate the peak output voltage in Fig. 17-16.

SOLUTION.
The input impedance is 500 kΩ. Since the source impedance is only 1 kΩ, essentially all of the 1-mV source signal reaches the noninverting input terminal, that is, $v_1 = 1$ mV.

The Thevenin or unloaded voltage gain is 100. Since there is no inverting input, the Thevenin output voltage is

$$v_{TH} = Av_1 = 100(1 \text{ mV}) = 100 \text{ mV}$$

The Thevenin output impedance is 50 Ω. With a load impedance of 10 kΩ, less than 1 percent of the signal is dropped across z_{out}. Therefore, the peak load voltage is almost equal to 100 mV. This output signal is in phase with the input signal.

17-5. THE OPERATIONAL AMPLIFIER

About a third of all linear ICs are *operational amplifiers* (op amps). An op amp is a high-gain dc amplifier usable from 0 to over 1 MHz. By connecting external resistors to an op amp, you can adjust the voltage gain and bandwidth to your requirements. There are over 2000 types of commercially available op amps. Almost all of these are monolithic

Figure 17-16. Example 17-6.

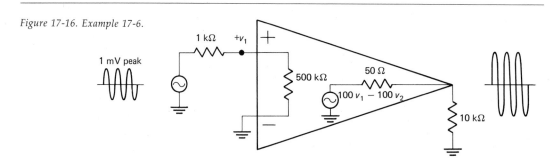

ICs with room-temperature dissipations under a watt. Whenever you need voltage gain, check available op amps. In many cases, an op amp will satisfy your requirements.

If you look at the schematic of a typical op amp, you will find most of the stages are diff amps; this allows operation down to zero frequency and provides common-mode rejection. Most op amps have two inputs and one output like Fig. 17-15.

First-generation op amps

The first quality IC op amp came out in 1965; it was the famous Fairchild μA709. Similar op amps followed: Motorola's MC1709, National Semiconductor's LM709, Texas Instruments' SN72709, and others. The last three digits in each of these types are 709. All these op amps behave the same; that is, they all have the same specifications. For this reason, we will refer to them as the 709.

The 709 family has different models like the 709, 709A, 709B, and 709C. For the same manufacturer, all these have the same schematic diagram, but the tolerances are larger for later members of the family; at the same time, the price of an op amp decreases. For instance, the 709 has the best tolerance and costs the most. At the other extreme, the 709C has the worst tolerance but costs the least.

When you look at the schematic diagram of a 709, be sure to look at the labels on each input terminal; these labels give you a great deal of information. Even better, if the data sheet includes a *pin diagram* like Fig. 17-17a or b, you can learn a lot about the op amp. For instance, pins 2 and 3 are the inverting and noninverting inputs. Pins 4 and 7 are for supply voltages. Pin 6 is the output, while pins 1 and 5 refer to lag networks to be externally connected.

The 709C has these typical values: $z_{in} = 250$ kΩ, $z_{out} = 150$ Ω, and $A = 45,000$ (approximately 93 dB). These values immediately tell us the 709C has a high input impedance, a low output impedance, and a high voltage gain.

Unbalance

We can expect data sheets to list CM_{rej} and other values that indicate the unbalance between halves of the diff amps. For instance, the 709 has a CM'_{rej} of 90 dB, equivalent to a CM_{rej} of approximately 30,000.

Op-amp data sheets list the *input offset voltage;* this is the differential input voltage needed to zero or *null* the dc output voltage. For example, the $V_{in(offset)}$ of a 709C is 7.5 mV; this is the maximum differential input voltage needed to null the output voltage of any 709C.

The *input offset current* is the *difference* between the two input currents for a zero output voltage. The 709C has an input offset current of 0.5 μA; when nulling the output voltage, you will have to provide up to 0.5 μA more current to one input than to the other.

Figure 17-17. Pin diagrams for linear IC amplifier.

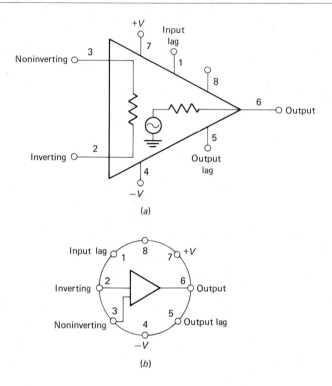

Need for dc return

Most op amps will not work properly unless you connect external dc returns. For instance, the input circuit of Fig. 17-12 has floating bases; it cannot work properly until a dc path is provided from each base to ground. It is the same with the typical op amp; the input bases are floating and need external dc returns.

The signal source driving the op amp provides the dc return unless it is a capacitively coupled source; in such a case, you have to add dc-return resistors (discussed in Sec. 5-13). Either way, the two dc currents of the first op-amp stage must flow through external dc resistances. The two currents are close in value, but not necessarily equal. When these currents flow through external dc returns, they may set up a slight unbalance.

The *input bias current* of an op amp is the average of the two input currents. As a general rule, the smaller the input bias current, the less the unbalance.

Risetime and slew rate

When pulses drive the op amp, you will want to know the risetime. It depends on the external resistors and capacitors you connect to the op amp. The data sheet lists risetimes for several combinations of external resistors and capacitors.

The *slew rate* symbolized dv_{out}/dt tells how fast the output voltage can change when the op amp is driven by a step input. A slew rate of 2 $V/\mu s$ means the output voltage can change no faster than 2 V during each microsecond. The slew rate of a 709C depends on the external resistors and capacitors connected to it; the data sheet shows a slew rate from 0.25 $V/\mu s$ to 12 $V/\mu s$ for different conditions.

There are other specifications on data sheets, most of them self-explanatory, like power dissipation, required supply voltages, and temperature effects.

17-6. BETTER OP AMPS

The 709 typifies the first generation of op amps. The advantages are high input impedance, low output impedance, and high voltage gain. The disadvantages include possible *latch-up,* no *short-circuit protection,* and the need for *external lag networks.* Latch-up means the output voltage can be latched or stuck at some value, regardless of the value of input voltage; latch-up may occur for certain values of common-mode input voltage. No short-circuit protection means that accidentally shorting the output terminals may destroy the op amp; this can easily happen with a 709 and other first-generation op amps. Finally, having to add external lag networks means extra work; ideally, these should be integrated in the op amp.

Second generation

The 741 is typical of second-generation op amps. It has no latch-up problems, includes short-circuit protection, and has its own integrated lag network. Unlike the 709, which has no internal capacitors, the 741 has an integrated 30-pF MOS capacitor. This capacitor is part of a lag network that rolls off the bel voltage gain at a rate of 20 dB per decade. The importance of this is discussed in Chap. 19.

Other improvements in the 741 are its higher input impedance (over a megohm typical), larger voltage gain (200,000 typical), offset nulling with an external 10-kΩ potentiometer, and lower input offset and bias currents.

Because it is inexpensive and easy to use, the 741 has become one of the most widely used op amps.

Third generation

The first- and second-generation op amps have input offset and bias currents in hundreds of nanoamps. These currents change with temperature; therefore, even though an op amp is balanced at one temperature, the output may not remain balanced when the temperature changes. In other words, the output voltage may *drift* slightly with temperature. National Semiconductor's LM101A is an example of a third-generation op amp; the main improvement is the input offset and bias currents. The LM101A has an input offset current of 20 nA max and an input bias current of 0.25 μA max. Other examples of third-generation op amps are Motorola's MC1539 and Sprague's 2139.

Fourth generation

Again, the major improvement occurs in the input offset and bias currents. National Semiconductor's LM108 has an input offset current of 0.2 nA max and an input bias current of 2 nA. These low values are an advantage with high-impedance sources because the voltage drifts with temperature are minimized. Motorola's MC1556 is another example of a fourth-generation op amp.

Along this line of reducing input offset and bias currents, the LM216 may be considered an improved fourth-generation op amp; it has an input offset current of 0.015 nA and an input bias current of 0.05 nA. When small currents like these flow through the input dc-return resistors, only small voltages are produced; therefore, with changing temperature, the output voltage is only slightly unbalanced.

Fifth generation

National Semiconductor's LM118 typifies the fifth generation in op amps. The outstanding features are a frequency range out of 15 MHz and a slew rate of 50 V/μs. The LM118 is internally compensated (has its own lag network), can be offset-nulled with a single potentiometer, and offers many other advantages.

Summary

As mentioned earlier, over 2000 types of op amps exist. Many of these are duplications of the same op amp. For instance, over a dozen manufacturers produce the 741. For a partial list of well-known types, look at Table 17-1. Some of these are dual op amps, two identical op amps in the same package (like the μA749 and the MC1437). Some are even *quad* op amps, four op amps on the same chip (like the LM3900 and the MC3401).

TABLE 17-1. OP-AMP TYPES

Fairchild: μA709, 741, 749, 776, 791
Motorola: MC1437, 1539, 1556, 1558, 1709, 1741, 3401
National: LM101, 101A, 108, 118, 709, 741, 3900
Texas Instruments: SN72709, SN72741

In the next chapter we use op amps to build excellent voltage amplifiers, current amplifiers, and other important circuits. At that time, you will understand why a third of all linear ICs are op amps.

17-7. OTHER LINEAR ICs

This section briefly examines other linear ICs. Our survey is not complete because there are too many special-purpose ICs. What we talk about are the main types.

Audio amplifiers (20 to 20,000 Hz)

Preamplifiers are audio amplifiers with less than 50 mW output power. Preamps are used at the front end of audio systems where they amplify signals from microphones, magnetic pickups, ceramic pickups, etc. Examples of integrated preamps are Motorola's MC1303 and National Semiconductor's LM381.

Medium-level audio amplifiers have output powers from 50 mW to 500 mW. These are useful near the output end of small systems like transistor radios and audio signal generators. A good example is Motorola's MFC4000P with an output power of 250 mW; another example is RCA's CA3020 with an output power of 500 mW.

Audio power amplifiers deliver more than 500 mW of output power; they are used in large radios, television receivers, high-fidelity systems, etc. One example is Motorola's MC1554 with an output power of 1 W. Other examples are Fairchild's μA706 (5 W), National Semiconductor's LM383 (5 W) and LM384 (10 W), etc.

Above 10 W, monolithic ICs have trouble dissipating the internal heat. For this reason, thick-film hybrid ICs may be used. A typical example is the Toshiba TH9013P, which delivers 20 W to an 8-Ω load.

Table 17-2 summarizes some of the available audio ICs.

TABLE 17-2. AUDIO ICs

Fairchild: μA706, 716, 745
Motorola: MC1303, 1306, 1554; MFC4000P, 9000P
National: LM377, 378, 380, 381, 382, 383, 384
RCA: CA3007, 3020, 3048, 3052

Wideband amplifiers (0 to 100 MHz)

A *wideband amplifier* has a flat response (constant bel gain) over a very broad range of frequencies. It also is known as a video amp, broadband amp, or linear-pulse amp. Wideband amps are not necessarily dc amps, but they often do have a response that extends down to zero frequency. They are used in applications where the range of input frequencies is very large. For instance, many oscilloscopes handle frequencies from 0 to over 10 MHz; instruments like these use wideband amps to increase the signal strength before applying it to the cathode-ray tube. As another example, the video section of a television receiver uses a wideband amp that can handle frequencies from near zero to about 4 MHz.

Table 17-3 is a sample of IC wideband amps. Many have voltage gains and bandwidths you can adjust by connecting different external resistors. For instance, the μA702 has a bel voltage gain of 40 dB and a break frequency of 5 MHz; by changing external components, you can get useful gain out of 30 MHz. The MC1553 has a bel voltage gain of 52 dB and a bandwidth of 20 MHz; these are adjustable by changing external components. The μA733 has a very wide bandwidth; it can be set up to give 20 dB gain and a bandwidth of 120 MHz.

TABLE 17-3.   WIDEBAND AMPLIFIERS

Fairchild: μA702, 733, 751
Motorola: MC1510, 1545, 1553, 1590
RCA: CA3001, 3011, 3012, 3020, 3021, 3023, 3035
Texas Instruments: SN2600, 2610, 5510, 7501, 7511

RF and IF amplifiers

A radio-frequency (RF) amplifier is usually the first stage in a radio or television receiver; intermediate-frequency (IF) amplifiers typically are the middle stages. Chapter 22 covers receivers, and at that time you will understand what RF and IF amplifiers do. The basic idea is this: RF and IF amplifiers are tuned or resonant so that they amplify only a *narrow band* of frequencies; this allows us to separate the signals from different radio or television stations.

Table 17-4 shows a few of the available IC amplifiers you can use for RF or IF stages. As mentioned in our general discussion, inductors and large capacitors are impractical to integrate on a chip. For this reason, you have to add L's and C's externally to get a resonant circuit.

TABLE 17-4.   RF AND IF AMPS

Fairchild: μA703, 717, 719
Motorola: MC1110, 1350, 1352, 1550
National: LM171, 271, 371, 703
RCA: CA3005, 3028, 3044, 3053

Voltage Regulators

Chapter 5 discussed rectifiers and power supplies. After filtering, we have a dc voltage with ripple. This dc voltage is proportional to the line voltage, that is, it will change 10 percent if the line voltage changes 10 percent. In many applications, a 10 percent change in dc voltage is too much. For this reason, we can use a *voltage regulator,* a device that delivers an almost constant dc output voltage even though the input voltage changes 10 percent or thereabouts. Table 17-5 is a sample of IC voltage regulators. The typical regulator can hold the output dc voltage to within 0.01 percent for normal

TABLE 17-5.   VOLTAGE REGULATORS

Fairchild: μA723, 7800
Motorola: MC1560, 1566, 1569
National: LM100, 109, 309

changes in line voltage and load resistance. Other features include positive or negative output, adjustable output voltage, and short-circuit protection.

Chapter 23 discusses voltage regulation further. At that time, you will see how to use some of these IC voltage regulators.

Summary

There are other linear ICs besides the ones we have listed.[3] Of growing importance are those for consumer markets like automotive, electrical appliance, and television. The ones we have described (audio, op amp, wideband, RF and IF, and voltage regulators) will be useful in later chapters.

Problems

17-1.   Calculate the ideal dc emitter current in each transistor of Fig. 17-18a.
17-2.   Calculate the value of dc voltage from each collector to ground in Fig. 17-18a.

[3] A good reference for ICs is *The Microelectronics Data Book,* Motorola Semiconductor Products Inc., 1969.

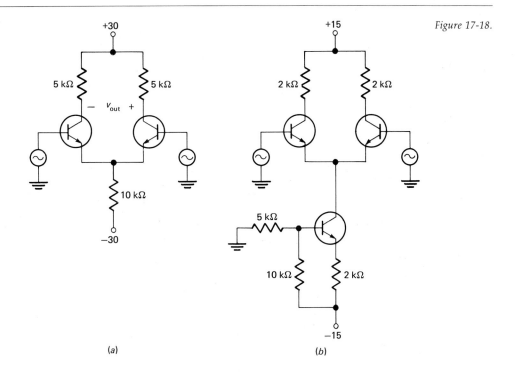

Figure 17-18.

(a)   (b)

Figure 17-19.

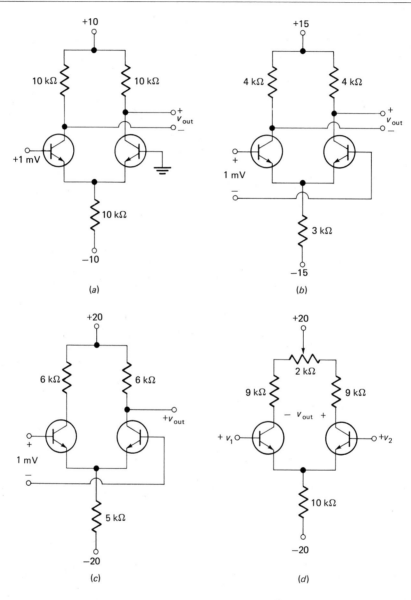

(a)

(b)

(c)

(d)

17-3. If each source in Fig. 17-18a has an R_S of 1 kΩ and if β equals 100, what is the value of dc emitter current (second approximation)?

17-4. What is the ideal value of dc emitter current in the diff amp of Fig. 17-18b?

17-5. Calculate the collector-to-ground voltage in Fig. 17-18b for the two upper transistors.

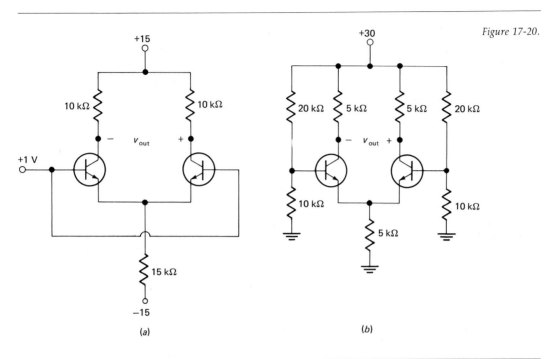

Figure 17-20.

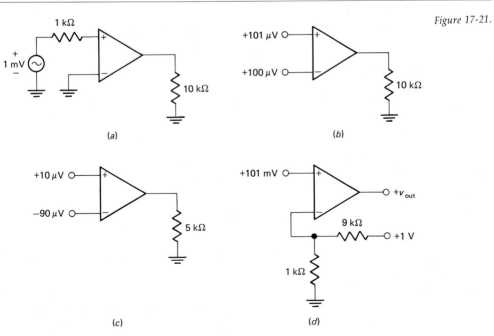

Figure 17-21.

17-6. In Fig. 17-18b suppose the collector resistors have a 1 percent tolerance. What is maximum dc output voltage if the two transistors are identical?

17-7. A sine wave with a peak of 1 mV drives the single-ended diff amp of Fig. 17-19a. How much output voltage is there?

17-8. What is the voltage gain A for the diff amp of Fig. 17-19b? With a differential input of 1 mV, what is the output voltage?

17-9. A sine wave with a peak of 1 mV drives the diff amp shown in Fig. 17-19c. How much ac output voltage is there? How much dc output voltage?

17-10. When you build a discrete diff amp, the two transistors almost never have close enough characteristics to get a reasonable balance between the two halves of the diff amp. For this reason, a potentiometer can be added as shown in Fig. 17-19d. To get an idea of the adjustment range, treat the transistors as identical and calculate the range of v_{out}.

17-11. In Fig. 17-19d the dc output voltage equals zero when the wiper is near the center of the potentiometer. What is the approximate value of voltage gain A? If v_1 equals 101 mV and v_2 equals 100 mV, how much output voltage is there?

17-12. What is the input impedance of the diff amp of Fig. 17-19a for a β of 75?

17-13. The common-mode input signal of 1 V in Fig. 17-20a produces an ac output voltage of 2 mV. Calculate the common-mode rejection ratio. Express the answer in decibels.

17-14. If a diff amp has a common-mode rejection ratio of 60 dB and a voltage gain of 200, how much common-mode output voltage do you get with a common-mode input voltage of 10 mV?

17-15. Figure 17-20b shows a diff amp using a single supply. Treat the voltage across the base-emitter diodes as zero and calculate the dc emitter current in each transistor.

17-16. Suppose the diff amp of Fig. 17-20b has a CM'_{rej} of 100 dB. The power supply delivers 30 V dc plus a ripple of 30 mV rms. How much common-mode output ripple is there?

17-17. The amplifier of Fig. 17-21a has a z_{in} of 1 MΩ, a z_{out} of 100 Ω, and an A of 10,000. Calculate the approximate output voltage.

17-18. As shown in Fig. 17-21b, the noninverting input equals 101 μV and the inverting input equals 100 μV. If A' equals 92 dB and z_{out} is 10 Ω, how much output voltage is there?

17-19. A positive voltage is applied to the noninverting input and a negative voltage to the inverting input in Fig. 17-21c. If A equals 100,000 and z_{out} is 10 Ω, how much output voltage is there?

17-20. In Fig. 17-21d, z_{in} is negligibly large, z_{out} negligibly small, and A equals 10,000. A sine wave with a peak of 101 mV is applied to the noninverting input. Another sine wave with exactly the same phase but a peak of 1 V is applied to the voltage divider driving the inverting input. What is the value of the output voltage?

18. Negative Feedback

Most attempts at perpetual motion *feed back* part of the output of a system to the input, the idea being for the system to run itself. In this chapter and the next, we discuss what happens when we feed back part of the output of an amplifier to the input. We will not get perpetual motion, but we will create remarkable new circuits.

18-1. THE FOUR PROTOTYPES

A *feedback amplifier* has two parts: an amplifier and a feedback circuit. In our diagrams we sometimes use a box like Fig. 18-1a to represent the amplifier. The signal travels from left to right.

The feedback circuit

Figure 18-1b shows a box for the feedback circuit. This may be another amplifier, a filter, a voltage divider, any circuit at all. In our drawings the signal will enter the right end of the feedback circuit and come out the left.

In this chapter the feedback circuit is always a resistive circuit with one or two resistors. Figure 18-1c shows one feedback circuit we use; it is nothing more than a voltage divider. In this particular case, a signal of 2 V driving the right end produces approximately 0.02 V at the left end. We will use the letter B to stand for the voltage gain of the feedback circuit. In Fig. 18-1c the voltage gain equals

$$B = \frac{0.02\ \text{V}}{2\ \text{V}} = 0.01$$

Figure 18-1. Building blocks.

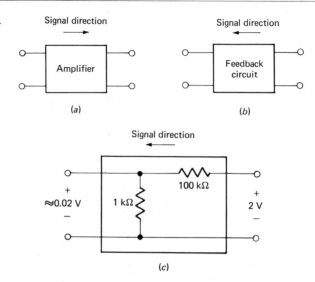

(a)

(b)

(c)

Four ways to connect

Given boxes with an input and output, there are *four* ways to connect two boxes. For instance, Fig. 18-2a shows an amplifier and feedback circuit connected *series-parallel* (SP); on the input side the amplifier and feedback circuit are in series, but on the output side they are in parallel.

Figure 18-2b is the *series-series* (SS) prototype, Fig. 18-2c the *parallel-parallel* (PP) prototype, and Fig. 18-2d the *parallel-series* (PS) prototype. As we will see, each prototype has its own distinct behavior. In other words, the prototypes are not interchangeable because they are fundamentally different.

Current and voltage feedback

In Fig. 18-2a and c, the voltage across the load resistor is the voltage to the feedback circuit; we are feeding back the output voltage. For this reason, a parallel-output connection is sometimes called *voltage feedback*. As we will prove later, voltage feedback tends to make the output voltage independent of R_L.

In Fig. 18-2b and d, the current through the load resistor flows into the feedback circuit. In this case, we are feeding back the output current. This is why a series-output connection is known as *current feedback*. We will prove this kind of feedback tends to make the output current independent of R_L.

EXAMPLE 18-1.

Figure 18-3a shows an SP feedback amplifier. With the given voltages, calculate each of these voltage gains:

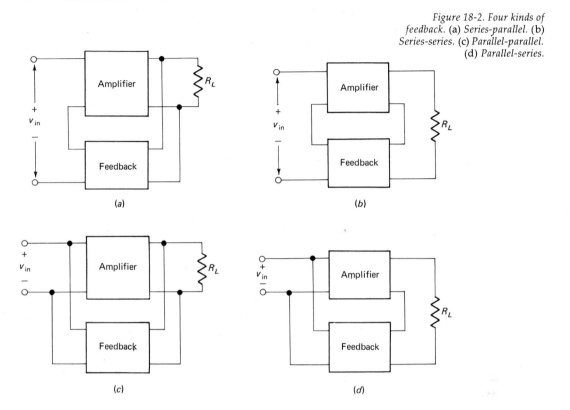

1. The gain of the amplifier
2. The gain of the feedback circuit
3. The gain of the feedback amplifier (source to output)

SOLUTION.
To keep the gains distinct, we use A for the internal amplifier, B for the feedback circuit, and A_{SP} for the feedback amplifier (overall circuit).

1. 1 mV goes into the amplifier and 10 V comes out. Therefore,

$$A = \frac{10}{0.001} = 10{,}000 \qquad \text{(internal gain)}$$

2. Because of the voltage feedback, the load voltage drives the feedback circuit. The gain of the feedback circuit is

$$B = \frac{0.2}{10} = 0.02$$

Figure 18-3. Examples 18-1 and 18-2.

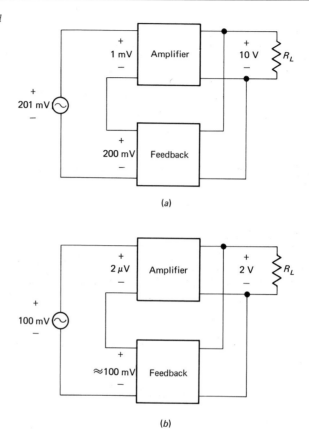

(a)

(b)

3. The input to the feedback amplifier (the source voltage) is 201 mV and the final output is 10 V. So, the gain of the feedback amplifier is

$$A_{SP} = \frac{10}{0.201} \cong 50 \qquad \text{(overall gain)}$$

to better than 1 percent.

The voltage gain with feedback is 50 compared to the voltage gain without feedback of 10,000. This giving up of gain is typical of all the feedback amplifiers we discuss in this chapter.

EXAMPLE 18-2.
The *sacrifice factor S* is the ratio of voltage gain without feedback to voltage gain with feedback, that is,

$$S = \frac{A}{A_{SP}} \qquad\qquad (18\text{-}1)^{***}$$

Calculate the value of S for the preceding example.

SOLUTION.
In the preceding example, A equals 10,000 and A_{SP} is 50. With Eq. (18-1), the sacrifice factor equals

$$S = \frac{10{,}000}{50} = 200$$

The sacrifice factor tells how much voltage gain we have sacrificed. In this example, an S of 200 means we have reduced the gain by a factor of 200.

EXAMPLE 18-3.
Calculate the values of A, A_{SP}, and S for the feedback amplifier of Fig. 18-3b.

SOLUTION.
2 μV goes into the internal amplifier and 2 V comes out. Therefore,

$$A = \frac{2}{2(10^{-6})} = 10^6$$

(This is equivalent to a bel voltage gain of 120 dB.) The voltage gain of the SP feedback amplifier is

$$A_{SP} = \frac{2}{0.1} = 20$$

And the sacrifice factor is

$$S = \frac{A}{A_{SP}} = \frac{1{,}000{,}000}{20} = 50{,}000$$

A large sacrifice factor is not uncommon in a feedback amplifier. Almost always, we will give up a great deal of the voltage gain. As we will see, by sacrificing voltage gain, we can improve many other amplifier qualities. In fact, a large sacrifice factor leads to characteristics impossible to attain without feedback.

18-2. IDEAL ANALYSIS OF AN SP AMPLIFIER

Let us make an ideal analysis to bring out a few of the big ideas about SP feedback. Later, we can improve the results with better approximations.

From here on, we concentrate on amplifiers with two inputs and one output such as an op amp or a wideband amp symbolized by the A block of Fig. 18-4. As we saw in the preceding chapter, an op amp has a high z_{in}, a low z_{out}, and a large A. Ideally, we can treat z_{in} and A as infinite, and z_{out} as zero.

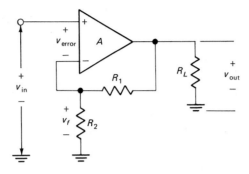

Figure 18-4. The SP negative-feedback amplifier.

Error voltage is ideally zero

Figure 18-4 shows an SP feedback amplifier. The input voltage to the feedback amplifier is v_{in}, the voltage out of the feedback circuit is v_f, and the voltage driving the internal amplifier is v_{error}. Summing voltages around the input loop gives

$$-v_{\text{in}} + v_{\text{error}} + v_f = 0$$

or
$$v_{\text{in}} - v_f = v_{\text{error}} \tag{18-2}$$

Voltage v_{error} is a differential input to the internal amplifier; therefore, with Eq. (17-6),

$$v_{\text{out}} = Av_{\text{error}}$$

or
$$v_{\text{error}} = \frac{v_{\text{out}}}{A} \tag{18-2a}$$

Now we can rewrite Eq. (18-2) as

$$v_{\text{in}} - v_f = \frac{v_{\text{out}}}{A}$$

With a typical op amp, A is very large, ideally infinite; therefore,

$$v_{\text{in}} - v_f \cong 0$$

or
$$v_{\text{in}} \cong v_f \quad \text{(ideal)} \tag{18-2b}***$$

This last result is important. Whenever you analyze an SP feedback amplifier like Fig. 18-4, you should realize the feedback voltage v_f almost equals the input voltage v_{in}. Furthermore, the error voltage v_{error} is small because A is very large. In symbols,

$$v_{\text{error}} = \frac{v_{\text{out}}}{A} \cong 0 \quad \text{(ideal)} \tag{18-2c}***$$

So, for ideal analysis, v_f approximately equals v_{in}, and v_{error} is almost zero.

Negative feedback

Equation (18-2) says

$$v_{error} = v_{in} - v_f \quad \text{(negative feedback)}$$

We call this *negative feedback* because the error voltage is the *difference* between input voltage and feedback voltage. The feedback voltage opposes the input voltage so that the internal amplifier has less voltage driving it. This means that we get less output voltage, equivalent to sacrificing voltage gain.

The next chapter talks about *positive feedback,* the idea being to return the feedback voltage so that it adds to the input signal. In symbols,

$$v_{error} = v_{in} + v_f \quad \text{(positive feedback)}$$

By doing this, we get more output voltage, equivalent to increasing the voltage gain.

For the remainder of this chapter we discuss negative feedback; it also is known as degenerative or inverse feedback. With negative feedback the value of S is greater than unity. On the other hand, positive feedback (also called regenerative or direct feedback) has an S value less than unity.

SP feedback voltage gain

What is the voltage gain A_{SP} in Fig. 18-4? That is, the formula for v_{out}/v_{in}? In Fig. 18-4 the z_{in} of the amplifier approaches infinity (ideally); since R_1 and R_2 form a voltage divider, the feedback voltage equals

$$v_f \cong \frac{R_2}{R_1 + R_2} v_{out} \quad \text{(neglecting } z_{in}) \qquad (18\text{-}2d)^{***}$$

or

$$\frac{v_{out}}{v_f} \cong \frac{R_1 + R_2}{R_2} = \frac{R_1}{R_2} + 1$$

Since $v_{in} \cong v_f$,

$$A_{SP} = \frac{v_{out}}{v_{in}} \cong \frac{v_{out}}{v_f}$$

or

$$A_{SP} \cong \frac{R_1}{R_2} + 1 \quad \text{(ideal)} \qquad (18\text{-}3)^{***}$$

This is important. It tells us the voltage gain of an SP negative-feedback amplifier is ideally determined by the ratio of two resistors. If we use precision resistors, we get precise values of voltage gain. In other words, in an SP negative-feedback amplifier where the sacrifice factor is high, the internal gain A is *swamped out.*

In the Appendix, we derive a more accurate formula for the voltage gain of an SP negative-feedback amplifier:

$$A_{SP} \cong \left(\frac{R_1}{R_2} + 1\right)\left(1 - \frac{1}{S}\right) \qquad (18\text{-}4a)$$

We almost never use this formula except for one thing. It tells us S must be large for Eq. (18-3) to be valid. If S is greater than 100, Eq. (18-3) is accurate to better than 1 percent. In a practical SP feedback amplifier, S is almost always greater than 100. For this reason, we will use Eq. (18-3) to calculate the voltage gain of an SP feedback amplifier.

Relation of A_{SP} to B

This is a minor point, but we should know the relation between A_{SP} and B. B is the voltage gain of the feedback circuit; this gain ideally equals

$$B \cong \frac{R_2}{R_1 + R_2} = \frac{v_f}{v_{\text{out}}} \cong \frac{v_{\text{in}}}{v_{\text{out}}} = \frac{1}{A_{SP}}$$

or

$$A_{SP} \cong \frac{1}{B} \qquad \text{(ideal)} \qquad (18\text{-}4b)\text{***}$$

This last formula is an alternative to Eq. (18-3); we sometimes use it to make equations more compact.

Input impedance

In Fig. 18-5a, z_{in} is the impedance between the noninverting and inverting input terminals, and $z_{\text{in}(SP)}$ is the input impedance of the SP feedback amplifier. The two impedances are not equal. To get the relation between these impedances, we use Fig. 18-5b. To begin with,

$$z_{\text{in}(SP)} = \frac{v_{\text{in}}}{i_{\text{in}}}$$

which we can rearrange to

$$z_{\text{in}(SP)} = \frac{v_{\text{in}}}{v_{\text{error}}/z_{\text{in}}} = \frac{v_{\text{in}}}{v_{\text{error}}} z_{\text{in}}$$

Since v_{error} ideally equals v_{out}/A,

$$z_{\text{in}(SP)} = \frac{v_{\text{in}}}{v_{\text{out}}/A} z_{\text{in}} = \frac{A}{A_{SP}} z_{\text{in}}$$

or

$$z_{\text{in}(SP)} = S z_{\text{in}} \qquad \text{(ideal)} \qquad (18\text{-}5)\text{***}$$

This says the input impedance of an SP feedback amplifier is S times the z_{in} of the internal amplifier. For instance, if the sacrifice factor equals 100 and z_{in} is 50 kΩ, the ideal value of $z_{\text{in}(SP)}$ is 5 MΩ.

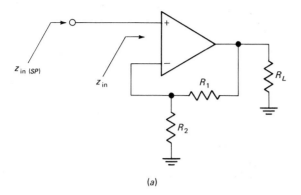

(a)

Figure 18-5. Deriving the formula for $z_{in(SP)}$.

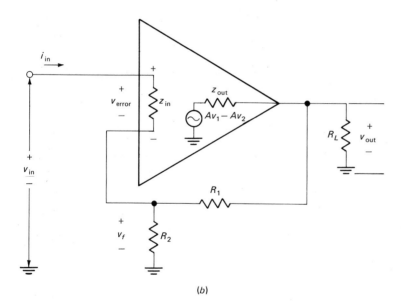

(b)

Common-mode input impedance

In a Darlington emitter follower, r_c' places an upper limit on the input impedance. A similar thing happens in an SP negative-feedback amplifier. To begin with, the *common-mode input impedance* $z_{in(CM)}$ is the impedance of both inputs to ground. This impedance is well into the megohms, depending on the design of the input stage. Data sheets for earlier-generation op amps do not include $z_{in(CM)}$, but those for later-generation op amps usually give you this value. As an example, the data sheet of an MC1556 shows a $z_{in(CM)}$ of 500 MΩ. This is typical for later-generation op amps.

Figure 18-6. Effect of common-mode input impedance.

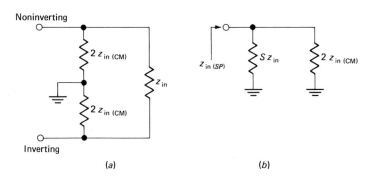

(a)   (b)

Since $z_{in(CM)}$ is measured with both inputs in parallel, each input has an impedance of $2z_{in(CM)}$ to ground (see Fig. 18-6a). When we include these impedances in the analysis, we get an input impedance of

$$z_{in(SP)} = Sz_{in} \parallel 2z_{in(CM)} \tag{18-6}$$

Figure 18-6b illustrates the shunting effect of $2z_{in(CM)}$. When the sacrifice factor is high, Sz_{in} approaches infinity, and $z_{in(SP)}$ approaches a value of $2z_{in(CM)}$. In other words, $2z_{in(CM)}$ places an upper limit on the input impedance of an SP negative-feedback amplifier.

Output impedance

The output impedance $z_{out(SP)}$ is the same as the Thevenin impedance looking back into the SP feedback amplifier. To get this impedance, we reduce all independent sources to zero as shown in Fig. 18-7. To find $z_{out(SP)}$, we apply a voltage v_{out} and work out the resulting current i_{out}. The ratio of v_{out} to i_{out} equals $z_{out(SP)}$.

In Fig. 18-7 the feedback voltage equals

$$v_f = Bv_{out}$$

This voltage drives the inverting input and produces a Thevenin output voltage of Av_f with the polarity shown. Therefore, the current i_1 equals

$$i_1 = \frac{v_{out} + Av_f}{z_{out}} = \frac{v_{out} + ABv_{out}}{z_{out}}$$

$$= \frac{v_{out}(1 + AB)}{z_{out}} \cong \frac{v_{out}(1 + A/A_{SP})}{z_{out}}$$

$$\cong S\frac{v_{out}}{z_{out}} \qquad \text{for large } S$$

or

$$\frac{v_{out}}{i_1} \cong \frac{z_{out}}{S}$$

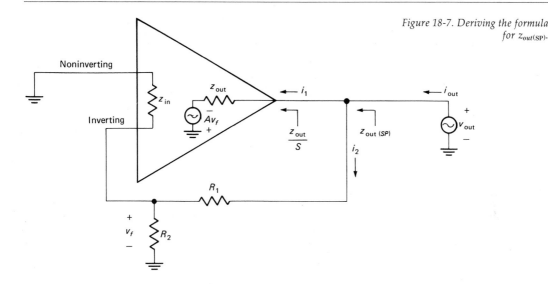

Figure 18-7. Deriving the formula for $z_{out(SP)}$.

This is the Thevenin output impedance looking directly into the output terminal of the amplifier (see Fig. 18-7).

The total Thevenin output impedance includes the effects of the voltage divider. In symbols,

$$z_{out(SP)} = \frac{z_{out}}{S} \parallel (R_1 + R_2)$$

Impedance z_{out} is usually under a kilohm to begin with. Since S is normally greater than 100, the first term is almost always much smaller than the second. For this reason, the practical equation to use is

$$z_{out(SP)} \cong \frac{z_{out}}{S} \quad \text{(ideal)} \qquad\qquad (18\text{-}7)^{***}$$

Summary

The important results are Eqs. (18-3), (18-5), and (18-7). When the S factor is greater than 100, A_{SP} equals $(R_1/R_2 + 1)$ to within 1 percent. Likewise, a large S factor means a very high input impedance and a very low output impedance. In fact, the SP negative-feedback amplifier ideally approaches the *perfect voltage amplifier,* one with *infinite input impedance, zero output impedance,* and *fixed voltage gain* as shown in Fig. 18-8. Whenever we see an SP negative-feedback amplifier, we can visualize it as a perfect voltage amplifier in preliminary analysis. This is important to remember because the remaining prototypes (SS, PP, and PS) do not act like voltage amplifiers.

Figure 18-8. The perfect voltage
amplifier.

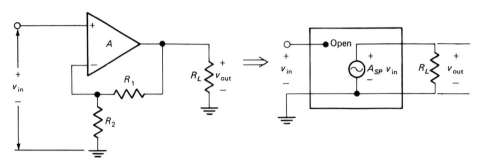

Figure 18-9. Examples 18-4 and
18-5.

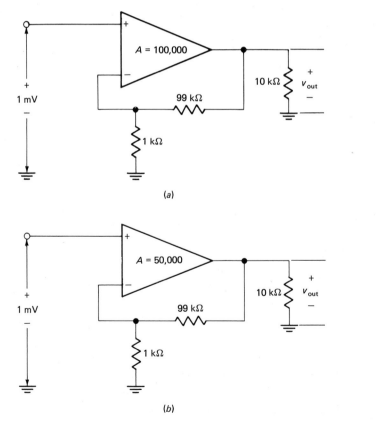

EXAMPLE 18-4.
Calculate A_{SP}, v_{error}, v_f, and S in Fig. 18-9a.

SOLUTION.
The overall voltage gain equals

$$A_{SP} = \frac{R_1}{R_2} + 1 = \frac{99{,}000}{1000} + 1 = 100$$

The output voltage is

$$v_{out} = A_{SP} v_{in} = 100(1 \text{ mV}) = 100 \text{ mV}$$

The error voltage is ideally zero. To a better approximation,

$$v_{error} = \frac{v_{out}}{A} = \frac{100 \text{ mV}}{100{,}000} = 1 \ \mu V$$

The feedback voltage equals

$$v_f = \frac{R_2}{R_1 + R_2} v_{out} = \frac{1000}{100{,}000} 100 \text{ mV} = 1 \text{ mV}$$

The sacrifice factor is

$$S = \frac{A}{A_{SP}} = \frac{100{,}000}{100} = 1000$$

Notice that the feedback voltage ideally equals the input voltage, and the error voltage is much smaller than the input voltage.

EXAMPLE 18-5.
Suppose we change the internal gain of Fig. 18-9a from 100,000 to 50,000. Recalculate A_{SP}, v_{error}, v_f, and S.

SOLUTION.
In Fig. 18-9b the voltage gain A_{SP} still equals 100 because the resistances R_1 and R_2 have the same values. The sacrifice factor S is so high that the more accurate formula, Eq. (18-4a), still gives a voltage gain close to 100.

Voltage v_{out} remains close to 100 mV; feedback voltage v_f still equals 1 mV, approximately.

The only quantities that change significantly are the error voltage and the sacrifice factor. First,

$$v_{error} = \frac{v_{out}}{A} = \frac{100 \text{ mV}}{50{,}000} = 2 \ \mu V$$

Second,

$$S = \frac{A}{A_{SP}} = \frac{50{,}000}{100} = 500$$

Here is what it boils down to. As long as the internal gain is much greater than the overall gain, the principal effect of a change in internal gain is a change in error voltage. In this example, internal gain drops in half and error voltage doubles. Physically what happens in going from Fig. 18-9a to 18-9b is this. Less internal gain means output voltage will *try* to decrease, but the slightest decrease in output voltage results in a smaller feedback voltage v_f. In turn, this produces a large increase in error voltage, which compensates for the decrease in gain A. The overall effect of a large change in A is a small change in v_{out} and a large compensating change in v_{error}.

EXAMPLE 18-6.
Figure 18-10 shows the complete schematic of an SP negative-feedback amplifier. Calculate the ideal values of $z_{in(SP)}$ and $z_{out(SP)}$.

SOLUTION.
As you may recall from the preceding chapter, a 741 is a second-generation op amp. The data sheet of a 741C indicates a typical A of 100,000, a z_{in} of 1 MΩ, and a z_{out} of 300 Ω. If you build the circuit as shown in Fig. 18-10, you will get a feedback voltage gain of

$$A_{SP} = \frac{R_1}{R_2} + 1 = 99 + 1 = 100$$

The sacrifice factor is

$$S = \frac{A}{A_{SP}} = \frac{100,000}{100} = 1000$$

The ideal input impedance of the feedback amplifier is

$$z_{in(SP)} = Sz_{in} = 1000(1 \text{ M}\Omega) = 1000 \text{ M}\Omega$$

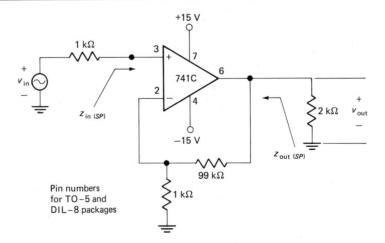

Figure 18-10. A complete SP negative-feedback amplifier.

Pin numbers
for TO–5 and
DIL–8 packages

and the ideal output impedance of the feedback amplifier is

$$z_{\text{out}(SP)} = \frac{z_{\text{out}}}{S} = \frac{300}{1000} = 0.3 \ \Omega$$

Because the ideal input impedance is so large, we have to use Eq. (18-6) to calculate the exact value of $z_{\text{in}(SP)}$. The data sheet of a 741 does not include the value of $z_{\text{in}(CM)}$; therefore, without measuring $z_{\text{in}(CM)}$, we can only estimate $z_{\text{in}(SP)}$ well into the megohms.

The output impedance $z_{\text{out}(SP)}$ is only 0.3 Ω; therefore, we can use very low values of R_L and still get an A_{SP} of 100. But we should mention that R_L is part of the r_C seen by the last stage of the op amp. As you recall from Chap. 10, the $I_C r_C$ of the last stage determines how large the signal can be before clipping occurs. R_L is part of r_C; therefore, a very low value of R_L forces the last stage to clip sooner than it would for larger values of R_L.

EXAMPLE 18-7.

Calculate the voltage gain A_{SP} for the feedback amplifier of Fig. 18-11.

SOLUTION.

This is a special case of SP negative feedback where R_1 is zero and R_2 is infinite. With Eq. (18-3),

$$A_{SP} = \frac{R_1}{R_2} + 1 = 0 + 1 = 1$$

The special case of Fig. 18-11 is known as a *voltage follower* because the output voltage equals or follows the input voltage. Since input impedance is very high and output impedance very low, the voltage follower reminds us of the emitter follower. The voltage follower offers a voltage gain extremely close to unity, and a much lower output impedance than is possible with an emitter follower.

EXAMPLE 18-8.

Figure 18-12*a* shows a widely used form of SP negative feedback with two discrete transistor stages. Calculate the approximate value of A_{SP}.

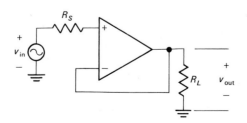

Figure 18-11. The voltage follower.

Figure 18-12. Discrete SP negative feedback amplifier.

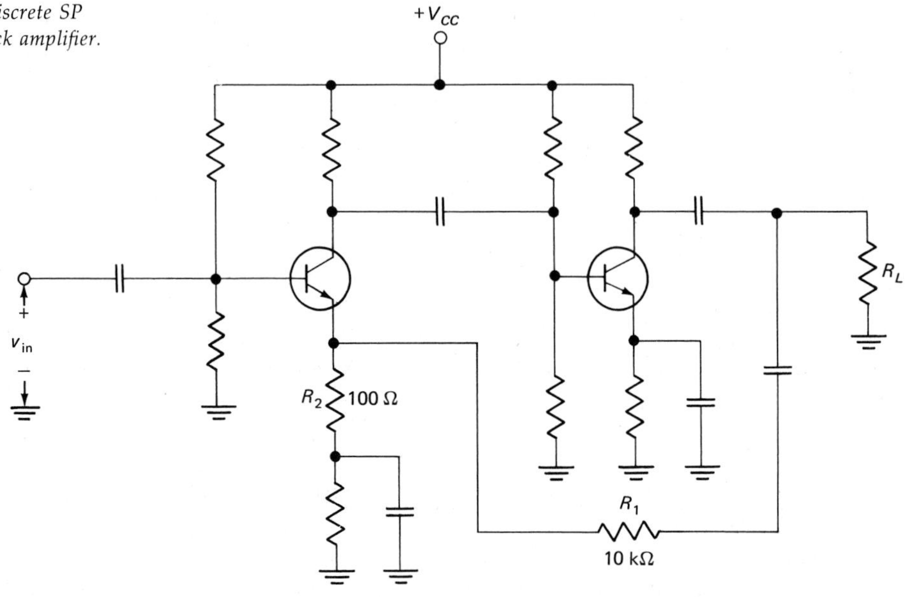

(a)

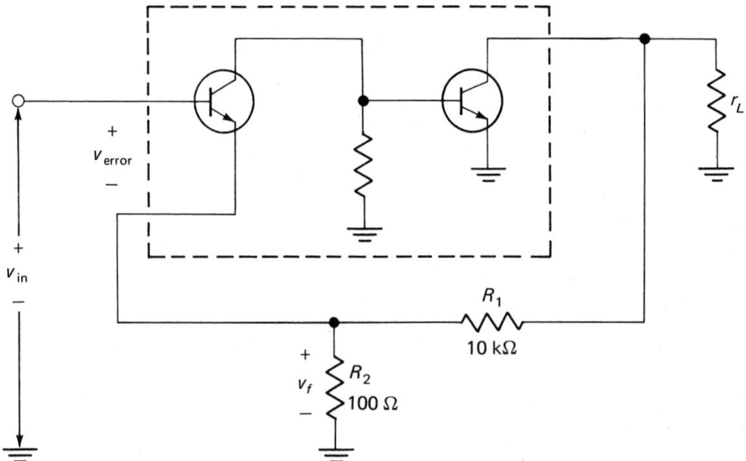

(b)

SOLUTION.

To see the feedback path, draw the ac equivalent circuit as shown in Fig. 18-12b. The box of gain A is inside the dashed lines. Since v_{error} appears across the r'_e of the first stage, the gain of the first stage is r_C/r'_e. The second stage has a similar gain. So, the total internal gain is

$$A = A_1 A_2$$

We can expect this internal gain to be 10,000 or more.

The voltage gain of the feedback circuit is

$$B \cong \frac{R_2}{R_1 + R_2} = \frac{100}{10{,}000 + 100} = 0.01$$

(This approximation is not as accurate as with the op amp because the first transistor loads the voltage divider.)

The feedback voltage gain is

$$A_{SP} \cong \frac{1}{B} \cong \frac{1}{0.01} = 100$$

The advantage of an SP two-stage feedback amplifier like Fig. 18-12a is that it swamps out the transistor parameters. Getting the exact value of internal gain A is unimportant; as long as A is high, A_{SP} equals $1/B$ to a close approximation.

18-3.   PP FEEDBACK

Figure 18-13a shows a *parallel-parallel* (PP) negative-feedback amplifier. The input voltage drives the *inverting* terminal of the amplifier; the noninverting terminal is grounded. The amplified-and-inverted output signal is then applied to the right end of a feedback resistor R.

Output voltage

The easiest way to find what the circuit does is to Millerize it, that is, split R into its two Miller components as shown in Fig. 18-13b. If the amplifier is a discrete circuit, we can expect a gain of 100 or more; if the amplifier is an op amp, the gain will be better than 10,000. For this reason, $A + 1$ is closely equal to A, and we can use the simplified circuit of Fig. 18-13c.

In Fig. 18-13c, R/A is in parallel with the z_{in} of the internal amplifier. Therefore, the input impedance of the PP feedback amplifier is

$$z_{\text{in}(PP)} = \frac{R}{A} \,\bigg\|\, z_{\text{in}} \tag{18-8a}$$

In a PP feedback amplifier, R/A will be much smaller than z_{in} so that

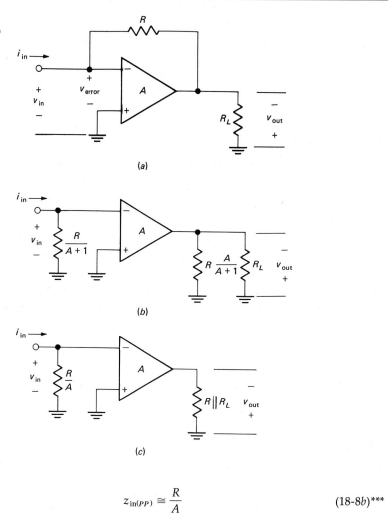

Figure 18-13. The PP negative-feedback amplifier. (a) Actual circuit. (b) Millerized circuit. (c) Simplified circuit.

Figure 18-13. The PP negative-feedback amplifier. (a) Actual circuit. (b) Millerized circuit. (c) Simplified circuit.

$$z_{in(PP)} \cong \frac{R}{A} \tag{18-8b}***$$

In other words, negligible current flows into the internal amplifier; almost all of i_{in} flows through the input Miller resistance.

The input voltage equals

$$v_{in} = i_{in} z_{in(PP)} \cong i_{in} \frac{R}{A}$$

and the output voltage is

$$v_{out} = A v_{in} \cong A i_{in} \frac{R}{A}$$

or
$$v_{\text{out}} \cong Ri_{\text{in}} \qquad \text{(ideal)} \qquad\qquad (18\text{-}9a)$$

This result is accurate to better than 1 percent in a practical PP feedback amplifier. If A is less than 100, or R/A more than 1/100 of z_{in}, you may wish to use this exact formula:

$$v_{\text{out}} = Ai_{\text{in}} \left(\frac{R}{A+1} \,\Big\|\, z_{\text{in}} \right) \qquad \text{(exact)} \qquad\qquad (18\text{-}9b)$$

Input impedance

When an op amp is used, we can easily keep R/A less than 1/100 of z_{in}. In this case, the $z_{\text{in}(PP)}$ closely equals

$$z_{\text{in}(PP)} = \frac{R}{A}$$

Ideally, A approaches infinity and $z_{\text{in}(PP)}$ goes to zero; therefore,

$$z_{\text{in}(PP)} \cong 0 \qquad \text{(ideal)}***$$

This is the crucial difference between SP and PP feedback. With an SP negative-feedback amplifier the input impedance tends toward infinity or an open circuit; with a PP negative-feedback amplifier the input impedance approaches zero or a short circuit. Symbolically,

Series input $\longrightarrow$ infinite z_{in}
Parallel input $\longrightarrow$ zero z_{in}

As we will see, similar results apply to the output side:

Series output $\longrightarrow$ infinite z_{out}
Parallel output $\longrightarrow$ zero z_{out}

These are equivalent to saying a series output connection acts like a current source, and a parallel output connection like a voltage source.

Output impedance

Like SP feedback, PP feedback makes the Thevenin output impedance approach zero. In other words, the output tends to act like a voltage source. The actual value of $z_{\text{out}(PP)}$ depends on the source impedance R_S connected to the input. By a derivation similar to that given for SP negative feedback, we can show

$$z_{\text{out}(PP)} = \frac{z_{\text{out}}}{1 + AB} \qquad\qquad (18\text{-}10a)$$

where
$$B = \frac{R_S}{R + R_S}$$
(18-10*b*)

The denominator of Eq. (18-10*a*) is so large in the typical PP feedback amplifier that $z_{\text{out(PP)}}$ approaches zero, much the same as in the SP feedback amplifier.

The ideal current-to-voltage transducer

Strictly speaking, a *transducer* is a device that changes an electrical quantity into a nonelectrical quantity, or vice versa. A microphone, a loudspeaker, a thermistor, and a photocell are examples. But the word *transducer* has also been used for devices that transform one kind of electrical quantity into another kind.

The PP negative-feedback amplifier ideally approaches the *perfect current-to-voltage transducer,* a device with *zero input impedance, zero output impedance,* and a *fixed relation between* i_{in} *and* v_{out}. Whenever we see a PP negative-feedback amplifier, we can visualize it as the perfect current-to-voltage transducer shown in Fig. 18-14*a*. The output voltage equals R times the input current.

How can we use PP negative feedback? A moving-coil ammeter has an internal resistance that may disturb the circuit whose current is being measured. For instance,

Figure 18-14. (a) *The perfect current-to-voltage transducer.* (b) *An electronic ammeter.*

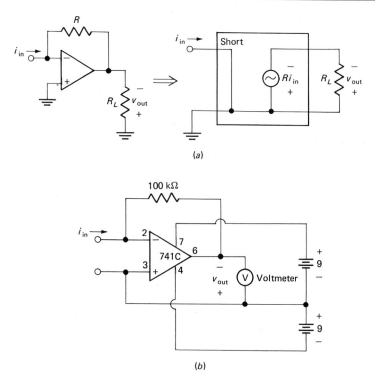

an ammeter with a full-scale deflection of 50 μA typically has a resistance of 2 kΩ, far from ideal; when an ammeter like this is placed in series with a branch, it adds 2 kΩ of resistance to the branch. With PP negative feedback, we can build an *electronic ammeter whose resistance approaches zero.*

Figure 18-14*b* shows a complete circuit for such an ammeter. The typical gain of a 741C is 100,000. Therefore,

$$z_{in(PP)} = \frac{R}{A} = \frac{100{,}000}{100{,}000} = 1\ \Omega$$

and
$$v_{out} = Ri_{in} = 10^5 i_{in}$$

The first equation tells us that we will add only 1 Ω of resistance to the branch whose current is being measured. The second equation tells us how sensitive the ammeter is; if i_{in} equals 50 μA, v_{out} equals 5 V, enough voltage to measure easily with an inexpensive voltmeter. Therefore, even though it uses inexpensive parts, the electronic ammeter of Fig. 18-14*b* is superior to a moving-coil ammeter as far as input impedance is concerned.

Where else can you use a current-to-voltage transducer? Many devices produce a current proportional to an input quantity. A photoemissive diode is an example. When light falls on it, this device produces a current of

$$I = KF$$

where K is a constant depending on the size and structure, and F is the amount of light hitting the diode. You can use the current I to drive a current-to-voltage transducer like Fig. 18-14*a*. With the output voltage Ri_{in}, you can drive a voltmeter, another amplifier, an oscilloscope, or other devices, depending on what you are trying to do.

EXAMPLE 18-9.
Suppose we have an oscilloscope with an input sensitivity of 10 mV/cm. By connecting the current-to-voltage transducer of Fig. 18-15*a* to the vertical input, we can measure current. If we want 1 μA of input current to produce 1 cm of vertical deflection, what value should R have?

SOLUTION.
To get 1 cm of deflection, 1 μA must produce 10 mV. With Eq. (18-9*a*),

$$R = \frac{v_{out}}{i_{in}} = \frac{10(10^{-3})}{10^{-6}} = 10\ \text{k}\Omega$$

Therefore, a 10-kΩ resistor in Fig. 18-15*a* allows us to measure microamp currents with an oscilloscope.

EXAMPLE 18-10.
Describe the action of Fig. 18-15*b*.

Figure 18-15. (a) *Using an oscilloscope to measure current.* (b) *Part of a transistor curve tracer.* (c) *Cascading a current-to-voltage transducer and a voltage amplifier.*

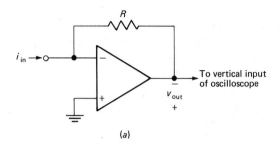

(a)

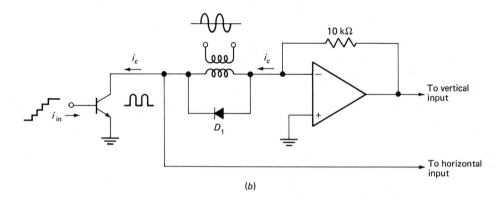

(b)

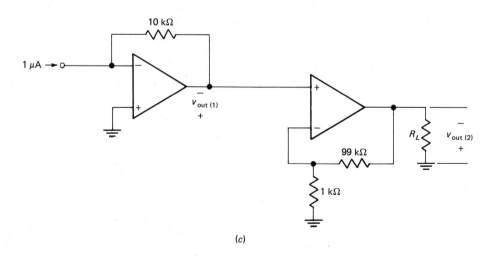

(c)

SOLUTION.
Figure 18-15*b* shows some ingredients for a transistor curve tracer. The sine wave is half-wave-rectified by diode D_1; this produces the half wave of collector voltage shown. The collector current flows through the current-to-voltage transducer. Therefore, the voltage applied to the vertical input of the oscilloscope corresponds to collector current, and the voltage to the horizontal input corresponds to collector voltage. By applying a synchronized *staircase* of current to the base, we will get collector curves. Each step produces the next higher curve.

EXAMPLE 18-11.
If 1 μA of input current i_{in} drives the system of Fig. 18-15*c*, what is the value of the output voltage?

SOLUTION.
The first stage is a current-to-voltage transducer with an output voltage of

$$v_{out(1)} = Ri_{in} = 10^4(10^{-6}) = 10 \text{ mV}$$

The second stage is a voltage amplifier with a gain of 100; therefore, the output voltage of the system is

$$v_{out(2)} = Av_{in} = 100(10 \text{ mV}) = 1 \text{ V}$$

18-4. SS FEEDBACK

Figure 18-16*a* shows a *series-series* (SS) negative-feedback amplifier. The input voltage drives the *noninverting* input terminal. The feedback voltage drives the inverting input terminal.

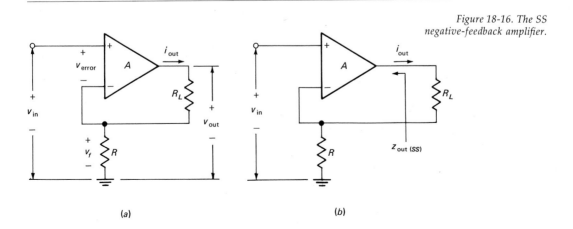

Figure 18-16. The SS negative-feedback amplifier.

(a)　　　　　　　　　(b)

Output current

What is the formula for i_{out} in Fig. 18-16a? First, note that i_{out} flows through the load resistor R_L and through the feedback resistor R. Therefore,

$$v_f = i_{out}R \quad (z_{in} \text{ negligibly high})$$

As we saw with SP feedback, a series input connection means v_f ideally equals v_{in}. Therefore,

$$v_{in} \cong i_{out}R$$

or

$$i_{out} \cong \frac{v_{in}}{R} \quad \text{(ideal)} \tag{18-11a***}$$

To get an exact formula, we proceed like this. The voltage gain of the feedback circuit is

$$B = \frac{R}{R_L + R} \quad (z_{in} \text{ negligibly high})$$

The output current is

$$i_{out} = \frac{v_{out}}{R_L + R} = \frac{Av_{error}}{R_L + R} = \frac{A(v_{in} - v_f)}{R_L + R}$$

or

$$i_{out} = \frac{A(v_{in} - Bv_{out})}{R_L + R} = \frac{Av_{in} - ABi_{out}(R_L + R)}{R_L + R}$$

Solving for i_{out} gives

$$i_{out} = \frac{A}{1 + AB} \frac{v_{in}}{R_L + R} \quad \text{(exact)} \tag{18-11b}$$

In the typical case, AB is much greater than unity so that

$$i_{out} \cong \frac{1}{B} \frac{v_{in}}{R_L + R} = \frac{R_L + R}{R} \frac{v_{in}}{R_L + R} = \frac{v_{in}}{R}$$

which agrees with the ideal formula found earlier.

Input impedance

Because of the series input connection, the input impedance approaches infinity. By a straightforward derivation like the one for SP feedback, we can prove

$$z_{in(SS)} \cong ABz_{in} \quad \text{(ideal)} \tag{18-12a}$$

where B equals $R/(R_L + R)$. As with SP feedback, the actual $z_{in(SS)}$ includes the shunting effect of $2z_{in(CM)}$, so that a better approximation is

$$z_{in(SS)} = ABz_{in} \parallel 2z_{in(CM)} \tag{18-12b}$$

Because of this, $2z_{in(CM)}$ is the upper limit on the value of $z_{in(SS)}$.

Output impedance

Look at Fig. (18-16b). Ideally, the output current of an SS negative-feedback amplifier is independent of the value of R_L. In other words, in Fig. 18-16b we can change R_L, but the value of i_{out} remains constant because it equals v_{in}/R (ideally). This implies the output of an SS feedback amplifier acts like a current source. In fact, the output impedance in Fig. 18-16b is

$$z_{out(SS)} = AR \qquad \text{for } A \gg 1 \tag{18-13}$$

For instance, if A is 10,000 and R is 100 Ω, $z_{out(SS)}$ equals 1 MΩ.

You can prove Eq. (18-13) easily. The output impedance equals the value of R_L that drops the output current in half. If you substitute $R_L = AR$ into Eq. (18-11b), i_{out} will reduce to half its normal value; this proves $z_{out(SS)}$ equals AR.

The ideal voltage-to-current transducer

Here is what counts. An SS negative-feedback amplifier ideally behaves like a *perfect voltage-to-current transducer*, a device with *infinite input impedance, infinite output impedance*, and a *fixed relation between v_{in} and i_{out}*. Figure 18-17a illustrates the idea. Apply an input voltage v_{in} to the transducer and out comes a current of v_{in}/R. Since R can be a precision resistor, the output current is precisely fixed by the ratio v_{in}/R.

Where can you use a voltage-to-current transducer? Figure 18-17b shows one application, a sensitive dc voltmeter. The load is an ammeter with a full-scale deflection of 100 μA. When v_{in} is 1 mV,

$$i_{out} = \frac{v_{in}}{R} = \frac{10^{-3}}{10} = 100 \ \mu A$$

This tells us we get full-scale deflection of the ammeter with only 1-mV input. In this way, we can accurately measure very small dc voltages.

If we change the 10-Ω resistor to a 100-Ω resistor, it will take a 10-mV input to get full-scale deflection. If we use a 1000-Ω resistor, it takes 100-mV input to get full-scale deflection, etc. By using a switch and precision resistors between 10 Ω and 1 MΩ, we will have a sensitive dc voltmeter with full-scale voltage ranges from 1 mV to 100 V.

The 741C data sheet indicates we can null the output with a 10-kΩ potentiometer between pins 1 and 5, as shown in Fig. 18-17b. This is how we can zero the voltmeter before taking a reading.

*Figure 18-17. (a) The perfect
voltage-to-current transducer. (b)
An electronic voltmeter.*

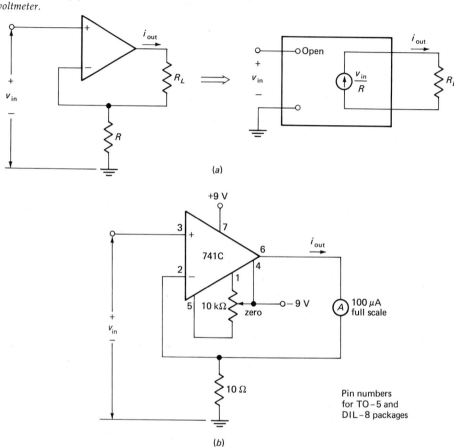

(a)

(b)

Pin numbers
for TO–5 and
DIL–8 packages

With a voltage-to-current transducer we can convert voltage waveforms to current waveforms. For instance, it is easier to generate a staircase of voltage than one of current. A staircase voltage, therefore, can drive a voltage-to-current transducer to produce a staircase of current. In this way, we can deliver a staircase of current to the transistor of Fig. 18-15b (curve tracer).

18-5. PS FEEDBACK

The fourth basic form of feedback is the *parallel-series* (PS) connection shown in Fig. 18-18a. The input current drives the *inverting* terminal; the noninverting terminal is grounded.

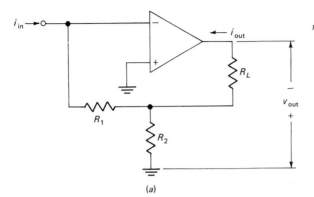

(a)

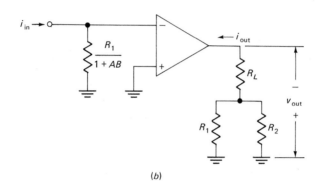

(b)

Figure 18-18. The PS negative-feedback amplifier. (a) Original circuit. (b) Millerized circuit.

Input impedance

The first step is to Millerize the circuit of Fig. 18-18a. The voltage gain we need is the gain from the inverting input terminal to the junction of R_1 and R_2. This gain includes voltage gain A of the internal amplifier and voltage gain B of the feedback circuit between the output of the op amp and the R_2 resistor. Since the error voltage approaches zero, we can visualize the left end of R_1 grounded. Therefore, the voltage gain of the feedback circuit from the output of the op amp to R_2 is

$$B = \frac{R_1 \parallel R_2}{R_L + R_1 \parallel R_2} \qquad (18\text{-}14a)$$

Now, we can split R_1 into its Miller components as shown in Fig. 18-18b. AB is always much greater than unity, at least 100 in a practical PS feedback amplifier. For this reason, the output Miller component is approximately R_1 as shown. The z_{in} of the amplifier is normally much greater than the input Miller component; therefore, to a close approximation,

$$z_{\text{in}(PS)} \cong \frac{R_1}{1 + AB} \quad \text{(ideal)} \tag{18-14b}$$

Since AB is usually much greater than 100, $z_{\text{in}(PS)}$ tends toward zero.

Current gain

In Fig. 18-18b the input voltage is

$$v_{\text{in}} = i_{\text{in}} z_{\text{in}(PS)} = i_{\text{in}} \frac{R_1}{1 + AB}$$

and

$$v_{\text{out}} = A v_{\text{error}} = A v_{\text{in}} = A i_{\text{in}} \frac{R_1}{1 + AB}$$

But v_{out} also equals

$$v_{\text{out}} = i_{\text{out}}(R_L + R_1 \parallel R_2)$$

By equating the two expressions for v_{out}, we can solve for the current gain:

$$\frac{i_{\text{out}}}{i_{\text{in}}} = \frac{A}{1 + AB} \frac{R_1}{R_L + R_1 \parallel R_2} \tag{18-15a}$$

Because AB is normally much greater than unity, the first factor on the right-hand side is closely equal to $1/B$. Since B is given by the expression of Eq. (18-14a), the practical formula for current gain is

$$\beta_{PS} = \frac{i_{\text{out}}}{i_{\text{in}}} \cong \frac{R_1}{R_2} + 1 \quad \text{(ideal)} \tag{18-15b}***$$

This is an easy formula to remember because it's identical to the formula for SP voltage gain. For convenience, we use the symbol β_{PS} to stand for the current gain of a PS negative-feedback amplifier.

Output impedance

Ideally, the output impedance of a PS negative-feedback amplifier tends toward infinity. This is apparent in Eq. (18-15b); since R_L does not affect the output current, the PS feedback amplifier acts like a current source. In fact,

$$z_{\text{out}(PS)} = A(R_1 \parallel R_2) \tag{18-16}$$

Since A is large, $z_{\text{out}(PS)}$ will be large.

The simplest way to prove Eq. (18-16) is this. Substitute $A(R_1 \parallel R_2)$ for R_L in the exact equation for current gain, Eq. (18-15a). The current gain will drop to half its normal value: this proves $z_{\text{out}(PS)}$ equals $A(R_1 \parallel R_2)$ because basic theory tells us the current drops in half when the load matches the output impedance.

The ideal current amplifier

The PS negative-feedback amplifier behaves ideally like a *perfect current amplifier,* a device with *zero input impedance, infinite output impedance,* and *fixed current gain* as shown in Fig. 18-19a. Inject a current i_{in} and out comes a current

$$i_{out} = \beta_{PS} i_{in}$$

where
$$\beta_{PS} \cong \frac{R_1}{R_2} + 1 \quad \text{(ideal)}$$

For instance, if R_1 equals 99 kΩ and R_2 is 1 kΩ, the current gain equals 100 to a very close approximation. If you need the exact answer, you can use Eq. (18-15a). If AB is greater than 100, the ideal current gain is accurate to better than 1 percent.

EXAMPLE 18-12.

Figure 18-19b shows part of a sensitive dc ammeter. How much input current do we need to get full-scale deflection?

Figure 18-19. (a) *The perfect current amplifier.* (b) *A sensitive electronic ammeter.*

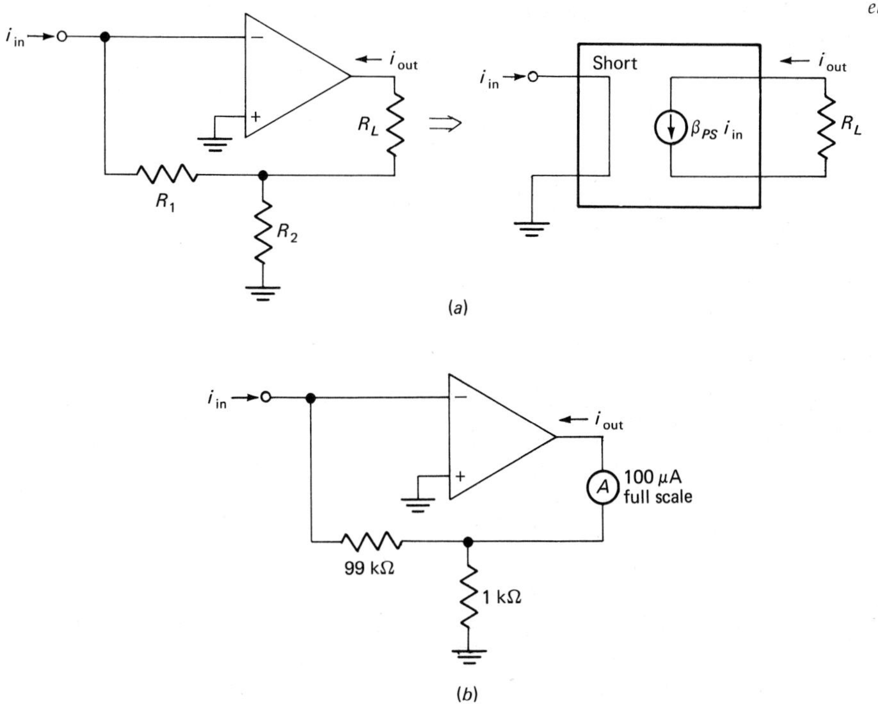

SOLUTION.

The current gain is

$$\beta_{PS} = \frac{R_1}{R_2} + 1 = 99 + 1 = 100$$

Since the ammeter has a full-scale current of 100 μA, it takes only 1 μA of input current to produce full-scale deflection. By switching in different R_1 and R_2 resistors, we can change the current ranges.

To make Fig. 18-19b practical, you should choose an op amp whose input bias and offset currents are low. As you recall, these currents flow through the source resistance driving the op amp. Because of this, you may get some error with high-impedance sources. If this problem arises in a particular application, all you do is choose a later-generation op amp whose input bias and offset currents are negligibly small (like an MC1556, LM108, or LM216). You may also need a voltage-nulling potentiometer to take care of the input offset voltage.

18-6.   BANDWIDTH OF A FEEDBACK AMPLIFIER

The internal amplifier has an upper break frequency f_2 produced by its dominant lag network. The break frequency of the entire feedback amplifier, however, does not equal f_2; it is much higher than f_2. This section tells you why negative feedback always *increases* the bandwidth.

A simple explanation

It is easy to understand why negative feedback increases the bandwidth. In Fig. 18-20 suppose the internal gain A equals 100,000. Because of the resistance values, the feedback gain A_{SP} equals 100. With a 10-mV input, the output voltage equals 1 V. This value is quite accurate because the sacrifice factor A/A_{SP} equals 1000.

As the frequency of the input signal increases, we eventually reach the break

Figure 18-20. An SP negative-feedback amplifier.

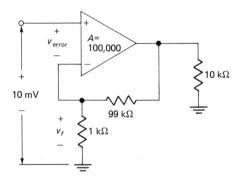

frequency f_2 of the *internal amplifier*. At this frequency the internal gain A is down 3 dB; in this case, A equals 70,700. The sacrifice factor is still high and equals 707. Because of this, the feedback voltage gain still closely equals 100 and the output voltage still equals 1 V. The internal gain is so high to begin with that even though it has gone down 3 dB, it still is high enough to ensure a large value of S.

As described in Example 18-5, as long as A is large, the only significant voltage change is in v_{error}. If A goes down by a factor of 2, v_{error} increases by a factor of almost 2; this compensates for the change in A.

With increasing frequency the internal gain A keeps decreasing until S is no longer large. When this happens, we begin to notice a decrease in v_{out}. When v_{out} is down 3 dB, we have reached the break frequency of the feedback amplifier. To keep this frequency distinct from others, we designate it as $f_{2(SP)}$.

Relation between break frequencies

By an advanced derivation, the two break frequencies are related as follows:

$$f_{2(SP)} = S_{mid} f_2 \qquad (18\text{-}17)^{***}$$

where

$$S_{mid} = \frac{A_{mid}}{A_{SP(mid)}} \qquad (18\text{-}18)$$

A_{mid} and $A_{SP(mid)}$ are the voltage gains well inside the bandwidth of the internal amplifier.[1] As an example, in Fig. 18-20 A_{mid} is 100,000 and $A_{SP(mid)}$ equals 100; therefore,

$$S_{mid} = \frac{100,000}{100} = 1000$$

If the internal amplifier has an f_2 of 200 Hz, Eq. (18-17) gives

$$f_{2(SP)} = 1000(200 \text{ Hz}) = 200 \text{ kHz}$$

What it means is this. When you sacrifice gain by a factor S_{mid}, you *stretch* the bandwidth of the feedback amplifier by the same factor. In the foregoing example we started with an internal gain A of 100,000 and a bandwidth of 200 Hz; we wind up with a feedback gain of 100 and a bandwidth of 200 kHz.

Special condition

There is a crucial condition that must be satisfied when you apply Eq. (18-17). The internal amplifier must have *one dominant lag network* so that the voltage gain A decreases 20 dB per decade above f_2. This 20-dB rolloff must continue to the break frequency $f_{2(SP)}$. In other words, the next important lag network in the internal ampli-

---

[1] Sometimes, A_{mid} is called the *open-loop gain* and $A_{SP(mid)}$ the *closed-loop gain*, where loop refers to the feedback path.

Figure 18-21. (a) The bel voltage gain of a 741C. (b) Bel voltage gain in Example 18-13. (c) Circuit for Example 18-13.

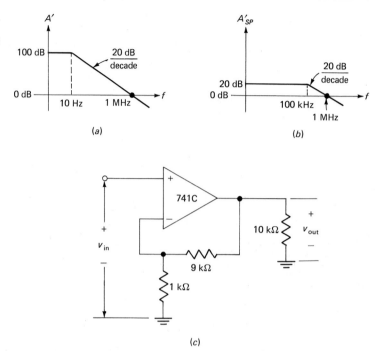

fier must have a break frequency greater than $f_{2(SP)}$. Unless this condition is satisfied, the neat relation given by Eq. (18-17) does not apply.

The proof of Eq. (18-17) is given in the Appendix for the interested reader.

EXAMPLE 18-13.

The data sheet of a 741C shows a bel graph like Fig. 18-21a. If the 741C is used in the SP negative-feedback amplifier of Fig. 18-21c, calculate the break frequency of the feedback amplifier.

SOLUTION.

In Fig. 18-21a the midband voltage gain of the 741C is 100 dB, equivalent to an ordinary gain of

$$A_{\text{mid}} = 100{,}000$$

The break frequency of the 741C is

$$f_2 = 10 \text{ Hz}$$

The bel graph indicates a rolloff of 20 dB per decade, meaning the 741C has one dominant lag network (integrated on the chip). This rolloff is constant out to the unity-gain frequency of

$$f_{unity} = 1 \text{ MHz}$$

In the feedback amplifier of Fig. 18-21c the midband feedback gain equals

$$A_{SP(mid)} = \frac{R_1}{R_2} + 1 = 9 + 1 = 10$$

which is equivalent to 20 dB. The sacrifice factor equals

$$S_{mid} = \frac{A_{mid}}{A_{SP(mid)}} = \frac{100,000}{10} = 10,000$$

With Eq. (18-17),

$$f_{2(SP)} = S_{mid}f_2 = 10,000(10 \text{ Hz}) = 100 \text{ kHz}$$

So, an S factor of 10,000 stretches the bandwidth by a factor of 10,000.

Figure 18-21b shows the bel voltage gain of the feedback amplifier. In the midband of the feedback amplifier, the bel gain is 20 dB. The break frequency of the feedback amplifier is 100 kHz. The rolloff rate is 20 dB per decade; therefore, the unity-gain frequency is still 1 MHz.

EXAMPLE 18-14.
By changing the ratio of R_1 to R_2 in Fig. 18-21c, we get different values of $A_{SP(mid)}$. Calculate the $f_{2(SP)}$ for each of the following values: $A_{SP(mid)} = 1000$, 100, 10, and 1.

SOLUTION.
Since A_{mid} is 100,000, Eq. (18-17) gives

$$f_{2(SP)} = \frac{100,000}{A_{SP(mid)}} f_2$$

We can use this to calculate $f_{2(SP)}$ for different $A_{SP(mid)}$. When $A_{SP(mid)}$ is 1000,

$$f_{2(SP)} = \frac{100,000}{1000} 10 \text{ Hz} = 1 \text{ kHz}$$

When $A_{SP(mid)}$ equals 100,

$$f_{2(SP)} = \frac{100,000}{100} 10 \text{ Hz} = 10 \text{ kHz}$$

When $A_{SP(mid)}$ is 10,

$$f_{2(SP)} = \frac{100,000}{10} 10 \text{ Hz} = 100 \text{ kHz}$$

When $A_{SP(mid)}$ equals 1,

$$f_{2(SP)} = \frac{100,000}{1} \; 10 \text{ Hz} = 1 \text{ MHz}$$

The point is this. We can trade off gain for bandwidth. By changing the external resistors in Fig. 18-21c, we can tailor-make a voltage amplifier to fit the needs of our particular application. The values in this example give us these choices of gain-bandwidth:

$$A_{SP(\text{mid})} = 1000, \; f_{2(SP)} = 1 \text{ kHz}$$

$$A_{SP(\text{mid})} = 100, \; f_{2(SP)} = 10 \text{ kHz}$$

$$A_{SP(\text{mid})} = 10, \; f_{2(SP)} = 100 \text{ kHz}$$

$$A_{SP(\text{mid})} = 1, \; f_{2(SP)} = 1 \text{ MHz}$$

Especially important, notice that the product of gain and break frequency is a constant value of 1 MHz.

18-7. THE GAIN-BANDWIDTH PRODUCT

The preceding example suggests the idea of the *gain-bandwidth product*. Since

$$f_{2(SP)} = S_{\text{mid}} f_2 = \frac{A_{\text{mid}}}{A_{SP(\text{mid})}} f_2 \tag{18-19}$$

we can multiply both sides by $A_{SP(\text{mid})}$ to get

$$A_{SP(\text{mid})} f_{2(SP)} = A_{\text{mid}} f_2 \tag{18-20}***$$

The right-hand side of this equation is the product of internal gain and the internal break frequency. For instance, the typical 741C has an A_{mid} of 100,000 and an f_2 of 10 Hz; therefore, it has a gain-bandwidth product of

$$A_{\text{mid}} f_2 = 100,000 (10 \text{ Hz}) = 1 \text{ MHz}$$

Since A_{mid} is dimensionless and f_2 is in hertz, the gain-bandwidth product always has the dimensions of hertz.

Gain-bandwidth product is constant

The left-hand side of Eq. (18-20) is the product of feedback gain and feedback break frequency. Therefore, no matter what the values of R_1 and R_2, the product of $A_{SP(\text{mid})}$ and $f_{2(SP)}$ must equal the gain-bandwidth product of the internal amplifier. If the internal amplifier has a gain-bandwidth product of 1 MHz, the product of $A_{SP(\text{mid})}$ and $f_{2(SP)}$ must equal 1 MHz. Example 18-14 illustrated this; each pair of $A_{SP(\text{mid})}$ and $f_{2(SP)}$ had a product of 1 MHz.

Equation (18-20) is often summarized by saying the *gain-bandwidth product is a con-*

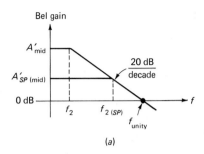

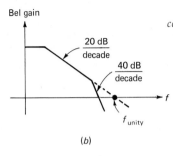

Figure 18-22. (a) Bel graph for constant gain-bandwidth product. (b) Second break frequency less than f_{unity}.

(a) (b)

stant. In other words, in Eq. (18-20) if the right-hand product is a constant, so too is the left-hand product. For this reason, even though $A_{SP(mid)}$ and $f_{2(SP)}$ change when we change external resistors, the product of these two quantities remains constant for a given internal amplifier.

Gain-bandwidth product equals unity-gain frequency

Figure 18-22a summarizes the idea of constant gain-bandwidth product. $A'_{SP(mid)}$ can have any value between 0 dB and A'_{mid}; for the special case of 0 dB, $A_{SP(mid)}$ equals unity, and Eq. (18-20) reduces to the useful relation

$$f_{unity} = A_{mid}f_2 \qquad (18\text{-}21)^{***}$$

This says the unity-gain frequency equals the gain-bandwidth product.

Figure 18-22a illustrates all the key ideas behind the gain-bandwidth product. The internal gain breaks at f_2; it drops 20 dB per decade. The feedback gain breaks at $f_{2(SP)}$; it too drops 20 dB per decade.[2] Both graphs superimpose beyond the $f_{2(SP)}$ break point until they cross the axis at a frequency of f_{unity}.

The gain-bandwidth product gives us a fast way of comparing amplifiers; the greater the gain-bandwidth product, the higher we can go in frequency and still have usable gain. The LM118, for example, with its gain-bandwidth product of 15 MHz provides gain out to higher frequencies than a 741C with a gain-bandwidth product of 1 MHz.

Constant only for 20 dB per decade rolloff

A final point. By looking at Fig. 18-22a, you can see the rolloff is 20 dB per decade until the graph crosses the horizontal axis. Beyond f_{unity} the internal amplifier will have other break frequencies. If the internal amplifier has a second break frequency *less* than f_{unity} (see Fig. 18-22b), the rolloff rate changes from 20 dB per decade to 40 dB per decade. In this case, the gain-bandwidth product is no longer constant out to f_{unity}.

[2] As discussed in Prob. 16-16, this equals 6 dB per octave.

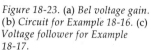

Figure 18-23. (a) Bel voltage gain. (b) Circuit for Example 18-16. (c) Voltage follower for Example 18-17.

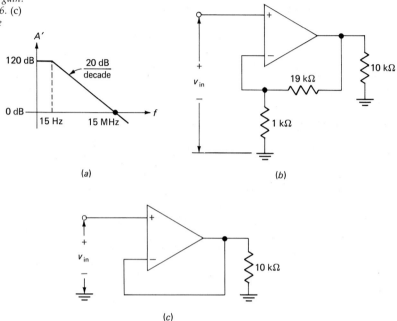

(a)

(b)

(c)

Therefore, do not try to apply the constant gain-bandwidth idea to every situation you encounter; *it applies only over the frequency range where the rolloff is 20 dB per decade.*

EXAMPLE 18-15.
Figure 18-23*a* shows the bel voltage gain of an op amp. What does the gain-bandwidth product equal?

SOLUTION.
We can get this in either of two ways. First, the midband gain is 120 dB, equivalent to

$$A_{\text{mid}} = 1{,}000{,}000$$

The internal break frequency is

$$f_2 = 15 \text{ Hz}$$

Therefore, the internal gain-bandwidth product equals

$$A_{\text{mid}}f_2 = 1{,}000{,}000(15 \text{ Hz}) = 15 \text{ MHz}$$

Second, we can get the same answer by reading the unity-gain frequency; this is how you normally would do it given the data sheet of an op amp. In Fig. 18-23*a*,

$$f_{\text{unity}} = 15 \text{ MHz}$$

which equals the gain-bandwidth product.

EXAMPLE 18-16.
If an op amp with the bel graph of Fig. 18-23a is used in Fig. 18-23b, what is the break frequency of the feedback amplifier?

SOLUTION.
The voltage gain of the feedback amplifier is

$$A_{SP(\text{mid})} = \frac{R_1}{R_2} + 1 = 19 + 1 = 20$$

Since f_{unity} equals 15 MHz, the gain-bandwidth product equals 15 MHz. With Eq. (18-20),

$$f_{2(SP)} = \frac{A_{\text{mid}}f_2}{A_{SP(\text{mid})}} = \frac{f_{\text{unity}}}{A_{SP(\text{mid})}}$$

$$= \frac{15 \text{ MHz}}{20} = 750 \text{ kHz}$$

EXAMPLE 18-17.
Suppose an op amp with the bel graph of Fig. 18-23a is used in the voltage follower of Fig. 18-23c. What is the break frequency of the voltage follower?

SOLUTION.
A voltage follower has unity gain (discussed in Example 18-7). Because of this, the break frequency equals f_{unity}; in this case, $f_{2(SP)}$ equals 15 MHz.

18-8. BREAK FREQUENCIES OF OTHER PROTOTYPES

Negative feedback also stretches the bandwidth of the PP, SS, and PS prototypes. In this section we state the formulas for the break frequencies; the proof of these formulas is similar to the derivation given in the Appendix for Eq. (18-17).

PP break frequency

The PP negative-feedback amplifier of Fig. 18-24b acts like a current-to-voltage transducer. For a *fixed* input current, the output voltage equals Ri_{in} in the midband of the feedback amplifier; as the frequency of the *fixed* input current increases, the output voltage breaks at a frequency of

$$f_{2(PP)} = (1 + A_{\text{mid}})f_2$$

*Figure 18-24. Break frequencies
with negative feedback. (a)
Voltage amplifier. (b)
Current-to-voltage transducer. (c)
Voltage-to-current transducer. (d)
Current amplifier.*

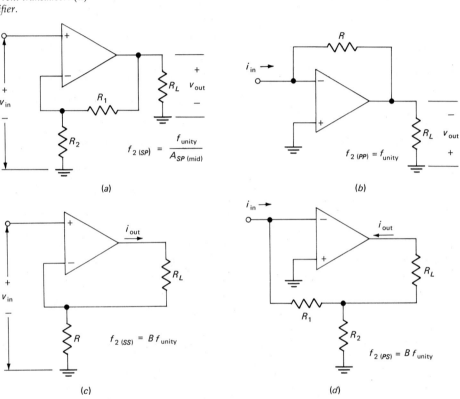

(a)

$$f_{2\,(SP)} = \frac{f_{unity}}{A_{SP\,(mid)}}$$

(b)

$$f_{2\,(PP)} = f_{unity}$$

(c)

$$f_{2\,(SS)} = B f_{unity}$$

(d)

$$f_{2\,(PS)} = B f_{unity}$$

Since A_{mid} is usually immense compared to unity, we can write

$$f_{2(PP)} \cong A_{mid} f_2 \qquad\qquad (18\text{-}22a)$$

The product of $A_{mid} f_2$ is the gain-bandwidth product of the internal amplifier; therefore, an alternative expression for the PP feedback break frequency is

$$f_{2(PP)} \cong f_{unity} \qquad\qquad (18\text{-}22b)$$

When A_{mid} is greater than 100, this is accurate to better than 1 percent.

As an example, if we use a 741C, we get a typical gain-bandwidth product of 1 MHz. If we drive the current-to-voltage transducer with a fixed amplitude of input current, we get a break frequency in output voltage of 1 MHz.

SS break frequency

Figure 18-24c shows an SS negative-feedback amplifier; it behaves like a voltage-to-current transducer. For a fixed amplitude of input voltage, the output current breaks at a frequency of

$$f_{2(SS)} = (1 + A_{mid}B)f_2$$

where
$$B = \frac{R}{R_L + R} \tag{18-23a}$$

In any practical SS feedback amplifier, $A_{mid}B$ is greater than 100; therefore, to within 1 percent accuracy, we can write

$$f_{2(SS)} \cong A_{mid}Bf_2 \tag{18-23b}$$

Since the gain-bandwidth product $A_{mid}f_2$ is equal to the unity-gain frequency, we can write

$$f_{2(SS)} \cong Bf_{unity} \tag{18-23c}$$

This is convenient because we often can read f_{unity} directly off a data sheet for an op amp.

As an example, the gain-bandwidth product of the typical 741C is 1 MHz. If R is 100 Ω and R_L equals 900 Ω in Fig. 18-24c, B equals 0.1; this means the break frequency $f_{2(SS)}$ will equal 100 kHz.

PS break frequency

In the PS negative-feedback amplifier of Fig. 18-24d we have a current amplifier. For a fixed amplitude of i_{in} the break frequency of i_{out} equals

$$f_{2(PS)} = (1 + A_{mid}B)f_2$$

where
$$B = \frac{R_1 \parallel R_2}{R_L + R_1 \parallel R_2} \tag{18-24a}$$

Again, the usual case is $A_{mid}B$ greater than 100, so that

$$f_{2(PS)} \cong A_{mid}Bf_2 \tag{18-24b}$$

to within 1 percent. Since $A_{mid}f_2$ equals f_{unity}, an alternative formula is

$$f_{2(PS)} \cong Bf_{unity} \tag{18-24c}$$

For example, if the R_1, R_2, and R_L of Fig. 18-24d have a B of 0.05, we will get a PS break frequency of 50 kHZ when we use an op amp with a gain-bandwidth product of 1 MHz.

18-9. OTHER BENEFITS OF NEGATIVE FEEDBACK

Negative feedback improves an amplifier in other ways: it reduces the effect of power-supply ripple, decreases nonlinear distortion, cuts down certain kinds of noise, etc. We are not ready to discuss these improvements; after covering harmonic theory, we will go into the effect of negative feedback on ripple, distortion, noise, and so on.

About all we can say at the moment is this. Unwanted signals generated inside an amplifier are fed back with a phase that tends to cancel the original disturbance. For instance, if power-supply ripple gets into a signal path, it is amplified and fed back 180° out of phase; this tends to reduce the power-supply ripple at the point where it gets into the signal path.

18-10. ANALOG-COMPUTER CIRCUITS

The PP negative-feedback amplifier is used a lot in analog computers. This section briefly looks at two of the mathematical operations you can perform with a PP negative-feedback amplifier.

Multiplication

Figure 18-25*a* shows a *multiplier* circuit. Because the error voltage into the op amp approaches zero, we can visualize the right end of R_1 at ground; therefore, the input current ideally equals

$$i_{in} \cong \frac{v_{in}}{R_1}$$

The current-to-voltage transducer produces an output voltage of

$$v_{out} \cong R_2 i_{in} \cong R_2 \frac{v_{in}}{R_1}$$

or
$$v_{out} \cong \frac{R_2}{R_1} v_{in} \quad \text{(multiplier)} \qquad (18\text{-}25)^{***}$$

In an analog computer, R_2/R_1 represents one of the numbers to be multiplied, and v_{in} stands for the other number. For instance, Fig. 18-25*b* shows the setup for multiplying 2 and 8. The output voltage equals

$$v_{out} = \frac{8000}{1000} 2 = 16 \text{ V}$$

Summing circuit

Figure 18-25*c* is a *summing circuit*. Since the inverting input terminal of the op amp is approximately at ground, each current equals

Figure 18-25. (a) *Analog multiplier.* (b) *Example of 2 times 8.* (c) *Two-input summing circuit.* (d) *Five-input summing circuit.*

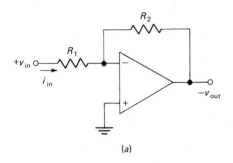

(a)

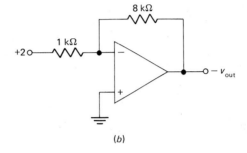

(b)

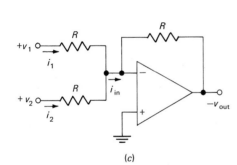

(c)

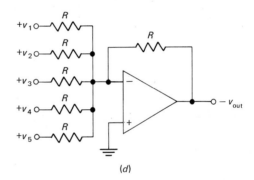

(d)

$$i_1 \cong \frac{v_1}{R}$$

$$i_2 \cong \frac{v_2}{R}$$

and $$i_{in} = i_1 + i_2$$

The current-to-voltage transducer produces an output voltage of

$$v_{out} \cong Ri_{in} \cong Ri_1 + Ri_2 \cong R\frac{v_1}{R} + R\frac{v_2}{R}$$

$$v_{out} \cong v_1 + v_2 \quad \text{(summer)} \qquad (18\text{-}26)^{***}$$

A summing circuit like Fig. 18-25c has other uses besides analog computers. Whenever we have two signals, we can add them and get a single-ended output. For instance, in Fig. 18-25c, v_1 may be a signal from one microphone and v_2 the signal from another. The summing circuit adds these two signals and produces an output of $v_1 + v_2$. In effect, we have taken the signals from two *separate* sources and put them together in a *single-ended* output. The output can then drive an audio amplifier if desired.

If you check the derivation for Eq. (18-26), you will see the same derivation applies to n inputs. That is, you can use as many input signals as you like, each with a separate R. In this case, the output voltage equals

$$v_{\text{out}} \cong v_1 + v_2 + v_3 + \cdots + v_n \tag{18-27}$$

For instance, Fig. 18-25d shows a five-channel summing circuit. With a circuit like this, you can combine five different telephone conversations into a single-ended output. These five signals can then travel over the same telephone wires and be separated at some distant point.

Problems

18-1. In Fig. 18-26a calculate the gain A of the internal amplifier, the gain B of the feedback circuit, and the gain A_{SP} of the feedback amplifier.

18-2. Calculate the sacrifice factor for the feedback amplifier of Fig. 18-26a.

18-3. How much output voltage is there in Fig. 18-26b? If the internal amplifier is a 741C with a typical gain of 100,000, what is the value of S?

18-4. If the internal amplifier of Fig. 18-26b has a gain of 1,000,000, what are the approximate values of v_{error} and v_f? If the gain decreases to 100,000, what are the approximate values of v_{error} and v_f?

18-5. What is the voltage gain of the feedback circuit in Fig. 18-26c? And the value of A_{SP}?

18-6. Calculate the output voltage in Fig. 18-26c. If the z_{in} of the internal amplifier equals 10 kΩ and A is 50,000, what is the ideal value of $z_{\text{in}(SP)}$?

18-7. The internal gain A of Fig. 18-26b equals 1,000,000 and the z_{in} is 100 kΩ. Calculate the ideal $z_{\text{in}(SP)}$. If the internal amplifier has a $2z_{\text{in}(CM)}$ of 10 MΩ from the noninverting input to ground, what is actual $z_{\text{in}(SP)}$?

18-8. Suppose the data sheet of an op amp lists an open-loop voltage gain of 100,000 (same as A), a z_{in} of 500 kΩ, and a common-mode input impedance of 100 MΩ. If this op amp is used in Fig. 18-26b, what is the actual value of $z_{\text{in}(SP)}$?

18-9. The third-generation op amp MC1556 has an internal gain A of 200,000 and an output impedance z_{out} of 1 kΩ.

   1. What is the value of $z_{\text{out}(SP)}$ when this op amp is used in Fig. 18-26b?
   2. If used in Fig. 18-26c, what is the value of $z_{\text{out}(SP)}$?

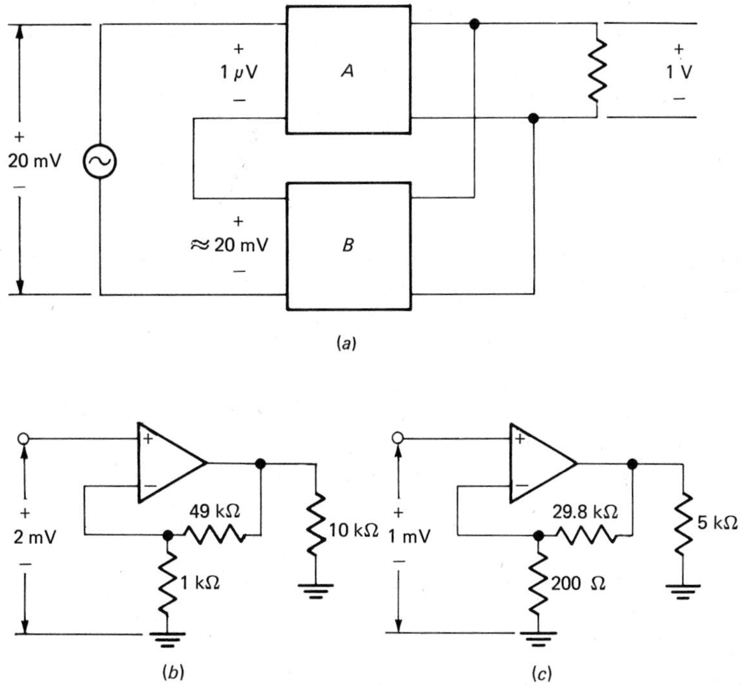

Figure 18-26.

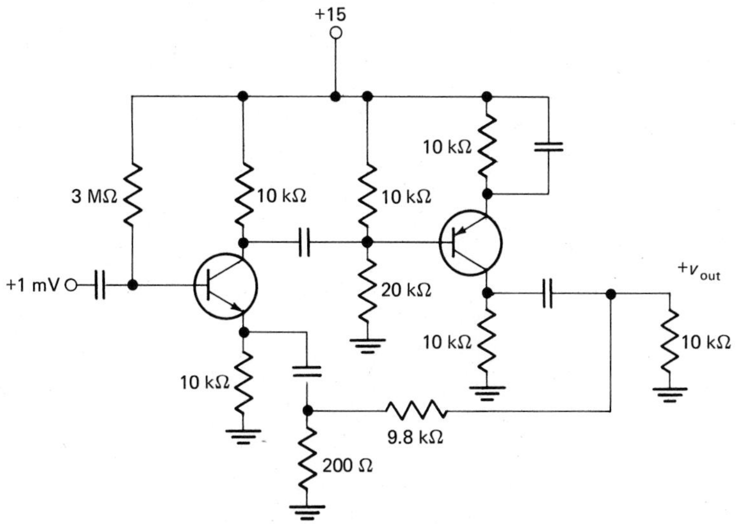

(a)

(b)

(c)

Figure 18-27.

18-10. What is the approximate output voltage in Fig. 18-27? If the internal gain is 10,000, what is the sacrifice factor? If the z_{in} of the internal amplifier equals 2500 Ω (the $\beta r'_e$ of the first stage), what does the ideal $z_{in(SP)}$ equal? If the z_{out} of the internal amplifier is 10 kΩ (same as collector resistor), what does the $z_{out(SP)}$ equal?

18-11. An input current of 1 mA drives the current-to-voltage transducer of Fig. 18-28a. What is the output voltage for each position of the switch?

18-12. Figure 18-28b shows a *photodiode* (the two arrows represent light) driving a current-to-voltage transducer. With 1 μA coming out of the photodiode, how much output voltage is there?

18-13. The 741C of Fig. 18-28c has a typical internal gain A of 100,000 and a z_{out} of 300 Ω. What is the value of $z_{in(PP)}$? If the source supplying the 1 mA has an impedance R_S of 1 MΩ, what is the value of $z_{out(PP)}$?

18-14. A 1-mA current source drives the current-to-voltage transducer of Fig. 18-28d. A voltmeter across the output has ranges of 1 mV, 10 mV, and 100 mV (full scale). We can use the circuit as an electronic ohmmeter. What is the value of $R_{unknown}$ that produces full-scale deflection for each given voltage range? If we

Figure 18-28.

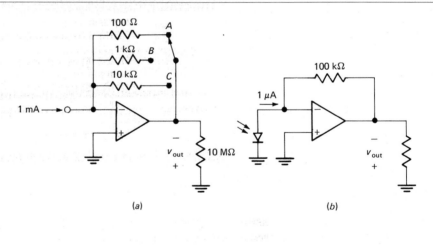

(a)   (b)

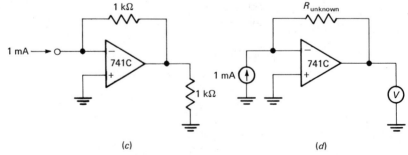

(c)   (d)

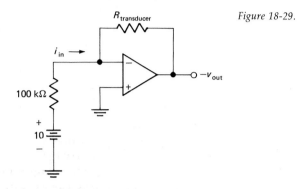

Figure 18-29.

change the current source to 1 μA, what is the value of $R_{unknown}$ that produces full-scale deflection for each range?

18-15. Many transducers are resistive types; a nonelectrical input quantity changes the resistance of the transducer. A carbon microphone is an example of a resistive transducer; the sound-wave input produces changes in resistance. Other examples are strain gages, thermistors, and photoconductive diodes. Suppose we use a resistive transducer in the circuit of Fig. 18-29; we can convert the changes in resistance to changes in output voltage.

1. What is the value of i_{in}?
2. If the quiescent value of the transducer resistance is 1 kΩ, what is the quiescent output voltage?
3. If the nonelectrical quantity driving the transducer produces a change of ±100 Ω in the transducer resistance, how much output ac voltage is there?

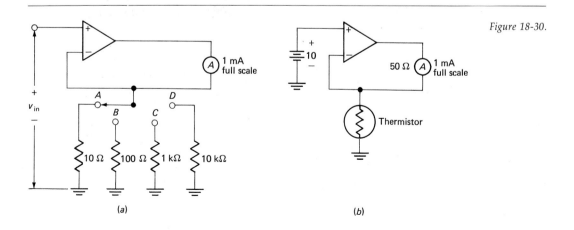

Figure 18-30.

(a)

(b)

18-16. Figure 18-30*a* shows a sensitive dc voltmeter. Calculate the input dc voltages that produce full-scale deflection of the ammeter in each switch position.

18-17. The ammeter of Fig. 18-30*a* has a resistance of 40 Ω. If the internal amplifier has a gain A of 1,000,000 and a z_{in} of 100 kΩ, what is the ideal value of $z_{in(SS)}$ in position A of the switch?

18-18. If the internal amplifier of Fig. 18-30*a* has a gain A of 100,000, what is the minimum value of $z_{out(SS)}$?

18-19. Figure 18-30*b* shows an electronic thermometer. At 0°C, the thermistor has a resistance of 20 kΩ. The resistance decreases 200 Ω for each degree rise, so that $R_{thermistor}$ equals 19.8 kΩ, 19.6 kΩ, 19.4 kΩ, and so on for T equals 1°C, 2°C, 3°C, etc. What does the ammeter read at 0°C? At 25°C? At 50°C?

18-20. The meter resistance in Fig. 18-31 is 50 Ω. The internal amplifier has a gain of 1,000,000. Calculate the values of $z_{in(PS)}$, $z_{out(PS)}$, and i_{out}.

18-21. To get a current gain of 200 in Fig. 18-31, what value should we use for R_1 if R_2 is kept at 100 Ω? If we decide to hold R_1 constant, what value should we use for R_2?

18-22. An SP feedback amplifier has a sacrifice factor of 1000. If the internal amplifier has a break frequency f_2 of 10 Hz, what is the feedback break frequency $f_{2(SP)}$?

18-23. The op amp of Fig. 18-32 has an internal gain A of 1,000,000 and a break frequency of 10 Hz. Work out the value of $f_{2(SP)}$ for each position of the switch.

18-24. Calculate the internal gain-bandwidth product for each of these:

1. $A_{mid} = 50,000$ and $f_2 = 100$ Hz
2. $A'_{mid} = 100$ dB and $f_2 = 20$ Hz
3. $A'_{mid} = 120$ dB and $f_2 = 15$ Hz

18-25. What is the value of f_{unity} for each amplifier given in the preceding problem if the internal gain rolls off 20 dB per decade until it crosses the horizontal axis?

18-26. If an amplifier has the bel voltage gain shown in Fig. 18-33*a*, what does its gain-bandwidth product equal? What is the value of f_{unity}?

Figure 18-31.

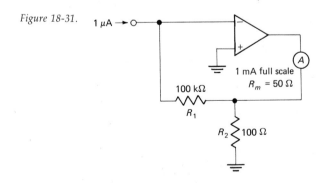

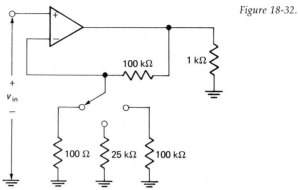

Figure 18-32.

18-27.   For an amplifier whose bel voltage gain looks like Fig. 18-33*b*, what is the value of its internal break frequency f_2? The value of its gain-bandwidth product?

18-28.   If a voltage follower uses an internal amplifier with the bel voltage gain shown in Fig. 18-33*b*, what will the $f_{2(SP)}$ equal?

18-29.   The SP feedback amplifier of Fig. 18-26*b* uses an internal amplifier whose bel voltage gain looks like Fig. 18-33*b*. What is the $f_{2(SP)}$ of the feedback amplifier?

18-30.   The internal amplifier used in Fig. 18-28*a* has the bel voltage gain shown in Fig. 18-33*b*. What is the value of $f_{2(PP)}$?

18-31.   The ammeter of Fig. 18-30*a* has a resistance of 50 Ω. The internal amplifier has a bel voltage gain like Fig. 18-33*b*. What does $f_{2(SS)}$ equal in position B of the switch? In position D?

18-32.   What does $f_{2(PS)}$ equal in Fig. 18-31 if the internal amplifier has a bel voltage gain like Fig. 18-33*b*?

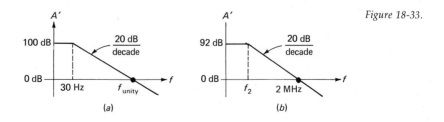

Figure 18-33.

19. Feedback Oscillators

The main use of positive feedback is in *oscillators,* circuits that generate an output signal without an input signal. In an oscillator, part of the output is fed back to the input; this feedback signal is the only input to the internal amplifier.

19-1. THE BASIC IDEA

Before we get to mathematical analysis, let us get the main idea of how an oscillator produces an output signal without an external input signal. To begin with, Fig. 19-1a shows a voltage source v driving the error terminals of the amplifier. The amplified signal Av drives the feedback circuit to produce feedback voltage ABv. This voltage returns to point x. If the phase shift through the amplifier and feedback circuit is correct, the signal at point x will be exactly in phase with the signal driving the error terminals of the amplifier. Stated another way, if the phase shift around the entire loop is $0°$, we have positive feedback.

Loop gain AB

The value of AB (the *loop gain*) is important. We will explain the action of an oscillator in a moment. For now, assume we connect points x and y, and remove voltage source v; the feedback signal drives the error terminals of the amplifier (Fig. 19-1b). If AB is less than unity, ABv is less than v, and the output signal will die out as shown in Fig. 19-1c because we are not returning enough voltage. On the other hand, if AB is greater than unity, ABv is greater than v, and the output voltage builds up (see Fig. 19-1d). Finally, if AB equals unity, there will be no change in output voltage; we will get a steady output like Fig. 19-1e.

Figure 19-1. (a) *Positive feedback.*
(b) *Oscillator.* (c) $AB < 1$. (d)
$AB > 1$. (e) $AB = 1$.

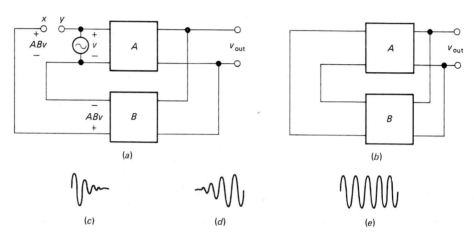

In a practical oscillator the value of loop gain AB is greater than unity when the power is first turned on. A small starting voltage is applied to the error terminals, and the output voltage builds up as shown in Fig. 19-1*d*. After the output voltage reaches a desired level, the value of AB automatically decreases to unity and the output amplitude remains constant (Fig. 19-1*e*).

The starting voltage

The starting voltage of an oscillator is built into every resistor inside the oscillator. As will be discussed in Chap. 21, every resistor generates noise voltages; these voltages are produced by the random motion of electrons in the resistor. The motion is so random it contains sinusoidal frequencies to over 1000 GHz (10^{12} Hz). In other words, each resistor acts like a small voltage source producing essentially all frequencies.

In Fig. 19-1*b* here is what happens. When you first turn on the power, the only signals in the system are noise voltages. These signals are very small in amplitude. All are amplified and appear at the output terminals. The amplified noise now drives the feedback circuit, usually a resonant circuit. Because of this, feedback voltage ABv will be maximum at the resonant frequency of the feedback circuit; furthermore, the phase is correct for positive feedback only at the resonant frequency.

In other words, the amplified noise is filtered so that only one sinusoidal component returns with exactly the right phase for positive feedback. When loop gain AB is greater than unity, the oscillations build up at this frequency (Fig. 19-1*d*). After a suitable level is reached, AB decreases to unity and we get a constant-amplitude output signal (Fig. 19-1*e*).

AB decreases to unity

There are two ways for AB to decrease to unity. First, the increasing signal will eventually force the output stage to clip; when this happens, the value of A decreases. The harder the clipping, the lower the voltage gain. In this way, the gain decreases to whatever value is needed to make AB equal to unity.

Second, we can put something in the feedback circuit to reduce the gain of the feedback circuit. Often, this something is a nonlinear resistor that reduces B when the output signal has reached the desired value. In this manner, the oscillator automatically makes AB equal unity after the oscillations have built up.

Summary

Here are the key ideas behind any feedback oscillator:

1. Initially, loop gain AB must be greater than unity at the frequency where the loop phase is $0°$.
2. After the desired level is reached, AB must decrease to unity by reducing either A or B.

19-2. THE LEAD-LAG NETWORK

Figure 19-2a shows a *lead-lag network,* so called because the phase angle leads for some frequencies and lags for others. There is one frequency where the phase shift exactly

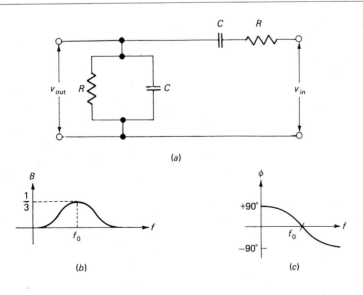

Figure 19-2. Lead-lag network. (a) *Circuit.* (b) *Voltage gain.* (c) *Phase angle.*

equals 0°; this important property allows the lead-lag network to determine the frequency of oscillation.

The main idea

At very low frequencies the series capacitor of Fig. 19-2a looks open and we get no output. At very high frequencies the shunt capacitor looks shorted and we get no output. In between these extremes the output voltage reaches a maximum value (see Fig. 19-2b). As will be proved, the frequency f_0 equals

$$f_0 = \frac{1}{2\pi RC} \qquad \text{(resonant frequency)} \qquad (19\text{-}1)$$

Figure 19-2c shows the phase angle of output voltage with respect to the input voltage. As we see, at very low frequencies the phase angle is positive and the circuit acts like a lead network. On the other hand, at very high frequencies the phase angle is negative and the circuit acts like a lag network. The frequency f_0 produces zero degrees phase shift.

Mathematical analysis

Here is the proof of Eq. (19-1). In Fig. 19-2a, the output voltage is

$$\mathbf{V}_{out} = \frac{\mathbf{Z}_2}{\mathbf{Z}_1 + \mathbf{Z}_2} \mathbf{V}_{in}$$

or

$$\frac{\mathbf{V}_{out}}{\mathbf{V}_{in}} = \frac{\mathbf{Z}_2}{\mathbf{Z}_1 + \mathbf{Z}_2}$$

where

$$\mathbf{Z}_1 = R - jX_C$$

and

$$\mathbf{Z}_2 = R \parallel (-jX_C)$$

With complex algebra we can get these formulas for the magnitude of gain and the phase angle:

$$B = \frac{1}{\sqrt{9 + (X_C/R - R/X_C)^2}} \qquad (19\text{-}2a)$$

and with the arctan function between 0 and 180°,

$$\phi = 90° - \arctan \frac{3}{X_C/R - R/X_C} \qquad (19\text{-}2b)$$

These formulas agree with Fig. 19-2b and c. That is, when the frequency is very low or very high, the value of B goes to zero; B reaches a maximum value of 1/3. Likewise, when f goes to zero, ϕ goes to 90°; when f goes to infinity, ϕ goes to $-90°$; at

$X_C = R$, the phase angle goes to zero. The special condition

$$X_C = R$$

implies

$$\frac{1}{2\pi fC} = R$$

or

$$f = \frac{1}{2\pi RC}$$

which proves Eq. (19-1).

Therefore, the lead-lag network of Fig. 19-2a acts like a resonant circuit; the voltage gain reaches a maximum at f_0, and the phase angle goes to zero at f_0. For this reason, f_0 is the *resonant frequency* of the lead-lag network. The lead-lag network is the key to how a *Wien-bridge oscillator* works; this is the most widely used oscillator for all low-frequency applications (less than 1 MHz).

EXAMPLE 19-1.
Over what frequency range can we tune the resonant frequency of Fig. 19-3?

SOLUTION.
When the capacitors are adjusted to 500 pF, the resonant frequency of the lead-lag network is

$$f_0 = \frac{1}{2\pi RC} = \frac{1}{2\pi(10^5)500(10^{-12})} = 3.18 \text{ kHz}$$

If we adjust the capacitors to 50 pF,

$$f_0 = \frac{1}{2\pi RC} = \frac{1}{2\pi(10^5)50(10^{-12})} = 31.8 \text{ kHz}$$

Therefore, a 10-to-1 change in capacitance produces a 10-to-1 change in resonant frequency.

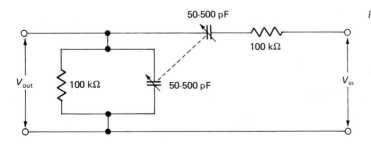

Figure 19-3. Tunable lead-lag network.

19-3. THE WIEN-BRIDGE OSCILLATOR

The Wien-bridge oscillator is the standard oscillator circuit for all low frequencies in the range of 5 Hz to about 1 MHz. It is almost always used in commercial audio generators and is usually the preferred circuit for other low-frequency applications.

The basic idea

Figure 19-4a shows a Wien-bridge oscillator; it uses positive and negative feedback. The positive feedback is through the lead-lag network to the noninverting input. The negative feedback is through the voltage divider to the inverting input.

When we first apply power, the *tungsten lamp* has low resistance and we do not get much negative feedback. For this reason, AB is greater than unity and oscillations can

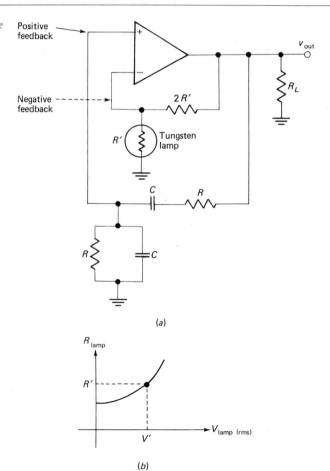

Figure 19-4. Wien-bridge oscillator. (a) *Circuit.* (b) *Lamp curve.*

(a)

(b)

build up at the resonant frequency f_0. As the oscillations build up, the tungsten lamp heats, and this increases its resistance. At a desired output level, the tungsten lamp has a resistance of R'. As we will prove, AB equals unity for this value.

Initial conditions

At first, the tungsten lamp has a resistance *less* than R'. The gain from the noninverting input to the output equals A_{SP}; therefore,

$$A_{SP} = \frac{R_1}{R_2} + 1 > 3$$

This gain is greater than 3 because R_1 equals $2R'$ and R_2 is less than R'. As proved earlier, the voltage gain B of the lead-lag network is 1/3 at the resonant frequency. Therefore, the voltage gain around the positive-feedback loop is initially greater than unity.

As the output voltage builds up, the resistance of the lamp increases as shown in Fig. 19-4b. At a certain voltage V' the tungsten lamp will have a resistance of R'. At this point, the gain from the noninverting input to the output is

$$A_{SP} = \frac{2R'}{R'} + 1 = 3$$

and the gain of the positive feedback loop becomes unity. When this happens, the output amplitude levels off and becomes constant. (In a practical oscillator the tungsten lamp does not become luminescent; this would waste signal power.)

Amplifier phase shift

In a Wien-bridge oscillator, the phase shift of the lead-lag network equals zero when the oscillations have a frequency of

$$f_0 = \frac{1}{2\pi RC}$$

Because of this, we can adjust the frequency by varying the value of R or C. This assumes the phase shift of the amplifier is negligibly small. Stated another way, the amplifier must have an SP break frequency well above the resonant frequency f_0.[1] In this way, the amplifier introduces no additional phase shift. If the amplifier did introduce phase shift, the neat calibration formula $f_0 = 1/2\pi RC$ would no longer hold.

Why called a Wien-bridge oscillator

Figure 19-5 shows another way to draw the oscillator. The lead-lag network is the left side of a bridge and the voltage-divider is the right side. This particular bridge is

[1] Specifically, $f_{\text{unity}}/3 \gg f_0$.

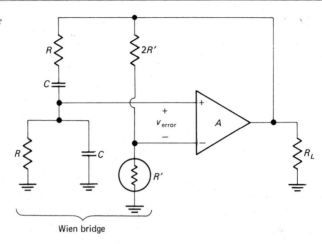

Figure 19-5. Wien-bridge oscillator.

Wien bridge

known as a *Wien bridge* and is used in applications besides oscillators. The error voltage is the output of the bridge. When the bridge approaches balance, the error voltage approaches zero.

The Wien bridge is an example of a *notch filter,* a circuit with zero output at one particular frequency. For the Wien bridge the notch frequency equals

$$f_0 = \frac{1}{2\pi RC}$$

Ideally, we get oscillations at f_0. But actually, we have to have some error voltage no matter how high the internal gain A. In the Wien-bridge oscillator the error voltage is so small the bridge is essentially balanced.

Discrete Wien-bridge oscillator

Figure 19-6 shows a Wien-bridge oscillator built with discrete components. The first stage is a FET because we need the high input impedance; this high input impedance is in shunt with R and therefore has negligible effect on the resonant frequency f_0. If we tried using a bipolar instead of a FET, the low input impedance would shunt R and throw off the calibration formula $1/2\pi RC$.

EXAMPLE 19-2.
Calculate the minimum and maximum frequency in the Wien-bridge oscillator of Fig. 19-7a.

SOLUTION.
The ganged rheostats can vary from 0 to 100 kΩ; therefore, the value of R goes from 1 kΩ to 101 kΩ. The minimum frequency of oscillation is

$$f_0 = \frac{1}{2\pi RC} = \frac{1}{2\pi(101)10^3(0.01)10^{-6}} = 158 \text{ Hz}$$

and the maximum frequency is

$$f_0 = \frac{1}{2\pi RC} = \frac{1}{2\pi(10^3)10^{-8}} \cong 16 \text{ kHz}$$

EXAMPLE 19-3.

Figure 19-7b shows the lamp resistance of Fig. 19-7a. Calculate the output voltage.

SOLUTION.

In Fig. 19-7a the output amplitude becomes constant when the lamp resistance equals 1 kΩ. In Fig. 19-7b this means the lamp voltage is 2 V rms. The current flowing through the lamp also flows through the 2-kΩ resistor, which means a 4-V rms signal across the 2-kΩ resistor; therefore, the output voltage equals the sum of 4 V and 2 V, or

$$V_{\text{out}} = 6 \text{ V rms}$$

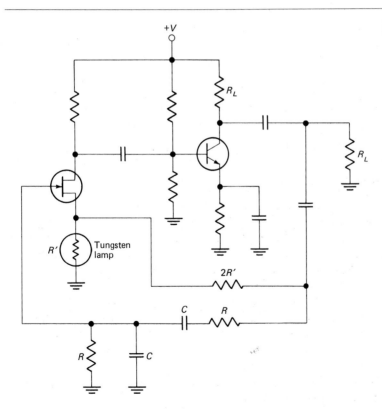

Figure 19-6. Discrete Wien-bridge oscillator.

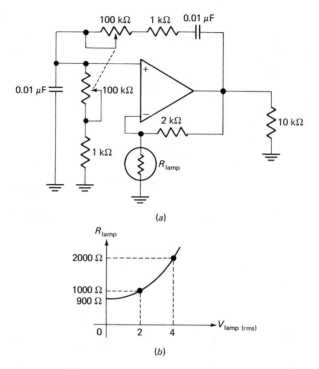

Figure 19-7. Examples 19-2 and 19-3.

(a)

(b)

19-4. PREVENTING UNWANTED OSCILLATIONS

When you build an amplifier with several stages of gain, you can easily get unwanted oscillations. This section briefly tells why they occur and how to prevent them.

Oscillations in negative-feedback amplifiers

Figure 19-8 shows a three-stage PP negative-feedback amplifier. In the midband of the internal amplifier the phase shift is 180° because there are an odd number of stages; therefore, the phase shift around the entire loop is 180° and the feedback is negative.

But *outside the midband* the lead and lag networks produce *additional phase shift*. Because of this, at some low and some high frequency the phase shift around the entire loop is 0° and the feedback becomes positive. For instance, each stage in Fig. 19-8 has a lead network formed by the coupling capacitors (C_2, C_3, and C_4). At some lower frequency these three lead networks can produce a phase shift of 180° (60° each if the networks are identical). Therefore, at this frequency the feedback is positive.

The only way to prevent oscillations outside the midband is to make sure loop gain AB is less than unity when the loop phase shift is 0°. The safest and most widely

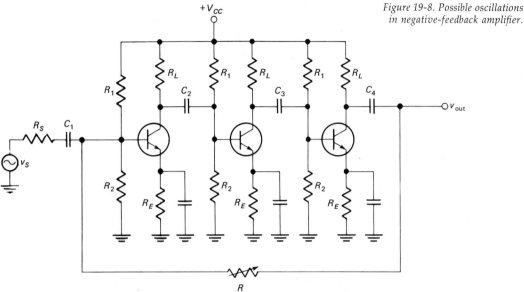

Figure 19-8. Possible oscillations in negative-feedback amplifier.

used method is this: make one of the lead networks and one of the lag networks dominant enough to produce a 20-dB rolloff until the loop voltage gain crosses the 0-dB axis. A 20-dB rolloff at the horizontal crossing means only one lead and one lag network are operating beyond their break frequencies; this implies *the loop phase shift is less than 0° when AB is unity.*

This is the reason we always used an op amp with a 20-dB rolloff in Chap. 18. With most later-generation op amps the dominant lag network is integrated on the chip and automatically provides a 20-dB rolloff until the bel voltage gain equals 0 dB.[2] With earlier-generation op amps like the 709 you have to add external resistors and capacitors to get this 20-dB rolloff; the manufacturer's data sheet tells you what sizes of R and C to use.

Low-frequency oscillations

Look at Fig. 19-9a. There is no feedback path from output to input; therefore, it cannot oscillate. Right? Wrong! There are subtle feedback paths that can make any high-gain amplifier produce unwanted oscillations.

---

[2] A rolloff of 20 dB per decade out to the f_{unity} of the internal amplifier is the simplest and safest way to avoid unwanted oscillations. Other methods are discussed in Graeme, J. G., G. E. Tobey, and L. P. Huelsman: *Operational Amplifiers,* McGraw-Hill Book Company, New York, 1971, pp. 165–197.

Figure 19-9. (a) High-gain
amplifier. (b) Power-supply
impedance. (c) Current feedback
through power-supply impedance.

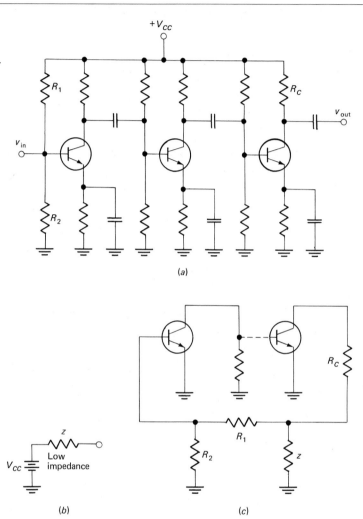

(a)

(b)

(c)

Motorboating is a putt-putt sound from a loudspeaker connected to a circuit like Fig. 19-9*a*. This sound represents very low-frequency oscillations, like a few hertz. The feedback path is *by way of the power supply*. Ideally, the supply looks like a perfect ac short. But, to a better approximation the power supply has a small nonzero impedance *z* as shown in Fig. 19-9*b*. Because of this, we can get current feedback through the supply impedance and through the voltage divider (R_1 and R_2) of the first stage.

For example, Fig. 19-9*c* shows the ac equivalent circuit with the first and last stage. As we see, the ac current out of the last stage flows through R_C and through power-supply impedance *z*. The voltage across *z* then can drive the R_1-R_2 voltage divider in the first stage. The frequency of oscillation is determined by the lead networks in the amplifier and the reactance of the power supply. At some frequency below the midband the phase shift around the entire loop equals 0°. If AB is greater than unity at this frequency, we will get oscillations.

What is the cure for motorboating if it arises? Here are some basic remedies:

1. Make one of the lead networks dominant enough to drop AB under unity before the loop phase shift equals 0°.
2. Use *RC decoupling* networks between stages as shown in Fig. 19-10a; the idea is to reduce the feedback between stages and reduce AB to less than unity.
3. Use a *regulated power supply* (see Chap. 23). This kind of supply has an internal impedance under 0.1 Ω (some as low as 0.0005 Ω); the current feedback then approaches zero.

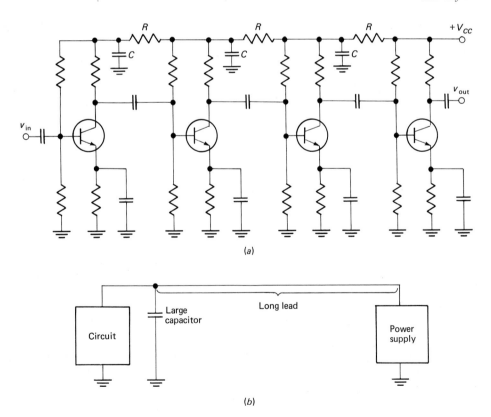

Figure 19-10. (a) Decoupling between stages. (b) Long supply lead may cause oscillations.

Methods 1 and 3 are best. Method 2 often succeeds only in lowering the frequency of oscillation.

A final practical point. Whenever you set up a circuit in the laboratory, watch out for the lead inductance between the power supply and the circuit (see Fig. 19-10b). A long lead may have enough inductance to result in oscillations. If you get oscillations, you can either shorten the lead or add a large capacitor across the circuit as shown.

High-frequency oscillations

You can get unwanted oscillations above the midband of the amplifier. The electric or magnetic fields around the last stage may induce feedback voltages in an earlier stage with the right phase for oscillations. If AB is greater than unity at the frequency where this happens, you will get oscillations.

Figure 19-11a illustrates the idea. The output circuit acts like one plate of a

Figure 19-11. Unwanted coupling from output stage to input. (a) Circuit. (b) Shielding each stage. (c) Baffle shield.

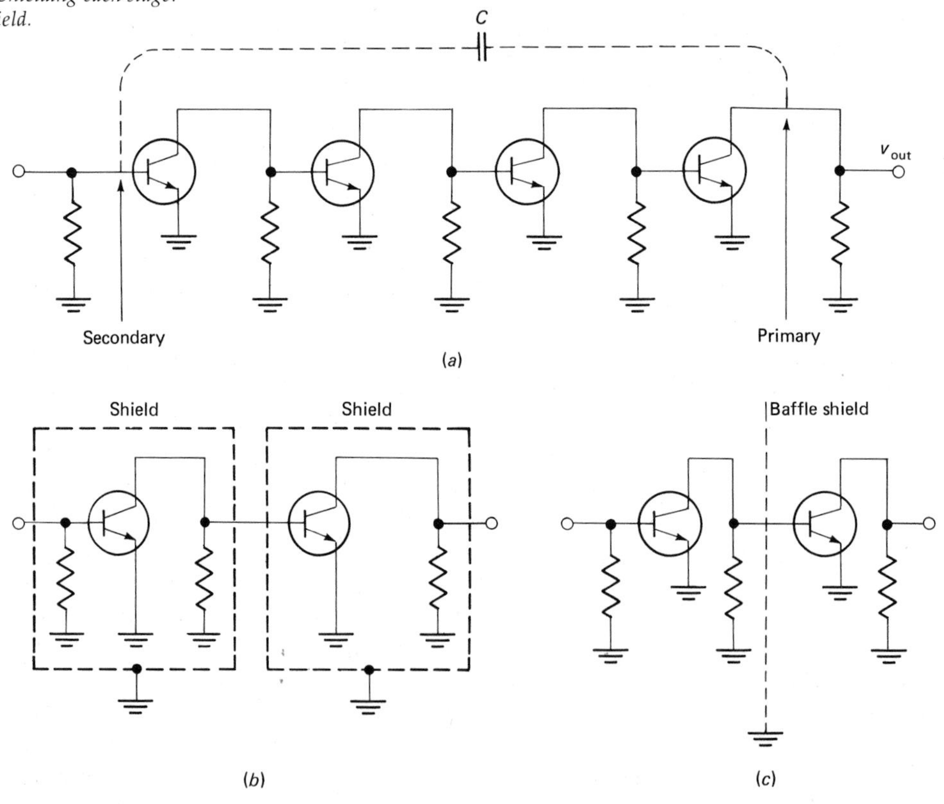

capacitor, the input circuit like the other plate. This capacitance between output and input is small (usually under 1 pF); but at a high frequency, the capacitance may feed back enough signal to produce oscillations.

Magnetic coupling is also possible. The output wire labeled *primary* in Fig. 19-11a can act like the primary winding of a transformer; the input wire labeled *secondary* can act like a secondary winding. As a result, ac current in the primary can induce a voltage in the secondary. If the feedback signal is strong enough and the phase correct, we get oscillations.

What is the cure for high-frequency oscillations? One of the approaches is to increase the distance between stages; this cuts down the coupling between them. If it is not practical to do this, you can enclose each stage in a *shield* or metallic container (see Fig. 19-11b). Shielding like this is common in many high-frequency applications; it blocks high-frequency electric and magnetic fields. If only the capacitive coupling is a problem, a *baffle shield* (a metallic plate) between stages may eliminate high-frequency oscillations (Fig. 19-11c).

Another subtle cause of high-frequency oscillations is the *ground currents.* In Fig. 19-11a the ac grounds do not look like perfect ac shorts at higher frequencies. If the ac ground currents of the last stage happen to flow through part of the chassis being used for an earlier stage, we may get enough unwanted feedback to produce oscillations. A problem like this is cured only by a proper layout of stages, the idea being to prevent ac ground currents of later stages from flowing through ground paths used by earlier stages.

19-5. *LC* OSCILLATORS

Although superb at lower frequencies the Wien-bridge oscillator is not suited to high frequencies (well above 1 MHz). For one thing, the R's and C's of the Wien bridge become too small; for another, the amplifier phase shift becomes a problem.

LC oscillators are standard for high-frequency applications up to 500 MHz. With an amplifier and an *LC* resonant circuit we can feed back a signal with the right amplitude and phase to sustain oscillations. As mentioned earlier, inductors and large capacitors are not practical to integrate on a chip. For this reason, ICs have no advantage over discrete transistors when it comes to *LC* oscillators.

CB oscillators

There are many possible ways to connect an amplifier and *LC* resonant circuit to get oscillations; we will discuss the basic *ac prototypes*. Figure 19-12a shows an *Armstrong oscillator*. In this circuit the collector signal drives an *LC* tank circuit. Because of the dot convention the feedback signal to the emitter is in phase with the collector signal. Since the phase shift from emitter to collector is 0°, the phase shift around the entire loop is 0° at the resonant frequency of the tank circuit. You may encounter varia-

Figure 19-12. Ac prototypes of CB oscillators. (a) *Armstrong.* (b) *Hartley.* (c) *Colpitts.* (d) *Clapp.* (e) *Crystal.*

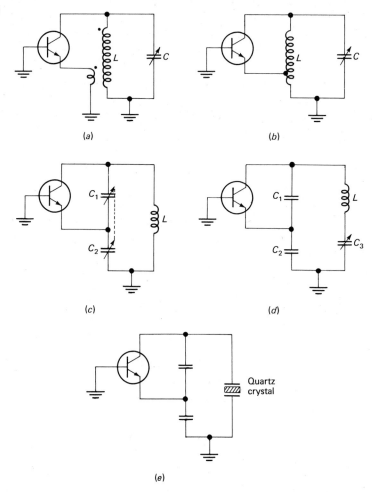

tions of the Armstrong oscillator; the way to recognize it is by the separate winding used for the feedback signal.

Figure 19-12*b* shows the ac prototype of a *Hartley oscillator*. Here the feedback signal comes from a tapped inductor. Because the feedback voltage drives the emitter, it sees a low impedance r'_e. For this reason, the tap is way down on the inductor. The Hartley oscillator is more practical than the Armstrong because it is easier to tune over a wide range. The Hartley oscillator is used a great deal in transistor radios and other entertainment receivers (discussed in Chap. 22).

The *Colpitts oscillator* of Fig. 19-12*c* is a superb circuit, widely used in commercial signal generators operating above 1 MHz. This kind of oscillator relies on a capacitive tap rather than inductive tap. That is, the voltage developed across C_2 is used to drive the emitter. By gang-tuning the two capacitors we can vary the frequency of oscillation.

Figure 19-12*d* shows the ac prototype of a *Clapp oscillator*. This circuit resembles a Colpitts because of the capacitive tap. The inductive branch, however, has a capacitor C_3. We will prove later that a Clapp oscillator produces a more stable frequency than a Colpitts oscillator.

When the accuracy and stability of the oscillator frequency are important, a *quartz-crystal oscillator* is used. In the ac prototype of Fig. 19-12*e* the feedback signal comes from a capacitive tap. As discussed later, the crystal acts like a large inductor in series with a small capacitor (similar to the Clapp).

CE prototypes

CE prototypes are similar to CB prototypes except we apply the feedback signal to the base. Figure 19-13 shows the ac prototypes for CE oscillators. In each of these the

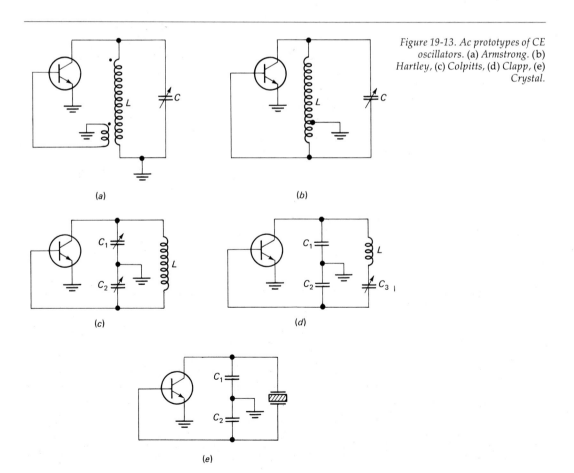

Figure 19-13. Ac prototypes of CE oscillators. (a) Armstrong. (b) Hartley, (c) Colpitts, (d) Clapp, (e) Crystal.

emitter is at ac ground. When the *LC* tank is resonant, it appears purely resistive to the collector. Above and below resonance the tank looks reactive. Only when the tank is at resonance do we get 180° phase shift from base to collector; there is an additional 180° phase shift through the tank to the base. For this reason, the phase shift around the loop is 0° at the resonant frequency.

The two main differences between the CB and CE prototypes are these. First, the input impedance of the base is approximately $\beta r'_e$; this higher input impedance produces less loading on the tank circuit. Second, the upper break frequency of a CB amplifier is much higher than the upper break frequency of a CE amplifier; this gives the CB prototype an advantage over the CE prototype when we are after very high oscillation frequencies.

Figures 19-12 and 19-13 show ac prototypes. Complete oscillator designs include biasing resistors for setting up a quiescent point. For an oscillator to start, *AB* must be greater than unity at the resonant frequency of tank circuit. For this reason, there must be some quiescent collector current to ensure we have voltage gain from base to collector; otherwise, the oscillator will not start. We discuss this later.

19-6. *LC* OSCILLATOR ANALYSIS

What is the resonant frequency of an *LC* oscillator? How large must *A* be for the oscillator to start? What output voltage do we get? These are questions answered in this section.

Resonant frequency

Oscillators operate class A, B, or C. As discussed in Chap. 12, when the collector current is not sinusoidal, you need a high-Q tank circuit to produce a sinusoidal output voltage. For this reason, most oscillators use tank circuits with a Q greater than 10. Basic circuit books prove this formula for the resonant frequency of a parallel tank circuit:

$$f_0 = \frac{1}{2\pi\sqrt{LC}} \sqrt{\frac{Q^2}{1+Q^2}} \tag{19-3a}$$

With this equation we can calculate the exact resonant frequency. Most of the time, however, we can use the approximation

$$f_0 \cong \frac{1}{2\pi\sqrt{LC}} \qquad \text{for } Q \gg 1 \tag{19-3b}$$

This is accurate to better than 1 percent when Q is greater than 10.

Given one of the ac oscillator prototypes, what values of *L* and *C* do you use in Eq. (19-3b)? Use the inductance and capacitance the *circulating* tank current flows through. For instance, in the Colpitts tank of Fig. 19-14a the circulating or loop current *I* flows

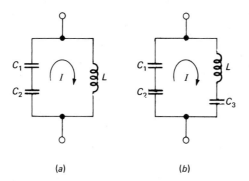

Figure 19-14. Circulating current. (a) In Colpitts tank. (b) In Clapp tank.

(a)   (b)

through L and through C_1 *in series with* C_2. The equivalent capacitance is

$$C = \frac{C_1 C_2}{C_1 + C_2} \quad \text{(Colpitts)} \tag{19-4a}$$

So, if C_1 and C_2 are 100 pF each, you would use 50 pF in Eq. (19-3b).

As another example, the Clapp-oscillator tank of Fig. 19-14b has a circulating current I that flows through the three capacitors in series. Therefore, the equivalent capacitance to use in the formula for resonant frequency is

$$C = \frac{1}{1/C_1 + 1/C_2 + 1/C_3} \quad \text{(Clapp)} \tag{19-4b}$$

As discussed later, in a Clapp oscillator C_3 is deliberately made *much smaller* than C_1 and C_2. Since the capacitors are in series with respect to circulating current, C_3 *dominates*, that is,

$$C \cong C_3 \quad \text{(Clapp)} \tag{19-4c}$$

For instance, if $C_1 = 1000$ pF, $C_2 = 5000$ pF, and $C_3 = 50$ pF,

$$C = \frac{10^{-12}}{1/1000 + 1/5000 + 1/50} \cong 50 \text{ pF}$$

The Armstrong and Hartley oscillators are straightforward; the equivalent L and C are the inductance and capacitance in each tank circuit (see Fig. 19-12a and b). The frequency of a crystal oscillator is discussed in Sec. 19-9.

Starting conditions

The required starting condition for an oscillator is

$$AB > 1$$

at the resonant frequency of the tank circuit. This is equivalent to

$$A > \frac{1}{B}$$

After you have reduced an oscillator circuit to prototype form, you have to calculate the small-signal gain A and the feedback gain B. In the prototypes of Figs. 19-12 and 19-13 the voltage gain is

$$A \cong \frac{r_C}{r_e'} \qquad \text{(small-signal gain)} \qquad (19\text{-}5a)$$

The value of B depends on the particular ac prototype. In any case, the ratio of r_C/r_e' must be greater than $1/B$, or the oscillator will not start.

Figure 19-15a shows the Colpitts tank circuit used in the CB oscillator prototype of Fig. 19-12c. The feedback voltage v_f is across C_2; the output voltage is across the series connection of C_1 and C_2. Therefore,

$$v_f = \frac{X_{C2}}{X_{C1} + X_{C2}} v_{\text{out}}$$

which means a feedback voltage gain B of

$$B = \frac{X_{C2}}{X_{C1} + X_{C2}}$$

Figure 19-15. Voltage gain of feedback network. (a) CB Colpitts. (b) CE colpitts. (c) Armstrong. (d) Hartley.

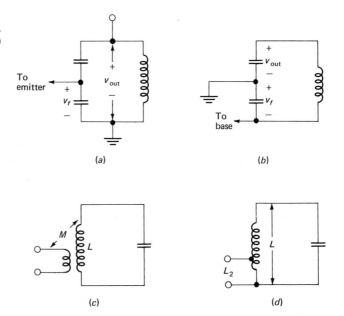

(a)

(b)

(c)

(d)

(The r_e' looking into the base alters this slightly, but the effect is minor.) By substituting $X_C = 1/2\pi fC$ and simplifying, we get

$$B = \frac{C_1}{C_1 + C_2} = \frac{1}{C_2/C_1 + 1}$$

The same formula supplies to the Clapp and crystal oscillators of Fig. 19-12d and e.

For a Colpitts, Clapp, or crystal oscillator to start, therefore, the following condition must be satisfied:

$$\frac{r_C}{r_e'} > \frac{C_2}{C_1} + 1 \qquad \text{(CB types)} \qquad (19\text{-}5b)$$

This is an approximation, but a good one. It also gives us the first clue for selecting the values of capacitors. In other words, we have to use a C_2/C_1 ratio that stays within the inequality. A typical value for C_2/C_1 is 10.

Figure 19-15b shows the CE Colpitts tank circuit. The output voltage is across C_1 only, and the feedback voltage across C_2. By a derivation similar to what we just went through, we can show that

$$B = \frac{C_1}{C_2}$$

and the starting condition is

$$\frac{r_C}{r_e'} > \frac{C_2}{C_1} \qquad \text{(CE types)} \qquad (19\text{-}5c)$$

The Armstrong oscillator has a feedback voltage gain of

$$B = \frac{M}{L}$$

where M is the mutual inductance between primary and secondary, and L is the inductance of the primary (see Fig. 19-15c). Therefore, for an Armstrong oscillator to start,

$$\frac{r_C}{r_e'} > \frac{L}{M} \qquad \text{(Armstrong)} \qquad (19\text{-}6)$$

The inductive tap on a Hartley is usually close to one end whether a CB or CE oscillator is involved; also, the mutual inductance is low in an air-core coil. For these reasons, we can approximate the feedback voltage as

$$v_f \cong \frac{X_{L2}}{X_L} v_{\text{out}} = \frac{L_2}{L} v_{\text{out}}$$

or

$$B \cong \frac{L_2}{L} \qquad \text{(negligible } M\text{)}$$

where L_2 is the tapped-off inductance in Fig. 19-15d, and L is the total inductance. So, the starting condition is

$$\frac{r_C}{r_e'} > \frac{L}{L_2} \quad \text{(Hartley)} \tag{19-7}$$

Output voltage

The exact formula for output voltage depends on the class of operation, the current gain-bandwidth product f_T, and other factors.[3]

Light feedback (small B) results in class A operation. With light feedback the value of A is only slightly larger than $1/B$. When you first turn on the power, the oscillations build up and the signal swings over more and more of the ac load line. With this increased signal swing, the operation changes from small signal to large signal. As described in Chap. 10, *large-signal voltage gain is less than small-signal voltage gain;* therefore, with light feedback the value of AB can decrease to unity without clipping. As an example, we can get light-feedback oscillations with a Colpitts oscillator by using a large C_2/C_1 ratio, provided AB is initially greater than unity. With increasing signal swing, AB drops to unity before clipping occurs.

With *heavy feedback* (large B), clipping occurs at either or both peaks, depending on the oscillator circuit, the amount of feedback, and other factors. This reduces the gain and decreases AB to unity. If the feedback is too heavy, you lose some of the output signal.

In practice, you can adjust the amount of feedback to maximize the output voltage. For instance, with a Colpitts oscillator you increase C_2/C_1 until you get maximum output voltage.

EXAMPLE 19-4.
In the Hartley oscillator of Fig. 19-16, $L = 20 \ \mu H$, $L_2 = 1 \ \mu H$, $C = 1000$ pF, $r_C = 5$ kΩ, and $r_e' = 50 \ \Omega$. Calculate the resonant frequency. Check the starting condition.

SOLUTION.
The resonant frequency approximately equals

$$f = \frac{1}{2\pi\sqrt{LC}} = \frac{1}{2\pi\sqrt{20(10^{-6})10^{-9}}} = 1.12 \text{ MHz}$$

The voltage gain A equals 100, and L/L_2 is 20; therefore,

$$A > \frac{L}{L_2} \tag{19-8}$$

$$100 > 20$$

So, the oscillator will start.

[3] For an accurate analysis, see Walston, J. A., and J. R. Miller, *Transistor Circuit Design*, McGraw-Hill Book Company, New York, 1963, pp. 307–320.

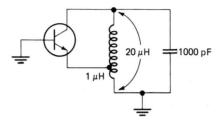

Figure 19-16. Example 19-4.

20 μH   1000 pF

1 μH

19-7.   THE COLPITTS OSCILLATOR

This section discusses the Colpitts oscillator.

The CB oscillator

Figure 19-17a shows a complete Colpitts CB oscillator. The base is ac grounded by the 0.02-μF capacitor. The *RF choke* ideally looks like a dc short but an ac open. The purpose of the RF choke is to prevent loss of signal power that would occur if you used a resistor. The final 2-kΩ load may be a discrete load or it may represent the input resistance of another stage.

When analyzing oscillators, you start by making a dc analysis. Figure 19-17b shows the dc equivalent circuit. At a glance, we recognize voltage-divider bias; the dc emitter current equals

$$I_E \cong \frac{10}{2000} = 5 \text{ mA}$$

This means an r'_e of about 5 Ω.

Figure 19-17c is the simplified ac equivalent circuit. If you work out the series capacitance, you get

$$C = \frac{C_1 C_2}{C_1 + C_2} = 909 \text{ pF}$$

The inductance is 15 μH; therefore, the oscillation frequency is

$$f_0 = \frac{1}{2\pi \sqrt{LC}} = \frac{1}{2\pi \sqrt{15(10^{-6})909(10^{-12})}}$$

or
$$f_0 = 1.36 \text{ MHz}$$

Basic circuit theory tells us the parallel resistance of the tank circuit in Fig. 19-17a equals

$$R_{\text{tank}} = Q_{\text{coil}} X_L = 50(2\pi)1.36(10^6)15(10^{-6})$$

or
$$R_{\text{tank}} = 6.4 \text{ k}\Omega$$

Figure 19-17. Colpitts CB oscillator. (a) Complete circuit. (b) Dc equivalent circuit. (c) Ac equivalent circuit. (d) Ac collector resistance.

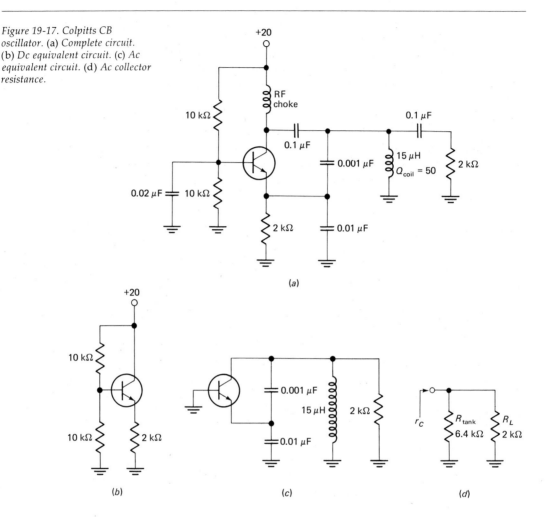

In other words, at resonance the tank circuit looks like a resistance of 6.4 kΩ. The collector sees R_{tank} in parallel with R_L (Fig. 19-17d). Therefore,

$$r_C = R_{\text{tank}} \parallel R_L \tag{19-9}$$

or

$$r_C = 6400 \parallel 2000 \cong 1.5 \text{ k}\Omega$$

Since r'_e is around 5 Ω, the ideal voltage gain is

$$A = \frac{r_C}{r'_e} = \frac{1500}{5} = 300$$

The actual voltage gain will be less than this because of the effects of α, r'_b, r'_c, and so on as discussed in Secs. 15-6 and 15-7. With Eq. (19-5b) the starting condition is satisfied because C_2/C_1 is only 10.

The optimum ratio of C_2/C_1 is found experimentally; you build the circuit and adjust the C_2/C_1 ratio to get maximum output voltage. In the case of Fig. 19-17a, a 10:1 ratio for C_2/C_1 produces a maximum output of approximately 20 V peak-to-peak.

The CE oscillator

Figure 19-18a shows a CE Colpitts oscillator. Again, voltage-divider bias sets up the quiescent point. (You can use any of the biasing circuits discussed in Chap. 7.) Voltage-divider bias has the advantage of setting up a stable quiescent point, and this leads to a reproducible output voltage when you mass-produce.

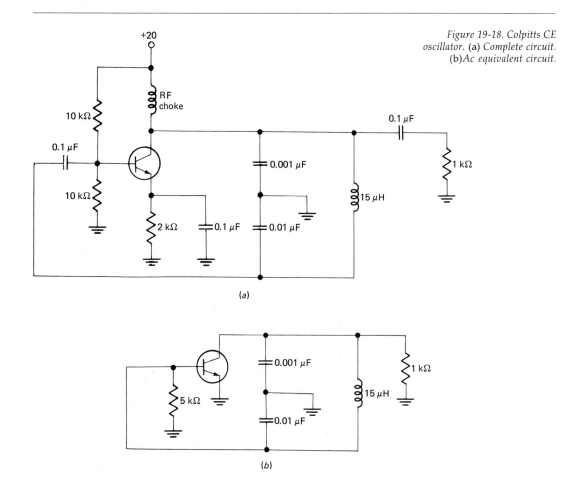

Figure 19-18. Colpitts CE oscillator. (a) Complete circuit. (b)Ac equivalent circuit.

Figure 19-18*b* is the ac equivalent circuit. The oscillation frequency, starting condition, etc., are similar to those for the CB oscillator of Fig. 19-17*a*. Either the CB or CE oscillator may work in a particular application. But often, you may get more output signal with a CB oscillator because the transistor can work to higher frequencies.

19-8. THE CLAPP OSCILLATOR

The Clapp oscillator is basically the same as a Colpitts except that we add a small capacitor in series with the inductor. The main idea behind a Clapp oscillator is to make the resonant frequency depend almost entirely on the value of L and C_3 in Figs. 19-12*d* and 19-13*d*. As shown earlier, the equivalent C in a Clapp oscillator is

$$C = \frac{1}{1/C_1 + 1/C_2 + 1/C_3}$$

By deliberate design, C_1 and C_2 are much larger than C_3; therefore, C_3 is the dominant capacitor and determines the value of C. In symbols,

$$C \cong C_3$$

Ideally, the resonant frequency in a Clapp oscillator is

$$f_0 = \frac{1}{2\pi \sqrt{LC_3}} \qquad \text{(Clapp)} \tag{19-10}$$

Why is this important? Because C_1 and C_2 are shunted by stray capacitances, transistor capacitances, and so on. Even though these additional capacitances are small, they affect the values of C_1 and C_2 slightly. In a Colpitts oscillator, therefore, the exact frequency depends on these extra capacitances. But in a Clapp oscillator C_3 is more important than C_1 and C_2. For this reason, the oscillation frequency of a Clapp oscillator is more stable and accurate than in a Colpitts oscillator. This is why the Clapp oscillator is occasionally used instead of the Colpitts.

19-9. QUARTZ CRYSTALS

Some crystals found in nature exhibit the *piezoelectric effect;* when you apply an ac voltage across them, they vibrate at the frequency of the applied voltage. Conversely, if you mechanically force them to vibrate, they generate an ac voltage. The main substances that produce this piezoelectric effect are *quartz, Rochelle salts,* and *tourmaline.*

Rochelle salts have the greatest piezoelectric activity: for a given ac voltage, they vibrate more than quartz or tourmaline. Mechanically, they are the weakest; they break easily. Rochelle salts have been used to make microphones, phonograph pickups, headsets, and loudspeakers.

Tourmaline shows the least piezoelectric activity but is the strongest of the three. It is also the most expensive. It is occasionally used at very high frequencies.

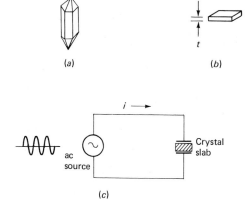

Figure 19-19. (a) Natural shape of quartz crystal. (b) A small slab of quartz. (c) Ac source driving mounted crystal slab.

Quartz is a compromise between the piezoelectric activity of Rochelle salts and the strength of tourmaline. Because it is inexpensive and readily available in nature, quartz is widely used for RF oscillators and filters.

The natural shape of quartz is a hexagonal prism with pyramids at the ends (see Fig. 19-19*a*). To get a usable crystal out of this, we have to slice a rectangular slab out of the natural crystal. Figure 19-19*b* shows this slab with a thickness *t*. The number of slabs we can get from a natural crystal depends on the size of the slabs and the angle of the cut.

There are a number of different ways to cut the natural crystal; these cuts have names like the *X* cut, *Y* cut, *XY* cut, and *AT* cut. For our purposes, all we need to know is the cuts have different piezoelectric properties. (Manufacturers' catalogs are usually the best source of information on different cuts and their properties.)

For use in electronic circuits the slab must be mounted between two metal plates as shown in Fig. 19-19*c*. In this circuit the amount of crystal vibration depends on the frequency of the applied voltage. By changing the frequency we can find resonant frequencies where the crystal vibrations reach a maximum. Since the energy for the vibrations must be supplied by the ac source, the ac current maximizes at each resonant frequency.

Fundamental and overtones

Most of the time, the crystal is cut and mounted to vibrate best at one of its resonant frequencies, usually the *fundamental* or lowest frequency. The higher resonant frequencies are called *overtones;* these frequencies are almost exact multiples of the fundamental frequency. As an example, a crystal with a fundamental frequency of 1 MHz has a first overtone of approximately 2 MHz, a second overtone of approximately 3 MHz, and so on.

The formula for the fundamental frequency of a crystal is

$$f = \frac{K}{t} \tag{19-11}$$

where K is a constant that depends on the cut and other factors, and t is the thickness of the crystal. As we see, the fundamental frequency is inversely proportional to thickness. For this reason, there is a practical limit on how high we can go in frequency. The thinner the crystal, the more fragile it becomes and the more likely it is to break because of vibrations.

Quartz crystals work well up to 10 MHz on the fundamental frequency. To reach higher frequencies we can use a crystal mounted to vibrate on overtones; in this way, we can reach frequencies up to 100 MHz. Occasionally, the more expensive but stronger tourmaline is used at higher frequencies.

Ac equivalent circuit

What does the crystal look like as far as the ac source is concerned? When the mounted crystal of Fig. 19-20a is not vibrating, it is equivalent to a capacitance C_m because it has two metal plates separated by a dielectric. C_m is known as the *mounting capacitance*.

However, when the crystal is vibrating, it looks like a tuned circuit. Figure 19-20b shows the ac equivalent circuit of a crystal vibrating at or near its fundamental frequency. Typical values of L are in henrys, C_s in fractions of a picofarad, R in hundreds of ohms, and C_m in picofarads. As an example, here are the values for one available crystal: $L = 3$ H, $C_s = 0.05$ pF, $R = 2000\ \Omega$, and $C_m = 10$ pF. Among other things, the cut, thickness, and mounting of the slab affect these values.

The outstanding feature of crystals compared with discrete LC tank circuits is their incredibly high Q. For the values just given, we can calculate a Q over 3000. Q's can easily be over 10,000. On the other hand, a discrete LC tank circuit seldom has a Q over 100. The extremely high Q of a crystal leads to oscillators with very stable values of frequency.

Figure 19-20. (a) Quartz crystal.
(b) Ac equivalent circuit.

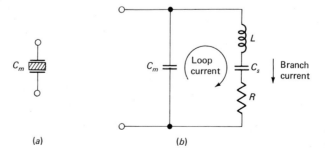

(a)   (b)

Series and parallel resonance

Besides the Q, L, C_s, R, and C_m of the crystal, there are two other characteristics we should know about. The *series resonant frequency* f_s of a crystal is the resonant frequency of the *LCR branch* in Fig. 19-20b. At this frequency the branch current reaches a maximum value because L resonates with C_s. The formula for this series resonant frequency is

$$f_s = \frac{1}{2\pi \sqrt{LC_s}} \tag{19-12a}$$

The *parallel resonant frequency* f_p of the crystal is the frequency where the circulating or *loop* current of Fig. 19-20b reaches a maximum value. Since this loop current must flow through the series combination of C_s and C_m, the equivalent C is

$$C_{\text{loop}} = \frac{C_m C_s}{C_m + C_s} \tag{19-12b}$$

and the parallel resonant frequency is

$$f_p = \frac{1}{2\pi \sqrt{LC_{\text{loop}}}} \tag{19-12c}$$

Two capacitances in series always produce a capacitance smaller than either; therefore, C_{loop} is less than C_s, and f_p is greater than f_s.

In any crystal, C_s is much smaller than C_m. For instance, with the values given earlier, C_s was 0.05 pF and C_m was 10 pF. Because of this, Eq. (19-12b) gives a value of C_{loop} just slightly less than C_s. In turn, this means f_p is only slightly more than f_s. When you use a crystal in an oscillator circuit like Fig. 19-21, the additional circuit capacitances appear in shunt with C_m. Because of this, *the oscillation frequency will lie between f_s and f_p*. This is the advantage of knowing the values of f_s and f_p; they set the lower and upper limits on the frequency of a crystal oscillator.

Crystal stability

The frequency of an oscillator tends to change slightly with time; this *drift* is produced by temperature, aging, and other causes. In a crystal oscillator the frequency

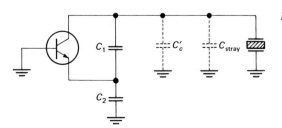

Figure 19-21. Effect of transistor and stray capacitance.

drift with time is very small, typically less than 1 part in 10^6 (0.0001 percent) per day. Stability like this is important in *electronic wristwatches;* they use quartz-crystal oscillators as the basic timing device.

By using crystal oscillators in precision temperature-controlled ovens, crystal oscillators have been built with frequency drifts less than 1 part in 10^{10} per day. Stability like this is needed in *frequency and time standards.* To give you an idea of how precise 1 part in 10^{10} is, a clock with this drift will take 300 years to gain or lose 1 s.

EXAMPLE 19-5.
A crystal has these values: $L = 3$ H, $C_s = 0.05$ pF, $R = 2000$ Ω, and $C_m = 10$ pF. Calculate the f_s and f_p of the crystal to three significant digits.

SOLUTION.
With Eq. (19-12a),

$$f_s = \frac{1}{2\pi \sqrt{LC_s}} = \frac{1}{2\pi \sqrt{3(0.05)10^{-12}}} = 411 \text{ kHz}$$

With Eq. (19-12b),

$$C_{\text{loop}} = \frac{(10 \text{ pF})(0.05 \text{ pF})}{10 \text{ pF} + 0.05 \text{ pF}} = 0.0498 \text{ pF}$$

With Eq. (19-12c),

$$f_p = \frac{1}{2\pi \sqrt{LC_{\text{loop}}}} = \frac{1}{2\pi \sqrt{3(0.0498)10^{-12}}} = 412 \text{ kHz}$$

If this crystal is used in an oscillator, the frequency of oscillation must lie between 411 and 412 kHz.

Problems

19-1. Calculate the voltage gain of a lead-lag network for each of these conditions:

1. $R/X_C = 0.1$
2. $R/X_C = 1$
3. $R/X_C = 10$

19-2. Work out the phase angle for conditions given in the preceding problem.

19-3. Calculate the minimum and maximum resonant frequency in position A of Fig. 19-22.

19-4. In Fig. 19-22 what are the minimum and maximum resonant frequency you can get in each switch position?

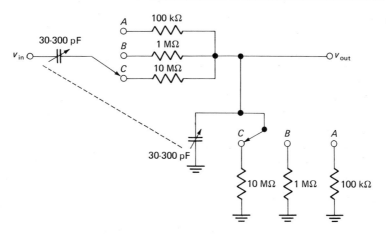

Figure 19-22.

19-5. The Wien-bridge oscillator of Fig. 19-23*a* uses a lamp with the characteristics of Fig. 19-23*b*. How much output voltage is there?

19-6. Position *D* in Fig. 19-23*a* is the highest frequency range of the oscillator. We can vary the frequency continuously by the ganged rheostats shown. What is the minimum and maximum frequency of oscillation on this range?

19-7. Calculate the minimum and maximum frequency of oscillation for each position of the ganged switch in Fig. 19-23*a*.

19-8. To change the output voltage of Fig. 19-23*a* to a value of 6 V rms, what change can you make?

19-9. In Fig. 19-23*a*, suppose we replace the 20-kΩ ganged rheostats by two carbon microphones. These microphones have resistances that fluctuate in step with the sound waves hitting them. Suppose both microphones pick up exactly the same sound. The quiescent frequency of oscillation is 500 kHz.

1. What is the frequency of oscillation when the microphones pick up no sound?
2. If the microphones pick up a sinusoidal sound wave, what happens to the oscillation frequency?

19-10. In Fig. 19-23*a* the break frequency of the amplifier with negative feedback is at least one decade above the highest frequency of oscillation. What is the break frequency?

19-11. Figure 19-24 shows one way to build a *phase-shift oscillator*. At the frequency of oscillation, the phase shift of the 741C is −270°; this is the sum of 180° plus an additional −90° produced by the internal lag network of a 741C. Each external lag network (10 kΩ and 0.001 μF) produces −45°; in this way, the loop phase shift is −360°, or equivalently, 0°.

Figure 19-23.

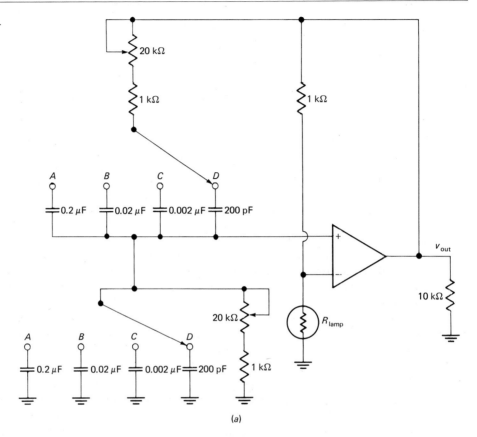

(a)

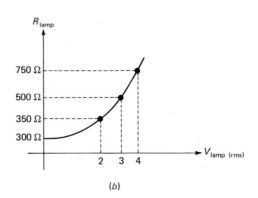

(b)

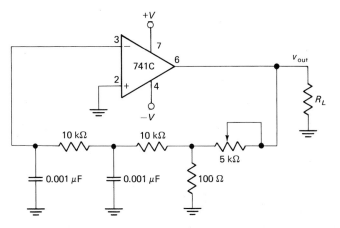

Figure 19-24.

In other words, we get oscillations at the break frequency of the external lag networks. Calculate the approximate frequency using $1/2\pi RC$. (The 5-kΩ rheostat has nothing to do with frequency; it allows you to change the output amplitude.)

19-12. Sometimes, you may need a rough estimate for the frequency of motorboating. Here is one way to get it. Figure 19-25a shows part of the last stage in amplifier. The 10 kΩ is the R_C of the stage and the 30 μF is the output capacitor of the power supply. The lag angle produced by this network will add to the lead angle produced by the lead networks in the rest of the amplifier. You can get oscillations when AB is greater than unity and the loop phase shift is 0°. As a rough estimate, this may occur at the break frequency of the lag network shown, that is, $f = 1/2\pi RC$. Estimate the motorboating frequency using the given R and C value.

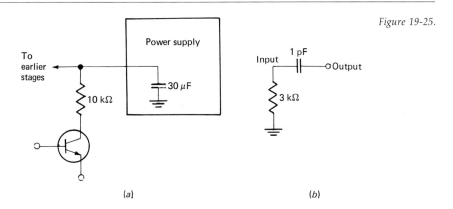

Figure 19-25.

(a)   (b)

Figure 19-26.

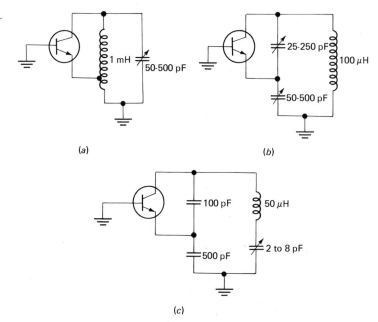

(a)

(b)

(c)

19-13. As discussed in Sec. 19-4, you can get high-frequency oscillations when enough output signal is capacitively coupled back to the input. Suppose the capacitance of Fig. 19-11a is 1 pF. If the input resistance is 3 kΩ as shown in Fig. 19-25b, you can estimate the frequency of oscillation by calculating the break frequency of Fig. 19-25b. What is this break frequency?

19-14. Figure 19-26a shows part of a Hartley oscillator. Calculate the minimum and maximum frequency of oscillation.

19-15. Work out the minimum and maximum frequency of oscillation for the Colpitts ac circuit shown in Fig. 19-26b. (C_2/C_1 always equals 2.)

19-16. Calculate the lowest and highest oscillation frequencies in the Clapp oscillator of Fig. 19-26c.

19-17. In the Clapp oscillator of Fig. 19-26c approximately what value of A do we need to start the oscillator?

19-18. In the Hartley oscillator of Fig. 19-26a, if the tapped inductance is 50 μH, what is the minimum value of A needed to start the oscillator?

19-19. What is the approximate value of dc emitter current in Fig. 19-27? The dc voltage from collector to emitter?

19-20. What is the approximate frequency of oscillation in Fig. 19-27? The value of B? For the oscillator to start, what is the minimum value of A?

19-21. A crystal has a fundamental frequency of 5 MHz. What is the approximate value of the first overtone frequency? The second overtone? The third?

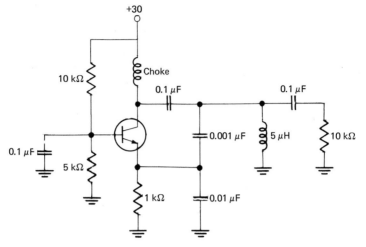

Figure 19-27.

19-22. A crystal has a thickness of t. If you reduce t by 1 percent, what happens to the frequency?

19-23. The ac equivalent circuit of a crystal has these values: $L = 1$ H, $C_s = 0.01$ pF, $R = 1000\ \Omega$, and $C_m = 20$ pF.

1. What is the series resonant frequency?
2. What is the Q at this frequency?

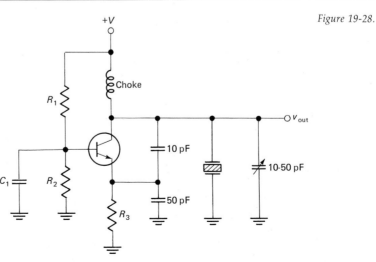

Figure 19-28.

19-24. The output voltage of an oscillator is not easy to calculate because it depends on the f_T of the transistor and many other factors. You can get a rough estimate as follows: the peak output voltage is V_{CEQ} or $I_{CQ}r_C$, whichever is smaller. If the r_C in Fig. 19-27 equals 1.5 kΩ, what is the peak output voltage?

19-25. To *pull* the crystal frequency means to change the frequency of oscillation slightly. Fig. 19-28 shows how it is done. The actual frequency of oscillation must lie between f_s and f_p. If the crystal has an f_s of 0.999 MHz and an f_p of 1.02 MHz, describe what you think the tuning capacitor is for.

19-26. The constant K of an X-cut crystal is 112.6 kHz in. What is the resonant frequency when t equals 0.1 in? When t equals 0.005 in?

20. The Frequency Domain

In our previous work, we have emphasized *time-domain analysis* by working out voltages from one instant to the next. But time-domain analysis is not the only method. This chapter discusses *frequency-domain analysis*.

20-1. THE FOURIER SERIES

Figure 20-1*a* shows a sine wave with a peak V_p and a period T. Ac-dc books concentrate on the sine wave because it is the most fundamental of all waves. Earlier chapters have likewise emphasized the sine wave. Now, we are ready to examine nonsinusoidal waves.

Periodic waves

The *triangular* wave of Fig. 20-1*b* traces its basic pattern during period T; after this, each cycle is a repetition of the first cycle. It is the same with the sawtooth of Fig. 20-1*c* and the half-wave signal of Fig. 20-1*d*; all cycles are copies of the first cycle. Waveforms with repeating cycles are called *periodic;* they have a period T in which the size and shape of every cycle is determined.

Harmonics

The sine wave is extraordinary. By adding sine waves with the right amplitude and phase, we can produce the triangular wave of Fig. 20-1*b*. With a different combination of sine waves we can get the sawtooth wave of Fig. 20-1*c*. And with still another

Figure 20-1. Periodic waves. (a) Sinusoidal. (b) Triangular. (c)Sawtooth. (d) Half-wave-rectified.

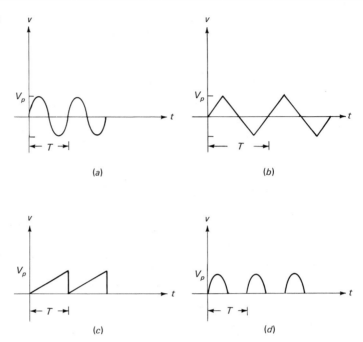

combination of sine waves we can produce the half-wave signal of Fig. 20-1*d*. In other words, *any periodic wave is a superposition of sine waves.*

These sine waves are harmonically related, meaning their frequencies are *harmonics* (multiples) of a *fundamental* (lowest frequency). Given a periodic wave, you can measure its period T on an oscilloscope. The reciprocal of T equals the fundamental frequency. Symbolically,

$$f_1 = \frac{1}{T} \qquad \text{(fundamental)} \qquad\qquad \text{(20-1)}***$$

The second harmonic has a frequency of

$$f_2 = 2f_1$$

The third harmonic has a frequency of

$$f_3 = 3f_1$$

In general, the *n*th harmonic has a frequency of

$$f_n = nf_1 \qquad\qquad \text{(20-2)}***$$

As an example, Fig. 20-2 shows a sawtooth wave on the left. This periodic wave is equivalent to the sum of the harmonically related sine waves on the right. Also

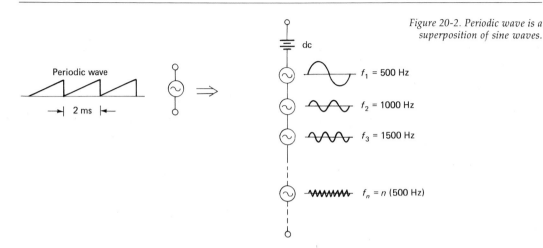

Figure 20-2. Periodic wave is a superposition of sine waves.

included is a battery to account for the average or dc value of the sawtooth wave. Since period T equals 2 ms, the harmonic frequencies equal

$$f_1 = \frac{1}{0.002} = 500 \text{ Hz}$$

$$f_2 = 2(500 \text{ Hz}) = 1000 \text{ Hz}$$

$$f_3 = 3(500 \text{ Hz}) = 1500 \text{ Hz}$$

and so on.

Formula for Fourier series

As a word formula, here is what Fig. 20-2 says:

Periodic wave = dc component + first harmonic
+ second harmonic + third harmonic + · · · + nth harmonic

In precise mathematical terms,

$$v = V_0 + V_1 \sin (\omega t + \phi_1) + V_2 \sin (2\omega t + \phi_2)$$
$$+ V_3 \sin (3\omega t + \phi_3) + \cdots + V_n \sin (n\omega t + \phi_n)$$

This famous equation is called the *Fourier series*.[1] It says a periodic wave is a superposition of harmonically related sine waves. Voltage v is the value of the periodic wave at any instant in time; we can calculate this value by adding the dc component and the instantaneous values of the harmonics.

---

[1] Fourier is pronounced foo' ree ay (ay as in say).

The first term in the Fourier series is V_0; this is a constant and represents the dc component. The coefficients V_1, V_2, V_3, . . . , V_n are peak values of the harmonics. The angles ϕ_1, ϕ_2, ϕ_3, . . . , ϕ_n are phase angles of the harmonics. Radian frequency ω equals $2\pi f_1$; as we see, each succeeding term in the Fourier series represents the next higher harmonic.

Theoretically, the harmonics continue to infinity, that is, n has no upper limit. Often, five to ten harmonics are enough to synthesize a periodic wave to within 5 percent. With the right combination of amplitudes (V_1, V_2, V_3, . . . , V_n) and angles (ϕ_1, ϕ_2, ϕ_3, . . . , ϕ_n) we can produce any periodic waveform.

EXAMPLE 20-1.
What frequencies do the harmonics of Fig. 20-3a have?

SOLUTION.
The periodic wave is a full-wave-rectified signal with a period of 1/120 s. Therefore, the reciprocal of the period equals

$$f_1 = 120 \text{ Hz}$$

and the higher harmonics have frequencies of

$$f_2 = 240 \text{ Hz}$$

$$f_3 = 360 \text{ Hz}$$

and so on.

EXAMPLE 20-2.
What frequencies do the first three odd harmonics of Fig. 20-3b have?

SOLUTION.
The period is 50 μs; therefore, the fundamental has a frequency of

$$f_1 = \frac{1}{50(10^{-6})} = 20 \text{ kHz}$$

The next odd harmonic has a frequency of

$$f_3 = 3(20 \text{ kHz}) = 60 \text{ kHz}$$

Figure 20-3. (a) Full-wave signal.
(b) Square wave.

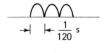

$\dashrightarrow$ $\vdash$ $\dfrac{1}{120}$ s

(a)

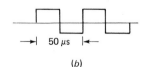

$\dashrightarrow$ 50 μs $\vdash$

(b)

The third odd harmonic has a frequency of

$$f_5 = 5(20 \text{ kHz}) = 100 \text{ kHz}$$

EXAMPLE 20-3.
Illustrate the superposition of harmonics to produce a square wave.

SOLUTION.
The square wave has only odd harmonics (discussed in the next section). When we add the first and third harmonics shown by the dashed lines of Fig. 20-4*a*, we get a new wave. Already, this looks more like a square wave than a sine wave.

By adding more harmonics, we can approach the square wave. Figure 20-4*b* shows the addition of the fifth harmonic. If we continue like this, we eventually get a flat top and vertical sides.

The example brings out these ideas:

1. Each harmonic must have exactly the right amplitude and phase to produce a given periodic wave. For instance, if the peak value of the third harmonic in Fig. 20-4*a* is too large, the dip in the new wave will be too large; it will be impossible to correct for this excessive dip with higher harmonics.
2. The Fourier theorem (superposing sine waves to get a periodic wave) seems

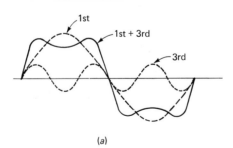

(a)

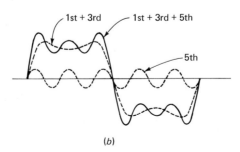

(b)

Figure 20-4. Adding sine waves to get a square wave.

plausible now. Limited as this example is, it does appear we can add sine waves to produce a periodic wave. The formal proof of the Fourier theorem is found in advanced books.

20-2. THE SPECTRUM OF A SIGNAL

The Fourier theorem is the key to frequency-domain analysis. Since we already know a great deal about sine waves, we can reduce a periodic wave to its sine-wave components; then, by analyzing these sine waves, we are indirectly analyzing the periodic wave. In other words, there are two approaches in nonsinusoidal circuit analysis. We can figure out what a periodic wave does at each instant in time, or we can figure out what each harmonic does. Sometimes, the first approach (time-domain analysis) is faster. But often, the second approach (frequency-domain analysis) is superior.

Spectral components

Suppose A represents the peak-to-peak value of a sawtooth wave. With advanced mathematics we can prove

$$V_n = \frac{A}{n\pi} \tag{20-3}$$

This says the peak value of the nth harmonic equals A divided by $n\pi$. For example, Fig. 20-5a shows a sawtooth wave with a peak-to-peak value of 100 V; the harmonics have these peak values:

$$V_1 = \frac{100}{\pi} = 31.8 \text{ V}$$

$$V_2 = \frac{100}{2\pi} = 15.9 \text{ V}$$

$$V_3 = \frac{100}{3\pi} = 10.6 \text{ V}$$

Figure 20-5. Sawtooth wave. (a) Time domain. (b) Frequency domain.

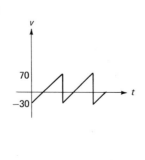

(a)

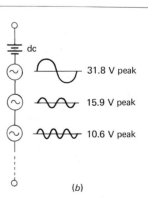

(b)

and so on. Figure 20-5b shows the first three harmonics for the sawtooth of Fig. 20-5a.

With an oscilloscope (a time-domain instrument) you see the periodic signal as a function of time (Fig. 20-5a). The vertical axis represents voltage and the horizontal axis stands for time. In effect, the oscilloscope displays the instantaneous value v of the periodic wave.

The *spectrum analyzer* differs from the oscilloscope. To begin with, a spectrum analyzer is a frequency-domain instrument; its horizontal axis represents frequency. With a spectrum analyzer we see the *harmonic peak values versus frequency*. For instance, if the sawtooth of Fig. 20-6a drives a spectrum analyzer, we will see the display of Fig. 20-6b. We call this kind of display a *spectrum;* the height of each line represents the harmonic peak value; the horizontal location gives the frequency.

Every periodic wave has a spectrum or set of vertical lines representing the harmonics. The spectrum normally differs from one periodic signal to the next. For instance, a square wave like Fig. 20-6c has a spectrum like Fig. 20-6d; this is different from the sawtooth spectrum of Fig. 20-6b.

Four basic spectra

For future reference, Fig. 20-7 shows four periodic waves and their spectra.[2] In each of these, A is the peak-to-peak value of the periodic wave. For convenience, we

[2] For other spectra, see *Reference Data for Radio Engineers,* Howard W. Sams & Co., Inc., New York, 1968, pp. 42-12 to 42-14.

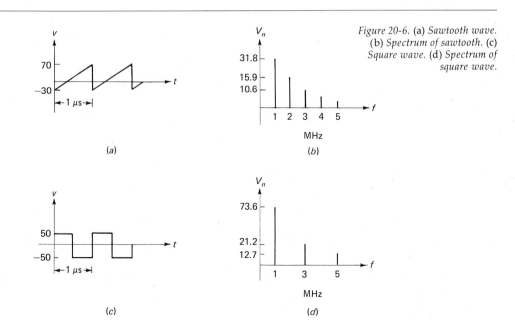

Figure 20-6. (a) *Sawtooth wave.* (b) *Spectrum of sawtooth.* (c) *Square wave.* (d) *Spectrum of square wave.*

Figure 20-7. Waveforms and spectra. (a) Sawtooth. (b) Square wave. (c) Full-wave-rectified. (d) Triangular.

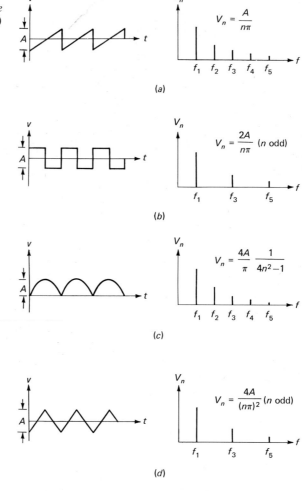

have shown only the harmonics to $n = 5$. In each case, the formula for harmonic peak values is given. For example, with a full-wave signal like Fig. 20-7c, we can calculate the peak value of any harmonic by using

$$V_n = \frac{4A}{\pi} \frac{1}{4n^2 - 1} \qquad (20\text{-}4a)$$

Or if we want the peak values of harmonics for the square wave of Fig. 20-7b, we use

$$V_n = \frac{2A}{n\pi} \qquad (n \text{ odd only}) \qquad (20\text{-}4b)$$

Half-wave symmetry

Of the waveforms in Fig. 20-7, two have *only odd harmonics*. Many other waveforms have this odd-harmonic property, and it helps to know a quick method for identifying such waveforms.

Any waveform with *half-wave symmetry* has the odd-harmonic property. Half-wave symmetry means you can invert the negative half cycle and get an exact duplicate of the positive half cycle. For example, Fig. 20-8a shows one period of a triangular wave. After inverting the negative half cycle, you have the *abc* half cycle (dashed line). This inverted half cycle is an exact duplicate of the positive half cycle. As a result, the triangular wave contains only odd harmonics. Similarly, the square wave of Fig. 20-8b has half-wave symmetry because the inverted negative cycle exactly duplicates the positive half; this means a square wave contains only odd harmonics.

In case of doubt, it helps to shift the inverted half cycle to the left; half-wave symmetry exists only if the shifted half cycle superimposes the positive half cycle. For instance, inverting the negative half cycle of a sawtooth wave produces the *abc* half cycle of Fig. 20-8c. When shifted left, the inverted half cycle exactly superimposes the positive half cycle; therefore, we are assured of half-wave symmetry and the odd-harmonic property.

The waveform of Fig. 20-8d does *not* have half-wave symmetry. When the negative half cycle is inverted and shifted left, it does not superimpose the positive half cycle. Therefore, Fig. 20-8d does not have the odd-harmonic property; it must contain at least one even harmonic.

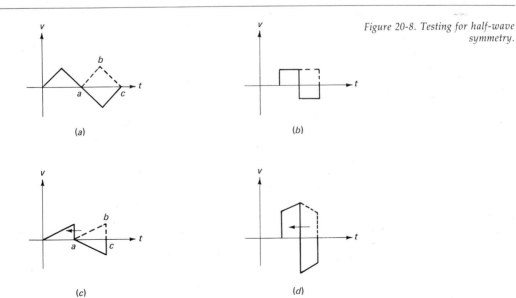

Figure 20-8. Testing for half-wave symmetry.

The dc component

The dc component is the average value of the periodic wave, defined as

$$V_0 = \frac{\text{area under one cycle}}{\text{period}} \qquad \text{(20-5)}^{***}$$

As an example, Fig. 20-9a shows a sawtooth wave with a peak of 10 V and a period of 2 s. The area under one cycle is shaded and equals

$$\text{Area} = \tfrac{1}{2}\,(\text{base} \times \text{height})$$
$$= \tfrac{1}{2}\,(2 \text{ s} \times 10 \text{ V}) = 10 \text{ V s}$$

Dividing by the period gives the average value of the sawtooth:

$$V_0 = \frac{10 \text{ V s}}{2 \text{ s}} = 5 \text{ V}$$

The area is positive when above the horizontal axis, but negative below. If part of a cycle is above and part below the horizontal axis, you algebraically add the positive and negative areas to get the net area under the cycle. Figure 20-9b shows a square wave that swings from $+70$ to -30 V. The first half cycle has an area above the horizontal axis; therefore, this area is positive. But the second half cycle is below the horizontal

Figure 20-9. Areas and average values.

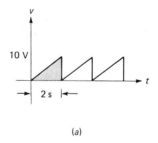

(a)

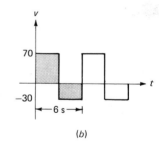

(b)

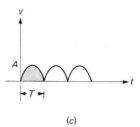

(c)

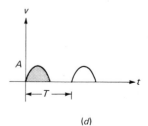

(d)

axis, so that the area is negative. Here is how to find the average value over the entire cycle:

$$\text{Positive area} = 3 \text{ s} \times 70 \text{ V} = 210 \text{ V s}$$

$$\text{Negative area} = 3 \text{ s} \times -30 \text{ V} = -90 \text{ V s}$$

$$\text{Net area under one cycle} = 210 - 90 = 120 \text{ V s}$$

Dividing the net area by the period gives

$$V_0 = \frac{120 \text{ V s}}{6 \text{ s}} = 20 \text{ V}$$

This is the average value for the entire cycle of Fig. 20-9b; it is the value a dc voltmeter would read.

Using the area-over-period formula, you can calculate the average value of other periodic waves, provided the waves have linear segments. For instance, sawtooth, square, and triangular waves are made up of straight lines. Because of this, you can use well-known geometry formulas to calculate areas; after dividing by period, you have the average value.

But what do you do with half-wave and full-wave signals like Fig. 20-9c and d? The waveforms are nonlinear and no simple geometry formulas are available for areas. The only way to calculate the areas is with calculus; using calculus, we can derive these average values:

$$V_0 = 0.636A \qquad \text{(full wave)} \qquad \qquad (20\text{-}6)^{***}$$

and

$$V_0 = 0.318A \qquad \text{(half wave)} \qquad \qquad (20\text{-}7)^{***}$$

Effect of dc component on spectrum

Figure 20-10a shows a triangular wave. Since it has half-wave symmetry, the spectrum contains only odd harmonics (Fig. 20-10b). If we add a dc component to the triangular wave, the wave moves up as shown in Fig. 20-10c. The only change in the spectrum is the appearance of a line at zero frequency (Fig. 20-10d). The height of this line represents the dc voltage. In general, adding a dc component to a waveform has no effect on the harmonics; the only spectral change is a new line at zero frequency.

EXAMPLE 20-4.

A square wave has a peak-to-peak value of 25 V and a period of 5 μs. Calculate the peak value and frequency of the ninth harmonic.

SOLUTION.

With Fig. 20-7b, the ninth-harmonic peak value is

$$V_n = \frac{2A}{n\pi} = \frac{2(25)}{9\pi} = 1.77 \text{ V}$$

Figure 20-10. Effect of a dc component on the spectrum.

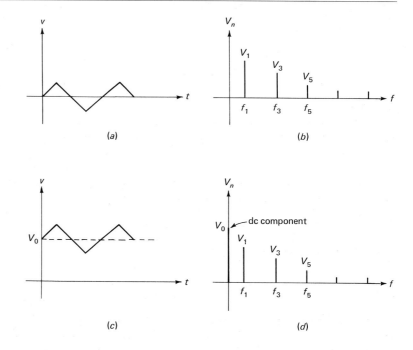

(a)   (b)

(c)   (d)

Since the period is 5 μs, the fundamental frequency equals

$$f_1 = \frac{1}{5(10^{-6})} = 200 \text{ kHz}$$

The frequency of the ninth harmonic is

$$f_9 = 9(200 \text{ kHz}) = 1.8 \text{ MHz}$$

EXAMPLE 20-5.
A 100-V-peak sine wave drives a full-wave rectifier as shown in Fig. 20-11a. The ideal output is a full-wave signal with a 100-V peak. Calculate the dc component and peak values of the first three harmonics of rectified voltage.

SOLUTION.
With Eq. (20-6),

$$V_0 = 0.636A = 0.636(100) = 63.6 \text{ V}$$

Referring to Fig. 20-7c,

$$V_n = \frac{4A}{\pi} \frac{1}{4n^2 - 1}$$

With this, we can calculate the peak values:

$$V_1 = 42.4 \text{ V}$$

$$V_2 = 8.49 \text{ V}$$

$$V_3 = 3.64 \text{ V}$$

EXAMPLE 20-6.

The input sine wave of Fig. 20-11a has a frequency of 60 Hz. Calculate the first three harmonic frequencies of the rectified voltage and show the input and output spectra.

SOLUTION.

The period of the input signal is 1/60 s. The full-wave signal has a period of half this value, or 1/120 s. Therefore, the fundamental frequency of the rectified output is

$$f_1 = 120 \text{ Hz}$$

and the next two harmonic frequencies are

$$f_2 = 240 \text{ Hz}$$

and

$$f_3 = 360 \text{ Hz}$$

The spectrum of a sine wave is a single line at the frequency of the sine wave. For this reason, the input spectrum contains only a single line with a peak value of 100 V and a frequency of 60 Hz (Fig. 20-11b).

Figure 20-11c shows the spectrum of rectified output voltage. The first line is the dc component with a value of 63.6 V (found in Example 20-5). The other lines are the harmonics of the rectified output signal; we have shown only the first three harmonics.

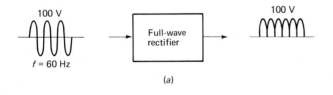

Figure 20-11. Full-wave rectification. (a) Circuit. (b) Input spectrum. (c) Output spectrum.

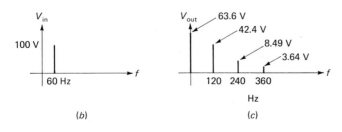

Especially important, the fundamental frequency in the output spectrum is 120 Hz, exactly double the input frequency. This doubling effect happens in all full-wave rectifiers.

20-3. FILTERS

Filters pass some sinusoidal frequencies but stop others. We first mentioned filters in our discussion of rectifiers; the idea was to remove all sinusoidal components, leaving only the dc component. Another example of a filter is the tuned class C amplifier. The collector current is a pulse waveform containing all harmonics of the fundamental frequency. The tuned tank removes all but one of the harmonics. This section describes filters and their effects on the spectrum of a signal.

Low-pass filter

A *low-pass filter* passes low frequencies but stops high ones. Figure 20-12a shows the ideal response of a low-pass filter; this is a graph of voltage gain versus frequency. As we see, the gain is unity from zero up to the *cutoff frequency* f_c. Beyond f_c, the voltage gain ideally drops to zero. The *passband* is the set of frequencies between zero and f_c; the *stopband,* the set of frequencies greater than f_c.

Any input sine wave with a frequency in the passband will appear at the output of the filter. But any sine wave whose frequency is greater than f_c cannot appear at the

Figure 20-12. (a) Ideal low-pass response. (b) Real low-pass response. (c) Low-pass filtering.

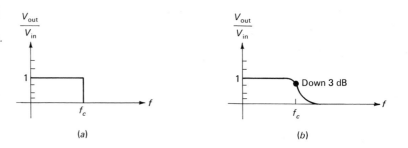

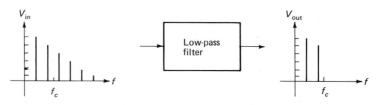

output of the low-pass filter. For instance, suppose a periodic signal has the spectrum shown in Fig. 20-12c. All sinusoidal frequencies up to f_c pass through the filter; all those beyond f_c are stopped. This is why the output spectrum contains only frequencies below f_c.

The ideal response of Fig. 20-12a is impossible to attain in practice, but we can get close. A real filter has a response like Fig. 20-12b. By definition, the gain is down 3 dB at the cutoff frequency. (The voltage gain in the passband may be slightly less than unity because of small losses in the filter.)

Other responses

Figure 20-13b shows the ideal response of a *high-pass* filter; the passband is the set of frequencies greater than f_c.

The *bandpass filter* has the ideal response shown in Fig. 20-13c; here, the passband is the set of frequencies between f_{c1} and f_{c2}.

The fourth basic response is the ideal *bandstop* of Fig. 20-13d; it stops all frequencies between f_{c1} and f_{c2}.

In a real filter the rolloff is not vertical as shown in the ideal responses of Fig. 20-13; the cutoff frequency is defined as the frequency where the voltage gain is down 3 dB from the passband value. The rolloff outside the passband may be 20 dB per decade, 40 dB per decade, or any multiple of 20 dB, depending on the construction of the filter.

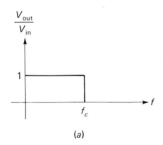

(a)

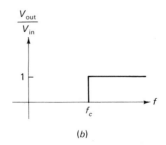

(b)

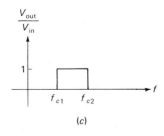

(c)

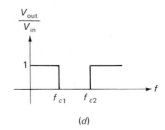

(d)

Figure 20-13. Ideal filter responses. (a) Low-pass. (b) High-pass. (c) Bandpass. (d) Bandstop.

Figure 20-14. (a) *Input spectrum.*
(b) *Ideal bandpass response.* (c)
Looking through the window. (d)
Final output spectrum.

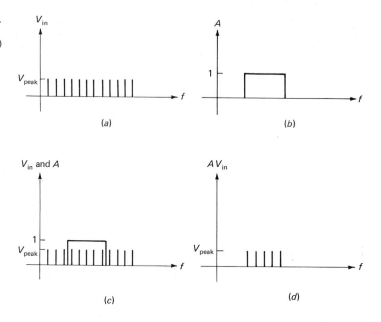

The bandpass filter (Fig. 20-13*c*) is often called a *window* because of its effect on a spectrum. As an example, Fig. 20-14*a* shows the spectrum of an input signal. Figure 20-14*b* is an ideal bandpass response. The easiest way to determine the output spectrum is to visualize the bandpass response superimposed on the input spectrum as shown in Fig. 20-14*c*. Since the gain is zero outside the passband and unity inside, only those components in the passband appear in the output spectrum of Fig. 20-14*d*. It's as though we are looking through a window at the spectrum of the signal.

In general, a filter is a frequency-domain device; it alters the spectrum of a signal. To determine the output spectrum, look at the input spectrum through the filter response, that is, with the filter response superimposed on the input spectrum; the output spectrum contains only those components in the filter passband.

Class C amplifier

Frequency-domain analysis explains the action of a class C amplifier better than time-domain analysis. As discussed in Chap. 12, an input sine wave drives the transistor class C, producing narrow pulses of collector current. Because this nonsinusoidal current is periodic, it contains harmonics of the fundamental frequency. When the duty cycle is very small, the spectrum resembles Fig. 20-14*a*. The tuned tank is a bandpass filter; its response is wide enough to pass only a single harmonic. If we tune the tank to the fundamental frequency, we get an output spectrum containing only the fundamental. If we tune the tank to a higher harmonic, we have a frequency multiplier.

Power-supply filter

Figure 20-15a shows the action of a full-wave rectifier and filter in the time domain. The input 60-Hz sine wave drives the rectifier to produce the 120-Hz full-wave signal. This full-wave signal is the input to a low-pass filter whose cutoff frequency is less than 120 Hz; therefore, the final output is a dc voltage because none of the sinusoidal components can get through the filter.

The filtering is easier to understand in the frequency domain. Figure 20-15b shows the input spectrum (a single line at 60 Hz) driving the rectifier. The output of the full-wave rectifier contains a dc component, a fundamental of 120 Hz, and higher harmonics. Because all sinusoidal frequencies are greater than the cutoff frequency of the low-pass filter, the output spectrum contains only the dc component. This dc component has an ideal value of

$$V_0 = 0.636A$$

where A is the peak value of the full-wave-rectified signal. The actual dc output voltage will be slightly less than this because of losses in the filter.

Filter prototype circuits

Filters often use inductances and capacitances. Figure 20-16a shows the prototype form of a low-pass LC filter. At lower frequencies the inductors look almost shorted and

Figure 20-15. Full-wave rectification. (a) Time-domain viewpoint. (b) Frequency-domain viewpoint.

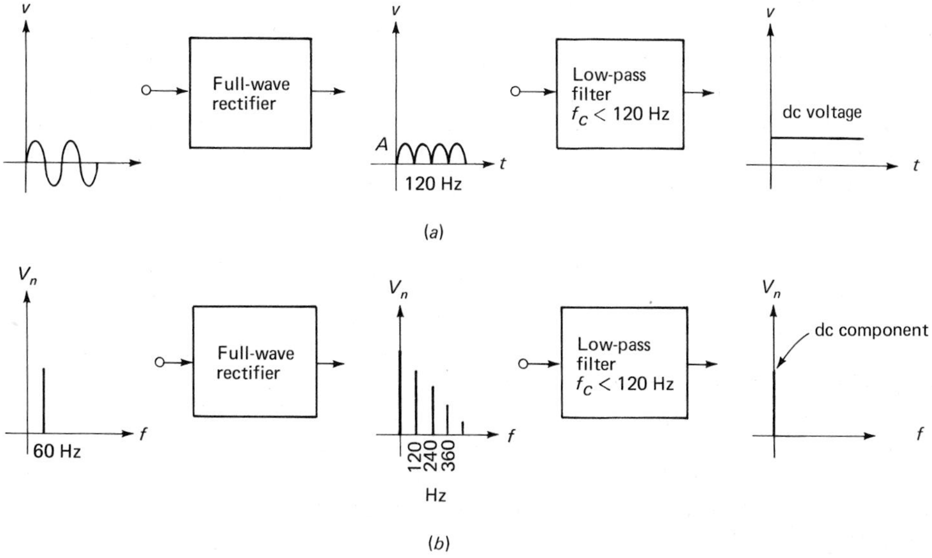

(a)

(b)

Figure 20-16. Filter prototypes.
(a) Low-pass. (b) High-pass. (c)
Series resonant bandpass. (d)
Parallel resonant bandpass. (e)
Parallel resonant bandstop. (f)
Series resonant bandstop.

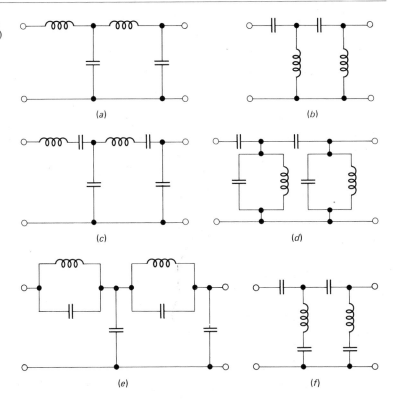

(a)

(b)

(c)

(d)

(e)

(f)

the capacitors almost open; this allows lower frequencies to reach the output. At higher frequencies the inductors approach open circuits and the capacitors short circuits; for this reason, higher frequencies are stopped.

Figure 20-16*b* is the prototype of a high-pass *LC* filter; the action is complementary to the low-pass. The capacitors look open and the inductors shorted at lower frequencies; this prevents lower frequencies from reaching the output. On the other hand, higher frequencies pass through to the output because the capacitors look shorted and the inductors open.

Many designs are possible with bandpass filters. Figure 20-16*c* shows a series resonant type, while Fig. 20-16*d* illustrates a shunt resonant type. Both of these prototypes pass sine waves when the circuits are resonant.

Figure 20-16*e* and *f* shows two types of bandstop filters. When the circuits resonate, the filter stops the sine wave from reaching the output terminals.

Filter analysis and design is covered in advanced books.[3]

[3] If you want more information, look for the two main methods: image-parameter design and modern-network synthesis. For an introductory discussion, see *Reference Data for Radio Engineers,* Howard W. Sams & Co., Inc., New York, 1968, pp. 7-1 to 9-9.

EXAMPLE 20-7.

Figure 20-17a shows a Wien bridge. How does the circuit affect the spectrum?

SOLUTION.

As discussed in Chap. 19, the bridge balances at a frequency of

$$f_0 = \frac{1}{2\pi RC}$$

In other words, when the input sine wave has a frequency of f_0, the output voltage drops to zero.

Figure 20-17b shows the response of a Wien bridge. As mentioned earlier, the Wien bridge is sometimes called a *notch filter* because of its effect on the spectrum. If you visualize Fig. 20-17b superimposed on an input spectrum, you will see it notches frequency f_0 out of the spectrum.

Figure 20-17c shows another kind of notch filter known as the *twin T* (also called the parallel T). When the load resistance R_L is much greater than R, the voltage gain is

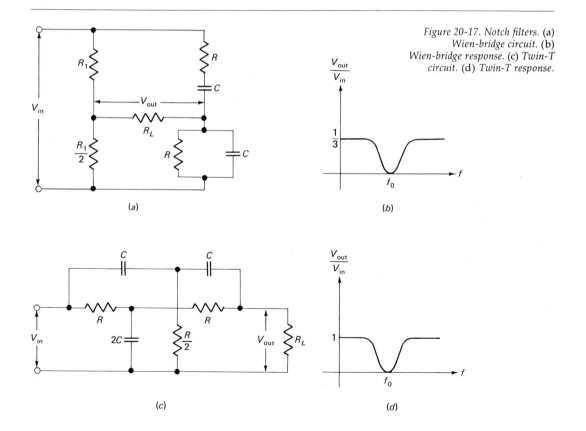

Figure 20-17. Notch filters. (a) Wien-bridge circuit. (b) Wien-bridge response. (c) Twin-T circuit. (d) Twin-T response.

(a)

(b)

(c)

(d)

approximately unity at low and high frequencies. As shown in Fig. 20-17d, the voltage gain drops to zero at frequency f_0. By using complex algebra, we can prove $f = 1/2\pi RC$.

20-4. HARMONIC DISTORTION

As discussed in Sec. 8-5, when the amplified signal is small, only a small part of the transconductance curve is used. Because of this, the operation takes place over an almost linear arc of the curve. Operation like this is called *linear* because changes in output current are proportional to changes in input voltage. Linear operation means the shape of the amplified waveform is the same as the shape of the input waveform. That is, we get no distortion when the operation is linear or small-signal.

But when the signal is large, we can no longer treat the operation as linear; changes in output current no longer are proportional to changes in input voltage. Because of this, we get nonlinear distortion. This section examines distortion from the viewpoint of the frequency domain.

Large-signal operation

When the signal swing is large, the operation becomes nonlinear. Figure 20-18 is an example of this. A sinusoidal V_{BE} voltage produces large swings along the transconductance curve. Because of the nonlinearity of the curve, the resulting current is no longer sinusoidal. In other words, the shape of the output current no longer is a true duplication of the input shape. Since the output current flows through a load resistance, the output voltage will also have nonlinear distortion.

Figure 20-19a shows nonlinear distortion from the time-domain viewpoint. The input sine wave drives an amplifier. If the operation is large-signal, the amplified output voltage is no longer a pure sine wave. Arbitrarily, we have shown more gain on one half cycle than the other; this kind of distortion is sometimes called *amplitude distortion.*

Figure 20-18. Nonlinear distortion.

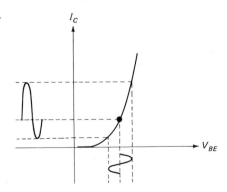

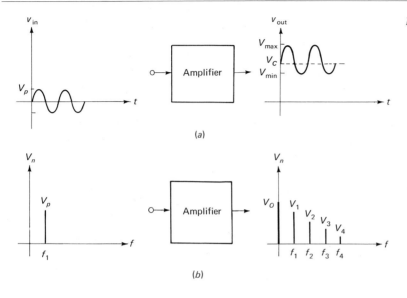

Figure 20-19. (a) *Amplitude distortion.* (b) *Harmonic distortion.*

The frequency domain gives us insight into amplitude distortion. Figure 20-19*b* shows how to visualize the same situation in the frequency domain. The input spectrum is a single line at f_1, the frequency of the input sine wave. The output signal is distorted but still periodic; therefore, it contains the dc component and harmonics shown. (Arbitrarily, we have stopped with the fourth harmonic.) The point is that a waveform with amplitude distortion contains a fundamental and harmonics; the strength of the higher harmonics is a clue to how bad the distortion is.

In fact, an alternative name for amplitude distortion is *harmonic distortion*. We use the term *amplitude distortion* when we visualize a signal in the time domain; we use *harmonic distortion* when we are thinking of the signal in the frequency domain; and when we are interested in the cause of the distortion, we use the term *nonlinear distortion*. All of these are synonyms for the kind of the distortion that occurs with *one input sine wave*. (The next chapter talks about distortion with two or more input sine waves.)

Formula for individual harmonic distortion

How are we going to compare the harmonic distortion of one amplifier with another? The larger the peak values of the harmonics, the larger the harmonic distortion. The simplest way to compare different amplifiers is by taking the ratio of the harmonics to the fundamental. Specifically, we define the percent of second harmonic distortion as

$$\text{Percent second harmonic distortion} = \frac{V_2}{V_1} \times 100 \text{ percent} \qquad (20\text{-}8a)$$

The percent of third harmonic distortion is

$$\text{Percent third harmonic distortion} = \frac{V_3}{V_1} \times 100 \text{ percent} \qquad (20\text{-}8b)$$

and so on for any higher harmonic. In general, the percent of nth harmonic distortion is

$$\text{Percent } n\text{th harmonic distortion} = \frac{V_n}{V_1} \times 100 \text{ percent} \qquad (20\text{-}8c)^{***}$$

As an example, suppose the output spectrum of Fig. 20-19b has $V_1 = 2$ V, $V_2 = 0.2$ V, $V_3 = 0.1$ V, and $V_4 = 0.05$ V.[4] Then, the harmonic distortions are

$$\text{Percent second harmonic distortion} = 10 \text{ percent}$$

$$\text{Percent third harmonic distortion} = 5 \text{ percent}$$

$$\text{Percent fourth harmonic distortion} = 2.5 \text{ percent}$$

Formula for total harmonic distortion

Data sheets usually give *total harmonic distortion,* all harmonics lumped together and compared to the fundamental. The formula for total harmonic distortion is

Percent total harmonic distortion
$$= \sqrt{(\text{percent second})^2 + (\text{percent third})^2 + \cdots} \qquad (20\text{-}9)^{***}$$

As an example, for $V_1 = 2$ V, $V_2 = 0.2$ V, $V_3 = 0.1$ V, and $V_4 = 0.05$ V, the individual harmonic distortions are 10, 5, and 2.5 percent; the total harmonic distortion is

Percent total harmonic distortion
$$= \sqrt{(10 \text{ percent})^2 + (5 \text{ percent})^2 + (2.5 \text{ percent})^2} = 11.5 \text{ percent}$$

Derivation of Eq. (20-9)

When a spectrum like Fig. 20-20 appears across a resistor, each sinusoidal component produces its part of the total power *independent of the other harmonics.* In symbols,

$$p = p_1 + p_2 + p_3 + \cdots$$

In terms of voltage,

$$\frac{V^2}{R} = \frac{V_1^2}{R} + \frac{V_2^2}{R} + \frac{V_3^2}{R} + \cdots$$

where V, V_1, V_2, V_3, etc., are *rms voltages.* Dropping the R,

$$V^2 = V_1^2 + V_2^2 + V_3^2 + \cdots$$

[4] Either rms or peak values can be used; $\sqrt{2}$ cancels in the ratio.

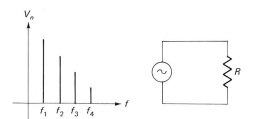

Figure 20-20. *Total power of harmonic components is superposition of individual powers.*

Taking the square root of both sides gives

$$V = \sqrt{V_1^2 + V_2^2 + V_3^2 + \cdots}$$

(20-10)

A formula like this is called a *quadratic sum* because we take the square root of the sum of squares.

Equation (20-10) gives the total rms voltage. If we want the rms voltage of the higher harmonics only, we can use

$$V_{\text{higher harmonics}} = \sqrt{V_2^2 + V_3^2 + V_4^2 + \cdots}$$

(20-11)

This result follows directly from Eq. (20-10); it says the rms value of all harmonics above the fundamental equals the quadratic sum of the harmonic rms voltages. For instance, if $V_2 = 0.2$ V rms, $V_3 = 0.1$ V rms, and $V_4 = 0.05$ V rms, the rms value of these harmonics lumped together is

$$V_{\text{higher harmonics}} = \sqrt{0.2^2 + 0.1^2 + 0.05^2} = 0.229 \text{ V rms}$$

Total harmonic distortion is defined as the rms value of higher harmonics divided by the rms value of the fundamental. In symbols,

$$\text{Percent total harmonic distortion} = \frac{V_{\text{higher harmonics}}}{V_1} \times 100 \text{ percent}$$

(20-12)

When we substitute the right-hand member of Eq. (20-11) into this equation, we get

$$\text{Percent total harmonic distortion} = \frac{\sqrt{V_2^2 + V_3^2 + V_4^2 + \cdots}}{V_1} \times 100 \text{ percent}$$

which reduces to

$$\text{Percent total harmonic distortion} = \sqrt{(\text{percent second})^2 + (\text{percent third})^2 + \cdots}$$

Therefore, we have proved Eq. (20-9).

Harmonic distortion sometimes useful

If you are trying to amplify speech or music, the less harmonic distortion the better. In the case of speech, total harmonic distortion must be less than 10 percent or thereabouts to preserve intelligibility. High-fidelity music needs less than 1 percent total harmonic distortion for good quality.

Don't get the idea harmonic distortion is always undesirable. There are many applications where we want as much harmonic distortion as possible. For instance, if we are building a frequency multiplier (Sec. 12-6), we want as much nth harmonic as possible. Therefore, we deliberately optimize the circuit to produce maximum distortion at the nth harmonic. The resonant tank circuit can then separate the nth harmonic from the others, giving us a pure output sine wave of frequency nf_1.[5]

EXAMPLE 20-8.
A spectrum analyzer shows the display of Fig. 20-21; the vertical scale is calibrated to read harmonic rms voltages rather than peak values (normal practice in a commercial analyzer).

1. Calculate the rms voltage for the periodic wave.
2. Calculate the rms voltage for the higher harmonics.
3. Calculate the total harmonic distortion.

SOLUTION.

1. With Eq. (20-10),

$$V = \sqrt{8^2 + 4^2 + 3^2} = 9.43 \text{ V rms}$$

If you measured the periodic wave with a true-rms voltmeter, you would read 9.43 V.

2. The rms value of the higher harmonics is found with Eq. (20-11):

$$V_{\text{higher harmonics}} = \sqrt{4^2 + 3^2} = 5 \text{ V rms}$$

If you notch the fundamental frequency out of the spectrum of Fig. 20-21, only the higher harmonics remain; measure these with a true-rms voltmeter and you will read 5 V.

3. Either Eq. (20-9) or (20-12) applies. Since we already have the rms value of the higher harmonics, Eq. (20-12) is more convenient:

[5] The *step-recovery diode* is one of the most efficient ways to generate high harmonics.

Figure 20-21. Example 20-8.

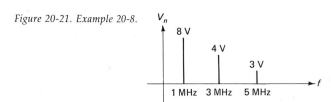

$$\text{Percent total harmonic distortion} = \frac{5}{8} \times 100 \text{ percent} = 62.5 \text{ percent}$$

Alternatively, you calculate each individual distortion to get

$$\text{Percent third harmonic distortion} = 50 \text{ percent}$$

$$\text{Percent fifth harmonic distortion} = 37.5 \text{ percent}$$

The quadratic sum of these distortions is

$$\text{Percent total harmonic distortion} = \sqrt{(50 \text{ percent})^2 + (37.5 \text{ percent})^2} = 62.5 \text{ percent}$$

Incidentally, a *distortion analyzer* is an instrument that measures the rms value of all higher harmonics lumped together. By taking the ratio of this value to the fundamental, the distortion analyzer reads the value of total harmonic distortion.[6]

20-5. THREE-POINT DISTORTION ANALYSIS

Often, we encounter the special case of no harmonics above the second. For instance, the square-law curve of a FET causes amplitude distortion where the highest harmonic is the second (proved in the next chapter). We may also encounter this special case in a bipolar amplifier; when the input signal is increased beyond small-signal operation, the first higher harmonic to appear is the second harmonic. In other words, there is a level of input signal between small-signal and large-signal operation where the only significant distortion component is the second harmonic. We will refer to this special case as *square-law distortion.*

The transconductance curve of a bipolar or FET will force the fundamental and second harmonic to line up as shown in Fig. 20-22. When we add the fundamental and second harmonic (dashed waves), we get the total distorted waveform (solid). The equation for the distorted waveform is

$$v = V_0 + V_1 \sin \omega t + V_2 \sin (2\omega t - 90°) \tag{20-13}$$

At $\omega t = 0°$, this equation reduces to

$$V_C = V_0 - V_2 \tag{20-13a}$$

At $\omega t = 90°$, Eq. (20-13) gives

$$V_{max} = V_0 + V_1 + V_2 \tag{20-13b}$$

And when $\omega t = 270°$, Eq. (20-13) reduces to

$$V_{min} = V_0 - V_1 + V_2 \tag{20-13c}$$

[6] Good examples of commercial distortion analyzers are the hp models 331A through 334A. For more information, consult the hp catalog.

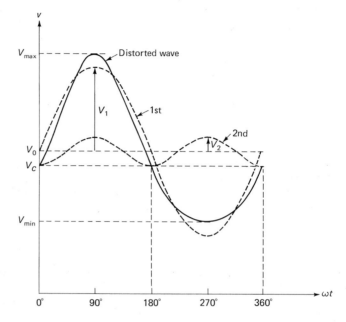

Figure 20-22. Adding fundamental and second harmonic to get distorted wave.

Solving the last three equations simultaneously gives

$$V_1 = \frac{V_{max} - V_{min}}{2} \tag{20-14}$$

$$V_2 = \frac{V_{max} + V_{min} - 2V_C}{4} \tag{20-15}$$

$$V_0 = V_C + V_2 \tag{20-16}$$

How do we use these results? If we see a distorted wave like Fig. 20-22 on an oscilloscope, we can measure V_{max}, V_{min}, and V_C. Then, we can use Eqs. (20-14) through (20-16) to calculate V_1, V_2, and V_0. In this way, we find the *dc component* and the *peak values* of the fundamental and second harmonics. This is known as a *three-point analysis* because we measure three voltages: V_{max}, V_{min}, and V_C. It gives exact values of V_1, V_2, and V_0 when there are no harmonics above the second; we can use it as a rough approximation even when there are smaller harmonics above the second.[7]

Note that V_C is the voltage level passing through the 0°, 180°, and 360° points on the distorted wave.

---

[7] For individual harmonics above the second, you normally would have to use a spectrum or wave analyzer. Alternatively, you can try a five-point analysis; see Seely, S.: *Electronic Circuits*, Holt, Rinehart and Winston, Inc., New York, 1968, pp. 120–123.

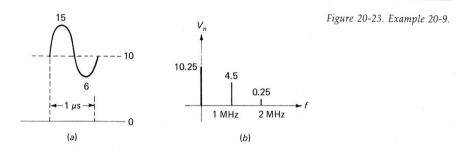

Figure 20-23. Example 20-9.

EXAMPLE 20-9.

Figure 20-23a shows a distorted output voltage. Assuming square-law distortion, calculate V_0, V_1, and V_2. (The 10-V level passes through the 0°, 180°, and 360° points on the distorted wave.)

SOLUTION.

Figure 20-23a is the kind of picture that can appear on an oscilloscope. We read these three values:

$$V_{max} = 15 \text{ V}$$

$$V_{min} = 6 \text{ V}$$

$$V_C = 10 \text{ V}$$

The rest is straightforward calculation. With Eqs. (20-14) through (20-16), we get

$$V_1 = \frac{15 - 6}{2} = 4.5 \text{ V}$$

$$V_2 = \frac{15 + 6 - 2(10)}{4} = 0.25 \text{ V}$$

and

$$V_0 = 10 + 0.25 = 10.25 \text{ V}$$

Figure 20-23b shows these spectral components.

20-6. OTHER KINDS OF DISTORTION

For nonlinear operation of an amplifier, an input sine wave produces a distorted output signal in the time domain; in the frequency domain, nonlinear operation is equivalent to a single-line spectrum producing an output spectrum with many lines (see Fig. 20-24a). This is what we mean by harmonic distortion; a pure input sine wave produces a fundamental and harmonics.

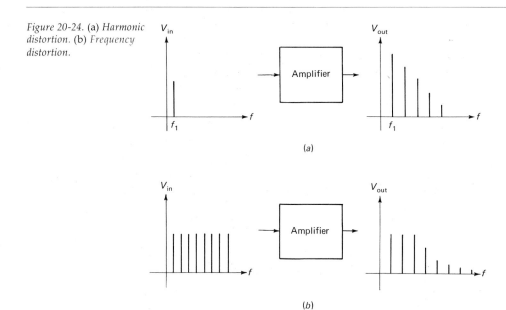

Figure 20-24. (a) Harmonic distortion. (b) Frequency distortion.

Frequency distortion

Frequency distortion is different. It has nothing to do with nonlinear distortion; frequency distortion can occur even with small-signal operation. The cause of frequency distortion is a change in amplifier gain with frequency. Figure 20-24*b* illustrates this kind of distortion. Arbitrarily, the input spectrum contains many equal-amplitude sinusoidal components. If the *break frequency* of the amplifier is *less than the highest sinusoidal frequency,* the higher frequencies in the output spectrum are attenuated as shown.

Frequency distortion, therefore, is nothing more than a change in the spectrum of the signal caused by amplifier filtering. This can affect the quality of speech or music signals. In other words, speech or music is a complex signal with many components in the spectrum. Unless all of these components are in the bandwidth of the amplifier, the components in the output spectrum will not have the correct amplitudes. Because of this, the speech or music will sound different from the original input signal.

Phase distortion

Phase distortion takes place when the phase of a harmonic is shifted with respect to the fundamental. As an example, the input signal of Fig. 20-25 shows the third harmonic peak in phase with the peak of the fundamental. If there is phase distortion, the third harmonic changes its phase with respect to the fundamental; arbitrarily, we have shown the third harmonic peak out of phase with the fundamental at the output.

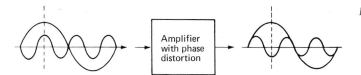

Figure 20-25. Phase distortion.

Frequency and phase distortion almost always occur together. In the midband of an amplifier the voltage gain and phase shift are constant (either 0° or 180°). Because of this, no frequency or phase distortion can occur if all sinusoidal components are in the midband of the amplifier. Outside the midband, the voltage gain drops off and the phase angle changes; therefore, we get frequency and phase distortion if the spectrum contains components outside the midband.

EXAMPLE 20-10.
An amplifier has a flat response from 20 Hz to 20 kHz. Harmonic distortion is negligible for input signals less than 0.1 V peak-to-peak. What kind of distortion occurs for each of these input signals:

1. The complex signal produced by middle C of a piano.
2. A wave whose spectrum has a fundamental of 12 kHz, plus *ultrasonic* components of 24 kHz and 36 kHz.

SOLUTION.
If either input signal has a peak-to-peak value greater than 0.1 V, we will get amplitude or harmonic distortion; this means harmonics will appear in the output not present in the input.

If both input signals have a peak-to-peak value less than 0.1 V, the only possible distortions are frequency and phase distortion. In this case,

1. Middle C has a fundamental frequency of 256 Hz. A piano produces harmonics up to 10 kHz or thereabouts. Because of this, the spectrum of middle C contains a 256-Hz fundamental plus all harmonics up to 10 kHz or so. Since all these sinusoidal components are in the midband of the amplifier, no frequency or phase distortion takes place.
2. The ultrasonic components (those with frequencies greater than 20 kHz) lie above the midband of the amplifier; therefore, they get less gain and a different phase shift than the fundamental. As a result, frequency and phase distortion occur.

20-7. DISTORTION WITH NEGATIVE FEEDBACK

This section proves negative feedback reduces harmonic distortion and other unwanted signals *generated inside* an amplifier.

Open loop

Figure 20-26a shows an amplifier before we connect the feedback circuit. An input sine wave produces an amplified fundamental plus a second harmonic. For the moment, we assume no harmonic distortion above the second. To account for the second harmonic in the output, we put a voltage source V_2 in series with the voltage source Av_{error}.

Closed loop

Figure 20-26b shows the same amplifier as part of an SP negative-feedback system. The feedback affects the second harmonic appearing across the output terminals; this is why we have labeled the second harmonic peak $V_{2(SP)}$. As we are about to prove, $V_{2(SP)}$ is much smaller than V_2.

In Fig. 20-26b, z_{out} is ideally small enough to neglect, so that

$$v_{out} = Av_{error} + V_2 = A(v_{in} - v_f) + V_2$$

or
$$v_{out} = A(v_{in} - Bv_{out}) + V_2 \qquad (20\text{-}17)$$

Expanding and solving for v_{out} gives

$$v_{out} = \frac{A}{1 + AB}\, v_{in} + \frac{V_2}{1 + AB}$$

Or since the sacrifice factor S equals $1 + AB$,

$$v_{out} = A_{SP}v_{in} + \frac{V_2}{S} \qquad (20\text{-}18)$$

The final result says the second harmonic is *reduced by the sacrifice factor*. For example, if the second harmonic has a peak value of 1 V with no feedback, then an SP feedback amplifier with a sacrifice factor of 1000 will have a second harmonic peak value of 1 mV.

Physically, here is why negative feedback reduces distortion. In Fig. 20-26a, the peak value of the second harmonic is V_2. When the amplifier is used in a negative-feedback system like Fig. 20-26b, the second harmonic is fed back to the input of the amplifier. Because of this, the error voltage contains both the input sine wave and a returning second-harmonic component. If you check the phase relations, you will find the amplified second-harmonic component arrives at the output 180° out of phase with the original distortion. Because of this, the resulting component $V_{2(SP)}$ is much smaller than V_2. Equation (20-18) exactly describes the amount of cancellation that takes place.

In Fig. 20-26b,

$$V_{2(SP)} = \frac{V_2}{S}$$

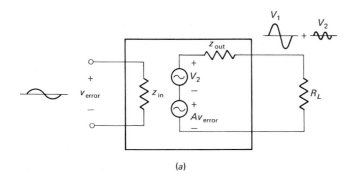

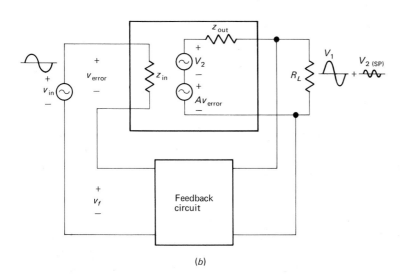

Figure 20-26. (a) Distortion without feedback. (b) Distortion with feedback.

(a)

(b)

We can divide both sides by V_1 to get

$$\frac{V_{2(SP)}}{V_1} = \frac{1}{S}\frac{V_2}{V_1}$$

or   Percent second harmonic distortion $= \dfrac{1}{S}\dfrac{V_2}{V_1} \times 100$ percent   (20-19)

This says the second-harmonic distortion with feedback equals the second-harmonic distortion without feedback reduced by the S factor. So, if an amplifier without feedback has a second-harmonic distortion of 5 percent, the same amplifier with feedback will have a second-harmonic distortion of 5 percent/S; if S is 1000, the second-harmonic distortion is 0.005 percent.

Total harmonic distortion

The derivation we have gone through applies to all higher harmonics. In Fig. 20-26a, we can replace V_2 by $V_{\text{higher harmonics}}$. This substitution applies to all derived equations so that

$$\text{Percent total harmonic distortion} = \frac{1}{S} \frac{V_{\text{higher harmonics}}}{V_1} \times 100 \text{ percent} \qquad (20\text{-}20)$$

This says the total harmonic distortion with feedback equals the total harmonic distortion without feedback reduced by the S factor.

As an example, if a data sheet gives a total harmonic distortion of 5 percent for a 10-V output signal, the same amplifier used in a negative-feedback system has a total harmonic distortion of 5 percent/S when the output signal is 10 V; if S is 100, the total harmonic distortion is only 0.05 percent.

Ripple and noise

In Fig. 20-26a, we are driving the amplifier with an input sine wave. Besides harmonic distortion, we may get some ripple and noise in the output; the ripple comes in with the supply voltage; the noise is generated by resistors, transistors, and other components in the amplifier. We can visualize ripple, noise, and other *internally generated* signals as an unwanted signal V_u appearing across the output terminals.

The derivation we went through for harmonic distortion applies to an unwanted signal internally generated in the amplifier. In Fig. 20-26a we can replace V_2 by V_u. This substitution therefore applies to all derived equations. So, we can rewrite Eq. (20-18) as

$$v_{\text{out}} = A_{SP} v_{\text{in}} + \frac{V_u}{S} \qquad (20\text{-}21)^{***}$$

This says the output voltage contains the original unwanted signal reduced by the S factor.

As an example, if the amplified ripple voltage appearing across the output of Fig. 20-26a is 0.1 V, then with a sacrifice factor of 1000 this ripple becomes 0.1 mV in the output of Fig. 20-26b.

Other kinds of feedback

By a derivation similar to what we have gone through for SP negative feedback, we can prove the other prototype forms of feedback reduce harmonic distortion and other internally generated signals. The S factor for each of the remaining prototype forms is identical to the stretching factor found for bandwidth improvement (Sec. 18-8).

Problems

20-1. What are the frequencies of the first three harmonics in the spectrum of Fig. 20-27*a*?

20-2. Work out the first four odd harmonic frequencies for the square wave of Fig. 20-27*b*.

20-3. Figure 20-27*c* shows a triangular wave. What is the fundamental frequency? The frequency of the twenty-fifth harmonic?

20-4. The first cycle of a sawtooth wave is shown in Fig. 20-27*d*. What frequencies do the first three harmonics have?

20-5. Calculate the peak values of the first three odd harmonics of Fig. 20-27*b*.

20-6. What is the peak value of the fundamental in Fig. 20-27*c*? The peak value of the tenth harmonic?

20-7. Draw the spectrum for the sawtooth wave of Fig. 20-27*d*. Include the dc component and the first four harmonics.

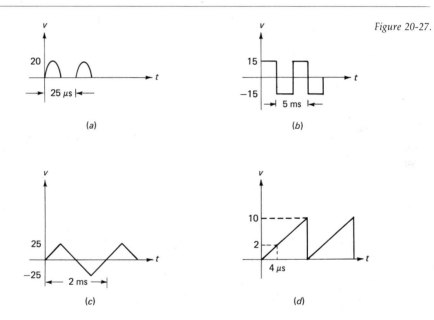

Figure 20-27.

(a)

(b)

(c)

(d)

20-8. What is the average voltage of the square wave in Fig. 20-28*a*?

20-9. Calculate the dc component of the pulse waveform shown in Fig. 20-28*b*.

20-10. Does the waveform of Fig. 20-28*c* have odd harmonics only? How about the staircase waveform of Fig. 20-28*d*?

20-11. If you shift the waveform of Fig. 20-28*a* vertically, can you find an average level that gives the waveform half-wave symmetry? If so, what is the new average value?

Figure 20-28.

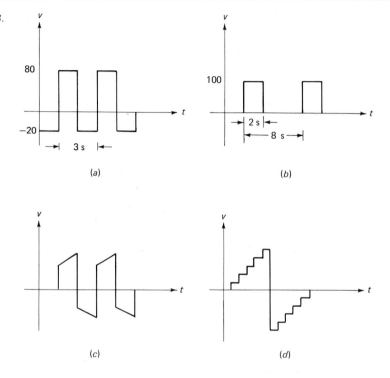

20-12. A bandpass filter has a lower cutoff frequency of 99 kHz and an upper cutoff frequency of 101 kHz; the gain of this filter is unity in the passband, but zero outside the passband. If a square wave drives this bandpass filter as shown in Fig. 20-29a, what is the peak value and frequency of the output signal?

20-13. The low-pass filter of Fig. 20-29b has a voltage gain of unity at zero frequency, a voltage gain of 0.005 at the fundamental frequency, and a voltage gain of zero for all other frequencies. What is the output dc voltage? How much ripple is there in the output?

20-14. What is the notch frequency of the twin-T filter shown in Fig. 20-30?

20-15. Suppose an amplifier has an input-output curve like Fig. 20-31a. What is the voltage gain when v_{in} equals 1 mV? When v_{in} equals 0.5 mV? Does this amplifier produce harmonic distortion?

20-16. Figure 20-31b shows the input-output curve of an amplifier. What is the voltage gain when v_{in} equals 1 mV? When v_{in} equals 0.5 mV? Does this amplifier produce nonlinear distortion?

20-17. A sine wave drives an amplifier. The output signal has a spectrum like Fig. 20-31c. Calculate the second harmonic distortion, the third harmonic distortion, and the total harmonic distortion.

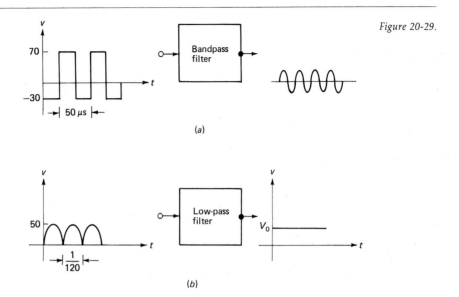

Figure 20-29.

(a)

(b)

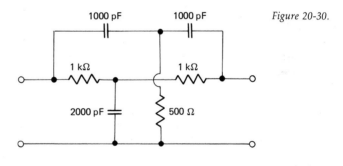

Figure 20-30.

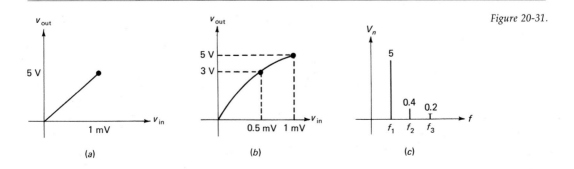

Figure 20-31.

(a)   (b)   (c)

20-18. An amplifier with a voltage gain of 500 has a total harmonic distortion of 0.2 percent for an input sine wave of 10 mV rms. If only the fundamental and third harmonics are present in the output, what is the rms value of the third harmonic?

20-19. The input voltage to an amplifier is changed in decade steps as follows: $v_{in} =$ 0.1 mV, 1 mV, 10 mV, 100 mV, and 1 V. The corresponding output voltages are $v_{out} = 10$ mV, 100 mV, 1 V, 9 V, and 12 V. If you want to avoid harmonic distortion, what is the largest input voltage you should use?

20-20. An amplifier produces only square-law distortion. Figure 20-32a shows the distorted output signal produced by a single input sine wave; 9.3 V is the voltage level through the 0°, 180°, and 360° points on the waveform. Calculate V_0, V_1, and V_2. Also work out the total harmonic distortion.

Figure 20-32.

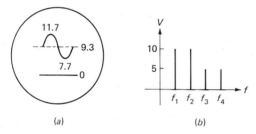

(a)   (b)

20-21. If you measure a V_{max} of 10 V, a V_{min} of 2 V, and a V_C of 5 V, what is the peak value of the second harmonic voltage?

20-22. The first two components of Fig. 20-32b have peak values of 10 V; the next two components have peak values of 5 V. Calculate the rms voltage of the periodic wave with this spectrum.

20-23. Calculate the total harmonic distortion for the signal whose spectrum is given by Fig. 20-32b.

20-24. An amplifier without feedback has output peaks of $V_1 = 10$ V and $V_2 = 0.5$ V. This amplifier is used in an SP negative-feedback system where the sacrifice factor S equals 500. If the fundamental output signal with feedback has a peak of 10 V, what is the peak value of the second harmonic?

20-25. A data sheet for an op amp says the total harmonic distortion equals 3 percent when the output signal is 20 V peak-to-peak. If you use this op amp in an SP negative-feedback system with a sacrifice factor of 5000, what will the total harmonic distortion equal when the output signal has a peak-to-peak value of 20 V?

20-26. Without feedback, the ripple across the output of an amplifier is 100 mV rms. The amplifier has an A of 50,000. If this amplifier is used in an SP negative-feedback amplifier with a B of 0.05, how much ripple is there across the output?

20-27. The electric field from a nearby motor induces an unwanted voltage in an amplifier. Without feedback, this unwanted signal appears across the output terminals with a value of 0.3 V rms. With SP negative feedback, this unwanted signal is greatly reduced. If A' equals 100 dB and A'_{SP} is 40 dB, how much unwanted signal is there across the output?

20-28. An input signal with a single spectral line at f_1 drives an amplifier without feedback; the output spectrum is shown in Fig. 20-32b.

1. Calculate the total harmonic distortion.
2. If the amplifier is used in an SP negative-feedback system where the sacrifice factor equals 5000, what does the total harmonic distortion equal when the desired output component has a peak value of 10 V?

21. Intermodulation and Mixing

When a sine wave drives a nonlinear circuit, harmonics of this sine wave appear in the output. If two sine waves drive a nonlinear circuit, we get harmonics of each sine wave and we get *new frequencies*. These new frequencies are *not* harmonics of either input sine wave. This chapter describes the theory and application of these new output frequencies.

21-1. NONLINEARITY

Figure 21-1 shows a graph of output voltage versus input voltage. Arbitrarily, the curve is concave up. In a practical amplifier, it may concave up or down, depending on the number of stages and other factors.

Small-signal operation

If the signal is small, the instantaneous operating point swings over a small part of the curve. In a case like this, the arc being used is almost linear. To a first approximation, the operation is linear. This allows us to write

$$v_{out} = Av_{in} \qquad (21\text{-}1a)$$

where A is a constant and v_{in} is the ac input signal. In Fig. 21-1, A represents the slope of the curve at point Q. In circuit terms, A is the small-signal voltage gain.

Figure 21-1. Nonlinear input-output curve.

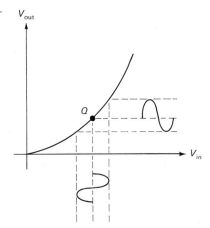

Large-signal operation

When the signal swing is large, we can no longer use the linear approximation. As we saw in the preceding chapter, large-signal operation produces nonlinear distortion; the output signal no longer is sinusoidal. Equation (21-1a) cannot be used for large-signal operation. Instead, we must use a *power series:*

$$v_{\text{out}} = Av_{\text{in}} + Bv_{\text{in}}^2 + Cv_{\text{in}}^3 + Dv_{\text{in}}^4 + \cdots \qquad (21\text{-}1b)$$

Note that each succeeding term has v_{in} raised to the next higher power. A power series like this applies to any nonlinear curve like Fig. 21-1.[1] For each combination of A, B, C, D, etc., we get a different nonlinear curve. In other words, if we have the values of A, B, C, D, and so on, we will know the exact relation between the input and output voltage.

The terms in Eq. (21-1b) have the following meanings:

Av_{in}—the linear term, the input signal amplified by a factor of A
Bv_{in}^2—the quadratic term; it leads to second harmonic distortion
Cv_{in}^3—the cubic term; it results in third harmonic distortion

The higher terms have similar meanings; the fourth-power term produces fourth harmonic distortion, the fifth-power term fifth harmonic distortion, and so forth.

21-2. LINEAR OPERATION

Figure 21-2 illustrates small-signal operation. Because the arc being used is so small, it closely approximates a straight line. If the input signal is sinusoidal, so too is the output signal.

[1] The proof is found in calculus books under the topic *Taylor series.*

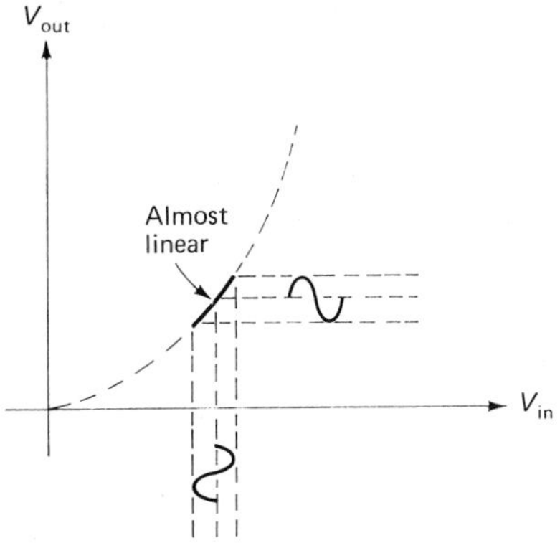

Figure 21-2. Small-signal operation.

Higher-order terms drop out

Equation (21-1b) applies to any nonlinear curve and includes small-signal operation as a special case. When the signal is small, all higher-power terms in Eq. (21-1b) drop out, leaving only the Av_{in} term. For instance, suppose an amplifier has this equation describing its ac output voltage:

$$v_{out} = 50v_{in} + 4v_{in}^2 + 3v_{in}^3$$

with all higher terms negligible. Then, an input peak of 0.1 V gives an output peak of

$$v_{out} = 50(0.1) + 4(0.1)^2 + 3(0.1)^3$$
$$= 5 + 0.04 + 0.003$$
$$\cong 5 \text{ V peak}$$

We have neglected the last two terms because they are small compared to the linear term; as a guide, we will neglect a higher-power term when it is less than 1 percent of the linear term.

Small-signal sinusoidal case

If the ac input signal is sine wave, we can express it by

$$v_{in} = V_x \sin \omega_x t$$

where V_x is the peak voltage and ω_x equals $2\pi f_x$. Figure 21-3a and b shows the input signal and its spectrum. If the signal is small, we get linear operation and the ac output voltage equals

$$v_{out} = Av_{in}$$
$$= AV_x \sin \omega_x t$$

This says the peak output voltage equals AV_x and the signal is sinusoidal with a radian

Figure 21-3. (a) Input signal. (b) Input spectrum. (c) Output signal. (d) Output spectrum.

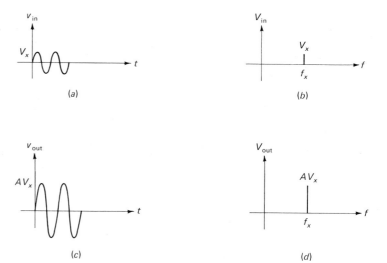

frequency of ω_x. Figure 21-3c and d shows the output signal and its spectrum. The output spectrum contains no harmonics; all that has happened is the spectral line at f_x has increased its height.

Small-signal operation with two input sine waves

What happens if two sine waves drive an amplifier? We can express the first sine wave by

$$v_x = V_x \sin \omega_x t$$

and the second by

$$v_y = V_y \sin \omega_y t$$

Figure 21-4a shows these sine waves; arbitrarily, v_x has the higher frequency. If the two signal sources are in series as shown in Fig. 21-4b,

$$v_{in} = v_x + v_y$$
$$= V_x \sin \omega_x t + V_y \sin \omega_y t \tag{21-1c}$$

By adding the ordinates of v_x and v_y (Fig. 21-4a) at each instant in time, we get the waveform of Fig. 21-4c. Experimentally, we can get this waveform by driving an oscilloscope with two sine-wave sources in series.[2] Each sinusoidal term in Eq. (21-1c)

---

[2] Stacking two sources in series is usually inconvenient in the laboratory because both sources may use the same ground (the third prong of the power plug). If you want to set up an experiment to look at the sum of two sine waves, use the op-amp summing circuit of Sec. 18-10.

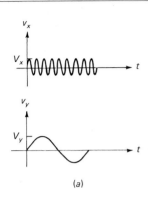

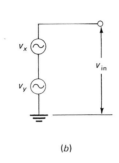

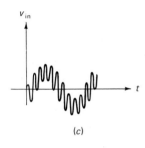

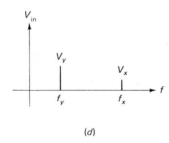

Figure 21-4. (a) *Two input signals.* (b) *Series connection of two signal sources.* (c) *Additive waveform.* (d) *Spectrum of additive waveform.*

produces a spectral line at the corresponding frequency. In other words, if we put a signal like Fig. 21-4c into a spectrum analyzer, we will see a spectrum like Fig. 21-4d. The height of each spectral line equals the peak value of the corresponding sine wave; the horizontal location gives the frequency. Since a spectrum analyzer is calibrated in cycle frequency f rather than radian frequency ω, the spectral lines appear at f_x and f_y.

If both signals are small, we get linear operation and the ac output voltage from an amplifier equals

$$
\begin{aligned}
v_{\text{out}} &= Av_{\text{in}} \\
&= A(v_x + v_y) \\
&= AV_x \sin \omega_x t + AV_y \sin \omega_y t
\end{aligned}
$$

What does this equation say? The ac output voltage is the sum of each input sine wave amplified by A. Figure 21-5a shows how this output signal looks in the time domain; it is nothing more than the signal of Fig. 21-4c amplified by A. Figure 21-5b shows the spectrum of the amplified signal; it is the original input spectrum with each line amplified by A. Especially important, no harmonics or other lines appear in the output spectrum for small-signal operation.

Figure 21-5. (a) Amplified additive signal. (b) Spectrum.

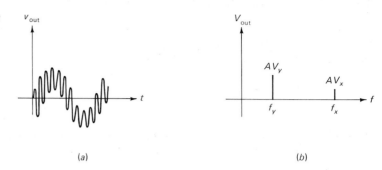

(a)   (b)

Additive waveforms

A waveform like Fig. 21-5*a* is called an *additive waveform* because it is the sum or superposition of two sine waves. You often see waveforms like this. As an example, if you are amplifying a sine wave with frequency f_y, an interference signal with frequency f_x may somehow get into the amplifier and be added to the desired signal; in this case, you will see the additive waveform of Fig. 21-5*a* at the output of the amplifier.

The interference signal can get into the amplifier in a number of ways. It may be excessive power-supply ripple being added to the desired signal. Or it may be a signal induced by nearby electrical apparatus, or possibly a transmitted radio signal. In any event, whenever you see an additive waveform like Fig. 21-5*a*, remember it represents the sum of two sine waves.

Linear operation with many sinusoidal components

What we have derived for two input sine waves applies to any number of input sine waves. In other words, if we have ten input sine waves, linear amplification results in ten output sine waves. The output spectrum will contain ten spectral lines with the same frequencies as the input spectrum; each component is amplified by A.

A good example of a multi-sine-wave input is the signal produced by middle C of a piano. This signal contains a fundamental with a frequency of 256 Hz plus harmonics to around 10,000 Hz (approximately 40 harmonics); the strength of the harmonics compared to the fundamental makes this note sound different from all others. If we *linearly* amplify this middle-C signal, the output spectrum contains only the original harmonics amplified by a factor A. Because of this, the output is louder but still has the distinct sound of middle C. A good high-fidelity amplifier therefore is a small-signal or linear amplifier; it does not add or subtract any spectral components; furthermore, it amplifies each component by the same amount.

21-3.   MEDIUM-SIGNAL OPERATION WITH ONE SINE WAVE

For typical values of A, B, C, etc., in Eq. (21-1b), the first higher-power term to become important is the quadratic term. That is, there is a level of input signal between the small-signal and large-signal cases where only the first two terms of Eq. (21-1b) are important:

$$v_{out} = Av_{in} + Bv_{in}^2 \qquad\qquad (21\text{-}2)$$

Because this special case lies between small-signal and large-signal operation, we call it the *medium-signal* case. Any transistor amplifier can operate medium-signal, with only a linear and quadratic term in the equation for ac output voltage. With further increase in signal, a bipolar amplifier goes into the large-signal case where the cubic and higher-power terms become important.

The FET amplifier

Unlike a bipolar amplifier, a FET amplifier operates medium-signal all the way to saturation and cutoff. In other words, when we increase the input signal, the FET amplifier continues to operate medium-signal; only the linear and quadratic terms appear in the expression for output voltage. Equation (21-2) applies to a FET amplifier for any signal level, provided the output signal is not clipped. This is a direct consequence of the parabolic or square-law transconductance curve. We do not have the time to prove it, but it is possible to derive Eq. (21-2) from the transconductance equation of a FET. Because of this, as long as the instantaneous operating point remains along the transconductance curve, Eq. (21-2) applies to a FET amplifier.

Quadratic term produces second harmonic

The quadratic term Bv_{in}^2 causes second harmonic distortion. Suppose the input voltage is a sine wave expressed by

$$v_{in} = V_x \sin \omega_x t$$

In the medium-signal case, the output voltage equals

$$\begin{aligned} v_{out} &= Av_{in} + Bv_{in}^2 \\ &= AV_x \sin \omega_x t + BV_x^2 \sin^2 \omega_x t \end{aligned} \qquad (21\text{-}3a)$$

A useful expansion formula proved in trigonometry is

$$\sin^2 A = \tfrac{1}{2} - \tfrac{1}{2} \cos 2A \qquad\qquad (21\text{-}3b)$$

where A represents angle. If we let $A = \omega_x t$, we can use Eq. (21-3b) to rearrange Eq. (21-3a); the result is

$$v_{out} = \tfrac{1}{2}BV_x^2 + AV_x \sin \omega_x t - \tfrac{1}{2}BV_x^2 \cos 2\omega_x t \qquad (21\text{-}4)$$

Figure 21-6. (a) x output in time domain. (b) Spectrum of x output.

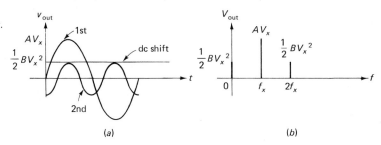

(a)

(b)

Each term in this expression is important. Briefly, here is what each means:

$\frac{1}{2}BV_x^2$—this has a constant value and represents a dc shift.

$AV_x \sin \omega_x t$—the linear output term; this is the amplified input sine wave.

$\frac{1}{2}BV_x^2 \cos 2\omega_x t$—this term has a radian frequency of $2\omega_x$, equivalent to a cycle frequency of $2f_x$; therefore, this term represents the second harmonic of the input sine wave.

Figure 21-6a shows each of these components in the time domain; if we add these components, we get the total waveform as it would appear on an oscilloscope. Figure 21-6b shows the spectrum; note the dc component, the fundamental with a frequency f_x, and the second harmonic with a frequency $2f_x$.

Summary

For the medium-signal case, an input sine wave v_x produces a dc shift, an amplified fundamental, and a second harmonic. In later discussion, we will refer to the right-hand member of Eq. (21-4) as the *x output*. We can visualize the x output either in the time domain (Fig. 21-6a) or in the frequency domain (Fig. 21-6b).

21-4.   MEDIUM-SIGNAL OPERATION WITH TWO SINE WAVES

When two input sine waves drive a medium-signal amplifier, something remarkable happens; in addition to the harmonics produced, *new frequencies* appear in the output. To be specific, here is what we will prove in this section. When two input sine waves with frequencies f_x and f_y drive an amplifier in the medium-signal case, the output spectrum contains sinusoidal components with these frequencies:

f_x and f_y: the two input frequencies
$2f_x$ and $2f_y$: the second harmonics of the input frequencies
$f_x + f_y$: a new frequency equal to the sum of the input frequencies
$f_x - f_y$: a new frequency equal to the difference of the input frequencies

For instance, if the two input frequencies are 1 kHz and 20 kHz, the output spectrum contains sinusoidal frequencies of 1 kHz and 20 kHz (the two input frequencies), 2 kHz and 40 kHz (second harmonics), 21 kHz (the sum) and 19 kHz (the difference).

The cross product

When two input sine waves drive an amplifier, we can express the input voltage as

$$v_{\text{in}} = v_x + v_y$$

where v_x and v_y are sine waves. For the medium-signal case, the output voltage equals

$$
\begin{aligned}
v_{\text{out}} &= Av_{\text{in}} + Bv_{\text{in}}^2 \\
&= A(v_x + v_y) + B(v_x + v_y)^2 \\
&= Av_x + Av_y + Bv_x^2 + 2Bv_xv_y + Bv_y^2
\end{aligned}
$$

By rearranging, we get

$$v_{\text{out}} = \underbrace{Av_x + Bv_x^2}_{x \text{ output}} + \underbrace{Av_y + Bv_y^2}_{y \text{ output}} + \underbrace{2Bv_xv_y}_{\text{cross product}} \tag{21-5}$$

The first two terms in this equation are the x output discussed in the preceding section, that is, the output from a medium-signal amplifier when only v_x drives the amplifier; therefore, we already know these two terms result in a dc shift, an amplified fundamental f_x, and a second harmonic $2f_x$.

The next pair of terms $Av_y + Bv_y^2$ is called the y output; because they are identical to the x output except for the y subscript, the y-output terms result in a dc shift, an amplified fundamental f_y, and a second harmonic $2f_y$. In other words, the y output is what we get out of a medium-signal amplifier when a sine wave of frequency f_y is the only input to the amplifier.

The unusual thing about Eq. (21-5) is the *cross product* $2Bv_xv_y$. If the cross product were *not* present in this equation, the output would be the superposition of the x output and the y output; this would allow us to take each input separately, find how the amplifier responds to it, and sum the x and y output to get the total output. But the cross product is there, and because of it the superposition theorem gives us only part of the total output.

In other words, if only the v_x input were present, we would get only the first two terms in Eq. (21-5). Likewise, if only the v_y input were present, we would get only the next two terms in Eq. (21-5). But when both inputs are present simultaneously, we get a cross product in addition to the x and y outputs.

Sum and difference frequencies

In Eq. (21-5), we already know the x-output terms represent a dc shift, an amplified fundamental f_x, and a second harmonic $2f_x$. Likewise, the y-output terms represent another dc shift, an amplified fundamental f_y, and a second harmonic $2f_y$. All that remains now is the meaning of the cross product.

When we substitute

$$v_x = V_x \sin \omega_x t$$

and

$$v_y = V_y \sin \omega_y t$$

into the cross product, we get

$$2Bv_x v_y = 2B(V_x \sin \omega_x t)(V_y \sin \omega_y t)$$
$$= 2BV_x V_y(\sin \omega_x t)(\sin \omega_y t) \qquad (21\text{-}6)$$

The product of two sine waves can be expanded by the trigonometric identity

$$\sin A \sin B = \tfrac{1}{2} \cos (A - B) - \tfrac{1}{2} \cos (A + B)$$

By letting $A = \omega_x t$ and $B = \omega_y t$, we can expand and rearrange Eq. (21-6) to get

$$2Bv_x v_y = BV_x V_y \cos (\omega_x - \omega_y)t - BV_x V_y \cos (\omega_x + \omega_y)t \qquad (21\text{-}7)$$

The first term on the right side is a sinusoidal component with a radian frequency of $\omega_x - \omega_y$, equivalent to a cycle frequency of $f_x - f_y$; therefore, this first term represents the difference frequency. The second term is also a sinusoidal component but has a cycle frequency of $f_x + f_y$; so, this second term represents the sum frequency. Figure 21-7a and b shows how these two terms look in the time domain. As we see, they are sinusoids with a peak value $BV_x V_y$, but one is the difference term while the other is the sum term.

Output spectrum

Equation (21-5) gives the output voltage from a medium-signal amplifier driven by two input sine waves. The first two terms are the x output; these terms represent a dc component, an amplified fundamental f_x, and a second harmonic $2f_x$; Fig. 21-8a shows the spectrum for this x output. In Eq. (21-5) the y-output terms represent another dc

Figure 21-7. (a) *Difference signal.* (b) *Sum signal.*

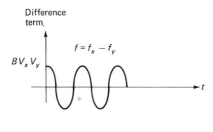

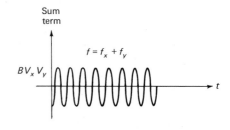

(a)

(b)

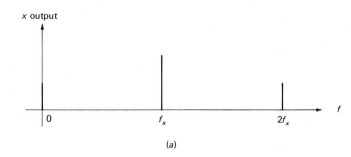

x output

0   f_x $2f_x$ f

(a)

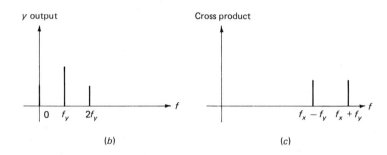

y output

0 f_y $2f_y$ f

(b)

Cross product

$f_x - f_y$ $f_x + f_y$ f

(c)

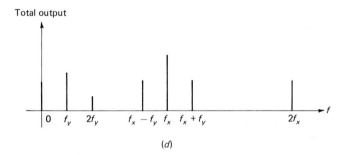

Total output

0 f_y $2f_y$ $f_x - f_y$ f_x $f_x + f_y$ $2f_x$ f

(d)

Figure 21-8. Spectra. (a) x output. (b) y output. (c) Sum and difference components. (d) Total output.

component, an amplified fundamental f_y, and a second harmonic $2f_y$; Fig. 21-8b illustrates the spectrum for the y-output terms. In Eq. (21-5) the cross product represents sum and difference terms; Fig. 21-8c shows the spectrum of this cross product.

The total output signal is the sum of the x output, the y output, and the cross product. The signal is too complicated to draw in the time domain but is easy to show in the frequency domain. Figure 21-8d is the output spectrum of an amplifier driven by two sine waves in the medium-signal case. As we see, the zero-frequency line is the sum of dc components. More important, the remaining lines represent sinusoidal signals; we have each input frequency and its harmonic, plus sum and difference frequencies.

Summary

We have found something new and useful. Put two sinusoidal signals into a medium-signal amplifier and out come six sinusoidal signals with frequencies f_x, $2f_x$, f_y, $2f_y$, a sum frequency of

$$\text{Sum} = f_x + f_y$$

and a difference frequency of

$$\text{Difference} = f_x - f_y$$

Sometimes, we write these new frequencies as

$$\text{New } f\text{'s} = f_x \pm f_y \qquad\qquad (21\text{-}7a)\,***$$

EXAMPLE 21-1.
Figure 21-9a shows an op-amp summing circuit driving a FET amplifier. What is the output spectrum of the FET amplifier?

SOLUTION.
If the operation of the FET is small-signal, the only significant output will be the amplified input sine waves. That is, the output spectrum will contain two spectral lines, one at 600 kHz and the other at 1 MHz.

If either input signal is large enough to produce medium-signal operation, the output spectrum will contain six spectral lines at frequencies of 400 kHz (difference), 600 kHz (one input), 1 MHz (another input), 1.2 MHz (second harmonic), 1.6 MHz (sum), and 2 MHz (second harmonic). Figure 21-9b shows these spectral lines. (The peak values are not important in this example.)

21-5. LARGE-SIGNAL OPERATION WITH TWO SINE WAVES

What happens when two sine waves drive a bipolar amplifier in the large-signal mode? We get an output spectrum containing each input frequency, all harmonics of these frequencies, and sum and difference frequencies produced by every combination of harmonics.

Derivation

For large-signal operation we need Eq. (21-1b) which says

$$v_{\text{out}} = Av_{\text{in}} + Bv_{\text{in}}^2 + Cv_{\text{in}}^3 + \cdots$$

This is an infinite series; there is no limit to the number of terms; each new term we add to the expression is the next higher power of v_{in}. With two input sine waves,

$$v_{\text{in}} = v_x + v_y$$

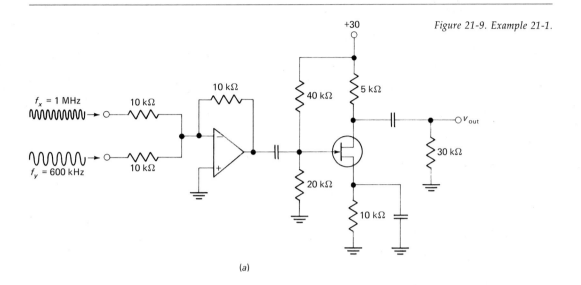

Figure 21-9. Example 21-1.

(a)

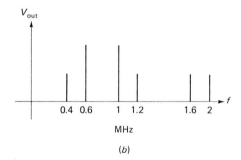

(b)

where v_x and v_y are sine waves. After substitution into the infinite series,

$$v_{\text{out}} = A(v_x + v_y) + B(v_x + v_y)^2 + C(v_x + v_y)^3 + \cdots$$

By applying the binomial theorem to each higher-power term, we can rearrange the expression to get

$$v_{\text{out}} = (Av_x + Bv_x{}^2 + Cv_x{}^3 + \cdots) + (Av_y + Bv_y{}^2 + Cv_y{}^3 + \cdots)$$
$$+ (2Bv_xv_y + 3Cv_x{}^2v_y + 3Cv_xv_y{}^2 + \cdots) \quad (21\text{-}8)$$

The first parenthesis in this equation is the *x output*:

$$x \text{ output} = Av_x + Bv_x{}^2 + Cv_x{}^3 + \cdots$$

This is the output we would get if v_x alone were driving the amplifier. With trigonometry we can prove $v_x{}^n$ produces the nth harmonic of f_x. Therefore, the *x* output contains

Figure 21-10. Spectra. (a) x output. (b) y output.

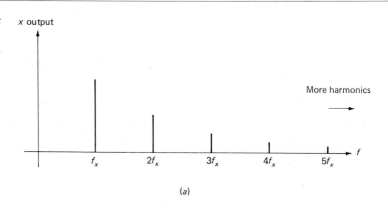

(a)

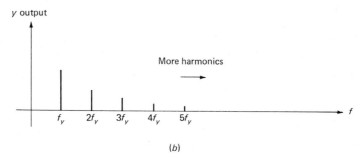

(b)

an amplified fundamental of frequency f_x, a second harmonic of frequency $2f_x$, a third harmonic of frequency $3f_x$, and so forth. Figure 21-10a shows the spectrum of the x output.

The second parenthesis in Eq. (21-8) is the *y output:*

$$y \text{ output} = Av_y + Bv_y{}^2 + Cv_y{}^3 + \cdot \cdot \cdot$$

This is the output we would get if only v_y were driving the amplifier. Again, the nth-power term introduces the nth harmonic; therefore, the spectrum of the y output contains spectral lines at f_y, $2f_y$, $3f_y$, and so forth as shown in Fig. 21-10b.

The third parenthesis in Eq. (21-8) is a collection of cross products from the binomial expansion of each power of $v_x + v_y$:

$$\text{Cross product} = 2Bv_xv_y + 3Cv_x{}^2v_y + 3Cv_xv_y{}^2 + \cdot \cdot \cdot$$

Following the last term are more terms of the form

$$Kv_x{}^mv_y{}^n$$

By advanced mathematics, we can prove each cross product produces a sum frequency of $mf_x + nf_y$ and a difference frequency of $mf_x - nf_y$. Or,

$$\text{New } f\text{'s} = mf_x \pm nf_y \qquad (21\text{-}9) \,***$$

where m and n can be any positive integers.

Orderly calculations

Equation (21-9) is the formula for every possible sum and difference frequency generated in a large-signal amplifier driven by two sine waves. Here is an orderly way to use this important equation:

Group 1. First harmonic of v_x combining with each harmonic of v_y gives frequencies of

$f_x \pm f_y$
$f_x \pm 2f_y$
$f_x \pm 3f_y$
etc.

Group 2. Second harmonic of v_x combining with each harmonic of v_y gives frequencies of

$2f_x \pm f_y$
$2f_x \pm 2f_y$
$2f_x \pm 3f_y$
etc.

Group 3. Third harmonic of v_x combining with each harmonic of v_y gives frequencies of

$3f_x \pm f_y$
$3f_x \pm 2f_y$
$3f_x \pm 3f_y$
etc.

and so on. In this way, we can calculate any higher sum and difference frequencies of interest.

A concrete example will help. Suppose the two input sine waves to a large-signal amplifier have frequencies of $f_x = 100$ kHz and $f_y = 1$ kHz. Then we can calculate sum and difference frequencies as follows:

Group 1.

100 kHz $\pm$ 1 kHz = 99 and 101 kHz
100 kHz $\pm$ 2 kHz = 98 and 102 kHz
100 kHz $\pm$ 3 kHz = 97 and 103 kHz
etc.

Group 2.

    200 kHz ± 1 kHz = 199 and 201 kHz

    200 kHz ± 2 kHz = 198 and 202 kHz

    200 kHz ± 3 kHz = 197 and 203 kHz

    etc.

Group 3.

    300 kHz ± 1 kHz = 299 and 301 kHz

    300 kHz ± 2 kHz = 298 and 302 kHz

    300 kHz ± 3 kHz = 297 and 303 kHz

    etc.

Continuing like this, we can find any sum and difference frequencies of interest. Figure 21-11a shows the spectrum for the first two groups of sum and difference frequencies.

In general, the spectrum of sum and difference frequencies looks like Fig. 21-11b. The total output spectrum contains these spectral lines plus the spectral lines of Fig. 21-10a and b.

Figure 21-11. Sum and difference spectra for large-signal operation.

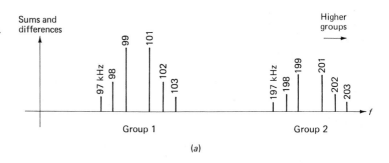

(a)

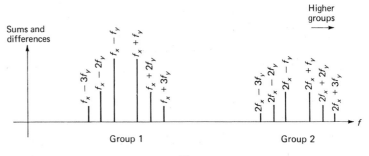

(b)

Summary

Unquestionably, the situation is complex in a large-signal amplifier. Put two sine waves into an amplifier and out come two amplified sine waves, their harmonics, and every conceivable sum and difference frequency produced by the input frequencies and their harmonics. Theoretically, an infinite number of spectral components exist. As a practical matter, the size of these components decreases for higher values of m and n.

EXAMPLE 21-2.

The two input sine waves to a large-signal amplifier have frequencies of $f_x = 10$ kHz and $f_y = 3$ kHz. Work out the first four sums and differences in group 1. Explain what to do with the negative frequency.

SOLUTION.

Group 1 has these frequencies:

   10 kHz $\pm$ 3 kHz = 7 and 13 kHz
   10 kHz $\pm$ 6 kHz = 4 and 16 kHz
   10 kHz $\pm$ 9 kHz = 1 and 19 kHz
   10 kHz $\pm$ 12 kHz = -2 and 22 kHz
   etc.

Whenever a negative frequency turns up, you can take the absolute value. In this example, -2 kHz becomes 2 kHz. The reason you can disregard the negative sign lies in the identity used to expand cross products:

$$\sin A \sin B = \tfrac{1}{2} \cos (A - B) - \tfrac{1}{2} \cos (A + B)$$

The first term on the right is the difference component. If A is less than B, you get a negative angle. But the cosine of a negative angle is equal to the cosine of a positive angle with the same magnitude. For instance,

$$\cos (-60°) = \cos 60°$$

Because of this, a negative difference frequency is equivalent to a positive frequency of the same magnitude.

21-6. INTERMODULATION DISTORTION

If we amplify speech or music in a nonlinear amplifier, the sum and difference frequencies in the output make the speech or music sound radically different. *Intermodulation distortion* is the change in the spectrum caused by the sum and difference frequencies.

Chord C-E-G

For reasons not yet understood, music sounds good when the notes have a mathematical relation. In Fig. 21-12*a*, middle C (point 1) has a fundamental frequency of

256 Hz. The next C on the right is an *octave* higher (point 2); this C has a fundamental frequency of 512 Hz, exactly double that of middle C. Play two notes that are an octave apart and the combination sounds pleasant.

Chords may sound even better. Play chord C-E-G (point 4) and it sounds very pleasant. Here is something interesting. Chord C-E-G has these mathematical properties:

C	E	G	NOTES IN CHORD
256	320	384	Fundamental frequency, Hz
1	1¼	1½	Ratio to middle C

The ratios suggest a chord sounds good when the fundamentals are quarter multiples of the lowest frequency.

Besides the fundamental frequencies shown, chord C-E-G contains harmonics to around 10 kHz. The amplitudes of these harmonics give the chord its distinct sound.

Figure 21-12. (a) Piano keyboard. (b) C–E–G spectrum.

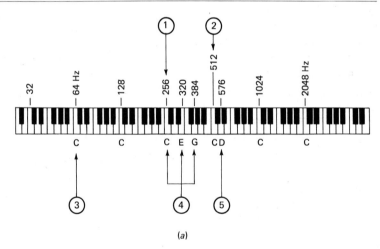

(a)

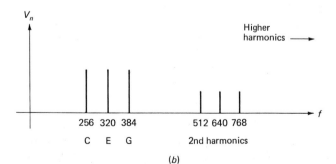

(b)

Figure 21-12*b* shows part of the spectrum for chord C-E-G. There are fundamental frequencies of 256-320-384, second harmonics of 512-640-768, third harmonics, fourth harmonics, and so on. The relative amplitudes of these harmonics must remain the same with respect to the fundamentals if we want to retain the exact sound. If this spectrum drives a *linear amplifier,* each spectral component receives the same gain, assuming all components are in the midband of the amplifier. In terms of the superposition theorem, the output is the sum of each amplified spectral component.

Sum and difference frequencies

If the spectrum of chord C-E-G drives a *nonlinear amplifier,* the output spectrum contains all the input spectral lines plus the sums and differences of every possible combination of these components. How will this sound? Terrible! The ear immediately detects sum and difference frequencies because these frequencies sound like mistakes.

For instance, the first two spectral lines of Fig. 21-12*b* have frequencies of 256 and 320 Hz; these components produce a difference frequency of 64 Hz and a sum frequency of 576 Hz. The 64-Hz component corresponds to the C note two octaves below middle C (point 3 in Fig. 21-12*a*). This extra C note is not unpleasant because it is harmonically related to middle C; however, it does represent a new component not in the original sound. Much worse is the sum frequency 576 Hz; this corresponds to the D note one octave above middle C (point 5 in Fig. 21-12*a*). This D note does not belong in chord C-E-G and sounds *discordant;* the effect is the same as if the pianist made a mistake.

Besides the sum and difference frequencies of the first two fundamentals, we get many other sum and difference frequencies produced by other combinations. As a result, nonlinear amplification produces many discordant notes.

Relation to nonlinear distortion

Nonlinear distortion causes harmonic and intermodulation distortion. Any device or circuit with a nonlinear input-output relation results in nonlinear distortion of the signal. In the time domain this means the shape of the periodic signal changes as it passes through the nonlinear circuit. In the frequency domain the result is a change in the spectrum of the signal. If only one input sine wave is present, only harmonic distortion occurs; if two or more input sine waves are involved, both harmonic and intermodulation distortion occur.

21-7.   FREQUENCY MIXERS

A *frequency mixer* is used in almost every radio and television receiver; it is also used in many other electronic systems.

The basic idea

Figure 21-13 shows all the key ideas behind a frequency mixer. Two input sine waves drive a nonlinear circuit. As before, this results in all harmonics and intermodulation components. The bandpass filter then passes one of the intermodulation components, usually the difference frequency $f_x - f_y$. Therefore, the final output of a typical mixer is a sine wave with frequency $f_x - f_y$.[3] In terms of spectra, a frequency mixer is a circuit that produces an output spectrum with a single line at $f_x - f_y$ when the input spectrum is a pair of lines at f_x and f_y.

A low-pass filter may be used in the place of a bandpass filter provided $f_x - f_y$ is less than f_x or f_y. For instance, if f_x is 2 MHz and f_y is 1.8 MHz, then

$$f_x - f_y = 2 \text{ MHz} - 1.8 \text{ MHz} = 0.2 \text{ MHz}$$

In this case, the difference frequency is lower than either input frequency; so we can use a low-pass filter if we wish. (Low-pass filters are usually easier to build than bandpass filters.)

But in applications where $f_x - f_y$ is between f_x and f_y, we must use a bandpass filter. As an example, if $f_x = 2$ MHz and $f_y = 0.5$ MHz, then

$$f_x - f_y = 2 \text{ MHz} - 0.5 \text{ MHz} = 1.5 \text{ MHz}$$

To pass only the difference frequency, we are forced to use a bandpass filter.

Usual size of input signals

In most applications, one of the input signals to the mixer will be large. This is necessary to ensure nonlinear operation; unless one of the signals is large, we cannot get intermodulation components. This large input signal is often supplied by an oscillator or signal generator.

The other input signal is usually small. By itself, this signal produces only small-

---

[3] The largest intermodulation components are the ones with the sum and difference frequencies $f_x \pm f_y$. Of these two, the difference is usually easier to filter.

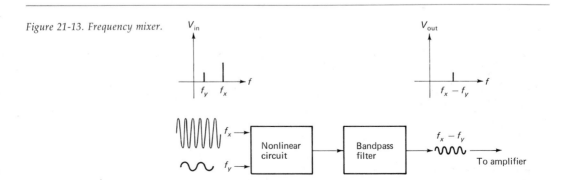

Figure 21-13. Frequency mixer.

signal operation of the mixer. One of the reasons this signal is small is because it often is a weak signal coming from an antenna (see Example 21-3).

The normal inputs to a mixer therefore are

1. A large signal adequate to produce medium- or large-signal operation of the mixer
2. A small signal that by itself can produce only small-signal operation

Transistor mixer

Figure 21-14a shows a *transistor mixer*. The two input signals produce an additive waveform at the junction of R and C_B. This additive waveform drives the base-emitter diode over a significant part of the transconductance curve. The resulting collector current contains harmonics and intermodulation components. With the LC tank tuned to the difference frequency, the output signal has a frequency of $f_x - f_y$.

Figure 21-14b shows an alternative way to couple the input signals. One signal drives the base, the other the emitter. The advantage is isolation of sources and sometimes a better impedance match.

Diode mixers

Instead of using a transistor for the nonlinear device, we can use a diode. We drive the diode with an additive waveform containing frequencies f_x and f_y. The diode current then contains harmonics and intermodulation components. With a filter, we can remove the difference frequency.

Incidentally, *heterodyne* is another word for mix, and *beat frequency* is synonymous with difference frequency. In Fig. 21-14a, we are heterodyning two input signals to get a beat frequency of $f_x - f_y$.

Conversion gain

Conversion gain refers to the power gain of the mixer. Specifically,

$$\text{Conversion gain} = \frac{p_{\text{out}}}{p_{\text{in}}} \qquad (21\text{-}10)\,\text{***}$$

where p_{out} is the output power of the difference signal and p_{in} is the input power of the smaller input signal. As an example, suppose the smaller signal in Fig. 21-14b delivers 10 μW to the emitter; if the output power of the difference signal is 40 μW, we have a conversion gain of

$$\text{Conversion gain} = \frac{40\ \mu\text{W}}{10\ \mu\text{W}} = 4$$

This is equivalent to a bel conversion gain of 6 dB.

Figure 21-14. Transistor mixers.

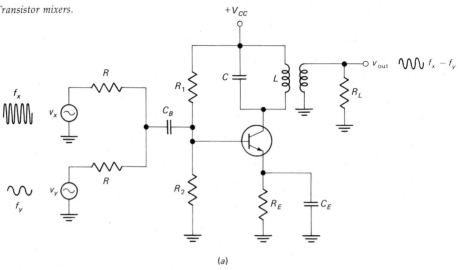

(a)

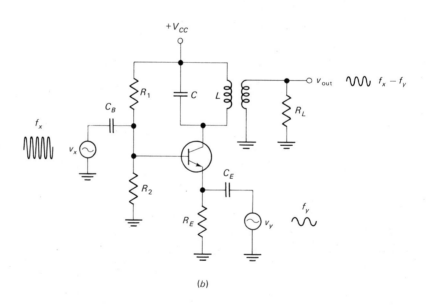

(b)

Also useful is the *conversion voltage gain* defined as

$$\text{Conversion voltage gain} = \frac{v_{\text{out}}}{v_{\text{in}}} \qquad (21\text{-}11)$$

Mixers in AM receivers

Figure 21-15*a* shows the front end of a typical AM broadcast receiver. The antenna delivers a weak signal to the radio-frequency (RF) amplifier. Although this increases signal strength, the f_y signal driving the mixer is still small. The other mixer input comes from a *local oscillator* (LO); this signal is large enough to produce nonlinear distortion. As a result, the output of the mixer is a signal with a frequency $f_x - f_y$.

The difference signal now drives several stages called the *intermediate-frequency* (IF) amplifiers. These IF amplifiers provide most of the gain in the receiver. The large signal coming out of the IF amplifiers then drives other circuits.

Here is a numerical example. When you tune to a station with a frequency of 1000 kHz, you are doing the following:

1. Adjusting a capacitor to tune the RF amplifier to 1000 kHz
2. Adjusting another capacitor to set the LO frequency to 1455 kHz

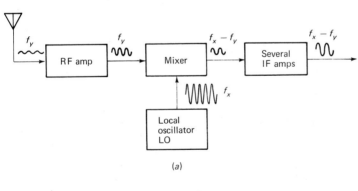

Figure 21-15. Front end of typical AM receiver.

As shown in Fig. 21-15b, the output of the mixer is 455 kHz. The IF amplifiers are fixed-tuned to this frequency; therefore, the 455-kHz signal receives maximum gain from the *IF strip* (group of amplifiers).

If you tune to a station with a frequency of 1200 kHz, the RF stage is tuned to 1200 kHz and the LO stage to 1655 kHz. The difference frequency is *still 455 kHz.* In other words, by deliberate design the frequency out of the mixer always equals 455 kHz no matter what station you tune to. We will discuss the reason for this in the next chapter. Briefly, it boils down to this: it is easier to design a group of amplifier stages tuned to a constant frequency than try to gang-tune these stages to each station frequency.

Another point. We used an IF frequency of 455 kHz because it is one of the common IF frequencies in commercial receivers. Other typical values are 456 kHz and 465 kHz. Many automobile radios use IF frequencies of 175 kHz or 262 kHz. Regardless of the exact value, any modern receiver has a group of IF amplifiers fix-tuned to the same frequency. The incoming signal frequency is converted down to this IF frequency.

EXAMPLE 21-3.
Figure 21-16 shows a mixer. The large input signal v_x has a frequency of 100 kHz. The small input signal v_y has a frequency of 100.5 kHz. If the output voltage has an rms value of 0.0125 V, what does the conversion voltage gain equal? What does the output lag network do?

SOLUTION.
With Eq. (21-11),

$$\text{Conversion voltage gain} = \frac{0.0125}{0.01} = 1.25$$

This is equivalent to approximately 2 dB. (If you build the circuit of Fig. 21-16, you should get similar results.)

The lag networks in the collector circuit are low-pass filters; they stop all components above a cutoff frequency of approximately 800 Hz. Since the difference frequency is 500 Hz, it passes through to the output.

21-8. SPURIOUS SIGNALS

A serious problem may arise in mixer applications. This section describes the problem and its cure.

Unwanted signals

The output filter of a mixer passes any signal whose frequency is in the passband. It is possible for small unwanted signals to get through the filter along with the desired

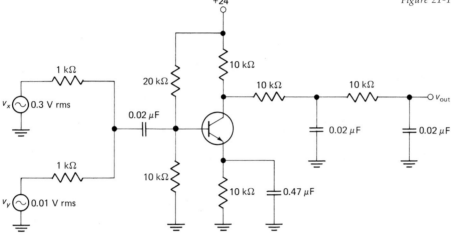

Figure 21-16. Bipolar mixer.

difference frequency. As we saw in Sec. 21-5, many difference frequencies are generated in a mixer; these are given by

$$\text{Difference frequencies} = mf_x - nf_y$$

where m and n take on all integer values. It is possible to find values of m and n that produce difference frequencies close to $f_x - f_y$. Such difference frequencies are called *spurious signals* because they can pass through the filter along with the desired difference frequency.

Here is an example. Suppose $f_x = 10.5$ MHz and $f_y = 2.5$ MHz. The desired difference is

$$f_x - f_y = 10.5 \text{ MHz} - 2.5 \text{ MHz} = 8 \text{ MHz}$$

There are many values of m and n that produce difference frequencies near this. For instance, if $m = 2$ and $n = 5$

$$mf_x - nf_y = 2(10.5 \text{ MHz}) - 5(2.5 \text{ MHz}) = 8.5 \text{ MHz}$$

This frequency is close to the desired frequency; if enough of this 8.5-MHz signal reaches the output, it may interfere with the desired signal. We can find other spurious frequencies close to the desired frequency. Because of this, the mixer output contains the desired signal plus many spurious signals.

Spurious signals occur for values of m and n greater than unity. For this reason, they are weak compared to the desired signal. Therefore, in some applications the bipolar mixer is acceptable even though it produces small spurious output signals.

*Figure 21-17. FET mixer. (a)
JFET. (b) Dual-gate MOSFET.*

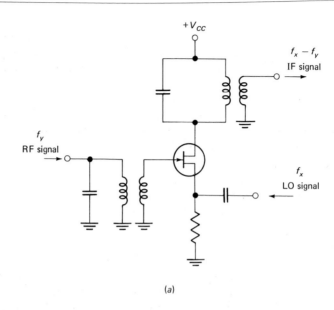

(a)

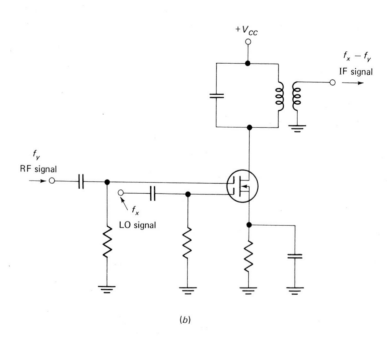

(b)

The FET mixer

But there are many applications where spurious signals out of a bipolar mixer cause problems.[4] The simplest way to eliminate spurious signals is to operate the mixer in the medium-signal case; the output voltage for this type of operation is

$$v_{out} = Av_{in} + Bv_{in}^2$$

As proved in Sec. 21-4, medium-signal operation produces only the input frequencies, their second harmonics, the sum frequency, and the difference frequency; there are no other difference frequencies; therefore, there can be no spurious signals with medium-signal operation.

Because the FET operates medium-signal all the way to cutoff and saturation, it is the ideal device to use in a mixer. With a FET mixer, spurious signals are almost eliminated. For this reason, the FET is superior to a bipolar transistor when it comes to mixer applications.

Figure 21-17a shows a JFET mixer. The RF signal drives the gate, and the stronger LO signal drives the source. For the FET to operate medium-signal, the LO signal must not drive the FET into saturation or cutoff; if this should happen, cubic and higher-power terms creep into the expression for output voltage. In other words, if the LO signal is large enough to cause clipping, the spurious signal content will increase. Ideally, the LO signal should cause the FET to swing over as much of the transconductance curve as possible without clipping; this results in the highest conversion gain with minimum spurious signals.

Figure 21-17b shows a dual-gate MOSFET mixer. The RF signal drives one of the gates, the LO signal the other gate. In a circuit like this, each input signal works into a high input impedance.

EXAMPLE 21-4.
Explain the action of Figs. 21-18 and 21-19.

[4] In a TV receiver, for instance, spurious signals out of a mixer are amplified along with the desired video signal. This results in a poor-quality picture when the spurious signals are large enough to interfere with the video signal.

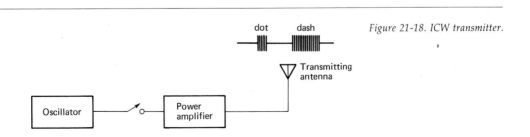

dot   dash

Transmitting antenna

Oscillator   Power amplifier

Figure 21-18. ICW transmitter.

Figure 21-19. ICW receiver.

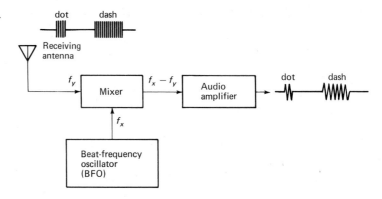

SOLUTION.

In Fig. 21-18, each time we close the switch, the antenna radiates a signal. Because of this, we can transmit short bursts of energy (dots) or long bursts (dashes).

For antennas of practical size, the radiated frequencies are far above the audio range (greater than 20 kHz). To hear the dots and dashes, therefore, we must first mix the incoming signal with a *beat-frequency-oscillator* (BFO) signal (see Fig. 21-19). The BFO signal has a frequency close to the received frequency, so close in fact that $f_x - f_y$ is an audio frequency. Because of this, the bursts of received energy are translated into bursts of audio frequencies, dots and dashes we can hear.

The system just described is called *interrupted-continuous-wave* (ICW) telegraphy.

21-9. NOISE

As mentioned earlier, noise contains sinusoidal components at all frequencies. Some of these noise components will mix with the LO signal and produce difference frequencies in the output of the mixer. Because of this, the remainder of the system amplifies the desired signal and unwanted noise out of the mixer.

In general, noise is any kind of unwanted signal not derived from or related to the input signal. This section describes noise.

Some types of noise

Where does noise come from? Electric motors, neon signs, power lines, ignition systems, lightning, etc., set up electric and magnetic fields. These fields can induce noise voltages in electronic circuits. To reduce noise of this type, we can shield the circuit and its connecting cables.

Power-supply ripple is also classified as noise because it is independent of the desired signal. As we know, ripple can get into signal paths through biasing resistors (also by induction). With regulated power supplies (Chap. 23) and shielding, we can reduce ripple to an acceptably low level.

If you bump or jar a circuit, the vibrations may move capacitor plates, inductor windings, and so on. This results in a noise called *microphonics*. In this respect, the transistor is far superior to the vacuum tube; because of its solid-state construction, a transistor has negligible microphonics.

Thermal noise

We can eliminate or at least minimize the effects of ripple, microphonics, and external field noise. But there is little we can do about *thermal noise*. Figure 21-20a shows the idea behind this kind of noise. Inside any resistor are conduction-band electrons. Since these electrons are loosely held by the atoms, they tend to move randomly in different directions as shown. The energy for this motion comes from the thermal energy of surrounding air; the higher the ambient temperature, the more active the electrons.

The motion of the billions of electrons is pure chaos. At some instants in time, more move up than down, producing a small negative voltage across the resistor. At other instants, more move down than up, producing a positive voltage. If amplified and viewed on an oscilloscope, this noise voltage would resemble Fig. 21-20b. Like any voltage, noise has an rms or heating value; as an approximation, the highest noise peaks are about four times the rms value.

The changing size and shape of the noise voltage implies components of many different frequencies. An advanced derivation shows the noise spectrum is like Fig. 21-20c. As we see, noise is uniformly distributed throughout the practical frequency range. The break frequency of approximately 10^{12} Hz is far beyond the capability of electronic circuits. For this reason, most people say that noise contains sinusoidal components at all frequencies.

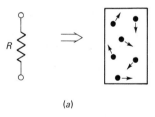

(a)

Figure 21-20. Thermal noise. (a) Random motion of electrons. (b) Appearance on oscilloscope. (c) Spectral distribution.

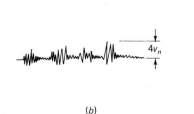

(b)

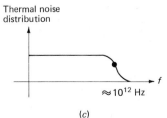

Thermal noise distribution

$\approx 10^{12}$ Hz

(c)

How much noise voltage does a resistor produce? This depends on the temperature, the bandwidth of the system, and the size of the resistance. Specifically,

$$v_n = \sqrt{4kTBR} \tag{21-12}$$

where v_n = rms noise voltage
k = Boltzmann's constant (1.37×10^{-23})
T = absolute temperature, Celsius + 273°
B = noise bandwidth, Hz
R = resistance, Ω

In Eq. (21-12), the noise voltage increases with temperature, bandwidth, and resistance.
At room temperature (25°C), the formula reduces to

$$v_n = 1.28(10^{-10}) \sqrt{BR} \tag{21-12a}$$

The noise bandwidth B is approximately equal to the 3-dB bandwidth of the amplifier, mixer, or system being analyzed. As an example, a transistor radio has an overall bandwidth of about 10 kHz. This is the approximate value of B in Eq. (21-21a). A television receiver, on the other hand, has a bandwidth of about 4 MHz; this is the value of B to use for noise calculations in a television receiver.

With Eq. (21-12) or (21-12a), we can get an estimate for the amount of noise produced by any resistor in an amplifier. Those resistors near the input of the amplifier are most important because their noise will be amplified most and will dominate the final output noise. We can get a rough estimate of the input noise level by calculating the noise generated by the Thevenin resistance driving the amplifier or system.[5]

EXAMPLE 21-5.
The Thevenin source resistance driving an amplifier is 5 kΩ (Fig. 21-21). Estimate the amount of noise this resistance delivers to the amplifier in the bandwidth shown.

SOLUTION.
The bandwidth is 100 kHz. We will assume the ambient temperature is 25°C. With Eq. (21-12a),

$$v_n = 1.28(10^{-10}) \sqrt{10^5(5000)}$$
$$= 2.86 \ \mu V$$

---

[5] There are many advanced books covering topics like noise, noise figure, and signal-to-noise ratio. For a start, see *Reference Data for Radio Engineers,* Howard W. Sams & Co., Inc., New York, 1968, Chap. 27.

Figure 21-21. Example 21-5.

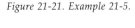
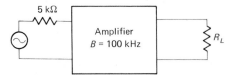

This gives us an estimate of the input noise level. If the desired signal is smaller than this, it will be masked or covered by the noise. Therefore, the desired signal has to be much greater than 2.86 μV to stand out from source noise and noises generated inside the amplifier.

Problems

21-1. In Eq. (21-1b), $A = 50$, $B = 10$, and $C = 5$; all other coefficients are zero. If v_{in} is a sine wave with a peak of 1 V, what values do the positive and negative output peaks have?

21-2. Same coefficients as given in Prob. 21-1. Calculate the positive and negative output peaks if v_{in} is a sine wave with a peak of 1 mV.

21-3. A linear dc amplifier has a voltage gain of 50 and a bandwidth of 30 kHz. The input spectrum has a single line with a peak of 15 mV and a frequency of 1 kHz. What are the peak values and frequencies of the output spectral lines?

21-4. The input spectrum to a linear amplifier looks like Fig. 21-22a. If all components are within the midband of the amplifier and the voltage gain is 75, what does the output spectrum contain?

21-5. An amplifier operates medium-signal with $A = 50$ and $B = 10$. The input spectrum is shown in Fig. 21-22b. Work out the value of

1. The dc shift
2. The peak output voltage of the fundamental
3. The peak output voltage of the second harmonic

21-6. Two sine waves drive a medium-signal amplifier. If the input frequencies are 56 kHz and 84 kHz, what frequencies does the output contain?

21-7. Three sine waves drive a medium-signal amplifier. If the input frequencies are 1 kHz, 4 kHz, and 9 kHz, what output frequencies are there? (Take two input frequencies at a time.)

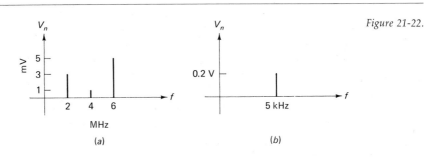

Figure 21-22.

(a)

(b)

Figure 21-23.

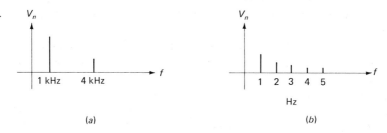

(a)   (b)

21-8.   Figure 21-23a shows the input spectrum to a large-signal amplifier. What harmonic frequencies does the output contain? What are the sum and difference frequencies of the first group?

21-9.   The difference frequency $f_x - f_y$ equals 2 MHz and the sum equals 16 MHz. What are the input frequencies?

21-10.   Figure 21-23b shows the output spectrum of an amplifier.

1.   If the amplifier is operating linearly, how many input sine waves are there?
2.   If the amplifier has harmonic distortion only, how many input sine waves are there?

21-11.   On the piano keyboard of Fig. 21-12a, what is the fundamental frequency of the lowest C note? The highest C note? The highest E note?

21-12.   Figure 21-24a shows the input spectrum to a medium-signal amplifier. What frequencies does the output spectrum contain?

21-13.   Figure 21-24b shows the sum and difference frequencies out of a medium-signal amplifier. What are the input frequencies?

21-14.   When tuned to channel 3, the mixer in a television set has a small input signal with a frequency of 63 MHz and LO signal with a frequency of 107 MHz. What is the difference frequency out of the mixer?

21-15.   AM station frequencies are from 540 to 1600 kHz. In an AM radio, the received signal is one input to a mixer, and the LO signal is the other input. The frequency of the LO signal is 455 kHz greater than the received signal. What

Figure 21-24.

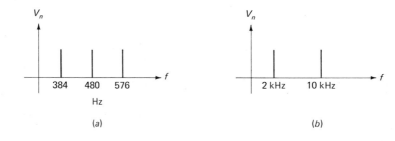

(a)   (b)

are the minimum and maximum LO frequencies? If a Hartley oscillator is used to generate the LO signal, what is the ratio of maximum to minimum tuning capacitance?

21-16. Suppose the transmitted frequency in Fig. 21-18 is 100 kHz. If the output of the audio amplifier in Fig. 21-19 sounds like a 2-kHz signal, what frequency does the BFO have?

21-17. A bipolar mixer has an input signal power of 3 μW and an output signal power of 1.5 μW. Express the conversion gain as an ordinary number and in decibels.

21-18. A FET mixer has an input signal voltage of 10 mV and an output signal voltage of 40 mV. Calculate the conversion voltage gain in decibels.

21-19. The data sheet of a diode mixer gives a conversion gain of -9 dB. If the input signal power to this mixer is 10 μW, how much output signal power is there?

21-20. Calculate the thermal noise voltage at room temperature for a bandwidth of 1 MHz and a resistance of

1. 1 MΩ
2. 10 kΩ
3. 100 Ω

21-21. How much thermal noise voltage is there at room temperature for a resistance of 10 kΩ and a bandwidth of 100 kHz? For a bandwidth of 1 kHz? 10 Hz?

21-22. The Thevenin resistance driving an amplifier is 1000 Ω. If the amplifier has a bandwidth of 1 MHz, approximately what is the input noise level at room temperature? If the amplifier has a voltage gain of 10,000, how large will the amplifier source noise be at the output?

22. Modulation

Radio, television, and many other electronic systems would be impossible without *modulation;* it refers to a low-frequency signal controlling the amplitude, frequency, or phase of a high-frequency signal.

22-1. AMPLITUDE MODULATION

When the low-frequency signal controls the amplitude of the high-frequency signal, we get *amplitude modulation* (AM).

A simple modulator

Figure 22-1a shows a simple modulator. A high-frequency signal v_x is the input to a potentiometer; therefore, the amplitude of the output signal depends on the position of wiper. If we move the wiper up and down sinusoidally, we get the AM waveform of Fig. 22-1b; the amplitude or peak value of the high-frequency signal is varying at a low-frequency rate.

The high-frequency signal is called the *carrier,* and the low-frequency signal the *modulating signal.* Hundreds of carrier cycles normally occur during one cycle of the modulating signal. For this reason, an AM waveform on an oscilloscope looks like the signal of Fig. 22-1c; the positive peaks of the carrier are so closely spaced they form a solid upper boundary known as the *upper envelope;* similarly, the negative peaks form the *lower envelope.*

Figure 22-1. Amplitude
modulation. (a) Voltage divider.
(b) AM waveform. (c) Envelope.

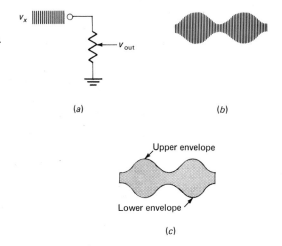

(a)

(b)

Upper envelope

Lower envelope

(c)

A transistor modulator

Figure 22-2 is an example of a transistor modulator. Here is how it works. The carrier signal v_x is the input to a CE amplifier. The circuit amplifies the carrier by a factor of A, so that the output is Av_x. The modulating signal is part of the biasing; therefore, it produces low-frequency variations in emitter current; in turn, this means variations in r'_e and A. For this reason, the amplified carrier signal looks like the AM waveform shown; the peaks of the output vary sinusoidally with the modulating signal. Stated another way, the upper and lower envelopes have the shape of the modulating signal.

Input voltages

For normal operation, Fig. 22-2 should have a small carrier. We do not want the carrier to influence voltage gain; only the modulating signal should do this. Therefore, the operation should be small-signal with respect to the carrier. On the other hand, the modulating signal is part of the biasing network. To produce noticeable changes in voltage gain, the modulating signal has to be large. For this reason, the operation is large-signal with respect to the modulating signal.

Input frequencies

Usually, the carrier frequency f_x is much greater than the modulating frequency f_y. In Fig. 22-2, we need f_x at least 100 times greater than f_y. Here is the reason. The capacitors should look like low impedances to the carrier and like high impedances to the modulating signal; in this way, the carrier is coupled into and out of the circuit, but the modulating signal is blocked from the output.

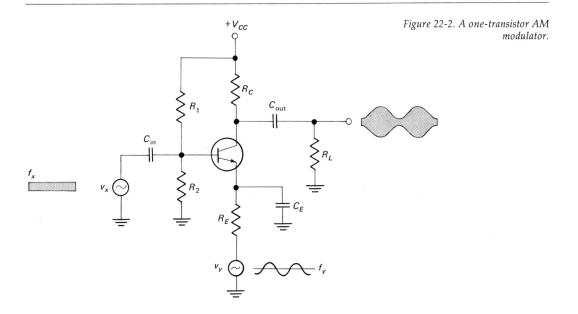

Figure 22-2. A one-transistor AM modulator.

EXAMPLE 22-1.
In Fig. 22-3, the input peak of the carrier is 10 mV. The input peak of the modulating signal is 8 V. Calculate the minimum, quiescent, and maximum voltage gain.

SOLUTION.
When the modulating voltage is zero, the voltage across the emitter resistor is 10 V. So, the emitter current is 1 mA and r'_e is approximately 25 Ω. The quiescent voltage gain therefore equals

$$A = \frac{r_C}{r'_e} = \frac{10{,}000 \,||\, 1500}{25} = 52 \qquad \text{(quiescent)}$$

At the instant the modulating voltage reaches a positive peak of 8 V, only 2 V is across the 10-kΩ emitter resistor. At this instant, the emitter current is 0.2 mA and r'_e is 125 Ω. The corresponding voltage gain is

$$A = \frac{1300}{125} \cong 10 \qquad \text{(minimum)}$$

At the negative peak of the modulating signal, the voltage across the emitter resistor is 18 V. The emitter current becomes 1.8 mA, r'_e decreases to 14 Ω, and

$$A = \frac{1300}{14} \cong 93 \qquad \text{(maximum)}$$

Figure 22-3. Example 22-1.

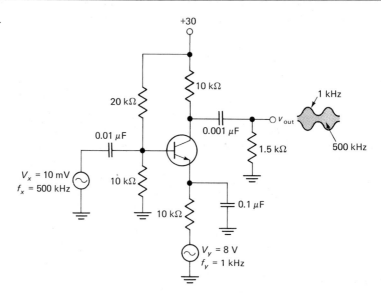

As far as the carrier is concerned, the CE amplifier changes its voltage gain from a low of 10 to a high of 93. Therefore, the output signal is an AM waveform as shown.

22-2. NEED FOR MODULATION

Without modulation, we would have no radio or television; we would also have to do without our modern telephone system. This section tells why.

Practical antenna length

Books on antenna theory describe why the length of an antenna should be at least a quarter wavelength. In terms of frequency, a quarter wavelength equals

$$L = \frac{7.5(10^7)}{f} \tag{22-1}$$

where L is in meters and f in hertz. If the antenna is shorter than this, it will not radiate signals efficiently.

Antennas would be immense at audio frequencies. For instance, to radiate 1000 Hz efficiently, we would need an antenna length of

$$L = \frac{7.5(10^7)}{1000} = 7.5(10^4) \text{ meters} \cong 47 \text{ miles}$$

This is much too long. For this reason, it is impractical to radiate audio frequencies directly into space. Instead, communication systems transmit radio frequencies (greater than 20 kHz).

AM radio

AM broadcast signals use carrier frequencies between 540 and 1600 kHz. In the studio, the audio signal modulates the carrier to produce an AM signal like Fig. 22-4a. (Actually, the envelope is more complicated than this because voice and music contain many sinusoidal components.) The transmitting antenna radiates this AM signal into space.

When received, the AM signal is processed by different stages and somewhere near the end of the receiver the original audio is retrieved (Fig. 22-4b). This audio signal is essentially the same as the original modulating signal. In this way, whatever goes into the microphone at the transmitting studio comes out the loudspeaker of the radio.

Telephone system

AM signals make modern telephone possible. If telephone systems had to use a single pair of wires for each conversation, the amount of wire would be impractical. In today's telephone systems, a single pair of wires carries hundreds of conversations. Each conversation has a different carrier frequency, and this is the key to how we can separate the conversation.

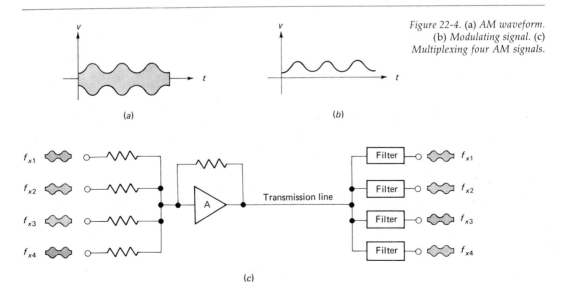

Figure 22-4. (a) AM waveform. (b) Modulating signal. (c) Multiplexing four AM signals.

Figure 22-4c shows the basic idea behind *multiplexing*, using the same transmission line to carry more than one signal. Four different AM signals are combined into a single-ended output by the op-amp summing circuit. These signals now travel down the transmission line (it may be miles long). At the receiving end, filters separate the different carrier frequencies, so that once again we have four separate AM signals. This approach can be extended to hundreds of AM signals and is used in modern telephone systems to reduce the amount of wire.

22-3. PERCENT MODULATION

A sinusoidal modulating signal produces a sinusoidal variation in voltage gain expressed by

$$A = A_0(1 + m \sin \omega_y t) \qquad (22\text{-}2)$$

where A_0 is the average gain and m is the *modulation coefficient*. As an example, if $A_0 = 100$ and $m = 0.5$, the graph of Eq. (22-2) looks like Fig. 22-5a. Here we see the voltage gain swinging from a maximum of 150 to a minimum of 50.

In general, the graph of voltage gain looks like Fig. 22-5b. The maximum value is $A_0(1 + m)$, and the minimum is $A_0(1 - m)$. For the special case of m equal to 1, we get the variation shown in Fig. 22-5c. When m is greater than 1, the maximum voltage gain is greater than $2A_0$ (see Fig. 22-5d).

Percent modulation is used to describe the amount of modulation that has occurred.

Figure 22-5. Modulator voltage gain. (a) $m = 0.5$. (b) *General case.* (c) $m = 1$. (d) $m > 1$.

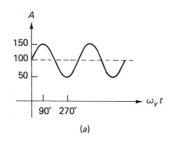

(a)

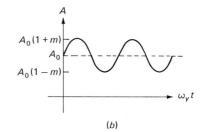

(b)

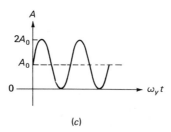

(c)

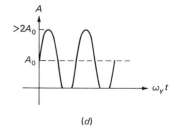

(d)

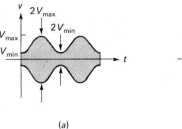 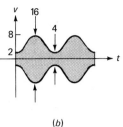

Figure 22-6. Calculating m.

Percent modulation equals

$$\text{Percent modulation} = m \times 100 \text{ percent} \qquad (22\text{-}3)$$

In Fig. 22-5a, m is 0.5; therefore, the percent modulation is 50 percent. In Fig. 22-5c, m is unity and percent modulation is 100 percent.

You can measure m as follows: Given a waveform like Fig. 22-6a, the maximum peak-to-peak voltage is $2V_{\max}$ and the minimum peak-to-peak voltage is $2V_{\min}$. By a straightforward derivation given in the Appendix,

$$m = \frac{2V_{\max} - 2V_{\min}}{2V_{\max} + 2V_{\min}} \qquad (22\text{-}4)^{***}$$

As an example, if we see a waveform like Fig. 22-6b, we calculate

$$m = \frac{16 - 4}{16 + 4} = 0.6$$

which is equivalent to 60 percent modulation.

22-4. AM SPECTRUM

The output voltage of an AM modulator looks like Fig. 22-7a and equals

$$v_{\text{out}} = Av_x$$

If the carrier is sinusoidal, we may write

$$v_{\text{out}} = AV_x \sin \omega_x t$$

where V_x is the peak value of the input carrier. With Eq. (22-2), the output voltage becomes

$$v_{\text{out}} = A_0(1 + m \sin \omega_y t)V_x \sin \omega_x t$$

or $\qquad\qquad v_{\text{out}} = A_0 V_x \sin \omega_x t + m A_0 V_x \sin \omega_y t \sin \omega_x t \qquad (22\text{-}5)$

Figure 22-7. AM signal and its components. (a) AM wave. (b) Unmodulated carrier. (c) Difference component. (d) Sum component.

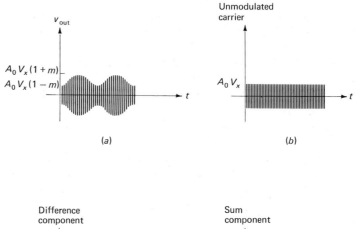

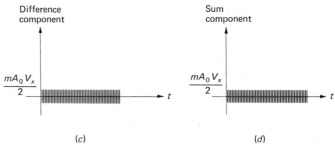

Unmodulated carrier

The first term in Eq. (22-5) represents a sinusoidal component with a peak of $A_0 V_x$ and a frequency of f_x. Figure 22-7b is the graph of the first term. We call this the unmodulated carrier because it is the output voltage when m equals zero.

Cross product

The second term in Eq. (22-5) is a cross product of two sine waves, similar to the cross products that occur in a mixer. As derived in the previous chapter, the product of two sine waves results in two new frequencies: a sum and a difference. Specifically, the second term of Eq. (22-5) equals

$$mA_0V_x \sin \omega_y t \sin \omega_x t = \frac{mA_0V_x}{2} \cos (\omega_x - \omega_y)t - \frac{mA_0V_x}{2} \cos (\omega_x + \omega_y)t \qquad (22\text{-}6)$$

The first term on the right is a sinusoid with a peak value of $mA_0V_x/2$ and a difference frequency of $f_x - f_y$. The second term is also a sinusoid with a peak $mA_0V_x/2$ but a sum frequency $f_x + f_y$. Figure 22-7c and d shows these sinusoidal components.

Spectral components

In the time domain, an AM signal like Fig. 22-7a is the superposition of three sine waves (Fig. 22-7b through d). One sine wave has the same frequency as the carrier, another has the difference frequency, and the third has the sum frequency.

In terms of spectra, here is what AM means. Figure 22-8a is the input spectrum to a modulator; the first line represents the large modulating signal with frequency f_y; the second line is for the small carrier with frequency f_x. Figure 22-8b is the output spectrum; here we see the amplified carrier between the difference and sum components. The difference component is sometimes called the *lower side frequency* and the sum is known as the *upper side frequency*.

The circuit implication of an AM signal is this. An AM signal is equivalent to three sine-wave sources in series as shown in Fig. 22-8c. This equivalence is not a mathematical fiction; the side frequencies really exist. In fact, with narrowband filters we can separate the side frequencies from the carrier.

EXAMPLE 22-2.

Figure 22-9a shows the inputs to a modulator. The modulating signal has a peak value of 10 V and a frequency of 1 kHz. The carrier has a peak of 0.05 V and a frequency of 200 kHz.

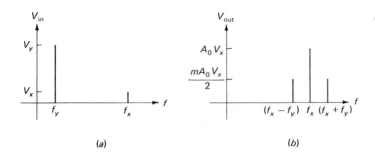

Figure 22-8. (a) Input spectrum to modulator. (b) AM spectrum. (c) AM is superposition of three sine waves.

(a)

(b)

(c)

Figure 22-9. (a) Input spectrum to AM modulator. (b) Output spectrum of AM modulator.

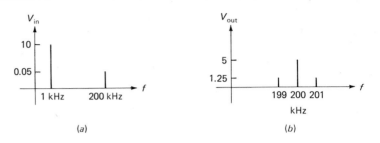

(a)   (b)

1. Find the frequencies in the output spectrum.
2. If A_0 is 100 and m is 0.5, calculate the peak values of the output spectral components.

SOLUTION.

1. In the output spectrum, the center component has the same frequency as the carrier; in this case, 200 kHz. The lower side has a difference frequency of 199 kHz, and the upper side a sum of 201 kHz.
2. Referring to Fig. 22-8b, the unmodulated carrier has a peak value of A_0V_x and the side components have peaks of $mA_0V_x/2$. So,

$$A_0V_x = 100(0.05) = 5 \text{ V}$$

and

$$\frac{mA_0V_x}{2} = \frac{0.5(100)0.05}{2} = 1.25 \text{ V}$$

Figure 22-9b shows the output spectrum.

22-5. FILTERING AM SIGNALS

In a multiplexer like Fig. 22-4c, each filter must pass only the desired spectral components: the carrier and two side frequencies of an AM signal.

Lower and upper sidebands

When the modulating signal is voice or music, the spectrum contains many sinusoidal components symbolized by the shaded region of Fig. 22-10a; the highest frequency is $f_{y(\max)}$. As an example, if the spectrum is for piano music, $f_{y(\max)}$ is approximately 10 kHz.

When voice or music modulates a carrier, the output spectrum looks like Fig. 22-10b. The shaded region to the left of the carrier is called the *lower sideband* because it contains all difference components. The region on the right of the carrier is the *upper sideband*; it contains all the sum components.

The *bandwidth of an AM signal* is the set of frequencies in the lower and upper

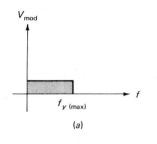

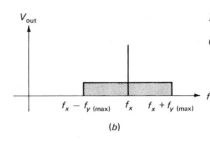

(a)

(b)

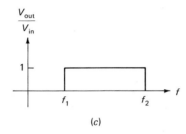

(c)

Figure 22-10. Sidebands of AM signal. (a) Input spectrum. (b) Output spectrum. (c) Window.

sidebands. Furthermore, the bandwidth value of an AM signal is

$$B_v = [f_x + f_{y(\max)}] - [f_x - f_{y(\max)}]$$

or $$B_v = 2f_{y(\max)} \qquad (22\text{-}7)^{***}$$

This says the bandwidth value of an AM signal is twice the highest modulating frequency.

Frequency distortion

We have to be careful when we filter AM signals. If the filter has a passband as shown in Fig. 22-10c, we must be sure to pass the desired spectral components while stopping undesired ones. That is, the window must pass only the components belonging to the desired AM signal.

If the filter passband is too narrow, we get frequency distortion. For instance, Fig. 22-11a shows three AM signals. We are trying to pass the center signal while stopping the adjacent signals. In this case, the window is too narrow and we lose the sinusoidal components at the edges of the AM spectrum. This is equivalent to losing the higher modulating frequencies.

Cross-talk

On the other hand, the filter passband may be too large as shown in Fig. 22-11b. We get not only the desired components but some components from adjacent signals.

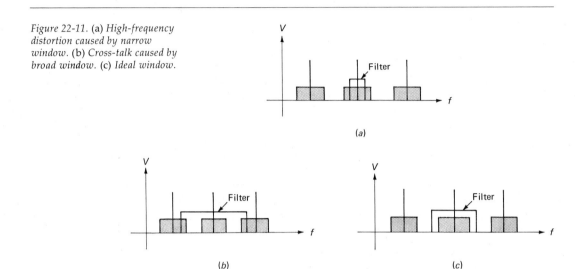

Figure 22-11. (a) *High-frequency distortion caused by narrow window.* (b) *Cross-talk caused by broad window.* (c) *Ideal window.*

This results in *cross-talk,* an interference between channels of information. In a telephone system, this is equivalent to higher modulating frequencies from adjacent signals interfering with the desired conversation. In AM radio it amounts to higher modulating frequencies from adjacent stations interfering with the modulating signal of the desired station.

Figure 22-11c illustrates ideal filtering; we get only the components of one channel of information. There is no frequency distortion, nor is there any cross-talk.

AM radio

AM radio transmission is an example of multiplexing because the same transmission medium (space) simultaneously transmits many channels of information (AM signals). The carrier frequencies are from 540 to 1600 kHz, spaced 10 kHz apart.

Figure 22-12 shows part of the AM broadcast spectrum. Especially notice the spectra do not overlap. To avoid overlapping spectra, each AM station must keep all spectral components within 5 kHz of the carrier. This implies the highest modulating frequency $f_{y(\max)}$ is less than 5 kHz.

EXAMPLE 22-3.
What is the bandwidth of the output signal in Fig. 22-13a?

SOLUTION.
The input audio spectrum contains all components up to 20 kHz. This spectrum drives a low-pass filter with a cutoff frequency of 5 kHz. Ideally, the output of the filter con-

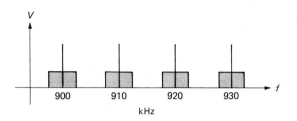

Figure 22-12. Part of the AM broadcast spectrum.

tains only spectral components up to 5 kHz. For this reason, the highest modulating frequency is 5 kHz (Fig. 22-13b).

Since the carrier frequency is 1 MHz, the output spectrum has a lower sideband from 995 to 1000 kHz and an upper sideband from 1000 to 1005 kHz (Fig. 22-13c). So, the bandwidth of the AM signal is from 995 to 1005 kHz; the bandwidth value is 10 kHz, twice the highest modulating frequency.

Incidentally, this example indicates why AM radio has low fidelity. Since the highest modulating frequency is 5 kHz, many of the higher harmonics in voice and music are not transmitted.

22-6. SUPPRESSED COMPONENTS

In some communication systems, part of the AM spectrum is suppressed.

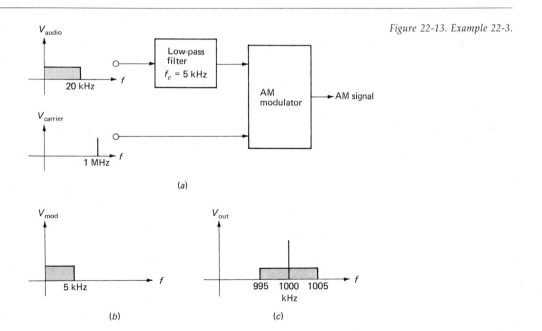

Figure 22-13. Example 22-3.

DSB-TC

Figure 22-14a shows the spectrum of an AM signal; a carrier component is between two sidebands. We classify this as *double sideband-transmitted carrier* (DSB-TC). In conventional AM radio, the transmitting antenna radiates a DSB-TC signal. The receiver picks up this signal and recovers the modulating signal.

DSB-SC

To transmit signals over great distances, the last stage in the transmitter must deliver large amounts of power to the antenna; even with a class C amplifier this implies a high power dissipation in the amplifying device. In some communication systems, this high dissipation is reduced by *suppressing the carrier*, that is, removing it as shown in Fig. 22-14b. This type of signal is classified as *double sideband-suppressed carrier* (DSB-SC). When the last stage of a transmitter amplifies a DSB-SC signal, the power dissipation is much less because there is no carrier power.

SSB-TC

Another way to reduce power dissipation is to use a *single sideband-transmitted carrier* (SSB-TC) signal illustrated by the spectrum of Fig. 22-14c. Either sideband may be suppressed. Since the last stage has to amplify only one sideband and the carrier, less power dissipation occurs in this stage.

Figure 22-14. Types of AM signals. (a) *DSB-TC.* (b) *DSB-SC.* (c) *SSB-TC.* (d) *SSB-SC.*

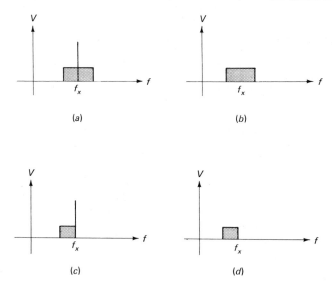

SSB-SC

Single sideband-suppressed carrier (SSB-SC) is the most drastic suppression of all. We take out one sideband and the carrier (see Fig. 22-14d). Besides greatly reducing the power dissipation of the last stage, a SSB-SC signal has half the bandwidth of a DSB signal. Because of this, twice as many channels can be multiplexed in a given frequency range.

EXAMPLE 22-4.

Figure 22-15a shows the spectrum of DSB-TC signal. If this signal drives a 1-kΩ resistor as shown in Fig. 22-15b, how much power does the resistor dissipate? If components are suppressed, how much power is there for each type of suppression?

SOLUTION.

1. The carrier component delivers a power of

$$p_{\text{carrier}} = \frac{(0.707 \times 5)^2}{1000} = 12.5 \text{ mW}$$

The lower-side component produces

$$p_{\text{lower}} = \frac{(0.707 \times 2.5)^2}{1000} = 3.125 \text{ mW}$$

and the upper-side component,

$$p_{\text{upper}} = \frac{(0.707 \times 2.5)^2}{1000} = 3.125 \text{ mW}$$

Therefore, the total power produced by the DSB-TC signal is

$$p = (12.5 + 3.125 + 3.125) \text{ mW} = 18.75 \text{ mW}$$

2. In Fig. 22-15b, assume we remove the carrier. Then, we have a DSB-SC signal. In

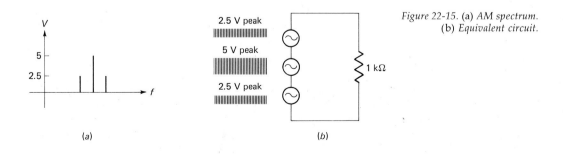

(a)   (b)

Figure 22-15. (a) *AM spectrum.* (b) *Equivalent circuit.*

this case, the two side components deliver a power of

$$p = (3.125 + 3.125) \text{ mW} = 6.25 \text{ mW}$$

3. If we suppress one of the side components, the resulting SSB-TC signal produces a total power of

$$p = (12.5 + 3.125) \text{ mW} = 15.625 \text{ mW}$$

4. If the carrier and one side component are suppressed, the SSB-SC signal delivers a total power of

$$p = 3.125 \text{ mW}$$

This example gives you some idea of why suppressed systems are sometimes used. We could just as easily be dealing in watts or kilowatts of power instead of milliwatts. Suppressing the carrier cuts the power dissipation by a factor of three; suppressing the carrier and one side component reduces the power by a factor of six. Aside from lowering the required $P_{D(\text{max})}$ rating of the last stage of a transmitter, suppressing components also reduces the size of the power supply. In some transmitters, the reduced size and weight can be vitally important (like the transmitter in a spacecraft).

22-7. AM MODULATOR CIRCUITS

The simple transistor modulator of Fig. 22-2 works but is not as useful as other modulators. This section describes some modulators used in discrete and integrated circuits.

DSB-TC modulator

Figure 22-16 shows a diff-amp modulator. The carrier drives one input and is amplified by a factor of R_C/r_e'. The modulating signal drives Q_3, which delivers a common-mode signal to the diff amp. Because of this, the modulating signal is suppressed across the output. All that appears is

$$v_{\text{out}} = Av_x$$

Since the modulating signal controls the emitter current in each half of the diff amp, the voltage gain varies as the modulating signal. Because of this, the output is a DSB-TC signal as shown.

The diff-amp modulator of Fig. 22-16 has this advantage over the one-transistor modulator of Fig. 22-2: no restriction exists on the carrier and modulating frequencies. In other words, the simple modulator of Fig. 22-2 will not work properly unless the carrier frequency is at least 100 times greater than the modulating frequency. The diff-amp modulator, however, works properly even when f_x/f_y is less than 100.

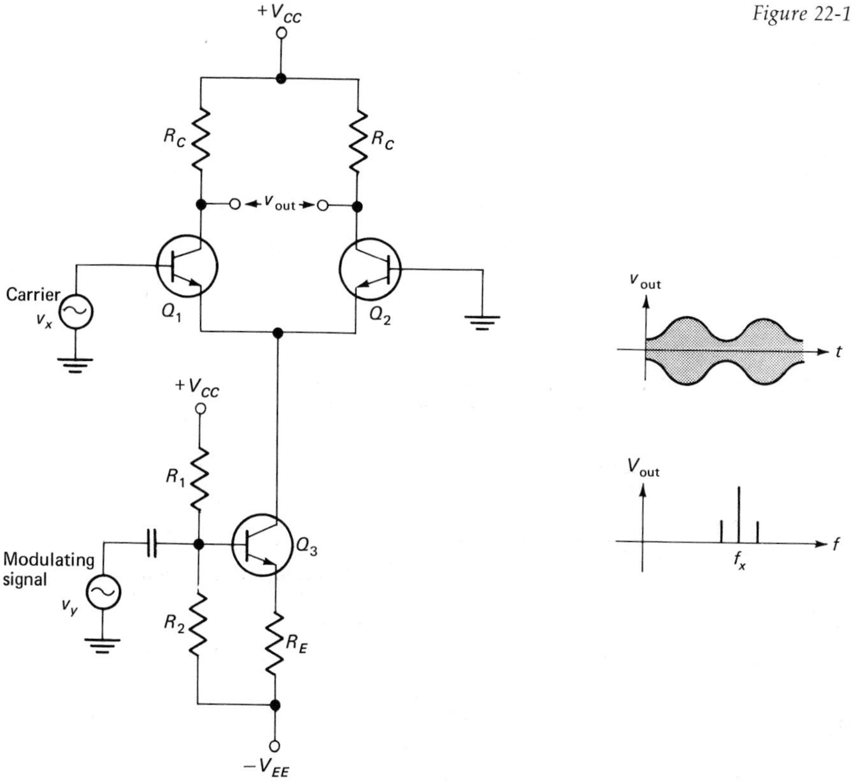

Figure 22-16. Diff-amp modulator.

For very high power requirements (kilowatts), DSB-TC modulators use a push-pull arrangement of vacuum tubes. This kind of modulation is called *plate modulation* and is discussed elsewhere.[1]

DSB-SC modulator

Figure 22-17a is a DSB-SC modulator. The carrier drives Q_3 and produces common-mode operation. For this reason, the carrier is suppressed across the output. The modulating voltage produces large-signal operation and out-of-phase AM collector currents. The difference of these currents is a DSB-SC signal.

The proof is too lengthy to show here. Briefly, here is what happens. The tuned output blocks the low-frequency modulating signal because the resonant circuit is tuned to the carrier frequency. Since the carrier is suppressed by common-mode opera-

[1] See *The Radio Amateur's Handbook*, American Radio Relay League, West Hartford, Conn.

Figure 22-17. Balanced modulator.
(a) Circuit. (b) Output spectrum.
(c) Output waveform.

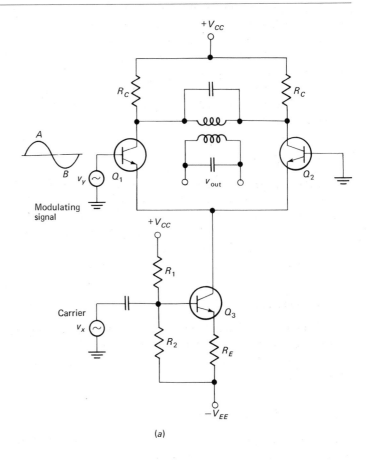

(a)

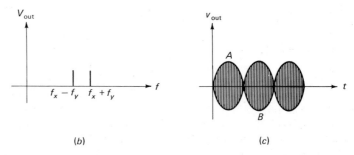

(b)

(c)

tion, all that reaches the final output are the lower and upper side frequencies. Because of this, the final output spectrum looks like Fig. 22-17b.

Figure 22-17c shows how a DSB-SC signal looks in the time domain. At first, this may appear like an ordinary AM signal. But look closely. The upper envelope resembles full-wave-rectified signal.

The modulator of Fig. 22-17a is often called a *balanced modulator* because each half of the diff amp is balanced (same gain) when no modulating signal is present. A balanced modulator suppresses the carrier.

SSB modulators

Given a DSB-TC or DSB-SC signal, we can produce SSB signals in either of two ways. First, we can use a bandpass filter to pass only one of the sidebands. Because the sidebands are so close together, we need an extremely sharp filter response. Usually, crystal filters are needed with this approach. Second, with the proper connection of modulators and phase shifters, we can make one of the sidebands cancel. The discussion of this phasing method is an advanced topic.[2]

IC modulators

Some IC modulators are available. These use diff-amp modulators similar to Fig. 22-16. Examples are Motorola's MC1595 and 1596, which can act like DSB-TC or DSB-SC modulators. By cascading crystal filters, we can get SSB signals with these IC modulators.

22-8. THE ENVELOPE DETECTOR

Once the AM signal is received, the carrier's work is over. Somewhere in the receiver is a special circuit that separates the modulating signal from the carrier. We call this circuit a *demodulator* or *detector.*

Diode detector

Figure 22-18a shows one type of demodulator. Basically, it is a peak detector (discussed earlier in Sec. 5-8). Ideally, the peaks of the input signal are detected, so the output is the upper envelope. For this reason, the circuit is called an *envelope detector.*

During each carrier cycle, the diode turns on briefly and charges the capacitor to the peak voltage of the particular carrier cycle. Between peaks, the capacitor discharges through the resistor. By making the *RC* time constant much greater than the period of the carrier, we get only a slight discharge between cycles. In this way, most of the car-

---

[2] If interested, see Pappenfus, E. W., W. B. Bruene, and E. O. Schoenike: *Single Sideband Principles and Circuits,* McGraw-Hill Book Company, New York, 1964, pp. 33–43.

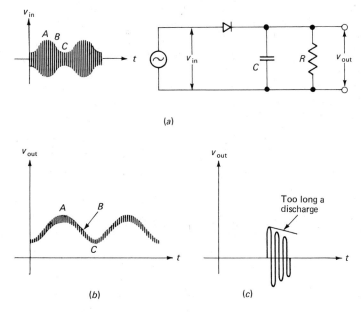

Figure 22-18. Envelope detector. (a) Circuit. (b) Output signal. (c) Detector time constant too large.

rier signal is removed. The output then looks like the upper envelope with a small ripple as shown in Fig. 22-18b.

Required RC time constant

But here is a crucial idea. Between points A and C in Fig. 22-18b, each carrier peak is smaller than the preceding one. If the RC time constant is too long, the circuit cannot detect the next carrier peak (see Fig. 22-18c). The hardest part of the envelope to follow occurs at B in Fig. 22-18b; at this point, the envelope is decreasing at its fastest rate. With calculus, the Appendix equates the rate of change of the envelope and the capacitor discharge to prove

$$f_{y(max)} = \frac{1}{2\pi RCm} \tag{22-8}***$$

where m is the modulation coefficient. With this equation, we can calculate the highest envelope frequency the detector can follow without attenuation. If the envelope frequency is greater than $f_{y(max)}$, the detected output drops 20 dB per decade.

The stages following the envelope detector are usually audio amplifiers with an upper break frequency less than the carrier frequency. For this reason, the small carrier ripple in Fig. 22-18b is reduced by these audio stages. Occasionally, we may add a low-pass filter on the output of the detector to remove the small carrier ripple.

Suppressed-carrier detection

If the received signal has a suppressed carrier, the envelope detector cannot recover the modulating signal. For instance, if we envelope-detect the DSB-SC signal of Fig. 22-17c, we get the upper envelope; this envelope has twice the frequency of the modulating signal.

What has to be done is this: the carrier frequency must be reinserted into the spectrum before detection. In other words, to receive a suppressed-carrier signal, we need an oscillator inside the receiver producing the same frequency f_x that was suppressed in the transmitter. This carrier component is added to the suppressed-carrier signal and the new signal is envelope-detected; the resulting output is the original modulating signal.

EXAMPLE 22-5.

In Fig. 22-18a, suppose the incoming signal has an m of 0.5. If R equals 6.2 kΩ and C is 0.01 μF, what is the highest envelope frequency we can detect without attenuation?

SOLUTION.

With Eq. (22-8),

$$f_{y(\max)} = \frac{1}{2\pi(6200)10^{-8}(0.5)} = 5.14 \text{ kHz}$$

This means the detector can follow any envelope frequency up to 5.14 kHz. Beyond this, we still get an output, but the amplitude falls off at 20 dB per decade.

22-9.   THE TUNED-RADIO-FREQUENCY RECEIVER

During the evolution of radio, the *tuned-radio-frequency* (TRF) receiver was used to receive AM signals. Even today, a few special applications use TRF receivers.

The basic idea

Figure 22-19a shows the block diagram of a TRF receiver. The idea is simple enough. The antenna picks up many AM signals, but the first four stages are tuned to only one of these signals. Therefore, the signal driving the detector is the AM signal from one station. The detected output then drives the audio amplifiers and the loudspeaker.

The TRF receiver relies on those first four tuned stages to separate one AM signal from the others. As mentioned, AM broadcast signals have bandwidths of 10 kHz; for this reason, the tuned RF stages in Fig. 22-19a have an overall bandwidth of approximately 10 kHz. Outside this bandwidth, the overall response should drop off as rapidly as possible.

Figure 22-19. TRF receiver. (a) Block diagram. (b) Four-stage response.

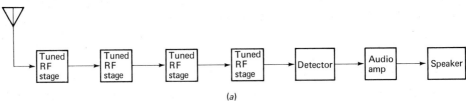

(a)

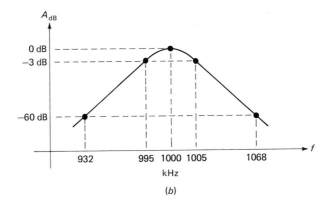

(b)

Selectivity

The *selectivity* of a receiver is its ability to separate the desired AM signal from all others received by the antenna. To have a high selectivity, a receiver must have a gain response that rolls off rapidly outside the 3-dB bandwidth. For a numerical definition, we will use

$$\text{Percent selectivity} = \frac{\text{BW}_{3\text{-dB}}}{\text{BW}_{60\text{-dB}}} \times 100 \text{ percent} \qquad (22\text{-}9)***$$

This formula compares the 3-dB bandwidth to the 60-dB bandwidth to give us a measure for how fast the gain rolls off. Maximum selectivity occurs when the 60-dB is almost as small as the 3-dB bandwidth. In this case, selectivity approaches 100 percent.

As an example, suppose Fig. 22-19*b* is the response of the four tuned RF stages of Fig. 22-19*a*. The receiver is tuned to a carrier frequency of 1000 kHz. The gain is down 3 dB at 995 and 1005 kHz, and down 60 dB at 932 and 1068 kHz. With Eq. (22-9),

$$\text{Percent selectivity} = \frac{1005 - 995}{1068 - 932} \times 100 \text{ percent} = 7.35 \text{ percent}$$

This is poor selectivity, but typical of a TRF receiver. Even a TRF receiver with six tuned stages has a selectivity less than 12 percent.

As mentioned, the best possible selectivity is 100 percent. A good communications receiver (not TRF) has a selectivity around 50 percent; this means the 60-dB bandwidth is only twice the 3-dB bandwidth. Small transistor radios have selectivities from 15 to 30 percent.

Disadvantages

What is wrong with TRF receiver? Besides the poor selectivity (usually less than 10 percent), the problem of tuning is serious. When we want to change stations in Fig. 22-19a, we have to retune the first four stages. Gang-tuning these stages is difficult. Even worse, the bandwidth of the tuned stages changes with retuning because the Q of inductors changes with frequency. For these and other reasons, the TRF receiver is seldom used.

22-10. THE SUPERHETERODYNE RECEIVER

The *superheterodyne* receiver (superhet) has a constant selectivity and is easier to tune over the frequency range.

The block diagram

Figure 22-20 is the block diagram of a superhet. Section 21-7 explained the mixing action near the front end of the superhet. As we recall, the received signal mixes with the LO signal to produce the IF signal. Because the LO of Fig. 22-20 operates 455 kHz above the RF signal, the IF spectrum has a center frequency of 455 kHz. The bandwidth of the IF signal is the same as the received signal. In AM radio, this means the RF and IF signals have bandwidths of 10 kHz.

The IF signal is amplified by *several* IF stages (see Fig. 22-20). The output of the last IF stage is envelope-detected to retrieve the modulating signal. This signal then drives the audio amplifiers and the loudspeaker. The output of the detector also feeds back a dc voltage to the IF amplifiers. This dc voltage is called the *AGC voltage* and is described in Sec. 22-11.

The major advantage of a superhet is the fixed-tuned IF stages. Because these stages are center-tuned to 455 kHz, the design is greatly simplified; furthermore, these IF amplifiers may be factory-tuned and left undisturbed; during normal use, only the RF and LO stages are ganged-tuned to different stations.

Double-tuned IF stages

To improve selectivity, the IF stages use double-tuned resonant circuits as shown in Fig. 22-21a. In a double-tuned stage like this, each resonant circuit is tuned to 455

Figure 22-20. Superheterodyne receiver.

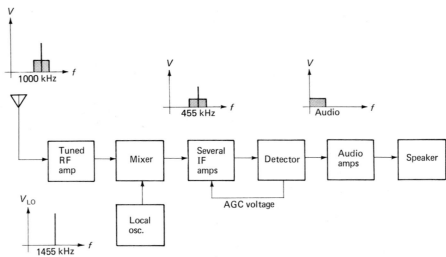

kHz. *Critical coupling* occurs when the coefficient of coupling equals

$$k = \frac{1}{\sqrt{Q_1 Q_2}}$$ (22-10)

where Q_1 is the Q of the primary resonant circuit and Q_2 the Q of the secondary resonant circuit.

When two resonant tanks are critically coupled, the response becomes *maximally flat* (see Fig. 22-21*b*). This means the response is flat over a broader range of frequencies than is possible with the single-tuned tank (compare Fig. 22-21*b* to 22-19*b*). Furthermore, the rolloff of a double-tuned stage is steeper than that of a single-tuned stage. For this reason, the double-tuned stage is more selective than the single-tuned stage.

As an example, Fig. 22-21*c* shows the overall response of three double-tuned IF stages. With a response like this, the selectivity is

$$\text{Percent selectivity} = \frac{460 - 450}{477 - 433} \times 100 \text{ percent} = 23 \text{ percent}$$

which is much better than the four-stage TRF receiver with a selectivity of only 7.35 percent (Fig. 22-19*b*). Furthermore, since the IF stages are fixed-tuned to 455 kHz, we get the same selectivity no matter what station we tune to.

Summary

The superhet is the standard for most communication systems. The central idea is to use a mixer to shift the received spectrum down to the IF frequency; at this constant

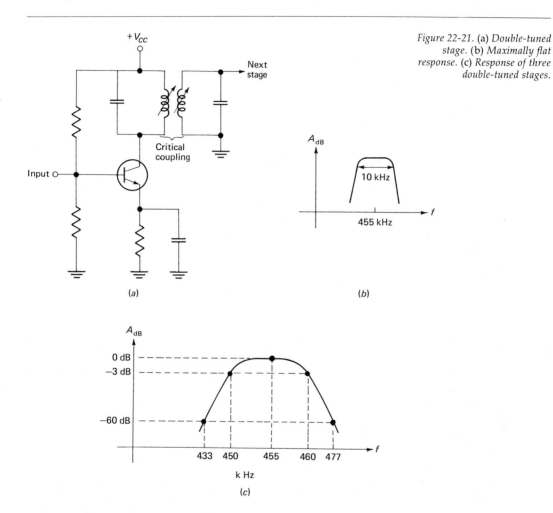

Figure 22-21. (a) *Double-tuned stage.* (b) *Maximally flat response.* (c) *Response of three double-tuned stages.*

and lower frequency, the IF stages can efficiently amplify the signal with a maximally flat response. The maximally flat IF response minimizes frequency and phase distortion of the recovered modulating signal. For these reasons, the superhet is widely used in radio, television, radar, and many other electronic systems.

EXAMPLE 22-6.
Some inexpensive receivers omit the RF stage as shown in Fig. 22-22*a*. What happens in such a receiver when the two given input AM signals are picked up by the antenna?

SOLUTION.
The AM signal centered on 645 kHz will mix with the LO signal to produce an IF signal with a center frequency of

Figure 22-22. Image-frequency problem. (a) Cross-talk produced by image frequency. (b) Tuned RF stage reduces effect of image signal.

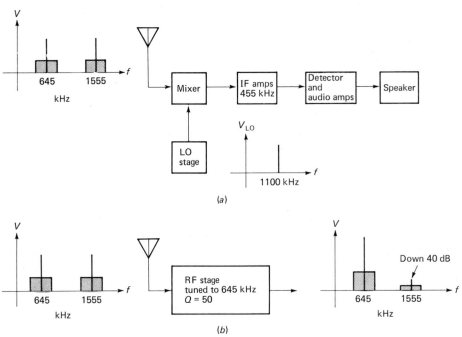

$$f_{out} = 1100 \text{ kHz} - 645 \text{ kHz} = 455 \text{ kHz}$$

Therefore, the lower of the two received AM signals produces a spectrum that will pass through the IF amplifiers.

The upper AM signal also mixes with the LO signal and produces an IF spectrum centered on

$$f_{out} = 1555 \text{ kHz} - 1100 \text{ kHz} = 455 \text{ kHz}$$

Therefore, the upper AM signal also produces a spectrum that passes through the IF amplifiers.

When both spectra are envelope-detected, we get audio information from both channels. Because of this, we hear both stations simultaneously (cross-talk).

This example brings out the need for a tuned RF stage. In the typical superhet, the desired RF spectrum is 455 kHz below the LO frequency, whereas the undesired RF spectrum is 455 kHz above the LO frequency; we call this undesired RF signal the *image signal*. The tuned RF stage in a superhet attenuates the image signal; by a straightforward calculation, we can show an RF stage with a Q of 50 provides the desired signal with 40 dB more gain than the image signal (Fig. 22-22b). In critical receiver applications, several tuned RF stages may be used.

22-11. AUTOMATIC GAIN CONTROL

When a receiver is tuned from a weak to a strong station, the loudspeaker will blare unless the volume is immediately decreased. Or the volume may change because of *fading*, a variation in signal strength caused by a change in the path between transmitting and receiving antennas. To counteract unwanted volume changes, most receivers use *automatic gain control* (AGC).

How it works

In Fig. 22-23, an AM signal drives the last IF stage Q_1. The amplified AM signal is envelope-detected to retrieve the audio signal. The average value of the detected output is labeled *AGC voltage;* this dc voltage is proportional to the unmodulated carrier. The stronger a received signal, the greater the AGC voltage.

The detected output drives audio amplifiers and also goes to a low-pass RC filter. The time constant of this AGC filter is typically 0.1 to 0.2 s. Because of this, the cutoff frequency is quite low, in the vicinity of 1 Hz.

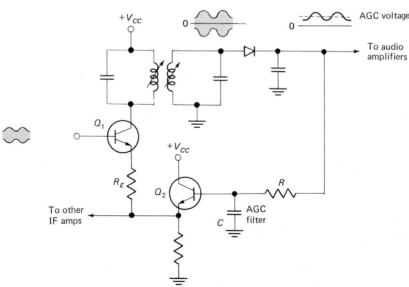

Figure 22-23. Automatic gain control.

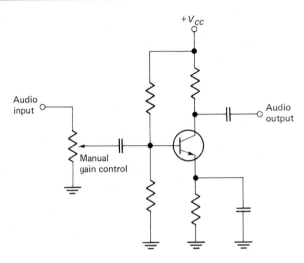

Figure 22-24. Manual gain control.

The dc voltage out of the AGC filter drives the base of emitter follower Q_2. This positive AGC voltage causes the following to happen: the dc emitter current in each IF stage decreases, the value of r'_e increases, and the voltage gain decreases. Especially important, the larger the AGC voltage, the lower the voltage gain of the IF amplifiers. Therefore, when we tune from a weak to a strong station, the AGC voltage increases; this automatically reduces the IF gain and prevents the loudspeaker from blaring.

Manual gain control

AGC action represents a coarse control of volume; we also need a manual control for fine adjustments. To avoid interacting with the AGC, the manual gain control must be in a stage following the detector. Figure 22-24 shows a typical manual control. Changing the position of the wiper changes the size of the audio signal into the amplifier. Since the audio stage is after the detector, the manual gain control has no effect on the AM signal driving the detector. This is why AGC control does not interact with manual control.

Other uses of AGC

The idea of automatically controlling gain has spread to television receivers, signal generators, and other electronic systems. AGC is a form of dc modulation because a dc voltage controls the gain.

Commercial signal generators often use AGC to flatten or level variations in signal strength. For instance, in Fig. 22-25 an oscillator drives amplifiers to produce an output signal. When we change the frequency of the oscillator, the signal strength changes

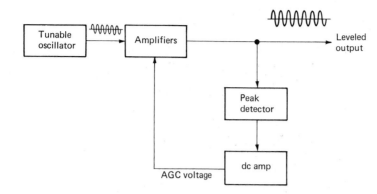

Figure 22-25. Signal generator with leveled output.

because of variations in coil Q, transistor parameters, etc. Without AGC, the final output signal may easily vary 20 dB over the tuning range of the oscillator. But with AGC, we can usually keep the variation under 1 dB.

The basic idea of Fig. 22-25 is similar to the AGC of a receiver; the output signal is peak-detected to get a dc voltage, which is amplified and used as the AGC voltage to the amplifiers. When the output signal tries to increase, more AGC voltage is fed back to reduce the gain of the amplifiers. In this way, the *leveled output* remains almost constant over the entire frequency range.

EXAMPLE 22-7.
Figure 22-26 shows a two-stage AGC amplifier, designed for the 50-kHz to 1-MHz range. For what output voltage does the AGC action become effective?

SOLUTION.
The first stage is voltage-divider biased with a base voltage of 3.2 V. Allowing 0.7 V across the base-emitter diode, we have 2.5 V at the top of the 2.2-kΩ emitter resistor. The AGC voltage at the bottom of this resistor is positive; therefore, the dc emitter current equals

$$I_E = \frac{2.5 - \text{AGC voltage}}{2200}$$

When the AGC voltage is small, the dc emitter current is large and so too is the voltage gain of the first stage. But for large AGC voltage, the emitter current goes down and so too does the gain.

The AGC action becomes very effective when the AGC voltage approaches 2.5 V. Near this value, the dc emitter current I_E approaches zero. For this condition, further increases in the input signal are almost completely offset by gain reduction.

What is the approximate output voltage corresponding to 2.5 V of AGC voltage? Allowing 0.7 V for the V_{BE} of emitter follower Q_3, and 0.3 V for the peak-detector

Figure 22-26. Example 22-7.

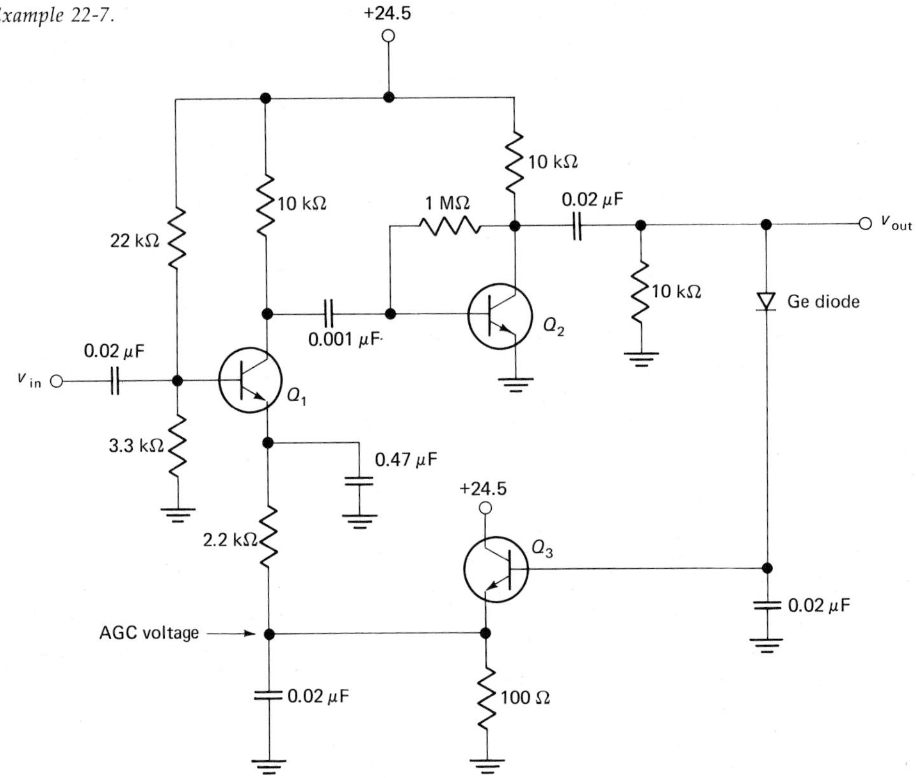

diode, we get

$$v_{out(peak)} = 2.5 + 0.7 + 0.3$$
$$= 3.5 \text{ V}$$

which is equivalent to 7 V peak-to-peak. This represents the upper limit (approximately) on the output voltage.

If you build the AGC amplifier of Fig. 22-26, here is what you will find. When the output voltage is much less than 7 V pp, the AGC voltage has no effect; for this weak-signal case, the input and output voltages are linearly related; a 100 percent increase in input voltage produces a 100 percent increase in output voltage.

But when the output voltage approaches 7 V pp, you will notice AGC action becomes quite effective. That is, when you increase the input signal, the gain is automatically reduced by almost the same factor; this results in an almost constant output. For instance, you may find that a 100 percent increase in input voltage produces less than a 1 percent change in output.

A final point about AGC amplifiers. Do not confuse an AGC amplifier with an overdriven amplifier. In an overdriven amplifier, an input sine wave produces an output square wave; when you increase the signal into the overdriven amplifier, you get no change in output amplitude. Similarly, in an AGC amplifier an increase in the input signal produces almost no increase in output voltage. But the crucial difference is this; the output of an AGC amplifier has the same shape as the input. In other words, an overdriven amplifier distorts the signal, but an AGC amplifier does not. For this reason, we use AGC amplifiers in radios, television receivers, signal generators, etc., where we want to limit the output signal with minimum distortion.

22-12. FREQUENCY MODULATION

When the modulation signal controls the frequency of the carrier, we get frequency modulation (FM). If the phase of the carrier is controlled, we have phase modulation (PM). FM is important in some communication systems; PM is rarely used. This section briefly describes the idea behind FM.[3]

The basic idea

The key idea in an FM modulator is to vary the frequency of a sine wave, usually by changing the capacitance in an *LC* oscillator. For instance, Fig. 22-27*a* shows the tuning capacitor of an oscillator. If the capacitance has a fixed value, the oscillator produces a *quiescent frequency* f_{xQ}. Now, imagine that we *rock* the capacitor back and forth, causing it to vary from a minimum to a maximum value. The number of round trips per second between minimum and maximum capacitance is called the *rocking frequency*. If we are fast enough, we might manually reach a rocking frequency of 2 Hz.

Figure 22-27*b* shows the output of an oscillator whose capacitance is rocked back and forth. When the capacitance is maximum, we get a minimum frequency $f_{x(\min)}$; for minimum capacitance, we get a maximum frequency $f_{x(\max)}$. The number of round trips per second between $f_{x(\min)}$ and $f_{x(\max)}$ equals the rocking frequency. Figure 22-27*b* shows the FM signal in the time domain. All that happens is the frequency of oscillation rocks back and forth between $f_{x(\min)}$ and $f_{x(\max)}$.

Frequency deviation

At any instant in time, the difference between the oscillator frequency f_x and the quiescent frequency f_{xQ} is called the *frequency deviation* f_d. Symbolically,

$$f_d = f_x - f_{xQ}$$

The maximum frequency deviation equals

---

[3] For a detailed discussion of FM, see books specializing in communication theory and circuits. An example is Shrader, R. L.: *Electronic Communication*, 2d ed., McGraw-Hill Book Company, New York, 1967.

Figure 22-27. Frequency modulation. (a) Manually controlled. (b) FM waveform. (c) Varactor controlled.

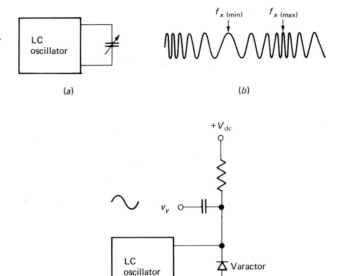

(a)

(b)

(c)

$$f_{d(\max)} = f_{x(\max)} - f_{xQ} \tag{22-11}$$

As an example, suppose $f_{x(\max)}$ equals 1075 kHz and f_{xQ} equals 1000 kHz. Then, the maximum frequency deviation is

$$f_{d(\max)} = 75 \text{ kHz}$$

In a practical FM modulator, we use a *varactor* in the place of a mechanically tuned capacitor. As discussed in Sec. 4-9, a reverse-biased diode acts like a capacitor; by varying the reverse bias, we can vary the capacitance. Figure 22-27c shows an FM modulator with a varactor determining the frequency of oscillation. The dc voltage sets up the quiescent value of capacitance; the modulating voltage v_y produces variations in this capacitance. Therefore, the modulating frequency equals the rocking frequency. For this reason, the modulating frequency equals the number of round trips per second between $f_{x(\min)}$ and $f_{x(\max)}$ in the FM wave of Fig. 22-27b.

If the modulating voltage is sinusoidal as shown in Fig. 22-27c, we may write

$$v_y = V_y \sin \omega_y t$$

where V_y is the peak voltage and ω_y the radian frequency. In a practical modulator, the frequency deviation is proportional to the modulating voltage. That is,

$$f_d = K V_y \sin \omega_y t$$

where K is a constant of proportionality that depends on the varactor and other factors. The maximum frequency deviation therefore equals

$$f_{d(\text{max})} = KV_y \qquad\qquad (22\text{-}12)$$

For instance, if K is 75 kHz/V, a peak modulating voltage V_y of 2 V produces

$$f_{d(\text{max})} = 150 \text{ kHz}$$

FM spectrum

It takes advanced mathematics to derive the spectrum of an FM signal. We will give the results without derivation. The spectrum of an FM signal contains the same frequencies as group 1 of a large-signal AM modulator. Specifically, an FM spectrum has these frequencies:

f_x
$f_x \pm f_y$
$f_x \pm 2f_y$
$f_x \pm 3f_y$
etc.

The carrier frequency f_x is in the center of the spectrum as shown by Fig. 22-28a; the distance between each component is f_y.

As an example, suppose f_x equals 1000 kHz and f_y is 15 kHz. Then the spectral components of an FM signal have frequencies of

1000 kHz
985 and 1015 kHz
970 and 1030 kHz
955 and 1045 kHz
etc.

The modulation coefficient of an FM signal equals

$$m = \frac{f_{d(\text{max})}}{f_y} \qquad\qquad (22\text{-}13)^{***}$$

If $f_{d(\text{max})}$ is 75 kHz and f_y is 15 kHz, then m equals 5. As shown in Fig. 22-28a, the peak values are given by $J_0(m)$, $J_1(m)$, $J_2(m)$, etc. These particular J values are called *Bessel functions*. If you need the peak values of the spectral components, you can look up the J values in reference books.[4]

---

[4] Among many such books, see *Reference Data for Radio Engineers*, Howard W. Sams & Co., Inc., New York, 1968, pp. 45-20 to 45-22.

Figure 22-28. (a) Spectrum of an FM signal. (b) Approximation for FM bandwidth.

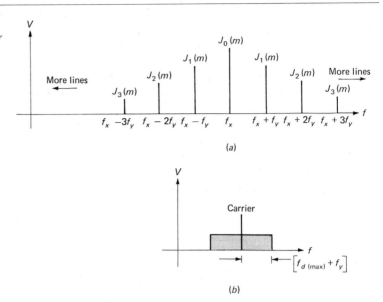

(a)

(b)

As an example, if m equals 1, we want the values of $J_0(1), J_1(1), J_2(1), J_3(1)$, etc. In any table of Bessel functions, we can read these values:

$J_0(1) = 0.765$ (carrier peak)
$J_1(1) = 0.440$ (two nearest side components)
$J_2(1) = 0.115$ (next nearest pair)
$J_3(1) = 0.0196$ (third nearest pair)

These J values represent the peak values of the spectral components relative to an unmodulated carrier peak of 1 V. To get the actual peak values, you multiply each J value by the peak value of the unmodulated carrier.

Bandwidth of an FM signal

In practice, we seldom bother looking up Bessel functions for an FM signal unless we are designing FM equipment or testing FM transmitters. But all of us should know the approximate *bandwidth of an FM signal* in case we ever have to filter FM signals.

In dealing with FM signals, many people use this convenient approximation: all significant side frequencies are within $[f_{d(max)} + f_y]$ of the carrier frequency as shown in Fig. 22-28b; when a component is further displaced than this, we can neglect it. In terms of bandwidth value,

$$B_v = 2[f_{d(max)} + f_y] \qquad (22\text{-}14)^{***}$$

As an example, suppose an FM broadcast signal has a carrier frequency of 100 MHz. By law, $f_{d(\max)}$ must be less than 75 kHz and f_y less than 15 kHz. Therefore, in the worst case,

$$f_{d(\max)} + f_y = 75 \text{ kHz} + 15 \text{ kHz} = 90 \text{ kHz}$$

The FM spectrum consists of a carrier frequency of 100 MHz and frequencies extending to approximately 90 kHz on each side of the carrier; this means a bandwidth value of about 180 kHz.

Advantage of FM over AM

FM has some advantages over AM. We will only mention the *noise advantage*. External noise from electric motors, lightning, ignition systems, etc., tends to amplitude-modulate transmitted signals. In an AM receiver, you will hear this interfering noise. But in an FM receiver, the amplitude of the signal is unimportant; only the changing frequency is detected. Because of this, FM receivers are very quiet as far as external noise is concerned.

Commercially, frequency modulation is used in FM radio; it is also used for the sound portion of a television signal.

Problems

22-1. In the transistor modulator of Fig. 22-3, suppose R_E is changed to 20 kΩ and V_y to 4 V. Calculate the quiescent, minimum, and maximum voltage gains. What are the corresponding peak carrier output voltages?

22-2. The input carrier to an AM modulator has a peak value of 20 mV. If the quiescent gain is 100, what is the quiescent peak output? If A changes ± 20 percent, what are the minimum and maximum output peak voltages?

22-3. How long is a quarter wavelength at each of these frequencies: 1 MHz (middle of AM broadcast band), 100 MHz (middle of FM broadcast band), and 200 MHz (channel 11 on TV)?

22-4. The lowest and highest carrier frequencies in AM radio are 540 kHz and 1600 kHz. Calculate the length of antennas for these carrier frequencies. (Use quarter wavelength.)

22-5. If an AM modulator has the gain variation of Fig. 22-29a, what is the value of m?

22-6. If you see an AM signal like Fig. 22-29b on an oscilloscope, what is the modulation coefficient?

22-7. A modulating signal with spectral components from 20 Hz to 20 kHz amplitude-modulates a carrier with a frequency of 1 MHz. Calculate the side frequencies for f_y equal to 20 Hz. What are the side frequencies when f_y is 20 kHz?

Figure 22-29.

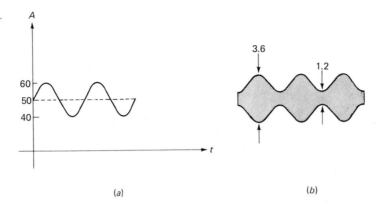

(a) (b)

22-8. An AM modulator has a quiescent voltage gain of 200 and a modulation coefficient of 0.3. If the peak carrier input V_x equals 10 mV, what are the peak values of the output spectral components?

22-9. Given an AM waveform like Fig. 22-29b, what are the peak values of the carrier and two side components?

22-10. Suppose the highest modulating frequency in an AM communication system is 20 kHz. What is the bandwidth value of the resulting AM signal?

22-11. The input frequencies to a DSB-SC modulator are 20 kHz and 1 MHz. What frequencies are in the output spectrum?

22-12. If the AM voltage of Fig. 22-29b appears across a 100-Ω resistor, how much carrier power is there? How much power is there in each sideband?

22-13. An AM signal has the spectrum shown in Fig. 22-30a. What is the bandwidth of the AM signal? The bandwidth value?

22-14. If the signal whose spectrum is shown in Fig. 22-30a is changed to a SSB-SC signal, what does the new spectrum look like?

22-15. Figure 22-30b shows the input spectrum to a SSB-TC modulator. What does the output spectrum look like?

22-16. If the spectrum of Fig. 22-30b drives a DSB-SC modulator, what is the bandwidth value of the output spectrum?

22-17. An AM signal drives an envelope detector whose R is 10 kΩ and C is 1000 pF. If the modulation coefficient is 30 percent, what is the highest detectable modulating frequency without attenuation?

22-18. A receiver has a 3-dB bandwidth of 20 kHz and a 60-dB bandwidth of 80 kHz. What is the percent selectivity?

22-19. A superhet has an IF frequency of 455 kHz. What is the LO frequency when the received frequency is 540 kHz? When the received frequency is 1600 kHz?

22-20. An IF stage is critically coupled with a Q_1 of 50 and a Q_2 of 75. What is the

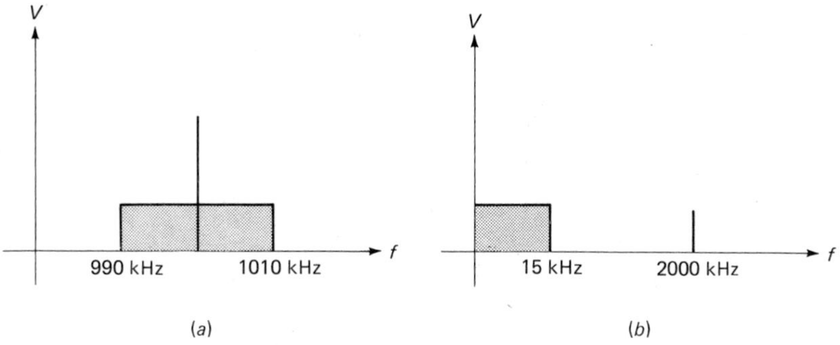

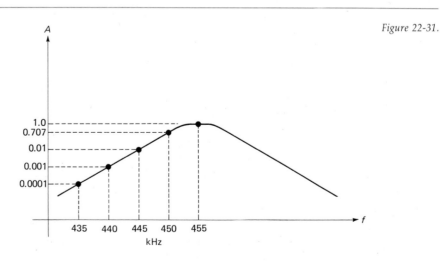

Figure 22-30.

value of k? In basic circuit books, you find this relation:

$$k = \frac{M}{\sqrt{L_1 L_2}}$$

If $L_1 = L_2 = 100$ μH, what is the value of M for critical coupling?

22-21. Figure 22-31 shows the response for the IF amplifiers in a receiver. Calculate the percent selectivity for this particular receiver.

22-22. The IF frequency equals 455 kHz in a receiver and the LO frequency is greater than the desired carrier frequency. What are the image frequencies for each of these cases:

1. A desired carrier frequency of 550 kHz
2. A LO frequency of 1355 kHz

Figure 22-31.

22-23. In the AGC amplifier of Fig. 22-23, assume three identical IF stages receive the AGC voltage. The r_C seen by each collector is 1 kΩ and r'_e is 25 mV/I_E. With no AGC voltage, I_E equals 1 mA in each stage.

1. What is the overall IF voltage gain when the AGC voltage is zero?
2. When the AGC voltage reduces the dc emitter current to 0.5 mA, what is the total IF voltage gain?
3. When I_E is 0.1 mA, what is the overall IF gain?

22-24. In an FM modulator, the highest oscillation frequency is 105.5 MHz. If the quiescent oscillation frequency is 105.3 MHz, what is the maximum frequency deviation?

22-25. In Eq. (22-12), K has a value of 10 kHz/V.

1. If V_y equals 3 V, what does $f_{d(\max)}$ equal?
2. If the maximum frequency deviation is 75 kHz, what is the peak modulating voltage?

22-26. The carrier frequency in an FM modulator is 20 MHz. If the modulating frequency is 100 kHz, what are the first three upper side frequencies in the FM spectrum? The first three lower side frequencies?

22-27. Calculate the modulation coefficient for an FM signal where the maximum frequency deviation is 50 kHz and the modulating frequency is 5 kHz.

22-28. In an FM modulator, the modulation coefficient equals 5 and the highest modulation frequency is 20 kHz. What is the approximate bandwidth value of the resulting FM signal?

23. Voltage Regulation

Ideally, the dc voltage out of a power supply should not change with variations in line voltage or load resistance. This chapter discusses ways to *regulate* dc voltage (hold it constant). The zener diode is the first step toward voltage regulation. By using negative feedback as well, we can build power supplies whose output voltage is almost immune to changes in line voltage or load resistance.

23-1. UNREGULATED SUPPLIES

The output of an unregulated supply is a dc voltage and ac ripple shown by the Thevenin circuit of Fig. 23-1a. V_S is the dc Thevenin voltage, v_{rip} is the ac Thevenin voltage, and Z is the Thevenin impedance.

Dc equivalent circuit

The total voltage delivered to a load resistance is the superposition of dc voltage and ac ripple. As far as the dc component is concerned, Z is purely resistive and the Thevenin circuit reduces to Fig. 23-1b. The value of R_S depends on the type of rectifier, the number of filter sections, and other factors. Typically, R_S is from tens to hundreds of ohms for unregulated supplies.

Figure 23-1. (a) *Power supply and Thevenin equivalent circuit.* (b) *Unloaded dc Thevenin circuit.* (c) *Loaded dc Thevenin circuit.*

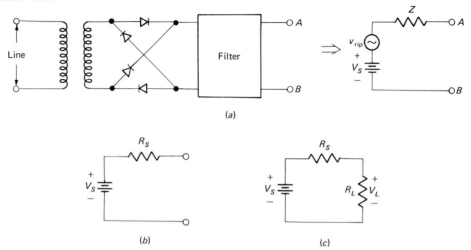

(a)

(b)   (c)

Relative change

In comparing one power supply with another, we need the *relative voltage change* defined as

$$\frac{\Delta V}{V} = \frac{V' - V}{V} \qquad\qquad (23\text{-}1)^{***}$$

where V' is the new value of voltage and V is the original value. For instance, if line voltage changes from 115 to 125 V rms, it has a relative change of

$$\frac{\Delta V}{V} = \frac{125 - 115}{115} = 0.087$$

or a percent change of 8.7 percent. As another example, if the dc voltage out of supply changes from 10 to 12 V, the relative change is

$$\frac{\Delta V}{V} = \frac{12 - 10}{10} = 0.2$$

which is equivalent to 20 percent.

Relative line change

In Fig. 23-1*a* the rectified voltage depends on line voltage; the greater the line voltage, the greater the rectified voltage. Neglecting the voltage drop across the diodes, the dc output voltage is directly proportional to line voltage. If line voltage increases 10 percent, so too does dc output voltage. Symbolically,

$$\frac{\Delta V_S}{V_S} = \frac{\Delta V_{\text{line}}}{V_{\text{line}}} \tag{23-2}$$

This is bad. Line voltage may vary from 105 to 125 V rms (more in some outlying areas). As a result, line changes can produce dc output changes approaching ± 10 percent.

Load changes

Figure 23-1c shows the dc Thevenin circuit loaded by R_L. The dc load voltage V_L is less than V_S because of the drop across R_S. By the voltage-divider theorem,

$$V_L = \frac{R_L}{R_S + R_L} V_S \tag{23-3}$$

In an unregulated supply V_L may be 5 or 10 percent less than V_S. Furthermore, changes in R_L produce changes in V_L. (In a complex piece of equipment, changes in R_L occur when we turn different circuits on and off.)

Summary

Here is the main idea. The dc voltage out of an unregulated power supply can change significantly when

Line voltage changes.
Load resistance changes.

The purpose of *voltage regulators* is to reduce the effects of these changes.

23-2. ZENER DIODES

Zener diodes are specially processed silicon diodes optimized for *breakdown operation*. By varying the doping level, a manufacturer can produce diode types with breakdown voltages from about 2 to 200 V. As we saw in Chaps. 4 and 5, the breakdown voltage stays almost constant even though the current is changing. In other words, the almost-vertical breakdown curve means the diode acts like a *constant-voltage device*.

Avalanche breakdown

Section 3-4 described the avalanche effect. Briefly, recall the following. When the reverse voltage reaches the breakdown value, the electric field is quite intense; minority carriers in this field are accelerated and can reach high enough velocities to dislodge valence electrons from outer orbits. The newly liberated electrons can then gain high enough velocities to dislodge other valence electrons. In this way, we get an avalanche effect. Avalance may occur for reverse voltages greater than 6 V or so.

Zener breakdown

Zener breakdown is different. When a diode is heavily doped, the depletion layer becomes very narrow. Because of this, the electric field across the depletion layer becomes intense. When a reverse voltage is applied, the field may be intense enough to dislodge valence electrons. If so, we have zener breakdown. Zener breakdown is sometimes called *high-field emission.*

Originally, people thought zener breakdown was the only breakdown mechanism in diodes. For this reason, the name came into general use for all breakdown diodes. In actuality, the zener effect dominates below 4 V, the avalanche effect over 6 V, and a combination of the two effects between 4 and 6 V.

Temperature coefficient

The *temperature coefficient* is the percent change in some quantity per degree Celsius. For instance, if a zener diode has a breakdown voltage of 10 V at 25°C and a temperature coefficient of 0.1 percent per degree, then its breakdown voltage progressively equals 10.01, 10.02, 10.03 V, and so on for each degree above 25°C.

When avalanche effect dominates, the temperature coefficient is positive; when zener effect is more important, the temperature coefficient is negative. Between 4 and 6 V where both effects are present, the temperature coefficient may be positive, negative, or zero depending on the current.

23-3. ZENER-DIODE REGULATOR

Figure 23-2*a* shows the prototype of a zener-diode regulator (note the schematic symbol for the diode). When the diode operates in the breakdown region, a *reverse current* I_Z flows through it. This current must satisfy Kirchhoff's current law:

$$I_Z = I_R - I_L \tag{23-4a}$$

The power dissipated by the zener diode is

$$P_Z = V_Z I_Z \tag{23-4b}$$

As long as P_Z is less than the maximum power rating on the data sheet, the zener diode is not destroyed. Furthermore, even though I_Z changes, V_Z remains almost constant in the breakdown region. Since R_L is directly across the zener diode, load voltage remains almost constant.

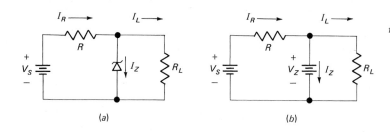

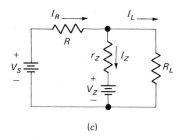

Figure 23-2. Zener-diode regulator. (a) *Actual*. (b) *Ideal*. (c) *Second approximation*.

Ideal approximation

To a first approximation, the zener diode acts like a constant-voltage source or battery as shown in Fig. 23-2b. Even though current through a battery changes, battery voltage remains constant. For this reason, load voltage ideally remains constant in spite of changes in source voltage V_S and other changes.

In Fig. 23-2b, load voltage ideally equals V_Z; therefore, load current equals

$$I_L = \frac{V_Z}{R_L} \qquad (23\text{-}4c)$$

Also, the voltage across R equals $V_S - V_Z$, so that

$$I_R = \frac{V_S - V_Z}{R} \qquad (23\text{-}4d)^{***}$$

Zener resistance

To a second approximation, we have to include zener resistance r_Z. In the breakdown region the voltage increases slightly when current increases. By taking the ratio of voltage change to current change, we get the zener resistance:

$$r_Z = \frac{\Delta V_Z}{\Delta I_Z} \qquad (23\text{-}5a)$$

The value of r_Z is included on data sheets for zener diodes. With r_Z, we can estimate how much voltage changes with current changes. For instance, by rearranging Eq. (23-5a),

$$\Delta V_Z = r_Z \Delta I_Z$$

If r_Z is 20 Ω, a change of 5 mA in zener current produces a change in voltage of

$$\Delta V_Z = 20(0.005) = 0.1 \text{ V}$$

Figure 23-2c shows the second approximation of a zener diode: a resistance r_Z in series with a battery V_Z. Zener current flows down, producing a load voltage of

$$V_L = V_Z + I_Z r_Z \quad \text{(second)} \qquad (23\text{-}5b) \text{***}$$

In this second-approximation formula, V_Z changes only slightly with temperature. Furthermore, when r_Z is small, changes in I_Z have little effect on load voltage; for this reason, Eq. (23-5b) ideally approaches

$$V_L = V_Z \quad \text{(ideal)} \qquad (23\text{-}5c) \text{***}$$

EXAMPLE 23-1.
Calculate the current through the zener diode and through the load resistance of Fig. 23-3.

SOLUTION.
The zener diode has a breakdown voltage of 10 V. The 3-kΩ resistor has 30 V across it; therefore, the current through this resistor is

$$I_R = \frac{V_S - V_Z}{R} = \frac{30}{3000} = 10 \text{ mA}$$

The load current equals

$$I_L = \frac{V_Z}{R_L} = \frac{10}{2000} = 5 \text{ mA}$$

To satisfy Kirchhoff's current law, zener current must equal

$$I_Z = I_R - I_L = 10 \text{ mA} - 5 \text{ mA} = 5 \text{ mA}$$

Figure 23-3. Examples 23-1 through 23-4.

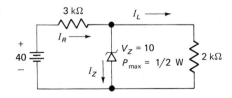

EXAMPLE 23-2.

Suppose the source voltage of Fig. 23-3 increases to 46 V (15 percent increase). Recalculate the three currents.

SOLUTION.

Ideally, the zener diode still has 10 V across it. Therefore, the voltage across the 3-kΩ resistor is 36 V and the current through this resistor equals

$$I_R = \frac{36}{3000} = 12 \text{ mA}$$

The load resistor has 10 V across it, so that load current still equals 5 mA. But zener current is

$$I_Z = 12 \text{ mA} - 5 \text{ mA} = 7 \text{ mA}$$

Comparing this to the previous example, we note I_Z has increased from 5 to 7 mA. So, a change in input voltage ideally has no effect on output voltage; the increase in source voltage is dropped across the series resistor.

EXAMPLE 23-3.

Suppose the load resistance of Fig. 23-3 changes to 4 kΩ. Calculate the three currents.

SOLUTION.

The zener diode has 10 V across it; therefore, the 3-kΩ resistor has 30 V across it and the current through it is

$$I_R = \frac{30}{3000} = 10 \text{ mA}$$

The load voltage is 10 V and load current is

$$I_L = \frac{10}{4000} = 2.5 \text{ mA}$$

The zener current equals

$$I_Z = 10 \text{ mA} - 2.5 \text{ mA} = 7.5 \text{ mA}$$

Comparing this to Example 23-1, we note a change in load resistance changes zener current but not zener voltage.

EXAMPLE 23-4.

If r_Z equals 15 Ω, what is load voltage in each of the three preceding examples?

SOLUTION.

In Example 23-1, I_Z is 5 mA. With Eq. (23-5b),

$$V_L = V_Z + I_Z r_Z = 10 + 0.005(15) = 10.075 \text{ V}$$

In Example 23-2, zener current equals 7 mA and load voltage is

$$V_L = 10 + 0.007(15) = 10.105 \text{ V}$$

In Example 23-3, I_Z equals 7.5 mA and

$$V_L = 10 + 0.0075(15) = 10.112 \text{ V}$$

These results demonstrate how small the changes in load voltage are. When source voltage changes from 40 to 46 V, load voltage changes only from 10.075 to 10.105 V; equivalently, a 15 percent increase in source voltage produces only a 0.3 percent increase in load voltage. Similarly, when load current decreases from 5 mA to 2.5 mA, load voltage increases from 10.075 to 10.112 V; in terms of percent, this means a 50 percent decrease in load current produces only a 0.37 percent increase in load voltage.[1]

23-4. THE ZENER FOLLOWER

Data sheets for zener diodes usually specify zener voltage and other characteristics only at one particular zener current I_{ZT}. As a rule, we try to design circuits to operate at or near this test current. For instance, a 1N821A has a zener voltage of 6.2 V and a temperature coefficient of 0.01 percent/C° when the zener current equals 7.5 mA; therefore, we try to operate the 1N821A at or near 7.5 mA.

Cascading an emitter follower

When the load resistance is directly across the zener diode (like Fig. 23-3), an increase of 1 mA in load current results in a decrease of 1 mA in zener current. In general,

$$\Delta I_Z = -\Delta I_L$$

As a guide, ΔI_Z should be less than 10 percent of I_{ZT}: this ensures we are operating the zener diode at or near its optimum current.

What do we do if ΔI_L is greater than 10 percent of I_{ZT}? We can add an *emitter follower* as shown in Fig. 23-4. The load voltage still equals the zener voltage (ideally), but the effect of load current is reduced by the β factor. Specifically,

$$\Delta I_Z = -\frac{\Delta I_L}{\beta} \tag{23-6}***$$

When necessary, we can use a Darlington pair instead of a single transistor to get a larger β.

We call Fig. 23-4 a *zener follower* because output voltage follows zener voltage. Source voltage V_S drives the zener regulator consisting of R and the zener diode. The

---

[1] For more discussion of zener diodes, see *Zener Diode Handbook,* Motorola Inc., Semiconductor Products Division, 1967.

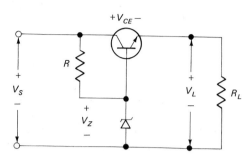

Figure 23-4. Zener follower.

voltage across the zener diode is the input to the emitter follower. Therefore, load volt-age ideally equals zener voltage. To a second approximation, load voltage differs from zener voltage by the V_{BE} drop of the transistor.

Series regulator

The zener follower is an example of a *series voltage regulator*. The collector-emitter terminals are in series with the load. Because of this, load current must pass through the transistor. Furthermore, the difference in voltage between the source and the load appears across the transistor. For this reason, the transistor must be able to dissipate a power of

$$P_D = V_{CE}I_C = (V_S - V_L)I_L$$

Series regulators are the most widely used type.

EXAMPLE 23-5.
If zener voltage in Fig. 23-4 is 15 V, what is the approximate load voltage?

SOLUTION.
The load voltage follows the zener voltage and ideally equals 15 V. To a second approximation,

$$V_L = 15 - 0.7 = 14.3 \text{ V}$$

23-5. SP REGULATION

In critical applications, zener voltages around 6 V are used because the temperature coefficient approaches zero. The highly stable zener voltage (sometimes called a *reference voltage*) can be amplified with an SP negative-feedback amplifier to get higher voltages with essentially the same stability as the reference voltage.

Figure 23-5. Discrete SP regulator. (a) *Actual circuit.* (b) *Redrawn to emphasize A and B blocks.*

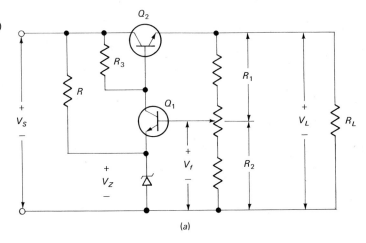

(a)

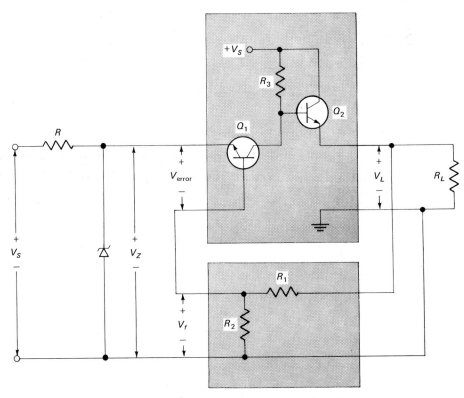

(b)

A discrete SP regulator

Figure 23-5a shows a discrete SP regulator. Transistor Q_2 acts like an emitter follower as before. Transistor Q_1 provides voltage gain in a negative-feedback loop. Here is the idea behind circuit operation. Suppose load voltage tries to increase. Feedback voltage V_f will increase. Since the emitter-to-ground voltage of Q_1 is held constant by the zener diode, more collector current flows through Q_1; this current also flows through R_3 and causes the base voltage of Q_2 to decrease. In response, the emitter voltage of Q_2 decreases; this offsets the original increase in load voltage.

Figure 23-5b shows the circuit redrawn so that we can recognize the amplifier and feedback sections. The output voltage is fed back to the input side. Since the circuit represents SP negative feedback, the SP voltage gain ideally is

$$A_{SP} = \frac{R_1}{R_2} + 1$$

The zener voltage V_Z is the input to the SP negative-feedback amplifier; therefore, the ideal output voltage is

$$V_L = A_{SP}V_Z \qquad (23\text{-}7)\,{}^{***}$$

Because of this, we can use a low zener voltage where the temperature coefficient approaches zero and still have a higher output voltage with an equally good temperature coefficient.

In Fig. 23-5a we can use the potentiometer to adjust output voltage to exactly the value required in a particular application. In this way, we can adjust for tolerance in zener-diode voltages and other circuit quantities.

IC SP regulator

In Fig. 23-5a, Q_1 provides internal voltage gain A. As we know, the higher A is, the more precise A_{SP}. Because of this, some applications require as much gain as possible to keep A_{SP} ideally constant. In a case like this, we can add more discrete stages. Better yet, we can replace Q_1 by an op amp.

Figure 23-6 shows an op amp in the place of Q_1. The output stage is an emitter follower. Zener voltage V_Z is the input to the op amp, so that the final load voltage equals $A_{SP}V_Z$. With an SP regulator like this, output voltage is as constant as the zener voltage. To increase load-current capability, a Darlington pair may be used in the place of Q_2.

Short-circuit protection

The SP voltage regulators of Figs. 23-5a and 23-6 are series regulators. As they now stand, both circuits have *no short-circuit protection*. If we accidentally place a short across the load terminals, we get an enormous current through Q_2. Either Q_2 will be destroyed or a fuse in the power supply will burn out, or both. To avoid this possibility, regulated supplies often include *current limiting*.

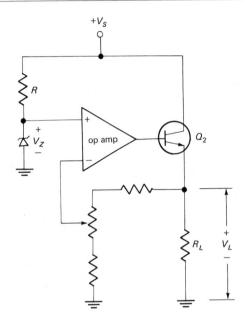

Figure 23-6. Op amp used to increase internal gain.

Figure 23-7 shows one way to limit load current to safe values even though the output terminals are shorted. For normal load currents, the voltage drop across R_4 is small and Q_3 is off; under this condition, the regulator works as previously described. If excessive load current flows, however, the voltage across R_4 becomes large enough to turn on Q_3. The collector current of Q_3 flows through R_3; this decreases the base voltage of Q_2 and reduces the output voltage to prevent damage.

In Fig. 23-7 current limiting starts when the voltage across R_4 is around 0.6 to 0.7 V. At this point, Q_3 turns on and decreases the base drive for Q_2. Since R_4 is 1 Ω, current limiting begins when load current is in the vicinity of 600 to 700 mA. By selecting other values of R_4, we can change the level of current limiting.

The current limiting of Fig. 23-7 is a simple example of how it is done. In more advanced approaches, Q_3 may be replaced by an op amp to increase the sharpness of the current limiting. Also, a PP current-to-voltage transducer (Sec. 18-3) can be used to monitor load current and provide the shutdown voltage to the emitter follower.[2]

23-6. IC REGULATORS

Monolithic IC regulators like the μA723 and the LM100 (two widely used industry standards) include a zener diode, feedback amplifier, short-circuit protection, and many

---

[2] For more discussion on regulated supplies, see *Dc Power Supply Handbook*, Hewlett-Packard Co., 1970.

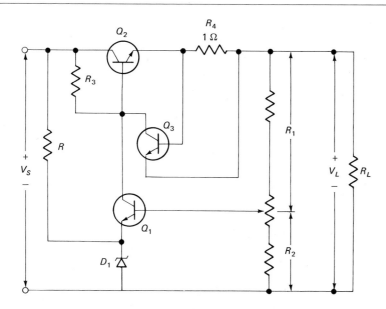

Figure 23-7. SP regulator including current limiting.

other features. You supply the unregulated dc input voltage; the IC regulator provides an almost constant output voltage. Monolithic ICs can provide load currents to about 150 mA. Beyond this, you have to add an *outboard* (external to the IC) power transistor. This transistor acts like an emitter follower and extends the load-current capability of the IC regulator.

Problems

23-1. The dc voltage out of a supply changes from 30 to 36 V. What does the relative change equal? If the supply voltage decreases from 30 to 27 V, what is the relative change? Express both answers as a percent.

23-2. In Fig. 23-1c the source voltage V_S equals 40 V. R_S is 50 Ω and R_L is 1000 Ω. What does V_L equal? If R_L decreases to 450 Ω, what does V_L equal?

23-3. A zener diode has a temperature coefficient of 0.01 percent/°C. If the zener voltage equals 8.2 V at 25°C, what does V_Z equal at 30°C?

23-4. To prevent the zener diode of Fig. 23-8a from burning out, what minimum value of R is required?

23-5. In Fig. 23-8b, what is the value of load current? Of zener current?

23-6. How much power does the zener diode of Fig. 23-8b dissipate?

23-7. Neglect r_Z and work out the value of zener current for each of these source conditions in Fig. 23-8c.

Figure 23-8.

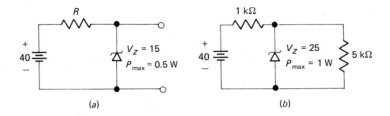

(a)   (b)

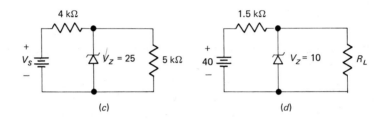

(c)   (d)

Figure 23-9.

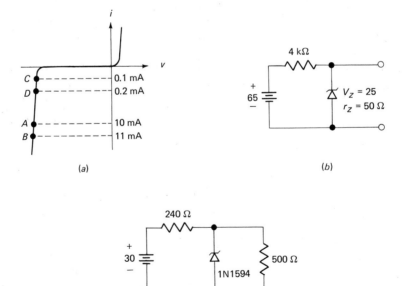

(a)   (b)

(c)

1. $V_S = 65$ V
2. $V_S = 70$ V
3. $V_S = 60$ V

23-8. In Fig. 23-8c, what is the minimum source voltage that keeps zener current greater than zero?

23-9. What is the approximate value of zener current in Fig. 23-8d for each of these conditions:

1. $R_L = 100$ kΩ
2. $R_L = 10$ kΩ
3. $R_L = 1$ kΩ

23-10. To keep zener current greater than zero, what is the minimum allowable load resistance in Fig. 23-8d?

23-11. Suppose the voltage between point A and B in Fig. 23-9a changes from 15 to 15.02 V. What is the value of r_Z between these points? If the voltage between C and D changes from 14.7 to 14.8 V, what is the value of r_Z between these points?

23-12. In Fig. 23-9b, what is the voltage across the zener diode when we take r_Z into account?

23-13. The 1N1594 of Fig. 23-9c has a V_Z of 12 V and an r_Z of 1.4 Ω. Calculate the ideal value of load voltage and the second-approximation value.

23-14. The load current in a zener follower increases by 500 mA. If the β equals 100, what is the change in zener current? If the zener resistance is 50 Ω, what is the change in load voltage?

23-15. In the SP regulator of Fig. 23-5a, R_1 is 1000 Ω and R_2 is 250 Ω. If V_Z equals 6.2 V, what does the output voltage equal ideally?

24. Vacuum Tubes

Most historians feel electronics began when Edison discovered a hot filament emits electrons (1883). Realizing the commercial value of Edison's discovery, Fleming developed the vacuum diode (1904). DeForest added a third electrode to get a vacuum *triode* (1906). Up to 1950, vacuum tubes dominated electronics; they were used in rectifiers, amplifiers, oscillators, modulators, etc.

Since 1950, solid-state devices like the bipolar transistor, FET, and IC have eliminated the vacuum tube from the mainstream of electronics. You will find vacuum tubes in older equipment, but new designs are solid-state; the exceptions are very high-power amplifiers, some microwave amplifiers, TV picture tubes, and a few other specialties. There is doubt about how long the tube can survive in the few applications where it is viable; research may give us practical kilowatt transistors, solid-state TV picture tubes, etc.

The point is that the tube is now a specialty item with a shrinking future.

24-1. VACUUM DIODES

Figure 24-1a shows the schematic symbol for a vacuum diode. The *heater* glows during operation because it reaches incandescent temperatures. This warms the nearby *cathode,* a specially coated electrode that releases electrons by *thermionic emission.* In other words, thermal energy can free outer-orbit electrons near the cathode surface. These emitted electrons are then in the vacuum between the cathode and the *anode.* If the anode of Fig. 24-1a is positive with respect to the cathode, it will attract and capture the emitted electrons.

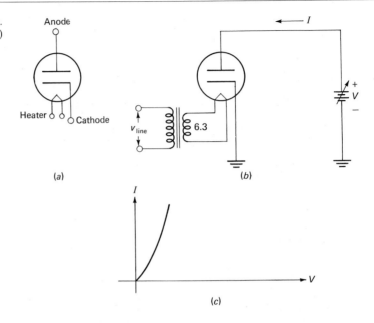

Figure 24-1. (a) *Vacuum diode.*
(b) *Getting the diode curve.* (c)
Diode curve.

Figure 24-1*b* is a circuit for measuring diode current. (The 6.3 V on the heater is a common value.) When anode voltage is positive, conventional current *I* flows. The greater *V* is, the larger *I* is. On the other hand, when *V* is negative, *I* is zero because the anode no longer captures electrons.

Figure 24-1*c* is a typical graph of *I* versus *V*. Unlike the semiconductor diode, the vacuum diode has no barrier potential. Because of this, current *I* appears as soon as *V* goes positive.

Ideally, we can approximate the vacuum diode as a switch; the switch is closed when the diode is forward-biased, and open when reverse-biased. For this reason, we can use vacuum diodes in any of the rectifier circuits discussed in Chap. 5.

As an example, Fig. 24-2 shows a full-wave peak rectifier using vacuum diodes instead of semiconductor diodes. If peak line voltage is 160 V, then each half of the secondary has a peak voltage of 240 V. Ideally, the first capacitor charges to 240 V, which is the approximate value of output dc voltage.

24-2. TRIODES

Figure 24-3*a* shows the schematic symbol for a *triode*, a vacuum tube with three main electrodes. The *control grid* is between the cathode and anode. Because of this, it can influence the current. When the grid is negative with respect to the cathode, it repels some of the electrons trying to flow from cathode to anode. The more negative the grid, the smaller the current.

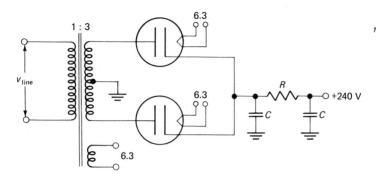

Figure 24-2. Full-wave peak rectifier with vacuum diodes.

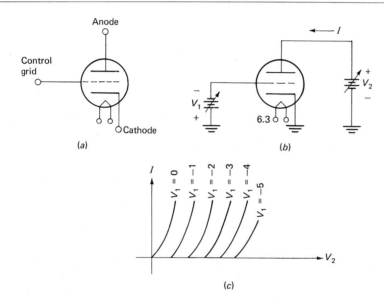

Figure 24-3. (a) Vacuum triode. (b) Getting triode curves. (c) Triode curves.

Triode curves

Figure 24-3*b* is a circuit for getting *triode curves* (similar to collector curves). By holding the grid voltage fixed, we can vary the anode voltage and measure the resulting anode current. For instance, if grid voltage V_1 is zero, we get the first curve of Fig. 24-3*c*. If we make V_1 equal to -1 V, it will take more anode voltage to overcome the effects of the negative grid. For this reason, the $V_1 = -1$ V curve has a knee as shown. Making the grid more negative increases the knee voltage and results in the other curves of Fig. 24-3*c*.

Figure 24-4. A triode amplifier.

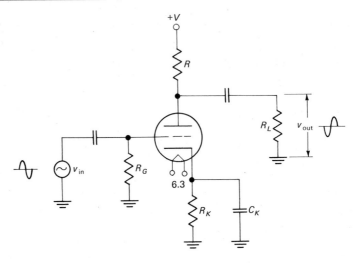

Self-bias

For linear operation the triode has to be biased at a quiescent point that allows ac fluctuations without clipping. As far as biasing is concerned, the triode is similar to an *n*-channel JFET. Because of this, we can use any of the biasing circuits discussed in Chap. 14 for *n*-channel JFETs.

Figure 24-4 shows *self-bias*, the common way to bias a triode. Anode current flows through R_K, producing a plus-minus voltage across it. This makes the cathode positive with respect to the grid, equivalent to the grid being negative with respect to the cathode. Using the triode curves, we can select appropriate values for R_K and R to set up a desired Q point, similar to setting up the Q point of a transistor circuit. Once we have a Q point, the amplifier is ready for class A operation.

Voltage gain

C_K is nothing more than a bypass capacitor; it ac grounds the cathode to prevent loss of voltage gain. If C_K is omitted or accidentally opens during operation, we get a swamped amplifier whose voltage gain ideally equals

$$\frac{v_{\text{out}}}{v_{\text{in}}} \cong \frac{r_L}{R_K} \qquad \text{(ideal)} \tag{24-1}$$

where r_L equals $R||R_L$ (the ac load seen by the anode).[1]

When C_K is present, the cathode is at ac ground. By an analysis similar to transistor analysis, we can prove the voltage gain of a triode amplifier equals

[1] Equation (24-1) is ideal; if you want a more accurate formula, see Prob. 24-6.

$$\frac{v_{\text{out}}}{v_{\text{in}}} = \frac{r_L}{r_p + r_L}\,\mu \qquad\qquad (24\text{-}2)$$

where μ and r_p are given on the data sheet of the tube.

As an example, suppose $R = 30$ kΩ and $R_L = 60$ kΩ. Then,

$$r_L = 30{,}000\,||\,60{,}000 = 20 \text{ k}\Omega$$

If the triode of Fig. 24-4 has a $\mu = 15$ and an $r_p = 5$ kΩ, the voltage gain equals

$$\frac{v_{\text{out}}}{v_{\text{in}}} = \frac{20{,}000}{5000 + 20{,}000}\,15 = 12$$

In Eq. (24-2), $r_L/(r_p + r_L)$ is less than unity for all values of r_L and r_p. As a result, the maximum possible voltage gain of a triode amplifier is

$$A_{\max} = \mu \qquad\qquad (24\text{-}3)$$

This is useful when examining data sheets for triodes; a glance at the value of μ immediately tells us the maximum possible gain we can get with the particular triode.

A study shows that two-thirds of available triode types have μ's less than 50; about 90 percent have μ's less than 75. As a guide, therefore, two-thirds of triode amplifiers have voltage gains less than 50; 90 percent of them have voltage gains less than 75.

24-3. TETRODES AND PENTODES

Like transistors and FETs, the triode has internal capacitances which become part of lag networks in an amplifier. In particular, the capacitance between the grid and the anode results in a large input Miller capacitance that limits the usefulness of a triode at high frequencies.

The *tetrode* of Fig. 24-5a uses a *screen grid* between the control grid and the anode. By applying a constant positive voltage to the screen grid, we can greatly reduce the capacitance between the control grid and the anode. This permits the tetrode to operate at higher frequencies than the triode.

The tetrode has a problem called *secondary emission* which limits its usefulness as a linear amplifier. What happens is this. Electrons arrive at the anode with a high enough velocity to dislodge outer-orbit electrons from the anode surface. Some of these emitted electrons are attracted by the positive screen grid; this reduces the anode current and leads to a distortion of the amplified signal.

The *pentode* of Fig. 24-5b uses a *suppressor grid* between the screen grid and the anode. This new grid is either grounded or connected to the cathode. Because of its low voltage, the suppressor grid forces secondarily emitted electrons to return to the anode; in this way, the undesirable effects of secondary emission are eliminated.

Figure 24-5c shows a typical set of pentode curves. As we see, these resemble JFET curves. For this reason, pentodes act like current sources. In fact, the ac analysis of a pentode amplifier is similar to a JFET amplifier and results in a voltage gain of

Figure 24-5. (a) Tetrode. (b)
Pentode. (c) Pentode curves.

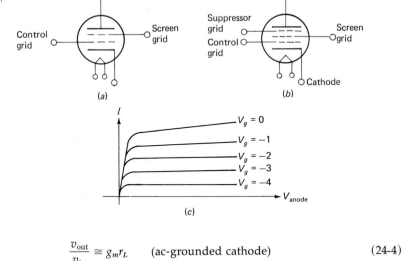

(a)

(b)

(c)

$$\frac{v_{out}}{v_{in}} \cong g_m r_L \qquad \text{(ac-grounded cathode)} \tag{24-4}$$

where g_m is the transconductance of the pentode. Similar to FETs, almost all pentodes have g_m's under 10,000 μmhos.

EXAMPLE 24-1.
In the pentode amplifier of Fig. 24-6, the dc anode current equals 8 mA and the screen-grid current equals 2 mA. Calculate the dc voltage from anode to ground and from cathode to ground. How much dc voltage is there from the control grid to ground?

SOLUTION.
The anode current flows through the 10-kΩ resistor, producing a dc voltage from anode to ground of

$$V = 250 - 0.008(10,000) = 170 \text{ V}$$

In a pentode, the dc cathode current is the sum of the anode current and screen-grid current. For this reason, the dc voltage from cathode to ground equals

$$V = 0.010(200) = 2 \text{ V}$$

Similar to a JFET, the controlling element has a negligible dc current. In other words, the control grid draws negligible current because it is negative with respect to the cathode. For this reason, the dc voltage is zero from the control grid to ground.

EXAMPLE 24-2.
Calculate the voltage gain from input to output in Fig. 24-6.

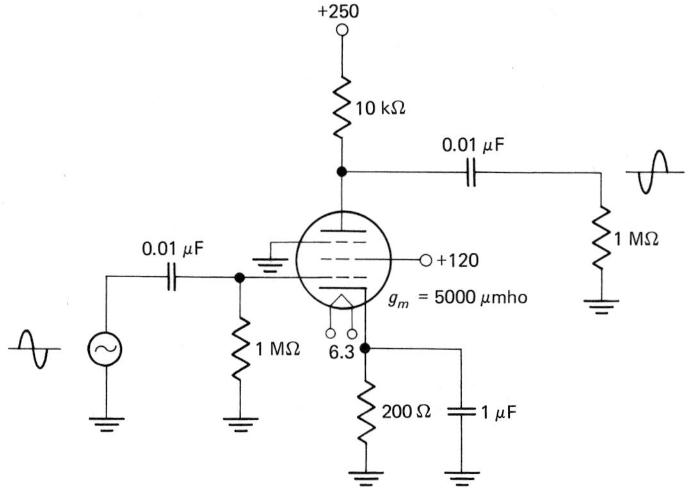

Figure 24-6. Pentode amplifier.

SOLUTION.

The ac load resistance seen by the anode equals

$$r_L = 10{,}000 || 1{,}000{,}000 \cong 10 \text{ k}\Omega$$

As shown in Fig. 24-6, the pentode has a g_m of 5000 μmhos; therefore, the voltage gain equals

$$\frac{v_{\text{out}}}{v_{\text{in}}} \cong g_m r_L = 0.005(10{,}000) = 50$$

In Fig. 24-6 the output sine wave is 180° out of phase with the input sine wave when the frequency is in the midband of the amplifier. As discussed in Chap. 16, coupling and bypass capacitors determine the lower break frequencies; internal and stray capacitances produce the upper break frequencies.

24-4. TRANSISTORS VERSUS TUBES

This section briefly examines some of the differences between transistors and tubes.

Input impedance

Vacuum tubes have very high input resistance because the control grid draws negligible dc current. If it weren't for FETs, we would need the vacuum tube for high-input-impedance applications. There are almost no applications, however, where the FET cannot replace the vacuum tube when input impedance is the main consideration.

Because the FET is more convenient, it is replacing the vacuum tube in electronic volt-meters, oscilloscopes, and other instruments where high input impedance is needed.

Voltage gain

You can get much more voltage gain with a bipolar transistor than with a vacuum tube. As mentioned, triode amplifiers normally have voltage gains less than 75. Similar to the way we compared bipolars and FETs, we can compare bipolars and pentodes (Sec. 14-9). That is,

$$\frac{A_{\text{bipolar}}}{A_{\text{pentode}}} = \frac{g_{m(\text{bipolar})}}{g_{m(\text{pentode})}}$$

The typical pentode has a g_m under 10,000 μmhos, but a bipolar has a g_m of about 40,000 μmhos when I_E is 1 mA, 80,000 μmhos for 2 mA, 160,000 μmhos for 4 mA, and so on. Therefore, given the same ac load resistance, we can get much more gain with a bipolar than with a pentode.

Voltages

Vacuum tubes run at much higher dc voltages than transistors. Typical power-supply voltages with receiver-type tubes are 100 to 300 V. More often than not, this is a disadvantage; lower voltages allow us to build portable, lightweight transistor equipment instead of heavier vacuum-tube equipment.

Power dissipation

Most power transistors have dissipations under 300 W. Vacuum tubes, on the other hand, can easily have dissipations in the kilowatt region. Therefore, the areas where the tube is still needed are high-power amplifiers, transmitters, linear accelerators, industrial control systems, microwave systems, etc. In the microwave systems, the kind of tubes we are talking about are *klystrons, magnetrons,* and *traveling-wave* tubes.

Heater

The transistor needs no heater; the vacuum tube does. This is one of the most important advantages of a transistor over a vacuum tube. The typical heater requires more than a watt of power. In a system using many tubes, the required heater power leads to a bulky power supply. Furthermore, there is the problem of getting rid of the heat.

The heater limits the tube's useful life to a few thousand hours. Transistors, on the other hand, normally last for many years. This is the reason transistors are usually soldered permanently into a circuit whereas tubes are plugged into sockets.

Miscellaneous

Low-power transistors are much smaller than vacuum tubes, roughly the size of a pea compared to a flashlight battery. This means transistor circuits can be more compact and lightweight. Furthermore, unlike the vacuum tube, a transistor is ready to operate as soon as power is applied.

Finally, transistors can be integrated along with resistors and diodes to produce incredibly small circuits.

In summary, solid-state devices have obsoleted the vacuum tube in most areas of electronics. Before long, the tube may be as anachronous as a zeppelin in the jet age.

Problems

24-1. In a triode amplifier, μ equals 25, r_p is 6 kΩ, and r_L is 18 kΩ. Calculate the voltage gain.

24-2. The relation between μ, r_p, and g_m is

$$\mu = g_m r_p$$

If a tube has a μ of 20 and an r_p of 5 kΩ, what does its g_m equal?

24-3. A pentode amplifier has a g_m of 6000 μmhos. If r_L equals 10 kΩ, what does the voltage gain equal?

24-4. If the turns ratio in Fig. 24-2 is changed to 1:5, what is the ideal value of output voltage?

24-5. In Fig. 24-6, suppose the dc anode current equals 10 mA and the screen-grid current equals 3 mA. What is the dc voltage from anode to ground? The dc voltage from cathode to ground?

24-6. If the 1-μF bypass capacitor is left out, the cathode no longer is at ac ground in Fig. 24-6. By a derivation similar to that given for the JFET, the voltage gain equals

$$\frac{v_{out}}{v_{in}} = \frac{r_L}{R_K + 1/g_m} \qquad \text{(no bypass)}$$

Calculate the voltage gain for this condition.

Appendix

Proof of Eqs. (5-5) and (5-6)

When the diode is off, the capacitor voltage equals

$$v = V_p \epsilon^{-t/R_L C}$$

The capacitor discharges for approximately T, and the final capacitor voltage equals

$$v_f \cong V_p \, \epsilon^{-T/R_L C} \tag{A-1}$$

A useful exponential identity is

$$\epsilon^x = 1 + x + \frac{x^2}{2!} + \frac{x^3}{3!} + \cdot \cdot \cdot$$

$$\cong 1 + x \qquad \text{for } x \ll 1$$

Since $T/R_L C$ is much smaller than unity in a peak detector, Eq. (A-1) becomes

$$v_f \cong V_p \left(1 - \frac{T}{R_L C}\right) = V_p - \frac{V_p T}{R_L C}$$

or

$$v_{\text{rip}} = V_p - v_f = \frac{V_p T}{R_L C}$$

or

$$\frac{v_{\text{rip}}}{V_p} = \frac{T}{R_L C}$$

When $R_L C > 100T$, v_{rip}/V_p is less than 0.01 (1 percent); this proves Eq. (5-5). Similarly, when $R_L C > 10T$, v_{rip}/V_p is less than 0.1 (10 percent); this proves Eq. (5-6).

Proof of Eq. (8-5)

The starting point for this derivation is the rectangular *pn* junction equation derived by Shockley:

$$I = I_S(\epsilon^{Vq/kT} - 1) \qquad (A\text{-}2)$$

where I = total diode current

$\quad I_S$ = reverse saturation current

$\quad V$ = total voltage across the depletion layer

$\quad q$ = charge on an electron

$\quad k$ = Boltzmann's constant

$\quad T$ = absolute temperature, C + 273

Equation (A-2) does *not* include the bulk resistance on either side of the junction. For this reason, the equation applies to the total diode only when the voltage across the bulk resistance is negligible.

At room temperature, q/kT equals approximately 40, and Eq. (A-2) becomes

$$I = I_S(\epsilon^{40V} - 1) \qquad (A\text{-}3)$$

(Some books use $39V$ but this is small difference.) To get r'_e, we differentiate I with respect to V.

$$\frac{dI}{dV} = 40I_S\epsilon^{40V}$$

With Eq. (A-3), we can rewrite this as

$$\frac{dI}{dV} = 40(I + I_S)$$

Taking the reciprocal gives r'_e:

$$r'_e = \frac{dV}{dI} = \frac{1}{40(I + I_S)} = \frac{25 \text{ mV}}{I + I_S} \qquad (A\text{-}4)$$

Equation (A-4) includes the effect of reverse saturation current. In a practical linear amplifier, I is much greater than I_S (otherwise, the bias is unstable). For this reason, the practical value of r'_e is

$$r'_e = \frac{25 \text{ mV}}{I}$$

Since we are talking about the emitter depletion layer, we add the subscript E to get

$$r'_e = \frac{25 \text{ mV}}{I_E}$$

Proof of Eq. (9-17a)

In Fig. 9-23a, the emitter of the second transistor is ac grounded; therefore, looking into the base of the second transistor (the one on the right), we have

$$z_{\text{in(base)2}} = \beta_2 r'_{e2}$$

This is the load resistance seen by the first emitter. Because of this, looking into the first base, we see

$$z_{\text{in(base)1}} = \beta_1 (r'_{e1} + \beta_2 r'_{e2})$$

By rearranging, this becomes

$$z_{\text{in(base)1}} = \beta_1 \beta_2 \left(r'_{e2} + \frac{r'_{e1}}{\beta_2} \right)$$

The quantity in the parentheses can be replaced by r'_e, which we interpret as the ac emitter resistance of a single equivalent transistor. In symbols,

$$r'_e = r'_{e2} + \frac{r'_{e1}}{\beta_2} \tag{A-5}$$

When the transistors are similar, the dc emitter current in the second transistor is β_{dc} greater than the dc emitter current of the first transistor; for this reason, r'_{e1} is approximately β_{dc} greater than r'_{e2}. If $\beta_{\text{ac}} = \beta_{\text{dc}}$, Eq. (A-5) simplifies to

$$r'_e \cong 2r'_{e2} \tag{A-6}$$

For instance, if r'_{e2} is 2.5 Ω, r'_{e1} will be approximately 250 Ω for a β_{dc} of 100. If the ac betas are equal to the dc betas,

$$r'_e = 2.5 + \frac{250}{100} = 5 \ \Omega$$

which is double the value of r'_{e2}.

Equation (A-6) is an excellent approximation for preliminary analysis of practical circuits.

Proof of Eq. (11-8)

In Fig. 11-11, the instantaneous power dissipation during the *on* time of the transistor is

$$\begin{aligned} p &= V_{CE} I_C \\ &= V_{CEQ} (1 - \sin \theta) I_{C(\text{sat})} \sin \theta \end{aligned}$$

This is for the half cycle when the transistor is conducting; during the *off* half cycle, $p = 0$ ideally.

The average power dissipation equals

$$p_{av} = \frac{area}{period} = \frac{1}{2\pi} \int_0^\pi V_{CEQ}(1 - \sin\theta) I_{C(sat)} \sin\theta \, d\theta$$

After evaluating the definite integral over the half-cycle limits of 0 to π, and dividing by the period 2π, we have the average power over the *entire cycle* for one transistor:

$$p_{av} = \frac{1}{2\pi} V_{CEQ} I_{C(sat)} \left[-\cos\theta - \frac{\theta}{2} \right]_0^\pi$$

$$= 0.068 V_{CEQ} I_{C(sat)} \tag{A-7}$$

This is power dissipation in each transistor over the entire cycle, assuming 100 percent swing over the ac load line.

If the signal does not swing over the entire load line, the instantaneous power equals

$$p = V_{CE} I_C = V_{CEQ}(1 - k \sin\theta) I_{C(sat)} k \sin\theta$$

where k is a constant between 0 and 1; k represents the fraction of the load line being used. After integrating

$$p_{av} = \frac{1}{2\pi} \int_0^\pi p \, d\theta$$

you get

$$p_{av} = \frac{V_{CEQ} I_{C(sat)}}{2\pi} \left(2k - \frac{\pi k^2}{2} \right) \tag{A-8}$$

Since p_{av} is a function of k, we can differentiate and set dp_{av}/dk equal to zero to find the maximizing value of k:

$$\frac{dp_{av}}{dk} = \frac{V_{CEQ} I_{C(sat)}}{2\pi} (2 - k\pi) = 0$$

Solving for k gives

$$k = \frac{2}{\pi} = 0.636$$

With this value of k, Eq. (A-8) reduces to

$$p_{av} = 0.107 V_{CEQ} I_{C(sat)} \cong 0.1 V_{CEQ} I_{C(sat)}$$

This final result says the average power dissipation over the entire cycle in each transistor has a maximum value of about one-tenth the quiescent voltage times the saturation current. This maximum power dissipation occurs when k is 0.636, that is, when approximately 63 percent of the ac load line is used.

Proof of Eq. (12-7a)

To begin with, assume the tank is tuned to the fundamental. Then, the fundamental sees an ac impedance of r_L; the harmonics, on the other hand, ideally see a short to ac ground. The load line for the fundamental has a saturation current of V_{CC}/r_L. Therefore, when 100 percent of the load line is used, the peak value of the fundamental current is

$$I_1 = \frac{V_{CC}}{r_L} \tag{A-9}$$

The actual current waveform driving the tank is a train of narrow pulses like Fig. 12-1a. To get the relation between the maximum current of the waveform and the peak current of the fundamental, you have to make a Fourier analysis.* This results in

$$I_{max} \cong \frac{\pi}{4k} \frac{V_{CC}}{r_L} \qquad \text{for } k \ll 1 \tag{A-10}$$

where I_{max} is the peak current in Fig. 12-1a, and k equals the pulse width W divided by period T.

The average value of I_{max} equals

$$I_{av} \cong \frac{2}{\pi} I_{max} = \frac{V_{CC}}{2kr_L} \tag{A-11}$$

When the duty cycle is low (as a guide, less than 10 percent), the collector voltage during the on time of the transistor approximately equals $V_{CE(sat)}$ for 100 percent swing. Because of this, we can get an approximate value of average dissipation during the on time as follows:

$$P_D' \cong V_{CE(sat)}I_{av} = V_{CE(sat)} \frac{V_{CC}}{2kr_L}$$

The dissipation over the entire cycle is

$$P_D \cong kP_D' = V_{CE(sat)} \frac{V_{CC}}{2r_L} \qquad \text{(Q.E.D.)}$$

Derivation of Eqs. (14-17) and (14-18)

Start with the transconductance equation

$$I_D = I_{DSS} \left[1 - \frac{V_{GS}}{V_{GS(off)}} \right]^2 \tag{A-12}$$

---

* See *Reference Data for Radio Engineers*, Howard W. Sams & Co., Inc., New York, 1968, p. 42-13, waveform G.

The derivative of this is

$$\frac{dI_D}{dV_{GS}} = g_m = 2I_{DSS}\left[1 - \frac{V_{GS}}{V_{GS(off)}}\right]\left[-\frac{1}{V_{GS(off)}}\right]$$

or

$$g_m = -\frac{2I_{DSS}}{V_{GS(off)}}\left[1 - \frac{V_{GS}}{V_{GS(off)}}\right]$$ (A-13)

When $V_{GS} = 0$, we get

$$g_{m0} = -\frac{2I_{DSS}}{V_{GS(off)}}$$ (A-14)

or by rearranging,

$$V_{GS(off)} = -\frac{2I_{DSS}}{g_{m0}}$$

This proves Eq. (14-18). By substituting the left member of Eq. (A-14) into Eq. (A-13),

$$g_m = g_{m0}\left[1 - \frac{V_{GS}}{V_{GS(off)}}\right]$$

This is the proof of Eq. (14-17).

Proof of Eq. (16-21a)

Current gain is measured under special conditions of r_C equals zero and R_S equals infinity (Fig. 16-18e), in other words, an ac short across the output and a current source driving the input. For these conditions, the Thevenin resistance in the base lag network becomes

$$R = \beta r_e'$$

and the capacitance is

$$C = C_e' + C_c'$$

The break frequency of the base lag network under these special conditions is called the *beta-cutoff frequency* f_β. It equals

$$f_\beta = \frac{1}{2\pi\beta r_e'(C_e' + C_c')}$$ (A-15)

By a derivation which we will discuss in a moment, f_β and f_T are related as follows:

$$f_\beta = \frac{f_T}{\beta}$$ (A-16)

With this relation, Eq. (A-15) can be rearranged to get

$$C'_e + C'_c = \frac{1}{2\pi f_T r'_e} \qquad \text{(Q.E.D.)}$$

A correct derivation for Eq. (A-16) is not easy. You have to include the delay time of carriers diffusing through the base. If interested, see Pettit, J. M., and M. M. McWhorter: *Electronic Amplifier Circuits*, McGraw-Hill Book Company, New York, 1961, pp. 26–35.

Proof of Eq. (16-28)

The key to the proof is showing that the impedance looking into the source is $1/g_m$ (see Fig. 16-23c and d). With Eq. (14-22), the voltage gain from gate to source is

$$\frac{v_s}{v_g} = \frac{r_S}{r_S + 1/g_m} = \frac{R_S}{R_S + 1/g_m} \qquad \text{(A-17)}$$

Visualize R_S as the load impedance on a source follower. Then the maximum voltage gain is unity. One way to find the Thevenin resistance looking back into the source follower is to find the value of R_S that reduces the voltage gain to one-half. By inspection, this occurs when

$$R_S = \frac{1}{g_m}$$

because Eq. (A-17) reduces to one-half. Therefore, the Thevenin resistance looking back into the source terminal is $1/g_m$. When viewed from the capacitor of Fig. 16-23c, the impedance becomes

$$R_{TH} = R_S \left\| \frac{1}{g_m} \right.$$

Proof of Eq. (18-4a)

In Fig. 18-4, think of v_{in} and v_f as single-ended inputs to the op amp. Then

$$v_{out} = Av_1 - Av_2 = Av_{in} - Av_f$$

Since B is the voltage gain of the feedback circuit,

$$v_{out} = Av_{in} - ABv_{out}$$

which can be written as

$$\frac{v_{out}}{v_{in}} = \frac{A}{1 + AB} = A_{SP} = \frac{A}{S}$$

Since $S = 1 + AB$, we can proceed as follows:

$$A_{SP} = \frac{A}{S} = \frac{AB}{SB} = \frac{AB + 1 - 1}{SB} = \frac{1}{B}\frac{S-1}{S}$$

$$= \frac{1}{B}\left(1 - \frac{1}{S}\right) \tag{A-18}$$

In Fig. 18-4, we assume z_{in} is so high it has negligible effect on the voltage divider, that is,

$$B \cong \frac{R_2}{R_1 + R_2} \qquad \text{(negligible } z_{in})$$

Substituting this into Eq. (A-18) and rearranging gives

$$A_{SP} \cong \left(\frac{R_1}{R_2} + 1\right)\left(1 - \frac{1}{S}\right)$$

Proof of Eq. (18-17)

In preceding proof, we showed that $S = 1 + AB$. Therefore,

$$A_{SP} = \frac{A}{S} = \frac{A}{1 + AB} \tag{A-19}$$

To get the relation between break frequencies, we now consider $\mathbf{A}_{SP}$ and $\mathbf{A}$ to be complex numbers. Provided $\mathbf{A}$ rolls off at a rate of 20 dB per decade, we can write

$$\mathbf{A} = \frac{A_{mid}}{1 + jf/f_2} \qquad \text{(one lag network)}$$

where f_2 is the break frequency of the internal amplifier. Substituting this into Eq. (A-19) gives

$$\mathbf{A}_{SP} = \frac{A_{mid}/(1 + jf/f_2)}{1 + A_{mid}B/(1 + jf/f_2)}$$

or

$$\mathbf{A}_{SP} = \frac{A_{mid}}{1 + jf/f_2 + A_{mid}B} = \frac{A_{mid}}{1 + A_{mid}B + jf/f_2}$$

$$= \frac{A_{mid}}{(1 + A_{mid}B)(1 + jf)/(1 + A_{mid}B)f_2}$$

or

$$\mathbf{A}_{SP} = \frac{A_{mid}}{1 + A_{mid}B}\frac{1}{1 + jf/(1 + A_{mid}B)f_2}$$

The first factor is the feedback-amplifier gain in the midband, that is,

$$A_{SP(mid)} = \frac{A_{mid}}{1 + A_{mid}B}$$

The second factor is a frequency-sensitive term. In fact, we may rewrite it as

$$\frac{1}{1 + jf/f_{2(SP)}} = \frac{1}{1 + jf/(1 + A_{mid}B)f_2}$$

This factor implies two things:

1. The bel voltage gain rolls off 20 dB per decade for f greater than $f_{2(SP)}$; therefore, the feedback amplifier also acts like a single lag network.
2. The break frequency of the feedback amplifier is

$$f_{2(SP)} = (1 + A_{mid}B)f_2$$

or with $S_{mid} = 1 + A_{mid}B$,

$$f_{2(SP)} = S_{mid}f_2 \qquad \text{(Q.E.D.)}$$

Proof of Eq. (22-4)

Start with an AM waveform of

$$v = Av_x$$

which can be rewritten as

$$v = A_0(1 + m \sin \omega_y t)V_x \sin \omega_x t$$

When both sine functions are maximum, we get the maximum output voltage:

$$V_{max} = A_0(1 + m)V_x$$

The minimum output voltage occurs when $\sin \omega_y t = -1$ and $\sin \omega_x t = 1$:

$$V_{min} = A_0(1 - m)V_x$$

The ratio of the last two equations is

$$\frac{V_{max}}{V_{min}} = \frac{1 + m}{1 - m}$$

The rest is algebra. Solving for m gives

$$m = \frac{V_{max} - V_{min}}{V_{max} + V_{min}}$$

Multiplying by 2,

$$m = \frac{2V_{max} - 2V_{min}}{2V_{max} + 2V_{min}}$$

Proof of Eq. (22-8)

The equation of the envelope is

$$v = V(1 + m \sin \omega_y t)$$

The derivative equals

$$\frac{dv}{dt} = m\omega_y V \cos \omega_y t$$

By physical reasoning, the steepest negative slope of the envelope occurs when $\omega_y t = 180°$; therefore, the maximum negative slope is

$$\frac{dv}{dt} = -m\omega_y V$$

The exponential discharge of the capacitor is given by

$$v = V\epsilon^{-t/RC}$$

which has a derivative of

$$\frac{dv}{dt} = -\frac{V}{RC} \epsilon^{-t/RC}$$

If we measure t from the instant when $\sin \omega_y t = 180°$, we can equate the rate of change of the envelope to the initial exponential discharge. Specifically,

$$-m\omega_y V = -\frac{V}{RC}$$

or

$$\omega_y = \frac{1}{RCm}$$

or

$$f_y = \frac{1}{2\pi RCm}$$

Since we equated rates of change at the steepest part of the envelope, the foregoing formula is for the maximum frequency we can detect.

Answers to Selected Odd-Numbered Problems

Chapter 1

1-1. 26 Ω. **1-3.** 1.26 Ω, 796 kHz. **1-5.** Between 13 Ω and 1.6 kΩ. **1-7.** 200 Ω. **1-9.** V_{TH} is 10 V and R_{TH} is 6.67 kΩ. **1-11.** V_{TH} equals 20 V and R_{TH} equals 10 MΩ.

Chapter 2

2-1. 8.48 Å, 13.2 Å. **2-3.** ¾ J, ¼ J.

Chapter 3

3-1. 0.63 V, 0.57 V, and 0.45 V. **3-3.** 64 nA and 2.05 μA.

Chapter 4

4-1. A is forward-biased, B is reverse-biased. **4-3.** A, A. **4-5.** Upper diode is forward-biased, lower diode is reverse-biased. **4-7.** 8 Ω. **4-9.** 23.3 Ω. **4-11.** 10 Ω. **4-13.** 2 mA. **4-15.** 2.5 mA; a half-wave signal with a 50-V peak. **4-17.** 1.5 mA. **4-19.** A half-wave signal with a peak of 8 V. **4-21.** 10 V. **4-23.** 0.34 mA. **4-25.** 24.8 mA and 7.44 V. **4-27.** 2500 MΩ, 2.5 MΩ.

Chapter 5

5-1. 148 V and 177 V. **5-3.** 160 mA. **5-5.** 2 to 1. **5-7.** 0.267 A (ideal) or 0.255 A (second approximation). **5-9.** 62.2 mA. **5-11.** 21.2 mA (ideal) or 20.5 mA (second approximation); half. **5-13.** 160 V, 80 V, 53.3 V. **5-15.** None. **5-17.** 12.7 V, 25.4 V. **5-19.** *a.* Negative clipped signal with positive peak of 20 V. *b.* Clipping level changes to −0.7 V. *c.* Positive peak changes to 19.5 V. **5-21.** 100 V; yes; no. **5-23.** 0.225 s; yes. **5-25.** 16 V. **5-27.** 38 V, 35 V. **5-29.** 48 A; yes. **5-31.** 60 V. **5-33.** 500 Ω. **5-35.** Balanced.

Chapter 6

6-1. 0.75 V, 0.65 V. **6-3.** 200,000; 9.8 million. **6-5.** 0.974. **6-7.** 4.7 V. **6-9.** 0.3 mA. **6-11.** 200 (approximate) or 199 (exact); 500 (approximate) or 499 (exact). **6-13.** 600 mA. **6-15.** 10 mA; 0.111 mA. **6-17.** 0.125 mA. **6-19.** 0.1 mA, 20.5 mA. **6-21.** 100; 5 mA; 233 mA. **6-23** 19.995 V; 50 V; no.

Chapter 7

7-1. 0.2 mA (ideal) or 0.186 mA (second approximation). **7-3.** 0.667 mA (ideal) or 0.62 mA (second approximation). **7-5.** Ideally: V_C is 8 V, V_E is 3.33 V, and V_{CE} is 4.67 V; to a second approximation, V_C is 8.14 V, V_E is 3.1 V, and V_{CE} is 5.04 V. **7-7.** 0.5 mA (ideal) or 0.465 mA (second approximation). **7-9.** Ideally, I_E and I_C equal 30 mA, and I_B equals 0.03 mA. To a second approximation, I_E and I_C equal 27.9 mA, and I_B equals 0.0279 mA. **7-11.** 10 V (ideal) or 10.4 V (second); 0 V (ideal) or 0.7 V (second); because the collector diode is not reverse-biased. **7-13.** Ideally, I_E is 2 mA, V_C is 18 V, and V_{CE} is 8 V; to a second approximation, I_E is 1.86 mA, V_C is 18.1 V, and V_{CE} is 8.84 V. **7-15.** 50 Ω (ideal) or 43 Ω (second approximation). **7-17.** 1.82 mA. **7-19.** R_E is 200 Ω (ideal) or 186 Ω (second approximation); V_C is 7.5 V. **7-21.** 5 mA (ideal) or 4.65 mA (second approximation); zero current because the Thevenin voltage driving the switch is zero; 5 V (ideal) or 5.35 V (second approximation); zero V. **7-23.** Ideally, 3.33 mA and 6.67 mA; to a second approximation, 3.22 mA and 6.43 mA. **7-25.** Ideally, $I_E = I_C = 30$ mA, $I_B = 0.2$ mA, $V_{BE} = 0$ V, $V_{CE} = 3$ V, and $V_{CB} = 3$ V; to a second approximation, $I_E = I_C = 26.5$ mA, $I_B = 0.177$ mA, $V_{BE} = 0.7$ V, $V_{CE} = 3.35$ V, and $V_{CB} = 2.65$ V. **7-33.** Ideally, 2.5 mA, 7.5 V, 1.25 V; to a second approximation, 1.8 mA, 8.2 V, 0.9 V. **7-35.** Ideally, 6.67 mA and 3.33 V; to a second approximation, 6.2 mA and 3.1 V. **7-37.** *a.* 1 V. *b.* 750 μV. *c.* 1000 μV. *d.* 0.5 V.

Chapter 8

8-1. 2.5 μF. **8-3.** 100 μF (approximate) or 105 μF (exact). **8-5.** 100 μF. **8-15.** Approximately 5 mA. **8-17.** 500 Ω. **8-19.** 25 Ω (ideal) or 26.9 Ω (second approximation).

8-21. 12.5 Ω (ideal) or 14.1 Ω (second approximation). **8-23.** 20.9 Ω (ideal). **8-25.** 200; 300. **8-27.** 0.1 mA; 0.5 mA.

Chapter 9

9-1. 74.1 μA. **9-3.** 0.167 mA, 4.17 mV, 25. **9-5.** 1.5-V peak if using I_E of 5 mA, or 1.29-V peak for I_E of 4.3 mA. **9-7.** 200. **9-9.** At the base, a sine wave with peak value of 5 mV superimposed on 20 V dc; at the emitter, a horizontal line at the 20-V level (ideal) or at 20.7-V level (second); at the collector, sine wave with peak of 1.5 V at $+10$-V level (ideal) or at $+8.6$-V level (second). **9-11.** 15 neglecting r'_e, or 14.1 including r'_e. **9-13.** 250 Ω (ideal) or 244 Ω (exact); 10 (ideal). **9-15.** 1.36. **9-17.** Base shorted to ground prevents FR bias. **9-19.** Roughly 1.25 kΩ with the second approximation of I_E; the answer is 469 Ω. **9-21.** *a*. 10 kΩ. *b*. 20 kΩ. *c*. 25 kΩ neglecting 10-kΩ R_E, or 24.4 kΩ including R_E. *d*. 50 kΩ. **9-23.** 0.526, 80, 42.1. **9-25.** Approximately 0.01 and 0.8. **9-27.** Ideally 0.426 for both. **9-29.** 69 kΩ, 10,000. **9-31.** 74.1, 192 for I_E of 4.3 mA. **9-33.** 10.3 kΩ. **9-35.** 100 Ω, 5 kΩ, 25 Ω; 200, 40.

Chapter 10

10-1. Ideally, $I_{CQ} = 10$ mA and $V_{CEQ} = 10$ V; to a second approximation, 9.3 mA and 11.4 V. **10-3.** Ideally, 25 mA and 5 V; to a second approximation, 23.3 mA and 5.35 V. **10-5.** Ideally, 10 mA and 10 V; to a second approximation, 9.3 mA and 10.7 V. **10-7.** Ideally, the quiescent point has coordinates of 50 mA and 5 V, the saturation current is 100 mA, and the cutoff voltage is 10 V. **10-9.** The cutoff voltage on the ac load line is 12 V; therefore, BV_{CEO} must be at least 12 V; $I_{C(\text{sat})}$ is 60 mA. **10-11.** Ideally, $I_{CQ} = 20$ mA, $V_{CEQ} = 16$ V, $I_{C(\text{sat})} = 60$ mA, and $V_{CE(\text{cutoff})} = 24$ V; to a second approximation the corresponding answers are 16.5 mA, 16.7 V, 58.2 mA, and 23.3 V. **10-13.** No. **10-15.** Approximately 170. **10-17.** Ideally, around 7:1; to a second approximation, about 8:1. **10-19.** Ideally, 2.5 W and 5 W; to a second approximation, 2.65 W and 5.3 W. **10-21.** Ideally, the total ac load power is 1.56 W with each ac load receiving half of this power; $P_{D(\text{max})}$ rating at least 3.12 W. To a second approximation, the total ac load power is 1.72 W and the required $P_{D(\text{max})}$ is 3.44 W. **10-23.** 43.5 W. **10-25.** Ideally, 0.5 W; to a second approximation, 0.611 W. **10-27.** 4 W; 8 W. **10-29.** 0.167 Ω. **10-31.** 0.0651 Ω. **10-33.** Ideally 20, 1000, and 125 Ω. **10-35.** 3.31, 248, 450 Ω, and 34.6 Ω. **10-37.** Roughly 40, 2000, and 62.5 Ω.

Chapter 11

11-1. 1.75 A. **11-3.** 0.1 A, 0.25 A, 0.8 A. **11-5.** 0.333 A. **11-7.** 20 mA. **11-9.** 0.68 V. **11-11.** 14.3 kΩ. **11-13.** 0.0667 mA, 10.5 kΩ, yes, 203 mA. **11-15.** 0.32 V, 9.68 V, germanium. **11-17.** Ideally, answers are 10 kΩ, 100, 1; to a second approximation, 10.2 kΩ, 100, 0.98. **11-19.** Ideally, 10 kΩ, 5000; to a second approximation, 10.5 kΩ, and 4760. **11-21.** 10 W worst-case; 6.8 W for 100 percent swing; 10 W for 63 percent swing.

11-23. 40 W, 8 W worst-case. **11-25.** 56.2 W, 11.2 W, 0.224 W. **11-27.** Approximately 900 Ω; a better answer obtained by compounding 1 percent ten times is 895 Ω. **11-29.** V_{BE} decreases 2.5 mV per degree rise. **11-31.** Exactly 1.215 kΩ, or nearest standard value is 1.2 kΩ. **11-33.** Ideal answers are 25 mA and 6 kΩ; second-approximation answers are 22.7 mA and 6.6 kΩ. **11-35.** Ideally, 20 kΩ, 30 mA, and 45 mA. **11-37.** 417 mW to $r_C + r_E$; 347 mW to r_C, 83 mW. **11-39.** Ideal answers are 100 Ω, 2.5 Ω, 20, and 100,000. **11-41.** Roughly, 50 W and 47.6 W; to a better approximation, 47.6 W and 45.4 W.

Chapter 12

12-1. 5 V, -5 V, -4.3 V, ideally 5 V. **12-3.** 200 mA, 20 V. **12-5.** 0.5 percent. **12-7.** 5 MHz. **12-9.** 5 percent, 50 kHz. **12-11.** 50, 5 kΩ. **12-13.** 5 kΩ, 25. **12-15.** 200 mW. **12-17.** 4 mW. **12-19.** 16 W when the entire load line is used. **12-21.** 500 kHz, 1 MHz, 1.5 MHz. **12-23.** Approximately 4 MHz, or more closely 3.98 MHz. **12-25.** 100 kΩ.

Chapter 13

13-1. 1.5(10^{11}). **13-3.** 2 V; 4.5 to 25 V. **13-5.** $0.009(1 + V_{GS}/3)^2$; 2.25 mA. **13-7.** About 6.12 mA. **13-9.** $0.012(1 + V_{GS}/5)^2$. **13-11.** Roughly -1.5 V for V_{GS}; about 6 V for V_P. **13-13.** 4.5 mA, 12.5 mA. **13-15.** 12 mA; 18 V; because V_{DS} is greater than V_P, which equals 6 V. **13-17.** 0.25 mA, 1 mA.

Chapter 14

14-1. Ideally, 5 mA, 25 V; to a second approximation, 4.5 mA, 25.5 V. **14-3.** 3 mA, 3.5 mA; 9 V, 5.5 V. **14-5.** $0.005(1 + V_{GS}/5)^2$; 3 mA (ideal), 3.5 mA (second approximation); 3.22 mA (third approximation). **14-7.** 6 mA, 18 V. **14-9.** 1.5 mA, 13 V. **14-11.** -8 V, -8 V, -17.8 V. **14-13.** 14.2 V, 8 V. **14-15.** 10 V, 10 V, 1 V, because its V_{GS} is only 1 V and its $V_{GS(th)}$ is 2 V. **14-17.** 2400 μmhos. **14-19.** Approximately 4875 μmhos or exactly 4600 μmhos. **14-21.** From 1500 μmhos to 3750 μmhos. **14-23.** 7.5. **14-25.** 0.8. **14-27.** 20. **14-29.** 7.5; 8.33. **14-31.** *a.* 2.04 V. *b.* 2.04 V. *c.* 0.5 mA. *d.* 10 MΩ.

Chapter 15

15-1. 0, 10, 20, 30, 40, and 50 dB. **15-3.** 12 dB, 42 dB, 62 dB. **15-5.** With five-place log tables, 34.7712 dB; or three significant digits, 34.8 dB. **15-7.** 26 dB, 400. **15-9.** 80 W. **15-11.** 31 dB. **15-13.** 6 dB, 26 dB, 46 dB, 66 dB. **15-15.** 74 dB. **15-17.** 40 dB, 34 dB, -6 dB; 68 dB; 2500. **15-19.** *a.* 17 dB (approximate) or 17.233 dB (five-place log tables). *b.* 46 dB. *c.* 63 dB. **15-21.** Neglecting effect of r_e' the total bel voltage gain is approximately 53 dB. **15-23.** 1 Ω, 50 kΩ. **15-25.** 50 μF, 500 pF. **15-27.** 25, approximately 100 pF and 4 pF. **15-29.** 339, 1.48 kΩ. **15-31.** 10^{-4}, 2 $\times$ 10^{-5} mhos. **15-33.** 100, 4 MΩ, 200 Ω.

Chapter 16

16-1. 15.9 MHz, 159 kHz, 1.59 kHz. **16-3.** 79.6 kHz. **16-5.** 16, 2.65 MHz, 2.65 MHz, −225°. **16-7.** *a.* 5, 3.18 MHz. *b.* 10, 1.59 MHz. *c.* 20, 796 kHz. *d.* Inversely. **16-9.** 7.96 MHz, drops to 3.98 MHz. **16-11.** *a.* 40, 15.9 MHz. *b.* 80, 7.96 MHz. *c.* 400, 1.59 MHz. *d.* 10 kΩ, 1 kΩ. **16-13.** 80 dB, 100 Hz, 40 dB, −90°. **16-15.** −45°, −135°, −180°. **16-17.** 2 V, 1.8 V, 70 kHz. **16-19.** 35 μs, 3.5 μs. **16-21.** 0.8 pF, 7.2 pF. **16-23.** Gate: 275 kHz; drain: 8.85 MHz. **16-25.** 1270 pF. **16-27.** Exact relation is $f_\beta = 1/2\pi\beta r'_e(C'_e + C'_c)$; approximate relation is $f_\beta = 1/2\pi\beta r'_e C'_e$ because C'_e is much greater than C'_c; the relation between f and f_T is $f_\beta = f_T/\beta$. **16-29.** Exact answers are 0.261 Hz and 8.84 Hz. **16-31.** 175 Hz. **16-33.** Approximately 68 Hz.

Chapter 17

17-1. 1.5 mA. **17-3.** 1.46 mA. **17-5.** 10 V (ideal); 10.3 V (second approximation). **17-7.** 200 mV peak. **17-9.** 240 mV peak, 8 V. **17-11.** 400; 400 mV. **17-13.** 100,000; 100 dB. **17-15.** 1 mA. **17-17.** 10 V. **17-19.** 10 V.

Chapter 18

18-1. 10^6, approximately 0.02, and 50. **18-3.** 100 mV, 2000. **18-5.** 0.00667, 150. **18-7.** 2000 MΩ, approximately 10 MΩ, exactly 9.95 MΩ. **18-9.** 0.25 Ω, 0.75 Ω. **18-11.** 100 mV, 1 V, 10 V. **18-13.** 0.01 Ω, 0.003 Ω. **18-15.** *a.* 0.1 mA. *b.* 100 mV. *c.* 20 mV peak-to-peak. **18-17.** 20,000 MΩ. **18-19.** 0.5 mA, 0.667 mA, 1 mA. **18-21.** 19.9 kΩ (exact), 20 kΩ (practical); 503 Ω (exact), 500 Ω (practical). **18-23.** 10 kHz, 2 MHz, 5 MHz. **18-25.** 5 MHz, 2 MHz, 15 MHz. **18-27.** 50 Hz, 2 MHz. **18-29.** 40 kHz. **18-31.** 1.33 MHz, 2 MHz.

Chapter 19

19-1. 0.0967, 0.333, 0.0967. **19-3.** 5.31 kHz and 53.1 kHz. **19-5.** 9 V rms. **19-7.** *A:* 37.9 Hz to 795 Hz, *B:* 379 Hz to 7.95 kHz, *C:* 3.79 kHz to 79.5 kHz, *D:* 37.9 kHz to 795 kHz. **19-9.** *a.* 500 kHz. *b.* The frequency of oscillation changes sinusoidally, going above 500 kHz on the positive half cycle and below 500 kHz on the negative half cycle (Chap. 22 discusses this *frequency modulation*). **19-11.** 15.9 kHz. **19-13.** Approximately 53 MHz. **19-15.** 1.23 MHz, 3.9 MHz. **19-17.** More than 6. **19-19.** 10 mA, 20 V. **19-21.** 10 MHz, 15 MHz, 20 MHz. **19-23.** 1.59 MHz and 9985. **19-25.** The tuning capacitor is in shunt with mounting capacitance C_m; therefore, the exact frequency of oscillation depends on the setting of the variable capacitor and is between 0.999 and 1.02 MHz.

Chapter 20

20-1. 40, 80, and 120 kHz. **20-3.** 500 Hz, 12.5 kHz. **20-5.** 19.1, 6.37, and 3.82 V. **20-7.** The spectrum contains a dc component of 5 V, and harmonics with peak values of

3.18, 1.59, 1.06, and 0.796 V; the first four harmonic frequencies are 50, 100, 150, and 200 kHz. **20-9.** 25 V. **20-11.** Yes; 0 V. **20-13.** 31.8 V, 0.106 V. **20-15.** 5000; 5000; no. **20-17.** 8 percent, 4 percent, and 8.94 percent. **20-19.** Of the given choices, 10 mV is the maximum input because the voltage gain still equals 100; somewhere beyond 10 mV but less than 100 mV, the voltage gain drops to less than 100 and nonlinear distortion appears. **20-21.** If no harmonics beyond the second are present, the peak value of the second harmonic voltage is 0.5 V; if harmonics beyond the second exist, 0.5 V is only rough estimate for the second harmonic peak voltage. **20-23.** 122 percent. **20-25.** 0.0006 percent. **20-27.** 0.3 mV.

Chapter 21

21-1. 65 V and −45 V. **21-3.** 0.75 V, 1 kHz. **21-5.** 0.2 V, 10 V, 0.2 V. **21-7.** Inputs: 1, 4, 9 kHz; second harmonics: 2, 8, 18 kHz; sums: 5, 10, 13 kHz; differences: 3, 5, 8 kHz. **21-9.** $f_x = 9$ kHz, $f_y = 7$ kHz. **21-11.** 32, 4096, 5120 Hz. **21-13.** $f_x = 6$ kHz, $f_y = 4$ kHz. **21-15.** 995 to 2055 kHz; roughly 4 to 1 or exactly 4.27 to 1. **21-17.** 0.5, −3 dB. **21-19.** 1.25 μW. **21-21.** 4.05 μV, 0.405 μV, 0.0405 μV.

Chapter 22

22-1. 26, 15.7, 36.5; 260 mV, 157 mV, 365 mV. **22-3.** 75, 0.75, and 0.375 m. **22-5.** 0.2. **22-7.** 1 MHz ± 20 Hz, 1 MHz ± 20 kHz. **22-9.** 2.4 V, 0.6 V. **22-11.** 1 MHz ± 20 kHz. **22-13.** 990 to 1010 kHz, 20 kHz. **22-15.** 2000 kHz carrier plus either a lower sideband extending down to 1985 kHz or an upper sideband extending to 2015 kHz. **22-17.** 53.1 kHz. **22-19.** 995 kHz, 2055 kHz. **22-21.** 33.3 percent. **22-23.** 64,000, 8000, 64. **22-25.** 30 kHz, 7.5 V. **22-27.** 10.

Chapter 23

23-1. 0.2, −0.1; or 20 percent, −10 percent. **23-3.** 8.2041 V. **23-5.** Ideally, 5 mA, 10 mA. **23-7.** 5 mA, 6.25 mA, 3.75 mA. **23-9.** 19.9 mA, 19 mA, 10 mA. **23-11.** 20 Ω, 1000 Ω. **23-13.** 12 V (ideal) or 12.0714 V. **23-15.** 31 V.

Chapter 24

24-1. 18.75. **24-3.** 60. **24-5.** 150 V, 2.6 V.

Index